MAGNETIC PROPERTIES OF LAYERED TRANSITION METAL COMPOUNDS

Physics and Chemistry of Materials
with Low-Dimensional Structures

VOLUME 9



*The titles published in this series are listed at the end of this volume.*

# MAGNETIC PROPERTIES OF LAYERED TRANSITION METAL COMPOUNDS

*Edited by*

L. J. DE JONGH

*Kamerlingh Onnes Laboratory,*
*Leiden University,*
*The Netherlands*

KLUWER ACADEMIC PUBLISHERS
DORDRECHT / BOSTON / LONDON

**Library of Congress Cataloging in Publication Data** CIP

Magnetic properties of layered transition metal compounds / edited by L.J. de Jongh.
p. cm. -- (Physics and chemistry of materials with low -dimensional structures. Series A, Layered structures)
Includes index.
ISBN-13: 978-94-010-7329-5 e-ISBN-13: 978-94-009-1860-3
DOI: 10.1007/978-94-009-1860-3
1. Thin films, Multilayered--Magnetic properties. 2. Metallic films--Magnetic properties. 3. Magnets. 4. Transition metal compounds--Magnetic properties. I. Jongh, L. J. de. II. Series.
QC176.9.M84M34 1989
530.4'175--dc20 89-2553
CIP

Published by Kluwer Academic Publishers,
P.O. Box 17, 3300 AA Dordrecht, The Netherlands.

Kluwer Academic Publishers incorporates the publishing programmes
of D. Reidel, Martinus Nijhoff, Dr W. Junk and MTP Press.

Sold and distributed in the U.S.A. and Canada
by Kluwer Academic Publishers,
101 Philip Drive, Norwell, MA 02061, U.S.A.

In all other countries, sold and distributed
by Kluwer Academic Publishers Group,
P.O. Box 322, 3300 AH Dordrecht, The Netherlands.

*Printed on acid-free paper*

# TABLE OF CONTENTS

# PREFACE

In the last two decades low-dimensional (low-d) physics has matured into a major branch of science. Quite generally we may define a system with restricted dimensionality $d$ as an object that is infinite only in one or two spatial directions ($d = 1$ and 2). Such a definition comprises isolated single chains or layers, but also fibres and thin layers (films) of varying but finite thickness. Clearly, a multitude of physical phenomena, notably in solid state physics, fall into these categories. As examples, we may mention:

- Magnetic chains or layers (thin-film technology).
- Metallic films (homogeneous or heterogeneous, crystalline, amorphous or microcristalline, etc.).
- 1-d or 2-d conductors and superconductors.
- Intercalated systems.
- 2-d electron gases (electrons on helium, semiconductor interfaces).
- Surface layer problems (2-d melting of monolayers of noble gases on a substrate, surface problems in general).
- Superfluid films of $^4He$ or $^3He$.
- Polymer physics.
- Organic and inorganic chain conductors, superionic conductors.
- 1-d or 2-d molecular crystals and liquid crystals.
- 1-d or 2-d ferro- and antiferro electrics.

The study of these systems has substantially advanced our fundamental knowledge of physics, apart from the technological applications (materials science) to which it may have led. No matter how widely different the above physical problems may appear at first sight, there are many parellellisms and characteristics that they all share, which is inherent in fact to the common intrinsic property of restricted dimensionality. It is for this very reason that we may justifiably consider low-d physics as an entity, as a recognizable field of science.

In this larger field then, the sub-branch of low-d magnetism has played a quite important role right from the outset. Theoretical work on magnetic chains and layers started as early as the twenties with the Ising chain, and was followed by studies of 2-d Ising magnets in the thirties and forties. About twenty-five years ago systematic experimental studies on quasi low-d magnetic systems were started in a few research groups. The maximum effort was attained in the seventies, the level having remained about constant since then. In most cases the systems studied are magnetic compounds consisting of weakly coupled chains or layers. Often the predicted behaviour for the isolated chain or layer has been found to be approximated to a surprisingly high degree. Ultrathin magnetic films of a few monolayers have also been studied. In general these systems should be distin-



*L. J. De Jongh (Ed.), Magnetic Properties of Layered Transition Metal Compounds* xi—xiii.

guished from the magnetic films considered in thin-film technology, which typically have a thickness of at least $10^3$—$10^4$ atomic layers.

It is beyond doubt that one of the most important contributions of low-d magnetism to fundamental physics has been in the field of phase transitions and critical phenomena. The study of the way in which cooperative phenomena are influenced by the dimensionality of the lattice, the symmetry of the interaction Hamiltonian, the quantum-mechanical nature of the spin, has led to such important advances as the concept of scaling, the ideas of universality and the theory of the renormalization group. The latter is a general theory of critical phenomena based upon principles of quantum-field theory, for which one of the principal proponents, K. G. Wilson, has been rewarded the Nobel Prize. In fact, the realization that continuum field theory and the theory of critical phenomena are different aspects of the same thing can be considered as one of the great advances in mathematical physics in recent years, the consequences of which are still being actively studied.

Lately, another unifying concept in physics has been introduced in terms of non-linear excitations or domain walls (solitons or kinks). Depending on the problem at hand, these excitations can be either static (topological) or dynamic. They occur as soon as the ground state of the system is degenerate, as in the conceptually most simple example of an Ising-type ferromagnet where the spin-up and spin-down ground states are distinct but have the same energy. Once more, low-d magnets have been of great importance, since they provided some of the most simple experimental system by which to study these physical phenomena of very general applicability. For instance, the absence of long-range magnetic order in an Ising type ferromagnetic chain becomes immediately transparent by realizing that the system is unstable to the thermal excitation of kinks, which are most effective in destroying the long-range correlations. For $d \geqslant 2$ the analogue of the kink in Ising-type systems is the boundary of the droplet formed e.g. by a fraction of spin-up spins in a background sea of spin-down spins. Thus the droplet picture of phase transitions has recently been placed on a more rigorous footing, providing at the same time a basis for the concept of scaling in the Kadanoff sense, as was used in the renormalization group theory.

Phase transitions have a very wide applicability, and this explains the usefulness and interest of the study of low-d magnets. Most physical problems concerning mutually interacting elements that form a spatial array can be 'translated into magnetic language' by describing it within a pseudospin formalism. Thus the same physics underlying magnetic systems also applies to interacting molecular rotators, structural transitions, quantum fluids, binary alloys or liquids, liquid—gas transitions, electric phase transitions and the melting of solids, to name but a few examples from solid state physics.

The advantage of studying a general physical problem in its 'magnetic form' is clearly that in magnetism a multitude of microscopic and macroscopic experimental tools are at hand to study the fundamental properties of a system.

We are therefore convinced that the knowledge gathered in the area of 2-d magnetism, as summarized in this volume, will be of value to many other 2-d problems as well, e.g. in surface physics. It was with such considerations in mind that the decision to edit this book was taken. Without attempting to be exhaustive, we have tried to provide an overview of the main streams in this field. In doing so we have tried to keep a balance between theoretical and experimental work. This illustrates the fact that low-d magnetism has been characterized by a long-standing, strong interaction between theory and experiment, which have both developed roughly in parallel and with a continuous cross-fertilization and mutual stimulation.

The interest and applicability of the concepts and knowledge of low-d magnets, have obtained a new impetus by the recent discovery of a series of layered copper oxide compounds with extremely high superconducting transition temperatures, for which A. Müller and G. Bednorz obtained the 1987 Nobel Prize. As will be mentioned below, the basic properties of the undoped materials, e.g. pure $La_2CuO_4$, are those of an insulating, primarily ionic, antiferromagnetic layer-type compound. Quite surprisingly, therefore, this class of materials not only has provided us with the highest-$T_c$ superconductors, but at the same time they belong to the best approximations of 2-dimensional antiferromagnetism. By now it has already become clear that the magnetic properties of these materials are probably also crucial for the superconductivity. It is hoped, therefore, that in this respect too the present review may serve as a useful guide.

In conclusion I wish to thank all the contributors to this volume for their enormous efforts in providing me with the various chapters. I know so well how little time each of them in principle has to spare for such ventures, which makes me all the more indebted to them for their enthusiasm and perseverance. I can only hope that they will be suitably satisfied with the final product. Last but not least, I wish to express my gratitude to Mrs. B. Cats-Houdijk and Mrs. E. van Rijsewijk-Salman for their invaluable help in preparing the manuscripts and the indices.

L. JOS DE JONGH Leiden, Autumn 1989

# INTRODUCTION TO LOW-DIMENSIONAL MAGNETIC SYSTEMS

L. J. DE JONGH

*Kamerlingh Onnes Laboratorium, Leiden University, P.O. Box 9506, 2300 RA Leiden, The Netherlands*

## 1. Experimental Realizations of 2-d Magnetic Systems

Several possible routes exist by which a 2-d magnetic system can be approximated. We mention the following.

- 'Magnetic separation' by crystallo-structural methods [1].
- Evaporation of mono- or oligoatomic layers of magnetic metals on a substrate [2].
- Langmuir—Blodgett technique: deposition on a substrate of mono- or oligoatomic layers of organic molecules which contain a magnetic atom [3].
- Physi-adsorbed layers of $^3$He on a substrate (2-d nuclear magnetism) [4].
- 2-d electron gases (electrons on helium, mosfets) [5].

The last two examples are rather special cases, and despite their obvious interest we shall not consider them further in this review. In the Langmuir—Blodgett technique monolayers of molecules carrying a magnetic atom, such as e.g. manganese stearate are considered [3]. These molecules are in the form of rods, with a hydrophobic and a hydrophilic end. A compact monolayer of such molecules floating on water can be obtained, with all the hydrophilic ends pointing into the water. If now a hydrophilic substrate such as glass or quartz is inserted into the liquid and subsequently lifted out, a monolayer film of the material is deposited onto the substrate. The main problem is that the magnetic samples obtained in this way are very small, so that only high-sensitivity experiments are feasible. The same holds in fact for the vacuum deposition of layers of magnetic metals. An additional problem there is that magnetic interactions in metals cannot usually be treated with any of the spin Hamiltonians considered in this review, which are based on localized moments and short-range interactions. For instance, the well-known RKKY (Rudermann—Kittel—Kasuya—Yosida) interaction in magnetic metals is oscillatory in character, and falls off as $r^{-3}$. For such reasons most of the efforts in the experimental studies of low-d magnetism have been on ionic insulating magnetic compounds, where a magnetic dimensionality of two may be approached by separating compact magnetic layers from one another as much as possible [1]. This branch of solid state chemistry and coordination chemistry has been well developed in the last twenty years or so, in parallel with the strong increase in the number of experimental and theoretical studies of low-d systems in general, and in low-d magnets in particular [1, 6—12]. Besides low-d magnets, the similar methods of 'lattice engineering' may be exploited to obtain low-d conductors. In the magnetic case the separation between the magnetic layers may be



*L. J. De Jongh (Ed.), Magnetic Properties of Layered Transition Metal Compounds* 1—51.

brought about by the insertion of nonmagnetic atoms, as shown in the two examples of 'model' lattices given in Figures 1 and 2. The first example is the series of compounds $(C_nH_{2n+1}NH_3)_2MX_4$ and $[NH_3(CH_2)_nNH_3]MX_4$, where $M^{2+}$ is a 3-d metal atom like Cr, Mn, Fe, Cu or Cd and X = Cl or Br. The length of the alkyl group $C_nH_{2n+1}$ and of the polyethylene group $(CH_2)_n$ can be varied widely, e.g. values of $n$ up to 16 have been reported. In Figure 1 the structures of $(C_2H_5NH_3)_2CuCl_4$ and of $(C_{10}H_{21}NH_3)_2CdCl_4$ have been drawn. They basically consist of $MCl_4^{2-}$ layers, well separated from one another by a bilayer of the hydrocarbon chains. The latter are non-intercalating and are weakly bonded by Van der Waals forces. For the diammonium series the metal halide layers are quite similar and the layers are held apart by the polyethylene chains. We mention that these compounds have attracted quite a lot of attention not only because of their magnetic properties, but also from the structural point of view. For example, the hydrocarbon layer structure in Figure 1 can be viewed as a smectic lipid bilayer embedded in a crystalline matrix [14]. Quite interesting melting transitions of the lipid chains have been reported. The crystal structural transitions in the compounds with smaller $n$ have also been studied extensively, similarly to the $A_2MX_4$

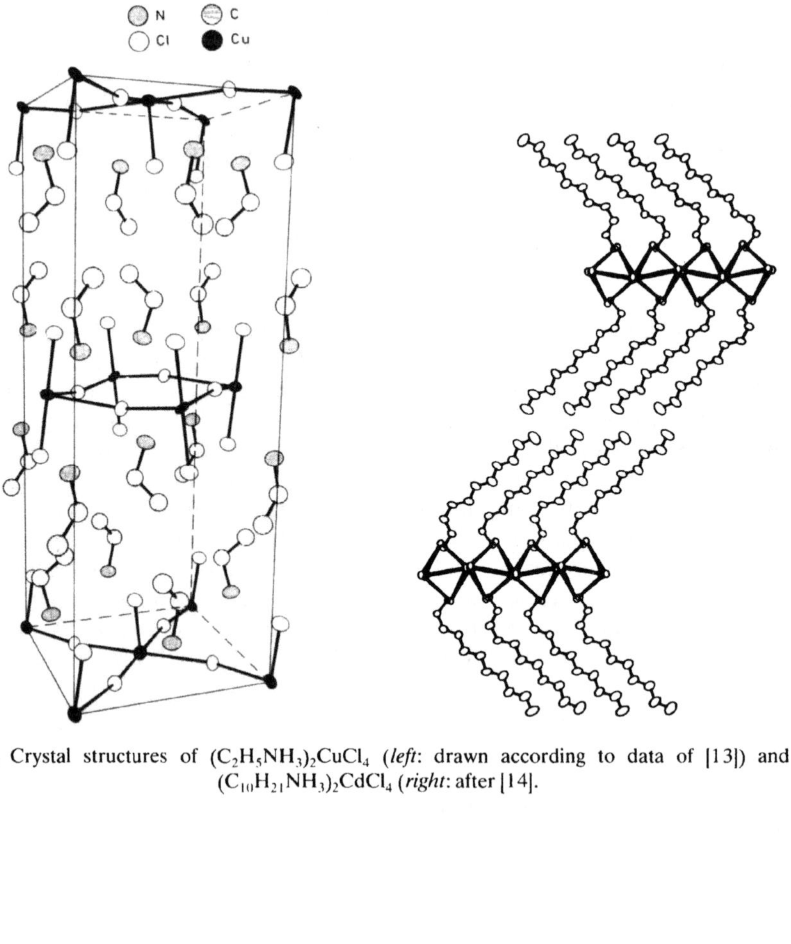


Fig. 1. Crystal structures of $(C_2H_5NH_3)_2CuCl_4$ (*left*: drawn according to data of [13]) and of $(C_{10}H_{21}NH_3)_2CdCl_4$ (*right*: after [14].

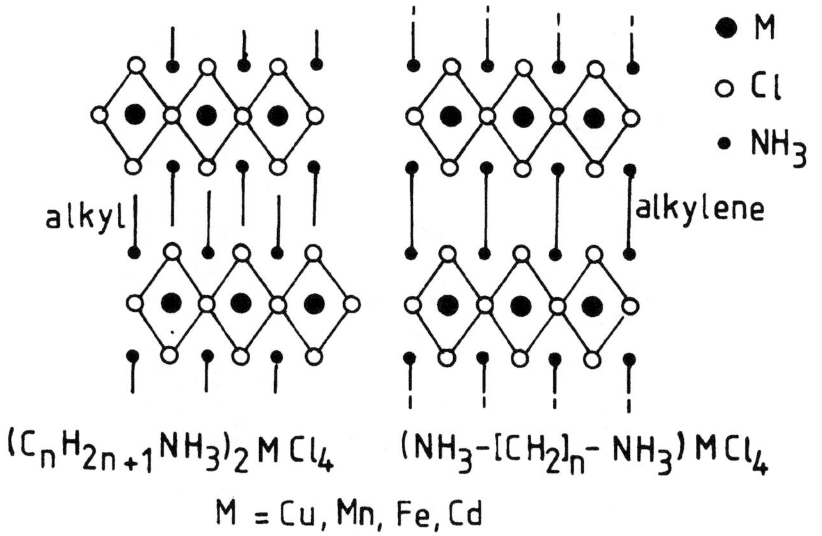


Fig. 2. Schematic comparison of the structures of the $(C_nH_{2n+1}NH_3)_2MCl_4$ and $[NH_3(CH_2)_nNH_3]MCl_4$ series of compounds (after [15]).

family (A = K, Rb; X = F; M = Mn, Fe, Co, Ni, Cu, Cd, Zn), the structure of which is exemplified by the $Rb_2MnF_4$ unit cell shown in Figure 3. This structure, often called the $K_2NiF_4$ structure, has become the 'fruit-fly' of research on 2-d magnetism, since so many materials having this structure have been studied and found to be excellent examples of 2-d magnets [1]. As may be seen in the same figure, one may consider this structure as being derived from the simple cubic perovskite structure found in $RbMnF_3$ by adding additional layers of nonmagnetic RbF which separate the magnetic MnF layers.

The same structures are also found in many metal oxides, and in fact the recently discovered high-$T_c$ superconductor $La_2CuO_4$ has basically the $K_2NiF_4$ structure, whereas the compound $LaCuO_3$ is a cubic perovskite. It is quite interesting to realize that (doped) $La_2CuO_4$ is not only an excellent quasi 2-d conductor, but at the same time the pure compound $La_2CuO_4$ is among the best 2-d antiferromagnets known to date, as we shall discuss further below.

Yet another way to arrive at low-d magnetic structures, in particular in $Cu^{2+}$ compounds, is the mechanism of cooperative Jahn—Teller ordering [16]. Consider the array of $CuF_6$ molecules in $KCuF_3$, the structure of which is a distorted version of the cubic perovskite unit cell shown in Figure 3. As a result of the Jahn—Teller effect the octahedral fluorine coordination of the $Cu^{2+}$ ions is tetragonally elongated. Since $Cu^{2+}$ is a $3d^9$ ion, which can be described by a single hole in the $3d$ shell, the hole will occupy the $d$-orbital that is perpendicular to the local axis of elongation.

Consequently, the way in which the elongated octahedra are ordered in the

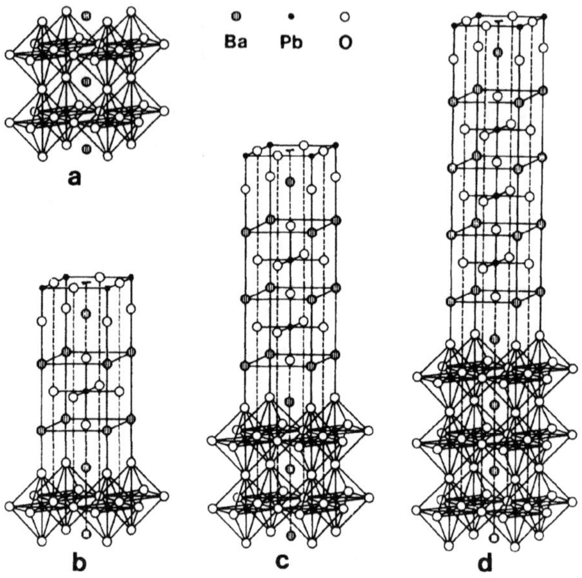


Fig. 3. Crystal-structure of (a) $AMX_3$ (cubic perovskite) (b) the mono-layer compounds $A_2MX_4$ (tetragonal $K_2NiF_4$ structure); (c) the bi-layer compounds $A_3M_2X_7$; (d) the triple-layer compounds $A_4M_3X_{10}$. In these materials M is a metal atom, A is e.g. K, Rb, La, $NH_4$, etc., and X = F, Cl, Br, O, etc.

crystal by means of the cooperative Jahn—Teller mechanism becomes crucial, since it determines the degree of overlap of the *d*-orbitals along the different crystallographic directions. The various possibilities are sketched in Figure 4. In the case of $KCuF_3$, the elongated axes of the fluorine octahedra order in anti-ferrodistortive fashion in the basal plane, as shown in Figure 5. This leads to a lack of overlap in this basal plane, leaving only the overlap along the *c*-axis. The resulting $Cu^{2+}$—$F^-$—$Cu^{2+}$—$F^-$—$Cu^{2+}$ superexchange along the *c*-axis yields indeed a strong antiferromagnetic interaction along the chains, with $J_{AF}/k_B \simeq -190$ K [17] (for a Cu—Cu distance of 3.93 Å), whereas the exchange coupling between the chains is ferromagnetic and very much weaker in strength, i.e. $J_F/k_B \simeq$ 10—20 K [18]. Similar antiferrodistortive arrangements are found for the $CuX_6$ layers in the copper halide (X = Cl, Br) compounds mentioned in the above. They are indeed characterized by ferromagnetic layers, with $J_F/k_B \simeq$ 10—20 K [1], and very weak coupling between layers due to the large interlayer spacings (cf. Figure 2). Precisely the same antiferrodistortive configuration is also found in $K_2CuF_4$, which basically has the $K_2NiF_4$ structure (Figure 3), and which has a ferromagnetic $J_F/k_B = +11$ K in the planes [19]. The weak ferromagnetic interaction is associated with the fact that in the basal plane of these structures the *d*-orbitals are at right angles to one another, as shown in Figure 5b (see e.g. ref. [20]).

On the other hand, it may also be seen from Figure 4 that when the elongated

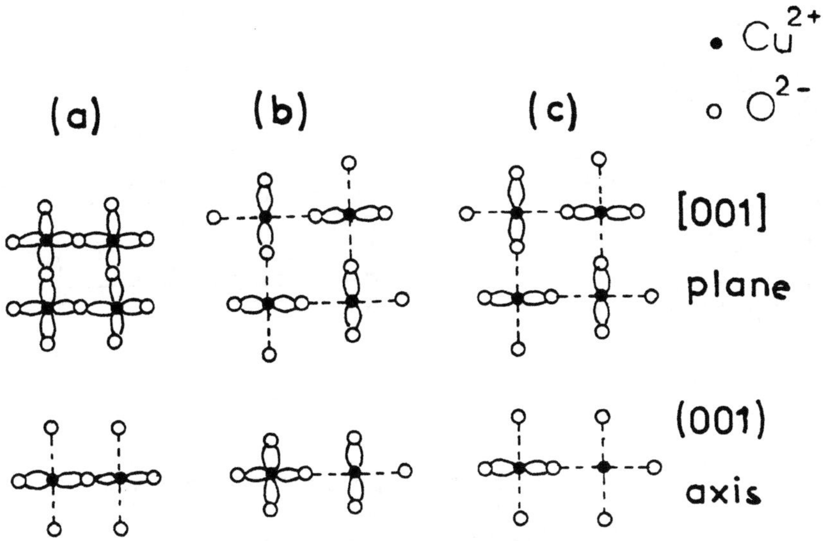


Fig. 4. Orbital ordering in structures of corner-sharing, elongated $CuL_6$ octahedra (L = F, Cl, Br, O). In (a) and (b) the ordering of the elongation axes is, respectively, ferrodistortive (all parallel), and antiferrodistortive (alternately at 90° rotation).

axes of the octahedra align parallel to the *c*-axis in what is called the ferrodistortive arrangement, then there will be little overlap along the *c*-axis, but a strong overlap in the basal plane, cf. Figure 5c. It is this mechanism, in fact, which is responsible for the extremely strong antiferromagnetic exchange ($J_{AF}/k_B \simeq -500$ K) found in $La_2CuO_4$ (and in the planes occurring in the related $YBa_2Cu_3O_7$ structure), making it an excellent example of a 2-d antiferromagnet [21]. Until the discovery of the 2-d antiferromagnetism in this material, the only good 2-d antiferromagnets with $Cu^{2+}$ known to the author were the compounds $Cu(C_5H_5NO)_6(BF_4)_2$ [22, 23] and $K_2BaCu(NO_2)_6$ [24]. The nearly-cubic packing of $CuO_6$ octahedra in the former material is shown in Figure 6. Here the superexchange paths connecting neighbouring Cu atoms involve two oxygens, i.e. Cu—O···O—Cu, and accordingly the antiferromagnetic interaction is much weaker, namely $J_{AF}/k_B \simeq -1$ K [22, 23]. Nevertheless, similar considerations to those discussed above apply and the ferrodistortive ordering of the octahedra leads to quasi 2-d antiferromagnetic behaviour in $Cu(C_5H_5NO)_6(BF_4)_2$. In the related compound $Cu(C_5H_5NO)_6(ClO_4)_2$, on the other hand, the cooperative Jahn—Teller ordering is antiferrodistortive and leads to antiferromagnetic chains, weakly coupled to one another, just as in $KCuF_3$. Finally, in the compounds $K_2BaCu(NO_2)_6$ and $K_2PbCu(NO_2)_6$, which have an fcc structure [25], the Jahn—

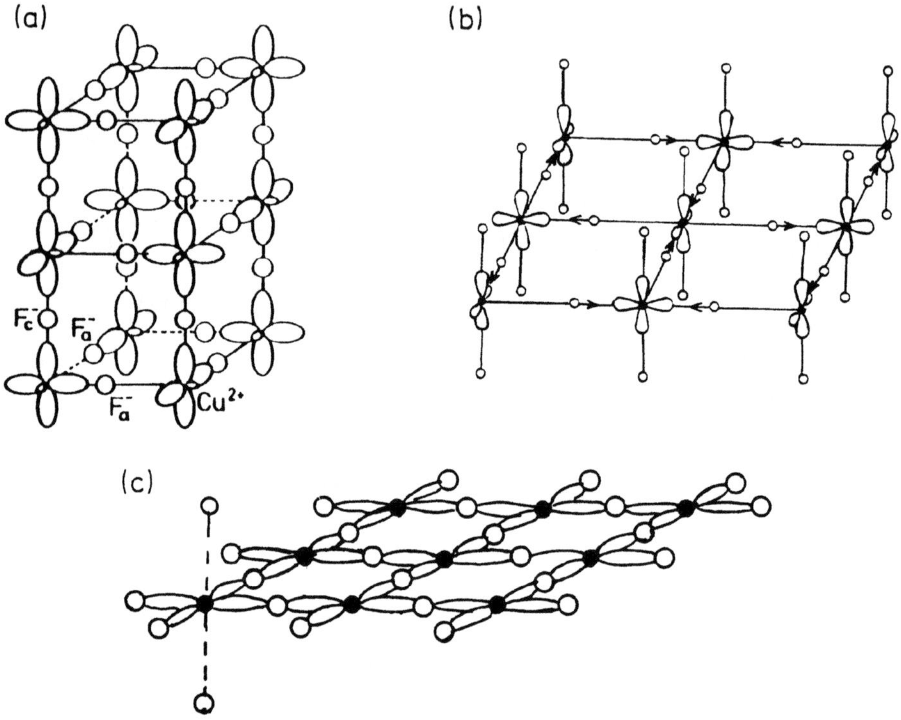


Fig. 5. Orbital ordering as found in (a) $KCuF_3$, (b) $K_2CuF_4$ and related layer-type copper halogenides; (c) $La_2CuO_4$ and related copper oxides. (For (b) and (c) only the basal planes of the layered structures are shown.)

Teller ordering is again ferro- and antiferrodistortive, respectively, leading to quasi 2-d and quasi 1-d behaviour [24], in agreement with the above considerations.

In concluding this section we point out that the perovskite structure also allows the growth of double layers, triple layers, etc., as exemplified by the series of oxides $SrRuO_3$, $Sr_2RuO_4$, $Sr_3Ru_2O_7$ and $Sr_4Ru_3O_{10}$.

The double antiferromagnetic layer is also found in $Rb_3Mn_2F_7$, $K_3Mn_2F_7$ and in $Rb_3Mn_2Cl_7$. It is interesting to note that the occurrence of similar multiple layers in the system $La_2CuO_4$, $La_3Cu_2O_7$ has recently also been reported [27].

In fact, as may be clear from the above discussion, the high-$T_c$ copper oxides are not exceptional in the above family of compounds. Although the copper—oxide bonds are more covalent than e.g. for the fluorides, the pure materials may still be viewed as basically ionic, magnetic insulators (the conductivity is caused by *doping* of the materials). After all, the value for $J_{AF}$ in $La_2CuO_4$ is only a factor of 2.5

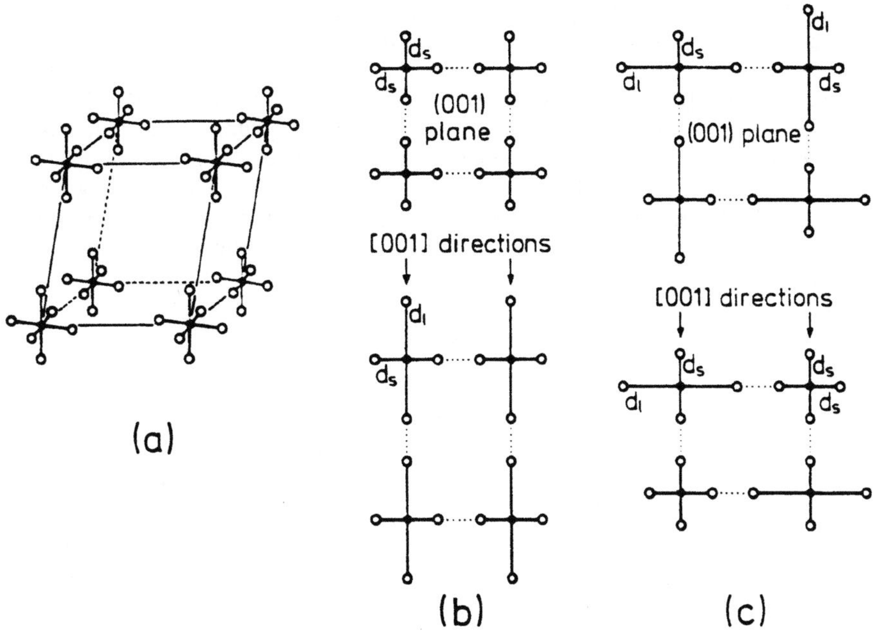


Fig. 6. (a) Stacking of the $CuO_6$ octahedra in the compounds $Cu(C_5H_5NO)_6X_2$. The cooperative Jahn—Teller ordering of the elongated octahedra may again be ferrodistortive (b), or antiferrodistortive (c), resulting in antiferromagnetic chains and antiferromagnetic layers, respectively (after [16]).

larger than is found for $KCuF_3$ along the $c$-axis, whereas the Cu—Cu distances involved are quite similar. Since $J_{AF}$ can be seen as an indication of the degree of covalency, the conclusion is that the two materials are likely to fall in the same class.

## 2. Magnetic Model Hamiltonians

In what follows the symbol d will be used for the dimensionality of the magnetic lattice, d = 1, 2, 3 corresponding to magnetic chains, layers, and 3-d arrays of magnetic moments, respectively. These three classes of systems may be further subdivided according to the type of magnetic interaction that is assumed to exist between the spins. We shall restrict ourselves mainly to interactions between nearest neighbours only. The various Hamiltonians that have been introduced have recently been reviewed by Boersma *et al.* [28], and are best discussed in terms of the general Hamiltonian:

$$\mathscr{H} = -2 \sum_{i<j} (J_x S_{ix} S_{jx} + J_y S_{iy} S_{jy} + J_z S_{iz} S_{jz}) - D \sum_i S_{iz}^2. \quad (1)$$

Taking first the case $D = 0$, we are left with the interaction between two neighbouring spins $\mathbf{S}_i$ and $\mathbf{S}_j$. Depending on the number $n$ of components ($x$, $y$, $z$) of the spins that are considered, one has a one-, two-, or three- component spin system. The number $n$ is called the spin-dimensionality, to be distinguished from the lattice-dimensionality. For given $n$ ($> 1$), one may in addition vary the number of *interacting* spin components by considering different combinations of $J_x$, $J_y$ and $J_z$, as illustrated in Table I. Of these models, the Heisenberg, XY and Ising models are probably the best known. Obviously, there are still more possibilities not listed in the table, for instance for $n = 3$ and $J_x \neq J_y \neq J_z$ one has the anisotropic Heisenberg (XYZ) model, etc.

TABLE I
Classification of model systems.

| Spin-dimensionality | Interaction | Model |
|---|---|---|
| $n=3$<br>$S_x^2+S_y^2+S_z^2=S(S+1)$ | $J_x=J_y=J_z$<br>$J_x=J_y;\ J_z=0$<br>$J_x=J_y=0;\ J_z$ | Heisenberg<br>XY<br>Z |
| $n=2$<br>$S_x^2+S_y^2=S(S+1)$ | $J_x=J_y$<br>$J_x=0;\ J_y$ | Planar<br>Planar Ising |
| $n=1$<br>$S_z^2=S(S+1)$ | $J_z$ | Ising |

Another variable quantity is the spin itself, since one may consider either quantum-mechanical spin operators ($S = 1/2, 1, 3/2, 2, 5/2, \ldots$), or classical spin vectors ($S = \infty$). In the latter case the conditions $\Sigma_\alpha S_\alpha^2 = S(S + 1)$ given in Table I should be replaced by $\Sigma_\alpha S_\alpha^2 = S^2$.

In the experimental examples that we shall consider, the anisotropic Ising or XY models are in most cases not realized by an anisotropy in the exchange interactions themselves. For metal atoms of the 3-d series the exchange between the true spins is quite isotropic, However, crystal-field effects in combination with spin—orbit coupling often leave a lower-lying multiplet as the only populated electronic level at temperatures corresponding to the magnetic ordering. In terms of such a ground-state multiplet, the magnetic properties may then be described in terms of an effective-spin formalism, and this may lead to quite anisotropic expectation values ($g$-values) for the magnetic moment in the ground state. The anisotropy in the expectation value of the magnetic moment may be taken into account in the form of an effective anisotropic exchange acting between (effective) isotropic spins. This situation is typically encountered in $Fe^{2+}$ or $Co^{2+}$ compounds, which often have a ground state doublet for which the magnetic behaviour, depending on the type of crystal-field symmetry, may be described by either the $S = 1/2$ Ising ($g_z \gg g_x, g_y$) or the XY model ($g_x \simeq g_y \gg g_z$). On the other hand $Mn^{2+}$ ($S = 5/2$)

and $Cu^{2+}(S = 1/2)$ compounds often quite closely approximate the isotropic Heisenberg model.

In some cases, as for $Ni^{2+}$, it may be convenient to add the crystal-field term to a Heisenberg-type of interaction, i.e. to consider the Hamiltonian

$$\mathscr{H} = -2J \sum_{i<j} \mathbf{S}_i \cdot \mathbf{S}_j - D \sum_i S_{iz}^2 . \tag{2}$$

For $D \to +\infty$ and $D \to -\infty$ the moments are forced to lie along and perpendicular to the $z$-axis, respectively. In this way the number of components is likewise reduced to $n = 1$ (Ising) and $n = 2$ (planar). For finite $D$-values one may speak of an Ising or XY type of anisotropy, the strength of which can be conveniently measured in units of $J$. Evidently there are many more possibilities to consider, of which we mention in particular the case of orthorhombic (biaxial) anisotropy. This is described by adding yet another term, $-E\,\Sigma_i S_{ix}^2$, to Hamiltonian (2). The case $D < 0$, $E > 0$, $E \ll D$ then corresponds to an easy plane (XY) type magnet with a small preference for the $x$-axis within this plane. In fact such a situation is more often than not found in the experimental compounds.

Next-nearest neighbour interactions will be neglected in most of what follows. In the materials to be considered the interactions are predominantly caused by superexchange interactions, which are of extremely short range ($J \propto r^{-n}$, $n > 12$) [29]. The dipolar interactions that experimentally, will always enter in addition, are of long-range ($\propto r^{-3}$) but are very much weaker in the examples to be discussed, so that they can mostly be neglected. On the other hand it should be noted that in case of e.g. $S$-state ions like $Mn^{2+}$, where the crystal field effects are very small, the anisotropy in the dipolar interactions may become the dominant source of the $DS_z^2$ anisotropy term in the Hamiltonian, and determine the preferred direction of the moments in the crystal.

Finally, we mention that dipolar interactions are particularly important for the rare-earth compounds. Since the $4f$-shell, which is responsible for the local magnetic moments of the (usually trivalent) rare earth ions, is located much to the interior, where it is surrounded by the $5s$ and $5p$ closed-shells, the exchange interactions between the $4f$ electrons of neighbouring rare-earth ions via intervening ligands are very indirect and therefore quite weak. Accordingly, the exchange part of the interaction is of the same order as, or often even very much weaker than the interionic dipolar interactions. Consequently, they provide excellent examples of dipolar magnets. The $Gd^{3+}$ and $Eu^{2+}$ ions have the ${}^8S_{7/2}$ ground state and may therefore be described by isotropic interaction models. For e.g. $Tb^{3+}$, $Dy^{3+}$, $Ho^{3+}$ and $Er^{3+}$ the electronic levels are again largely split by spin—orbit coupling and crystal-field effects (the former is now predominant). Similar to the above description, this may lead to a lower-lying doublet as the ground-state relevant for the magnetic ordering phenomena. Within the effective-spin $S = 1/2$ formalism both the Ising and XY type of anisotropy are found to occur in these materials ($g \gg g_\perp$ and $g_\perp \gg g$, respectively). Some recent examples are discussed in [30] and references cited therein. For nearest-neighbour-only models, however, as considered throughout this book, the rare-earth ionic compounds are obviously less suitable, which is the rason why we shall concentrate on the ionic compounds of the 3-d series in what follows.

## 3. Survey of the Predicted Thermodynamic Behaviour

One of the principal advances in the study of phase transitions has been brought about by the theoretical and experimental investigations of the ways in which the lattice-dimensionality $d$ and the spin-dimensionality $n$ influence the critical behaviour of many-body systems. In the diagram of Table II the various combinations obtained from $n = 1, 2, 3$ and $d = 1, 2, 3$ are summarized. The first question concerns the presence of a transition to long-range order at a temperature $T_c$. In Table II the existence of a finte $T_c > 0$ has been indicated by a cross, whereas a circle indicates that $T_c = 0$, i.e. the system cannot sustain long-range magnetic order at any nonzero temperature. It is seen that, regardless of the value of $n$, the 3-d systems all have $T_c > 0$, whereas 1-d systems have $T_c = 0$. For an Ising-type ferromagnetic chain the absence of long-range order is most transparent from the argument by Landau [31], which shows that the ordered chain is unstable for the thermal excitation of kinds which separate it into spin-up and spin-down regions. Although the creation of such a kink costs an exchange energy $\mathrm{d}U = JS^2$, this is overcompensated by the entropy term $\mathrm{d}S = k\ln N$ (the kink may be placed at $N$ different sites) as soon as the number $N$ of spins in the chain is large enough. As is clear from Table II, the 2-d systems are most interesting since they form the borderline, so to speak, between ordering and nonordering systems. The 2-d Ising model still has a transition to long-range order in the 'conventional' sense, as evidenced by the famous Onsager solution for the free energy in zero field [32]. On the other hand, the 2-d Heisenberg model does not order at any finite temperature, as has been definitely proved by Mermin and Wagner [33] In between one has the 2-d XY model, which has been marked by a cross *and* a circle in Table II, because this model is distinct from the other ones. It has no transition to long-range order in the conventional sense, i.e. a temperature below which an order-parameter such as the spontaneous magnetization can be defined. Yet there exists a finite temperature at which the susceptibility diverges in an exponential fashion. The possibility of such a special transition was postulated by Stanley and Kaplan, first for the 2-d Heisenberg model [34] and in later work also for the 2-d planar model [35]. Their assumption was based upon analyses of the

TABLE II

Absence (○) or presence (×) of a transition to a conventional long-range ordered state at a finite temperature.

| | $d = 1$ | $d = 2$ | $d = 3$ |
|---|---|---|---|
| Ising | ○ | × | × |
| XY | ○ | ⊗ | × |
| Heisenberg | ○ | ○ | × |

high-temperature series expansions for the susceptibility of these models, so that the nature of the low-temperature phase was left unspecified. They pointed out, however, that the absence of spontaneous magnetization and the presence of a susceptibility divergence need not be incompatible.

For a certain period of time this transition temperature was called the Stanley—Kaplan temperature ($T_{SK}$). In later work Berezinskii [36] and Kosterlitz and Thouless [37] investigated the nature of the low-temperature phase, and found it to correspond with a state of topological order in which, at not too low temperatures, pairs of vortex and antivortex arrangements of spins may be found. The transition itself is marked by the unbinding of vortex—antivortex pairs. The transition occurs only for the planar models, not for the 2-d Heisenberg model (which thus has $T_c = 0$). These matters are discussed more fully in the following chapters by Pokrovsky and Uimin and by Navarro.

At the advent of the Berezinskii—Kosterlitz—Thouless theory, there were no good examples of 2-d planar systems in existence. But the XY Hamiltonian also applies to a superfluid [37—39], and thus the first experiments bearing upon this peculiar transition were performed on ultrathin layers of superfluid helium [40]. Subsequently, the theory has been extended to describe other 2-d critical phenomena, such as the melting of a 2-d solid, etc. For such reasons the model wandered away from magnetism for some time, and the transition is nowadays known by most people as the Kosterlitz—Thouless transition. However, as will be discussed in particular in the chapter by Regnault and Rossat-Mignod, some quite good examples of this special transition have recently been found in quasi 2-d magnetic systems.

In what follows are shall briefly review some theoretical predictions for the thermodynamic behaviour of 2-d magnetic systems, in particular as compared to those for the more conventional 3-d systems and for magnetic chains. We shall include a few experimental data on representative examples; much more information on the experiments can be found in the subsequent chapters. As for the theory, with the exception of the Onsager solution for the 2-d Ising model, the results are mainly from high-temperature series expansions, spin-wave theory and from the $n$-vector model theories. These three main streams of theory will each be treated more extensively in the following chapters.

The thermodynamic behaviour of low-d magnets differs fundamentally from their 3-d counterparts. In order to discuss this properly, it is of interest first to review very briefly the results of molecular (MF) theory, which is the simplest theory to describe cooperative phenomena (for more extensive treatments, see e.g. [41 and 42]). Nevertheless, it has been successful in describing the qualitative behaviour of the magnetic phase transitions in 3-d magnets. For 2-d and 1-d magnets, however, it appears to be completely inadequate, and it is of importance to understand the basic reasons for this failure. The discussion will be based upon the Heisenberg Hamiltonian with the Zeeman-term added:

$$\mathcal{H} = -2J \sum_{\langle ij \rangle} \mathbf{S}_i \cdot \mathbf{S}_j - g\mu_B \mathbf{H} \cdot \sum_{\ell} \mathbf{S}_{\ell}.$$

For simplicity we assume a ferromagnetic interaction ($J < 0$) that is limited to the $z$ nearest neighbours of the reference spin ($z$ is often called the magnetic coordination number). In the simplest MF approximation that we shall consider here, one retains only the $z$-components of the spins, i.e. takes $S_i^z = \langle S^z \rangle$ for all nearest-neighbour sites of the reference spin $j$, and writes:

$$\mathcal{H}_{\mathrm{MF}} = -\left\{ 2 \sum_i J \langle S^z \rangle + g\mu_{\mathrm{B}} H \right\} S_j^z.$$

Here $\langle S^z \rangle$ denotes the expectation value for $S^z$. The many-body problem is now reduced to the single-ion problem of a paramagnetic system of spins in an effective field $H_{\mathrm{eff}} = H + H_{\mathrm{MF}}$, so that the Hamiltonian becomes:

$$\mathcal{H}_{\mathrm{MF}} = -g\mu_{\mathrm{B}} H_{\mathrm{eff}} S_j^z$$

where the molecular field is given by:

$$H_{\mathrm{MF}} = 2 \sum_i J \langle S^z \rangle / g\mu_{\mathrm{B}} = 2zJ \langle S^z \rangle / g\mu_{\mathrm{B}} = \lambda M_{\mathrm{s}}$$

here $M_{\mathrm{s}} = g\mu_{\mathrm{B}} \langle S^z \rangle$ is the spontaneous magnetization (per spin), and $\lambda = 2zJ/g^2\mu_{\mathrm{B}}^2$ is called the molecular field parameter. The problem is now easily solved by elementary statistical mechanics. The partition function is:

$$Z = \mathrm{Tr}(e^{-\mathcal{H}_{\mathrm{MF}}/k_{\mathrm{B}}T}) = \sum_i e^{-E_i/k_{\mathrm{B}}T}.$$

Taking the example of spin $S = \frac{1}{2}$, one has $E_i = \pm \frac{1}{2} g m_{\mathrm{B}} H_{\mathrm{MF}}$ so that

$$Z = 2\cosh(g\mu_{\mathrm{B}} H_{\mathrm{MF}}/2kT).$$

The magnetic energy $U_{\mathrm{m}}$, the entropy $S_H$ at constant field and the magnetization $M_T$ at constant temperature are all given as the first derivatives of the free energy, $F = -kT \ln Z$, by the relations

$$U_{\mathrm{m}} = \langle \mathcal{H} \rangle = -\partial \ln Z / \partial \left( \frac{1}{kT} \right) = -Z^{-1} \partial Z / \delta \left( \frac{1}{kT} \right)$$

$$S_H = -(\partial F/\partial T)_H; \qquad M_T = -(\partial F/\partial H)_T,$$

from which the specific heat at constant field $C_H$ and the isothermal susceptibility $\chi_T$ are obtained as the second derivatives:

$$C_H = (\partial U/\partial T)_H = -T(\partial S/\partial T)_H = -T(\partial^2 F/\partial T^2)_H$$

$$\chi_T = (\partial M/\partial H)_T = -(\partial^2 F/\partial H^2)_T.$$

The behaviour of these thermodynamic quantities is as sketched in Figure 7.

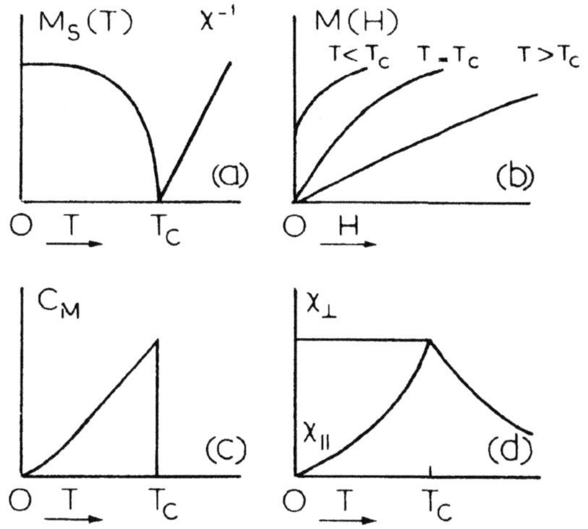


Fig. 7. Sketch of the (static) thermodynamic behaviour according to Molecular Field (MF) theory. (a) Temperature dependence of the order parameter (spontaneous or sublattice magnetization) $M_s$, and of the reciprocal of the ferromagnetic susceptibility, $\chi^{-1}$. (b) Field-dependence of the magnetization isotherms at various temperatures. (c) Temperature dependence of the specific heat. (d) Temperature dependence of the susceptibility of a (weakly anisotropic) antiferromagnet.

What, now, are the essential differences between MF theory and more sophisticated theoretical treatments for the Ising, XY and Heisenberg models [41—44]? This may best be illustrated by considering the temperature dependence of the pair-correlation functions [44]:

$$\Gamma_r(T) = \langle S_0^z S_r^z \rangle / \tfrac{1}{3}S(S+1) \qquad (r = 1, 2, \ldots)$$

which give the degree of correlation between the reference spin $\mathbf{S}_0$ and another spin $\mathbf{S}_r$ at a distance $r$ from $\mathbf{S}_0$. For convenience we only consider the $z$-components of the spins. The brackets denote the expectation value and the factor $\frac{1}{3}S(S+1)$ is a normalization constant. In fact all the thermodynamic behaviour can be related to the correlation functions. Although we are here considering the static behaviour, the extension to time-dependent correlation functions to describe the frequency (energy) dependent effects is straightforward. The latter are important for such experimental studies as NMR, ESR, Mössbauer effect and neutron scattering (see the chapters by Benner and Boucher, by Hirakawa and Ikeda, and by Regnault and Rossat-Mignod).

The qualitative behaviour of the $\Gamma_r(T)$ for a system that shows a transition to long-range order (LRO) at some finite temperature $T_c$ is as sketched in Figure 8. Except for the infinite-range correlation function $\Gamma_\infty$, which vanishes at $T_c$, all the $\Gamma_r$ remain finite in some temperature range above $T_c$. This reflects the so-called short-range order (SRO) that is present in the magnetic system above $T_c$. As $T_c$ is approached from above, the correlation length that characterizes the degree of correlation will progressively increase and finally diverge at $T_c$ itself, signalling the

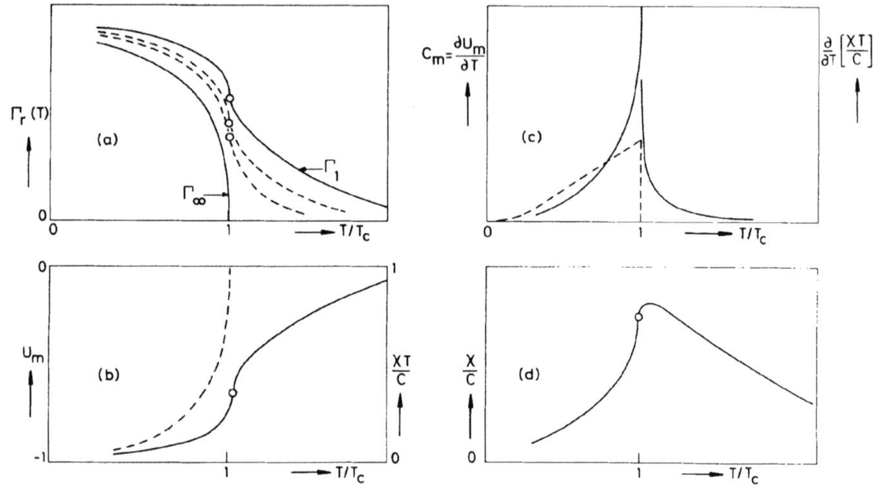


Fig. 8. (a) Qualitative temperature dependence of the static correlation functions $\Gamma_r$. Plotted are $\Gamma_1$ and $\Gamma_\infty$ (solid curves) and two functions with intermediate $r$ (broken curves). (b) Behaviour of the magnetic energy. $U_m \simeq -|\Gamma_1|$, as well as of the product of the antiferromagnetic parallel susceptibility and the temperature ($C$ denotes Curie's constant), $\chi_\parallel T/C \simeq 1-|\Gamma_1|$. The broken curve gives the behaviour of the square of the spontaneous magnetization ($M_s^2 \sim \Gamma_\infty$), to which $U_m$ should be equal according to MF theory. (c) Temperature dependence of the magnetic specific heat, $C_m = \partial U_m/\partial T$, as well as of the temperature derivative of $\chi_\parallel T/C$. The broken curve again denotes the MF result. (d) Antiferromagnetic parallel susceptibility $\chi_\parallel/C$ versus the relative temperature $T/T_c$ (in (a), (b) and (d) the position of $T_c$ has been indicated by the open circles). (After Fisher [44]).

establishment below $T_c$ of nonzero correlations between spins that are arbitrarily far apart. For a magnetic system the order parameter associated with this long-range order (LRO) is the spontaneous magnetization $M_s$ (or sublattice magnetization if an antiferromagnet is concerned). It follows that $M_s^2 \propto \Gamma_\infty$. The finite range correlations $\Gamma_r$, on the other hand, show an inflection point at $T_c$. For the 2-d Ising model exact results are known, from which it appears that, near to $T_c$ the $\Gamma_r$ are given by:

$$\Gamma_r(T) = A + B|T-T_c| \ln |T-T_c|$$

It immediately follows that the derivatives $\partial\Gamma_r(T)/\partial T$ should exhibit logarithmic singularities at $T_c$. For systems with only nearest-neighbour exchange, the magnetic energy will be simply related to $\Gamma_1(T)$ as:

$$U_m(T) = -NzJ\langle S_0^z S_1^z\rangle$$

so that the temperature behaviour of $U_m(T)$ is just that of $-\Gamma_1(T)$. Now, since the magnetic specific heat is the temperature derivative of the energy, it follows that $C_m(T)$ will display a logarithmic singularity on both sides of $T_c$ for the 2-d Ising model considered, as shown schematically in Figure 8. Also the magnetic susceptibility is directly related to the correlation functions. For the parallel susceptibility $\chi_{\parallel}$ one has:

$$\chi_{\parallel} T/C_0 = 1 + \sum_{r=1}^{\infty} \Gamma_r(T)$$

with the Curie constant, $C_0 = N_0 g^2 \mu_B^2 S(S+1)/3k$. For a ferromagnet all the terms in the above series are positive and the susceptibility diverges for $T \rightarrow T_c$ according to some power law (see below). In an antiferromagnet all the terms are alternately positive and negative. As shown by Fisher [45], one can write, to a good approximation:

$$\chi_{\parallel}^{AF} T/C_0 \simeq 1 - D_0|\Gamma_1(T)| = 1 - D_1|U_m(T)|$$

where $D_0$ and $D_1$ are constants that depend at most weakly on temperature near $T_c$. It follows that for the antiferromagnet the product $\chi_{\parallel}^{AF} T/C_0$ behaves qualitatively as the magnetic energy. Obviously, the behaviour of $\chi_{\parallel}$ itself as a function of $T$ may be derived from this. For the 2-d Ising model one concludes that $\chi_{\parallel}^{AF}$ will have a behaviour near $T_c$ of the same form as $\Gamma_1(T)$, whereas the derivative $\partial\chi_{\parallel}^{AF}/\partial T$ will show the same singularity at $T_c$ as the specific heat (and thus may be used experimentally to define $T_c$).

For the 3-d Ising, XY, and Heisenberg models, which also have a transition to LRO at a finite $T_c$, the singularities are not of the same form as for the 2-d Ising model (see below). However, qualitatively, the same line of argument also applies to these 3-d transitions. There is just one important remark to be made here, namely that the effects of SRO diminish as one passes from 2-d Ising (2-d I) to 3-d Heisenberg (3-d H), 3-d XY and 3-d Ising (3-d I), in that order. Referring to Figure 8 we stipulate that by SRO-effects we mean that $\Gamma_r \neq 0$ in some range above $T_c$, which as we have seen implies that the specific heat should have a 'high-temperature tail', i.e. $C_m(T) \neq 0$ for $T > T_c$, and also that the maximum in $\chi_{\parallel}^{AF}(T)$ occurs at some temperature $T_{max} > T_c$, whereas at $T_c$ itself $\partial\chi_{\parallel}/\partial T$ is a maximum.

It is now a trivial matter to understand the difference with MF theory since, in that approach, at least in its simplest form, *SRO is not taken into account at all.* In MF theory the magnetic interaction of a given spin with its $z$ nearest neighbours is replaced by an average interaction with all the other spins in the lattice. The molecular field associated with this interaction is proportional to $M_s$, so that in fact only the $\Gamma_\infty$ is taken into account. Accordingly, in MF theory one has $-U_m \propto \Gamma_\infty \propto M_s^2$. The specific heat will be zero above $T_c$ and show the behaviour sketched in Figures 7 and 8. For the $\chi_{\parallel}^{AF}$ it follows that $T_{max}$ and $T_c$ will coincide (cf. Figure 7). Thus the MF model is at the very end of the sequence 2-d I, 3-d H, 3-d XY, 3-d I, MF, which indicates a continuous decrease of the relative importance of SRO effects. This also applies to the relative importance of the specific heat tail above $T_c$, which reflects the amount of entropy spent in the SRO processes. It may also be inferred from the trends in the predicted $T_c$ values for these models when compared to the MF value $\theta = 2zJS(S+1)/3k$. From the compilation in Table III it can be concluded that the deviations from MF theory increase when:

(i) the dimensionality of the lattice is lowered;
(ii) the spin dimensionality is increased (I → XY → H);
(iii) the spin value is decreased; and
(iv) the number of nearest neighbours is decreased.

If we consider that the MF model discussed above applies to classical spins, in the approximation where only the $z$-components are taken into account (Ising approximation), the trends (ii) and (iii) are self-explanatory. The trend (iv) follows from the fact that MF theory is known to become exact for sufficiently long range of the magnetic interaction $J$, i.e. for sufficiently large number of interacting neighbours. Lastly, the trend (i) can be understood by considering that MF theory also becomes exact for lattice dimensionality $d \geqslant 4$ (critical behaviour, see below), that for $d = 1$ there is no transition to LRO ('$T_c = 0$') for I, H and XY, whereas for $d = 2$ the XY and H models also have no LRO.

Clearly, MF theory fails completely for the low-d systems which show no LRO, since it predicts a transition to LRO regardless of the lattice-dimensionality, the value of $T_c$ being only dependent on the lattice through the number of nearest neighbours $z$.

Furthermore, also for systems that do show LRO, the MF theory becomes inadequate when we come close to the critical point. In particular, it cannot accurately describe the critical behaviour, by which is meant the singularities occurring in the thermodynamic function at $T_c$. It is by now well established that in the majority of cases the critical behaviour of a thermodynamic function $f(t)$ is described by a power law in the relative temperature $t$, where $t = 1 - T_c/T$ for $T$

TABLE III
Critical temperatures of $S = \frac{1}{2}$ ferromagnetic models.

| | | $T_c/\theta$ | | $T_c/\theta$ | $(z)$ |
|---|---|---|---|---|---|
| Heisenberg | f.c.c. | 0.68 | Ising triangular | 0.607 | (6) |
| XY | f.c.c. | | Ising quadratic | 0.567 | (4) |
| Ising | f.c.c. | 0.816 | honeycomb. | 0.506 | (3) |
| Heisenberg | s.c. | 0.56 | | | |
| XY | s.c. | 0.673 | | | |
| Ising | s.c. | 0.752 | | | |

Critical temperatures for varying spin-value (ferro-magnets).

| | | $T_c/\theta$ | | | $T_c/\theta$ |
|---|---|---|---|---|---|
| Heisenberg | $S = \frac{1}{2}$ | 0.68 | Ising | $S = \frac{1}{2}$ | 0.816 |
| f.c.c. | $S = 1$ | 0.75 | f.c.c. | $S = 1$ | 0.851 |
| | $S = 2$ | 0.77 | | $S = 2$ | 0.864 |
| | $S = \infty$ | 0.80 | | $S = \infty$ | 0.874 |

$> T_c$ and $t = 1 - T/T_c$ for $T < T_c$. The exponents appearing in the power laws are called critical exponents or critical indices. Thus one has for instance:

| | | | |
|---|---|---|---|
| specific heat: | $C_m \propto t^{-\alpha}$ | for | $T \to T_c^+$ |
| | $\propto t^{-\alpha'}$ | for | $T \to T_c^-$ |
| ferromagnetic susceptibility | $\chi \propto t^{-\gamma}$ | for | $T \to T_c^+$ |
| | $\propto t^{-\gamma'}$ | for | $T \to T_c^-$ |
| Spontaneous magnetization | $M_s \propto t^{\beta}$ | for | $T \to T_c^-$. |

The extent of the critical region, i.e. the minimum value of $t$ for which the function in question is indeed described, say within a few percent, by the singular term, depends on the function of interest. Roughly speaking one should have $t < 10^{-1}$, $t < 10^{-2}$ and $t < 10^{-4}$—$10^{-6}$ for $\chi$, $M_s$ and $C_m$ respectively [1, 46]. This simply reflects the strength of the singularity, i.e. the values of the critical indices, which are in the order $\gamma > \beta > \alpha$. Values (approximate) for these indices for the 3-d models, the 2-d I model and the MF model are listed in Table IV. Clearly, the deviations from the MF model again increase in the order 3-d I, 3-d XY, 3-d H and 2-d I. We note here also the existence of the so-called 'scaling laws' between the critical indices [41—43]. These laws follow from certain requirements of homogeneity of the thermodynamic functions, and lead to the equivalence of the (primed) exponents below $T_c$ to those above $T_c$. Furthermore they lead to so-called 'scaling relations' between the exponents, such as $\alpha + 2\beta + \gamma = 2$, since they predict that only two of the critical exponents are independent, so that from the knowledge of two arbitrary indices, all the others can be derived.

TABLE IV

Comparison of theoretical values for the critical exponents $\alpha$, $\beta$ and $\gamma$ according to different models.

| | MF | $d=3$ | | | $d=2$ |
|---|---|---|---|---|---|
| | | $n=1$ | $n=2$ | $n=3$ | $n=1$ |
| $\alpha$ | 0 (discont.) | $\simeq$ 1/8 | 0 (log) | $\simeq$ −0.08 | 0 (log) |
| $\beta$ | $\frac{1}{2}$ | $\simeq$ 0.31 | $\simeq$ 0.33 | $\simeq$ 0.35 | 1/8 |
| $\gamma$ | 1 | 1.25 | $\simeq$ 1.33 | $\simeq$ 1.38 | 7/4 |

We further remark that the presence of a finite discontinuity in $f(t)$ at $T_c$, or of a logarithmic singularity, can both be comprised in the power-law divergence in the limit that the index is zero. For a logarithmic divergence this can be easily seen by noting that:

$$\lim_{\alpha \to 0} \frac{t^{\alpha} - 1}{\alpha} = \ln t.$$

The case $\alpha = 0$ further comprises a finite discontinuity at $T_c$, as predicted for the specific heat by MF theory [41].

We conclude this section with the following remarks. As we have seen, the MF theory is a very rough approximation that fails completely for 1-d and 2-d systems and is only qualitatively correct for 3-d Ising systems. By incorporating short-range order effects the MF approach may in principle be somewhat improved; however, it still remains inadequate for the description of 2-d cooperative phenomena in systems with short-range interactions.

One therefore has to take recourse to the theory for the Ising, XY and Heisenberg models [43]. Unfortunately, few exact results are known for these models, and most of these refer to chain systems (for a brief survey, see [1]. Since the chain models cannot sustain LRO, the temperature dependence of the specific heat and the antiferromagnetic susceptibilities take the form of broad, non-anomalous curves, whereas the ferromagnetic susceptibility diverges at $T = 0$. The same is true, in all probability, for the 2-d Heisenberg model [22, 47, 48]. The 2-d Ising model stands on its own, in that it is one of the very few many-body problems for which exact solutions are available, as mentioned above. Not surprisingly, it has become one of the cornerstones of the theory of phase transitions and critical behaviour, and has contributed enormously to our understanding of cooperative phenomena in general.

Also, as regards the theory for the 2-d XY models, we have witnessed impressive progress in the last two decades, notably for the continuum versions of this model. We refer the reader to the next two chapters for further information.

Concerning the 3-d models, our knowledge of the thermodynamic behaviour still is based on approximate methods. High-temperature series expansions have played a very important role here, and for that reason they are extensively reviewed in the chapter by Navarro. In the low-temperature range predictions can be obtained by considering the possible magnetic excitations of the systems. In the Ising models these are single or multiple spin flips ('magnons' and multi-magnon bound states, or droplets). The associated low-temperature series expansions for the Ising case are also reviewed by Navarro. For the XY and Heisenberg models the excitations are on the one hand the harmonic spin waves, which represent low-amplitude deviations from the ground state and on the other the nonlinear excitations like domain walls (kinks, solitons, skyrmions), which can be topological or nontopological. These are highly localized excitations that connect different ground states of the system, e.g. the spin-up and spin-down domains in a ferromagnet [49, 50]. Spin-wave theory and its application to low-d systems has been briefly summarized in [1]. The theory will be treated in detail in the chapter by Arts and de Wijn. Furthermore, a discussion of the nonlinear excitations can be found in Pokrovsky and Uimin's chapter.

We finally mention that our knowledge of the *critical* behaviour in 3-d systems has of course benefitted enormously from the development of the renormalization group theories in the last two decades. Since several reviews of this field have already appeared, we refer the reader to these papers for further information [51]. When combining the knowledge obtained from all the different approaches mentioned above, there emerges a quite reasonable overall picture of how phase transitions and cooperative phenomena are influenced by lattice- and spin-dimensionality ($d$ and $n$), range and sign of the interaction, spinvalue and the like. It is amusing to note that in the theory of the renormalization group the parameters $d$ and $n$ appear as continuous variables, which may take any value from 0 to $\infty$. As mentioned already, this has lead to the theoretical prediction that for $d \geqslant 4$ all critical exponents take their mean field values. Experimentally, it is of course difficult to visualize noninteger values for $d$ or $n$ in real magnets. However, an attempt to measure a 4-d critical exponent in an experimental system under suitable conditions may be found in [52].

## 4. Lattice- and Spin-Dimensionality Crossovers in Quasi 2-d Magnetic Systems

A crucial problem in the field of low-dimensional magnetism is the question as to what extent the thermodynamic properties of the experimental approximations of theoretical 1-d and 2-d model systems are affected by the weak interchain and interlayer interactions that evidently will be present. Intuitively, one would expect a crossover to 3-dimensional behaviour to occur at low enough temperature. In fact, a molecular field-type argument for estimating the 3-d ordering temperature $T_c^{3-d}$ for systems of weakly coupled chains or layers explains the basic principle. In this argument one puts $kT_c^{3-d} \simeq \xi_{1d} J' S^2$ and $kT_c^{3-d} \simeq \xi_{2d}^2 J' S^2$ for weakly coupled chains and layers, respectively, where $\xi_{1d}$ and $\xi_{2d}$ are the (temperature dependent) correlation lengths in the chain and in the layer (evaluated at $T_c$), respectively, and $J'$ is the weak interchain or interlayer interaction, respectively. The $T_c^{3d}$ is thus found by equating the thermal energy to the energy of interacting chain, or layer segments, respectively. It follows that the 3-d order is basically driven by the divergence of the correlation length in the low-dimensional system as the temperature is lowered (to zero or to $T_c^{2-d}$). By the same argument it follows immediately that a long-range order in the planes, as for quasi 2-d Ising systems, will automatically entail an order in three dimensions as soon as $J' \neq 0$. In fact the ordering temperature will be (slightly) enhanced with respect to the $J' = 0$ case.

A quite analogous problem is the anisotropy crossover (or spin-dimensionality crossover), which is particularly important for quasi 2-d systems, since it has the consequence that even in an ideal 2-d system (i.e. $J' = 0$) long-range 2-d order can still occur as soon as a finite amount of Ising-type anisotropy is present. The reason is that the 2-d Ising model has a phase transition, and by a similar molecular field-type argument as just given it may be argued that any finite amount of Ising-type anisotropy in a 2-d Heisenberg or XY system should thus lead to a $T_c^{2-d}$. Obviously, if, besides this small amount of anisotropy, an interlayer interaction $J' \neq 0$ is present, which is so weak that in the absence of anisotropy it would have given rise to a $T_c^{2-d}$, it follows that for the combination of both deviations $T_c^{3-d}$ will necessarily coincide with $T_c^{2-d}$. Since the crossover entails a change in the character of the order parameter, and thus in the effective spin dimensionality, the anisotropy crossover may also be called a spin dimensionality crossover.

We may put the above in a more abstract form by considering that the total

Hamiltonian of the quais low-d magnetic system will be composed of different contributions, namely:

$$\mathscr{H} = \mathscr{H}_0 + \mathscr{H}_1 + \mathscr{H}_2 + \mathscr{H}_3 + \dots \tag{3}$$

Here by $\mathscr{H}_0$ we denote for instance the Hamiltonian for an ideal isotropic Heisenberg chain or layer system.

$$\mathscr{H}_0 = -2J \sum_{\langle i,j \rangle}^{(d)} \mathbf{S}_i \cdot \mathbf{S}_j \tag{4}$$

where $d = 1, 2$ indicates that the summation is for nearest neighbours along the chains or within the layers only. For the next term we may consider the weak interchain or interlayer interaction term: ($|J'/J| \ll 1$):

$$\mathscr{H}_1 = -2J' \sum_{\langle i,j \rangle}^{(3-d)} \mathbf{S}_i \cdot \mathbf{S}_j \tag{5}$$

the summation now extending in the two (for chains, $d = 1$) or one (for layers, $d = 2$) remaining direction(s). As the third and fourth terms we may consider Ising and XY type anisotropy terms:

$$\mathscr{H}_2 = -D \sum_i S_{iz}^2 \tag{6}$$

$$\mathscr{H}_3 = -E \sum_i S_{ix}^2. \tag{7}$$

Here the $D$-term would favour an ordering along the $z$-axis or within the XY plane according to whether $D > 0$ or $D < 0$, respectively. In the latter case the $E$-term would then single out the $x$-axis as the preferential axis within the easy (XY) plane, and thus constitute an Ising-type anisotropy in an XY-type system. As the fifth term we may introduce the long-range dipolar interaction (between classical moments $\boldsymbol{\mu}_i = g\mu_B \mathbf{S}_i$):

$$\mathscr{H}_4 = \sum_{\langle i,j \rangle} \left[ \frac{\boldsymbol{\mu}_i \cdot \boldsymbol{\mu}_j}{r_{ij}^3} - \frac{3(\boldsymbol{\mu}_i \cdot \mathbf{r}_{ij})(\boldsymbol{\mu}_j \cdot \mathbf{r}_{ij})}{r_{ij}^5} \right] \tag{8}$$

where $\mathbf{r}_{ij}$ is the vector that connects the moments $\boldsymbol{\mu}_i$ and $\boldsymbol{\mu}_j$. In case $J$ and $J'$ are very small, the dipolar term may indeed become the predominant term. Furthermore, we should add that in many cases this term leads to an anisotropy in the total interaction that may approximately be taken into account by a term as in $\mathscr{H}_2$ (although strictly speaking there is a difference).

The various types of crossover that may occur can now be understood on basis of the so-called 'Principle of Universality', as formulated by Kadanoff [53], by Griffiths [54], and by others [55]. Applied to our present problem, it states that the behaviour of the system at low enough temperatures or close enough to $T_c$ (which ever of the two is appropriate) will be determined by the weakest term in the Hamiltonian. Let us illustrate the meaning of this with an example. Consider a quasi 2-d system with $\mathcal{H} = \mathcal{H}_0 + \mathcal{H}_1 + \mathcal{H}_2$, for which $|J| \gg |D| \gg |J'|$. Then it follows that at high enough temperature the system will behave as a 2-d Heisenberg magnet. As the temperature is lowered, it will first show an anisotropy-crossover to 2-d Ising-type behaviour (if $D > 0$), and finally a lattice-dimensionality crossover to 3-d Ising-type behaviour. Particularly interesting is the case when the ratio of the above parameters is such that the crossovers occur in the critical region, i.e. close enough to $T_c$ for the behaviour of the thermodynamic functions to be described by power laws, as discussed in the above. In that case the crossovers are seen as changes in the critical exponents, since the values of these depend on both lattice- and spin-dimensionality.

Perhaps the earliest and most famous theoretical calculation of a lattice-dimensionality crossover is the behaviour of the specific heat of the 2-d Ising model on the quadratic lattice, with inequivalent exchange constants along the X and Y directions, so that it can be viewed as a 2-d array of weakly coupled Ising chains as the ratio of these exchange constants is varied from 1 to 0. As already mentioned, this solution, obtained by Onsager [32], is one of the few analytical results in the theory of phase transitions. The curves are shown in Figure 1 of Navarro's chapter for some ratios $R = J'/J$. As $R$ decreases to values of $10^{-2}$ and lower, it is seen that down to very low temperatures the behaviour of the coupled chain system is indistinguishable from that of the isolated Ising chain, which is the curve with the broad maximum (identical to the specific heat of a two-level Schottky anomaly). Since the specific heat for the ideal $d = 1$ system is non-anomalous, whereas the $d = 2$ system has a logarithmic divergence at the long-range ordering temperature, the crossover from $d = 1$ to $d = 2$ behaviour is very clearly seen in this example by the appearance of the small spike that reflects, so to speak, the 2-d ordering anomaly. If one were to analyse the critical behaviour of these spikes, one would find the same logarithmic divergence as for the pure $d = 2$ system, if only $T_c$ is approached closely enough. Evidently, as $R$ is lowered and the transition occurs at very low temperature, there is hardly any entropy left in the system, so that an (experimental) determination even of the spike itself would become quite difficult.

Similar behaviour is found for the $d = 1$ to $d = 3$ crossover in the specific heat of 3-d assemblies of weakly coupled Ising chains. For these systems no analytic solutions are available. The crossovers were studied by Navarro and the author [56], using high-temperature series expansion results of Stanley and coworkers for the region above $T_c$, complemented by low-temperature expansions in the range $T/\theta \lesssim 0.4$. These calculations are further described in the chapter of Navarro. A beautiful experimental study of $d = 2 \rightarrow d = 3$ dimensionality crossover in the specific heat is provided by the work of Bloembergen [48] on very weakly coupled

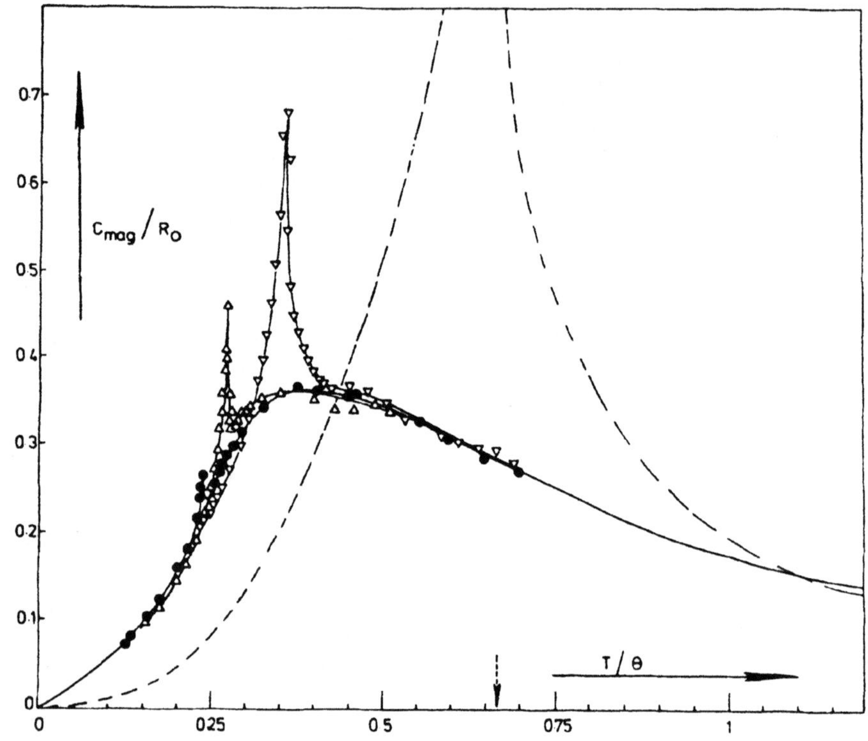
0.7
0.6
0.5
0.4
0.3
0.2
0.1
0
$c_{mag}/R_o$
T / θ
0
0.25
0.5
0.75
1

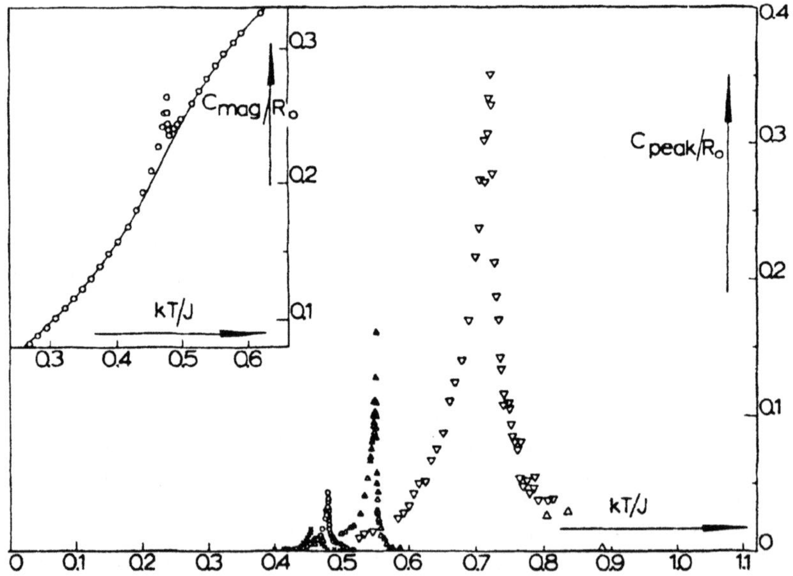
$C_{mag}/R_o$
0.3
0.2
0.1
kT/J
0.3
0.4
0.5
0.6
$C_{peak}/R_o$
0.4
0.3
0.2
0.1
0
kT/J
0
0.1
0.2
0.3
0.4
0.5
0.6
0.7
0.8
0.9
1.0
1.1

Heisenberg ferromagnetic layer systems, as shown in Figure 9. The value for the interlayer interaction $|J'/J|$ is $8 \times 10^{-3}$ for $Rb_2CuCl_4$. For the compounds $(C_nH_{2n+1}NH_3)_2CuCl_4$ it is $8 \times 10^{-4}$ for $n = 2$; $6 \times 10^{-5}$ for $n = 1$, and about $10^{-5}$ for $n = 5$ [1]. Since the values for the spin anisotropy are extremely small and roughly the same for all these materials, the variation seen in the size of the $\lambda$-anomaly in the specific heat may be associated with the variation in $|J'/J|$.

Another excellent demonstration of lattice-dimensionality crossover in the specific heat ($d = 1 \rightarrow d = 3$) has been given by Kopinga and de Jonge and coworkers [57], in their work on a series of linear chain $Mn^{2+}$ compounds. The low-temperature magnetic specific heats shown in Figure 10 are for $[(CH_3)_2NH_2]MnCl_3$ (DMMC), for $(C_5H_5NH)MnCl_3$ (PMCA), and for $[(CH_3)_4N]MnCl_3$ (TMMC). These compounds approximate quite well the $S = 5/2$, Heisenberg antiferromagnetic chain model, with intrachain exchange constants $J/k = -5.8$ K, $-6.5$ K and $-6.7$ K, respectively. The solid curve gives the theoretical prediction for this model, which yields a linear term in the low-temperature limit, as also follows from simple spin-wave theory. The interchain interactions for these materials are given by $R = |J'/J| = 1 \times 10^{-3}$, $4 \times 10^{-4}$ and about $10^{-4}$—$10^{-5}$, respectively. Clearly, the value of $T_c$ is seen to decrease progressively to zero with decreasing $R$. Concomitantly, the area under the 3-d ordering specific heat anomaly decreases.

The more general question as to what extent the thermodynamic quantities are affected by the interlayer or interchain interactions has been considered by Stanley and the author [58], making use of the work of Citteur and Kasteleyn [59], and of Liu and Stanley [60]. The basic idea in [59 and 60] is to develop the thermodynamic functions in expansions with respect to the parameter $R = J'/J$, and atempt to calculate the coefficients in the expansions by combinatorial methods. For instance, for the reduced susceptibility, $\bar{\chi} = \chi T/C$, the expansion reads:

$$\bar{\chi}(R) = \bar{\chi}_0(0) + \bar{\chi}_1(0)R + \bar{\chi}_2(0)R^2/2! + \cdots + \bar{\chi}_n(0)R^n/n! + \cdots \quad (9)$$

where

$$\bar{\chi}_n(0) \equiv (\partial^n \bar{\chi}_d(R)/\partial R^n)_{R=0}, \quad (10)$$

and $d = 1, 2$. The abovementioned authors were able to obtain the following rigorous results. For the susceptibility it is found that

$$\bar{\chi}_{3d} = \bar{\chi}_d + 2z'R(J/kT)\bar{\chi}^2{}_d + \cdots \quad (11)$$

or

$$\Delta\bar{\chi}/\bar{\chi}_d \simeq 2z'R(J/kT)\bar{\chi}_d. \quad (12)$$

Fig. 9. *Upper part*: Magnetic heat capacity of the compounds $Rb_2CuCl_4$ ($\nabla$), $(C_2H_5NH_3)_2CuCl_4$ ($\Delta$) and $(CH_3NH_3)_2CuCL_4$ ($\bullet$) in the neighbourhood of $T_c$. The dashed curve represents the $S = \frac{1}{2}$, fcc Heisenberg ferromagnet. $\theta \equiv T_{MF} = 2zJS(S + 1)/3k$. *Lower part*: Singular part of the magnetic heat capacity of the compounds $Rb_2CuCl_4$ ($\nabla$), $(C_nH_{2n+1}NH_3)_2CuCl_4$ with $n = 2(\Delta)$, $n = 1(\circ)$ and $n = 5(\times)$, obtained after subtraction of the continuous magnetic background (cf. insert). (Both figures are from [48].)

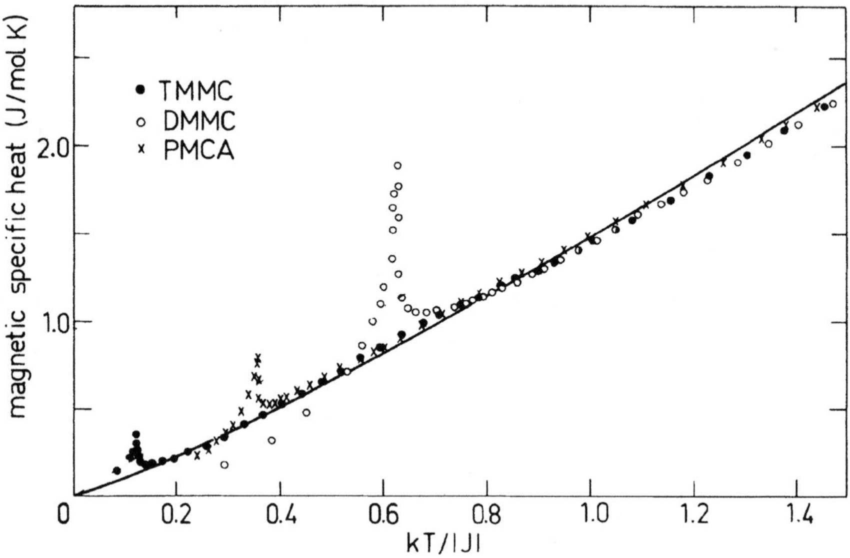


Fig. 10. Low-temperature magnetic specific heat data for three $Mn^{2+}$ chain compounds (see text), demonstrating the $d = 1 \rightarrow d = 3$ lattice dimensionality crossover for decreasing values of $R = J'/J$. Solid curve is the prediction for the $S = 5/2$, antiferromagnetic Heisenberg linear chain. The temperature is given in units of $J/k$, i.e. the intrachain exchange constant for each of the compounds. (After [57].)

Here we have introduced the notation $\bar{\chi}_{3d} = \bar{\chi}(R)$ for the reduced susceptibility of the 3-d array of weakly coupled chains or layers. Furthermore $\bar{\chi}_d$ is the reduced susceptibility of the ideal isolated chain or layer ($d = 1$ or 2), and $\Delta\bar{\chi} = \bar{\chi}_{3d} - \bar{\chi}_d$.

For the reduced magnetization $\bar{M} = M/N_0 g\mu_B S$ it is found similarly that:

$$\Delta\bar{M}/\bar{M}_d \simeq 2z'R(J/kT)\bar{\chi}_d \tag{13}$$

whereas for the specific heat the first-order term in the $R$-expansion is zero, so that $\Delta C_m \propto R^2$. (This is to be expected, since $\Delta C_m$ should not depend on the sign of $R$). We remark that the first result for the susceptibility can also be obtained by treating the interlayer (chain) interaction in the molecular field approximation. In that case we may write $\chi_{3d} = \chi_d/(1 - \lambda'\chi_d)$, or $\chi_{3d} \simeq \chi_d(1 + \lambda'\chi_d) \approx \chi_d + \lambda'\chi_d^2$, where $\lambda'$ is the molecular field parameter corresponding to the interaction $J'$. With $\lambda' = 2z'J'/N_0 g^2\mu_B^2 = 2z'J'/kC$ one retrieves Equation (11) above).

From these few results one may already draw a number of important conclusions. Let us consider the approximate relation (12) for the susceptibility, which may be rewritten as:

$$\Delta\chi/\chi_d = \left(\frac{\chi_d\theta}{C}\right)\left\{\frac{3z'R}{zS(S+1)}\right\} \tag{14}$$

where $z$ and $z'$ are the numbers of magnetic neighbours associated with $J$ and $J'$. We may now distinguish between low-dimensional ferromagnets and antiferromagnets. In the latter case an upper bound to $\chi_d$ can be given by $\chi_d < C/2\theta$, since $C/2\theta$ is the value of the susceptibility reached at $T = \theta$ in the molecular field theory. It follows that, apart from a numerical factor of the order of unity, one has $\Delta\chi/\chi_d \simeq R$, *independent* of temperature. One may conclude that for low-dimensional antiferromagnets the susceptibility is scarcely affected by interlayer or interchain couplings, so long as these do not exceed $R \simeq 10^{-2}$. As may be seen in Figure 11 and in following chapters, this conclusion is nicely born out by the experiments. In Figure 11 experimental antiferromagnetic susceptibilities are compared with theory for the Heisenberg (upper part) and the Ising model (lower part). For the Heisenberg model, the relevant quantity is the perpendicular susceptibility ($\chi_\perp$), since this is the quantity that is measured in the limit of zero anisotropy. The figure [10, 61] shows single-crystal $\chi_\perp$ data for the monolayer compound $K_2MnF_4$ [62] and for the simple cubic perovskite $RbMnF_3$ [63], as well as powder susceptibility data for the bilayer compound $Rb_3Mn_2F_7$ [61] (cf. Figure 3). The latter data are for $T > T_c$ where one has $\chi_\perp = \chi_{powder}$ within the limits of experimental error. The values for $T_c/\theta$ are 0.70, 0.58 and 0.44 for the s.c., the bilayer and the monolayer compounds, respectively. The theoretical curves in the figure have been obtained from high-temperature series expansions ($T/\theta > 0.5$) and from 2-d spin-wave theory ($T/\theta < 0.5$) [1, 62, 64]. It is seen that the data agree well with these predictions, in particular we note that the $K_2MnF_4$ data appear to be described by 2-d theory over the whole temperature range.

The same is true for the $\chi_\parallel$ data for $K_2CoF_4$, which nicely fit the 2-d Ising predictions at all temperatures, including $T_c$ (for the Ising model, the parallel susceptibility is the quantity to consider). We note that the deviations seen for $T > \theta$ are to be ascribed to the failure of the effective spin $S = \frac{1}{2}$ description for $Co^{2+}$ at high temperatures. The conclusion is that, since the antiferromagnetic susceptibilities show smooth (nondiverging) behaviour as a function of temperature for all lattice dimensionalities, crossover behaviour is not to be expected except for very large $R$ values. We add that this does not exclude anisotropy crossover; in fact the $\chi_\parallel$ that is observed experimentally in weakly anisotropic Heisenberg antiferromagnets (not plotted here) can be interpreted as such a phenomenon, since it is due to a weak Ising-type anisotropy.

For ferromagnets ($J > 0$), on the other hand, the chain or layer susceptibility $\chi_d$ will both diverge at low temperatures (either at a finite $T_c$ or at $T = 0$), so that there will be a strong crossover effect near $T_c$, no matter how small is $R$ (cf. Equation (12)). Since the sign of $\Delta\chi$ in this case is determined by the sign of $J'$, it follows that an antiferromagnetic coupling will yield a sudden rapid decrease of $\chi_{3d}$ from the diverging behaviour of the ideal ferromagnetic chain or layer. The decrease corresponds, of course, to the antiferromagnetic ordering that is established between the ferromagnetic chains or layers, due to the action of $J'$. Some examples of $d = 1$ to $d = 3$ crossover in ferromagnetic Ising-type chain systems [65] are shown in Figure 12, where these effects are clearly distinguished (obviously, the smaller $|R|$ is, the lower the temperature at which the crossover occurs).

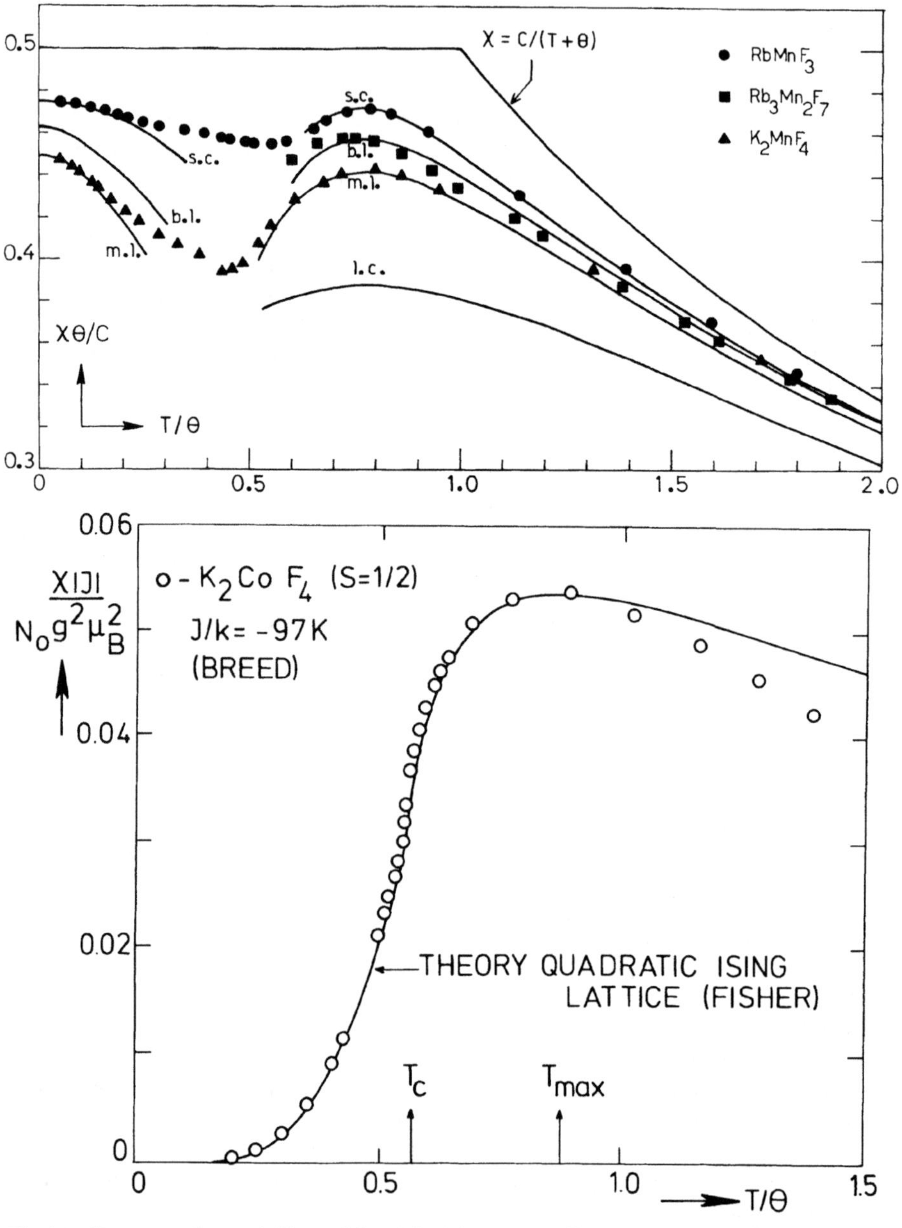


Fig. 11. *Upper part*: Susceptibilities of $S = 5/2$ Heisenberg antiferromagnets. Here $\theta$ and $C$ are the Curie—Weiss temperature and the Curie constant. Top curve is the molecular-field prediction for $\chi_\perp$. The other solid curves are theoretical predictions for the $S = 5/2$ Heisenberg antiferromagnet on the simple cubic (s.c.), bilayer (b.l.) monolayer (m.l.) and linear chain (l.c.) lattices, as discussed in the text. (After [1, 10, 61—64].) *Lower part*: Parallel susceptibility of $K_2CoF_4$ in reduced units, compared to the prediction for the $S = \frac{1}{2}$, quadratic layer Ising antiferromagnet. (From [1].)

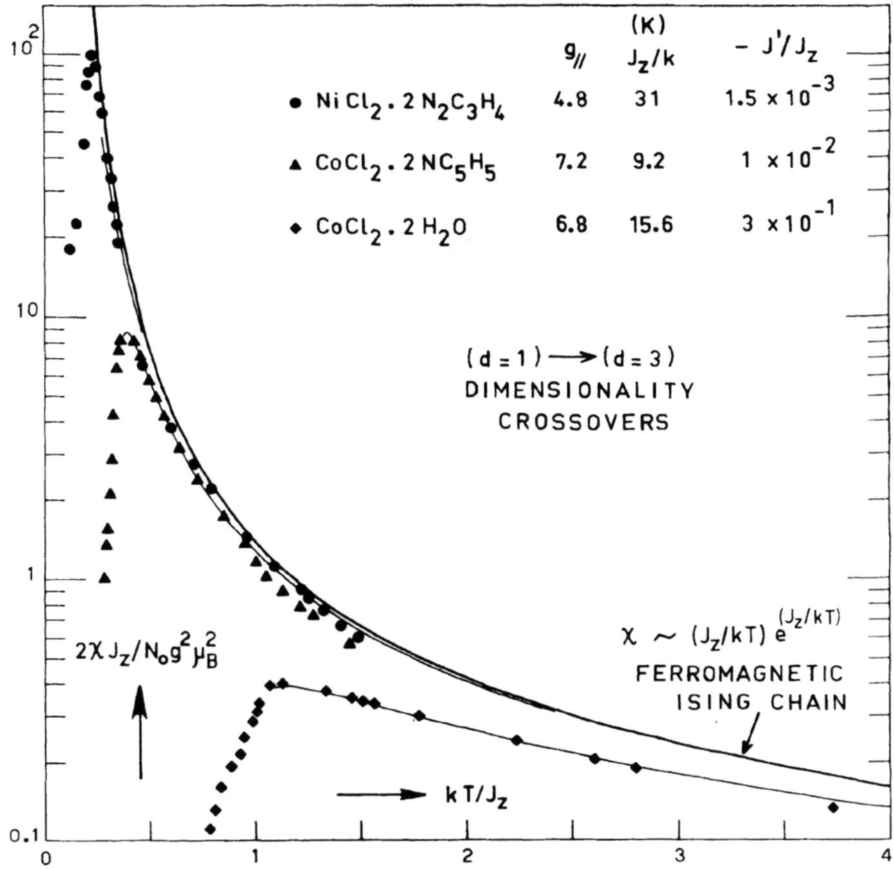


Fig. 12. Susceptibilities of ferromagnetic chain compounds. The thick solid curve is the exact result for the ferromagnetic $S = \frac{1}{2}$ Ising chain. The thin curves are calculated for an s.c. array of ferromagnetic Ising chains with antiferromagnetic interaction couplings $J'$, using the indicated values for $J'/J_z$. (From [65].)

For $J' > 0$, on the other hand, the effect of $J'$ will be an increase of the susceptibility with respect to $\chi_d$. A nice example is found [1, 58, 66] in the series of copper compounds $(C_nH_{2n+1}NH_3)_2CuCl_4$, which are approximations of the 2-d Heisenberg ferromagnet. The three compounds in Figure 13 have similar values for the anisotropy, so that their main difference is in the values of $R$. All three compounds have a planar anisotropy of about $3 \times 10^{-3}$; the $n = 10$ compound has a slightly lower Ising-type anisotropy ($6 \times 10^{-5}$), than the other two ($2 \times 10^{-4}$). The effect of this is a slight displacement of the $\chi$-curve with respect to the other two. This has been corrected for in constructing Figure 13, in the way described in [66], cf. Figure 8 in that paper. Since, for all compounds, the Ising anisotropy is a stronger perturbation than the interlayer coupling, it follows that

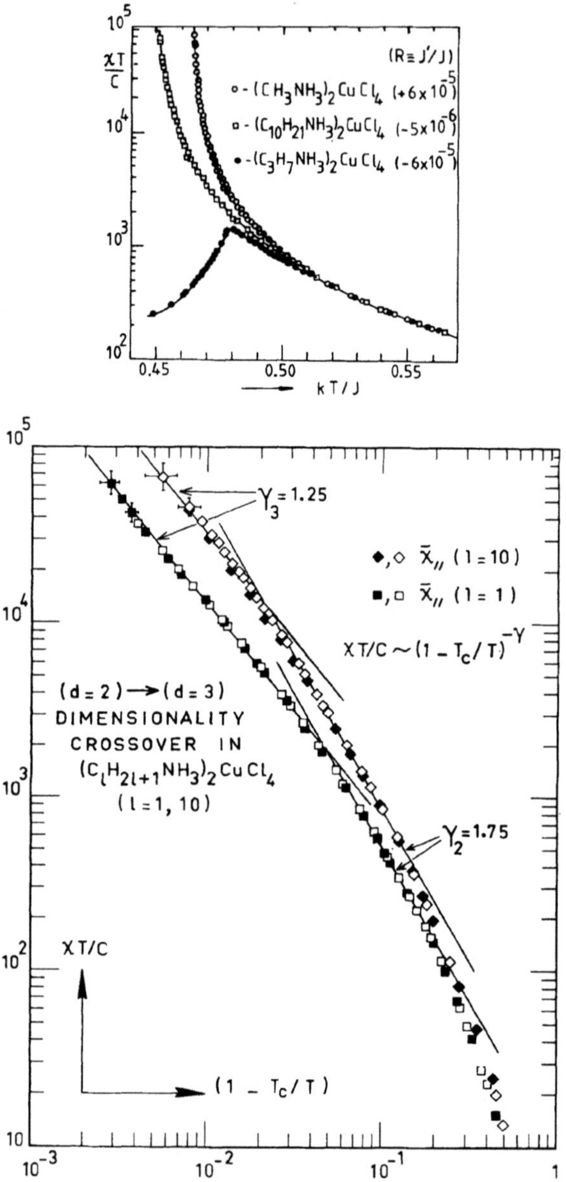


Fig. 13. *Upper part*: Susceptibility behaviour in the neighbourhood of $T_c^{3-d}$ for the compounds $(C_nH_{2n+1}NH_3)_2CuCL_4$ with $n = 1$, 3 and 10. These materials have $kT_c^{3-d}/J = 0.464$, 0.478 and 0.445, respectively. The interlayer coupling constants are: $R = +(5.5 \times 10^{-5})$, $-(6 \times 10^{-5})$ and $\pm(3 \times 10^{-6})$. *Lower part*: Critical behaviour of the ferromagnetic susceptibilities for the compounds with $n = 1$ and 10. (After [58] and [66].)

the lattice-dimensionality crossovers are preceded by an anisotropy crossover, i.e. the lattice-dimensionality crossover will be from 2-d Ising to 3-d Ising behaviour.

Now the compounds $(CH_3NH_3)_2CuCl_4$ and $(C_3H_7NH_3)_2CuCl_4$ both have $|R| \simeq 6 \times 10^{-5}$, however they have $R > 0$ and $R < 0$, respectively. The compound $(C_{10}H_{21}NH_3)_2CuCl_4$ on the other hand has $|R| \lesssim 5 \times 10^{-6}$, i.e. more than an order of magnitude smaller, so that in the temperature region shown, it may serve as an estimate of the ideal $\chi_{2d}$, with respect to which the susceptibilities of the other two materials should start to deviate as $T$ is lowered, one upward the other downward. As seen in the figure this is precisely what occurs. For the two ferromagnetic compounds the crossover occurs in fact in the critical region, where it can be seen (Figure 13) by a change in the value of the critical index $\gamma$ from the 2-d to the 3-d Ising prediction (cf. Table II). Furthermore, we mention that by comparing the $\chi$-data from many different ferromagnetic layer-type copper compounds, with varying values for the anisotropy and for $J'$, an experimental estimate of the 'ideal' $\chi_{2d}$ of the 2-d Heisenberg ferromagnet (with $S = \frac{1}{2}$) could be obtained [66] down to $kT/J \simeq 0.5$. These data were later shown by Kosterlitz and Santos [67] to agree reasonably well with their theoretical formula for this model, predicting an exponential temperature dependence of the form $\tilde{\chi}_{2d} \propto T^2 \exp(c/T)$ (with $c$ a constant).

As shown in Figure 9 the same series of copper compounds has allowed a beautiful demonstration of the dimensionality crossover in the specific heat, as discussed by Bloembergen. With decreasing value of $R$, the value of $T_c$ is seen to decrease, and the spike in the heat capacity that reflects the $\lambda$-anomaly corresponding to the 3-d order is observed to become smaller and smaller. For $R \simeq 6 \times 10^{-5}$ it is in fact scarcely observable. Thus the experiments at the same time provide a reasonably accurate experimental determination of the specific heat of the ideal 2-d Heisenberg ferromagnet, which should correspond to the broad anomalous curve underlying the data (as mentioned, the anisotropies are very small in these materials).

Lastly we discuss the crossover behaviour for the magnetization. For the direct magnetization of low-d antiferromagnets, that is the magnetization curves in uniform fields, the similar arguments as for the susceptibility apply, that is they will not be affected by $J'$ as long as $|R| \lesssim 10^{-2}$. On the other hand, in case of the spontaneous magnetization of a ferromagnet a crossover to 3-d behaviour will occur for $T \to T_c$, as follows from Equation (13). This is also the case for the sublattice magnetization of the antiferromagnet, since this being a staggered quantity, the $\chi_d$ in Equation (13) should then correspond to the staggered susceptibility, so that the whole problem is analogous to the ferromagnetic case. Experimentally the sublattice magnetization of many quasi 2-d antiferromagnets with the $K_2NiF_4$ structure have been studied with neutron diffraction, e.g. $K_2NiF_4$ [68], $K_2MnF_4$, $Rb_2MnF_4$ [69] etc., as well as by NMR methods. As will be discussed in the following chapters, in all these cases 2-d behaviour is found over the whole temperature range below $T_c$. As seen in Figures 14, and 15, close to $T_c$ the critical behaviour is found to be of the 2-d Ising-type, in agreement with the universality hypothesis. In these materials the weak, Ising-type anisotropy is of the

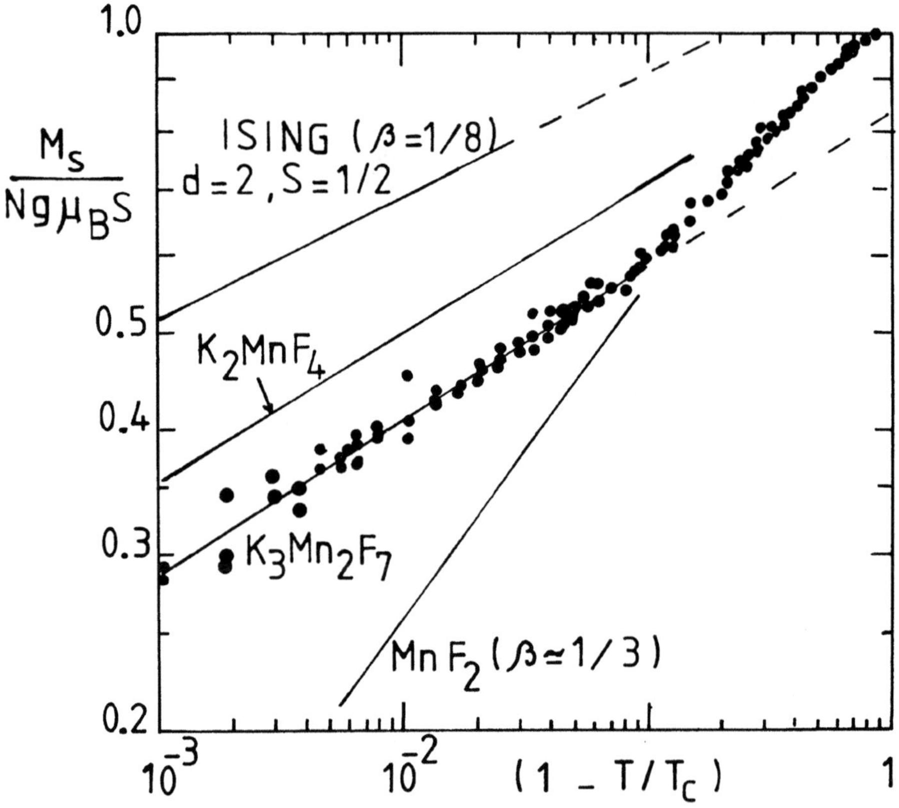


Fig. 14. Comparison of the behaviour of the sublattice magnetization of three antiferromagnetic manganese compounds (Heisenberg-type, $S = 5/2$), namely for $MnF_2$ (3-d; rutile structure), for $K_3Mn_2F_7$ (double layer, cf. Fig. 3); and for $K_2MnF_4$ (monolayer, cf. Fig. 3). (After ref. [75].)

order of $10^{-2}$—$10^{-4}$ [1], so that it constitutes a stronger perturbation than the interlayer couplings ($10^{-4}$—$10^{-6}$). The long-range ordering is thus primarily induced by the anisotropy-crossover from 2-d Heisenberg to 2-d Ising behaviour, which preceeds the 2-d Ising to 3-d Ising dimensionality crossover that should occur close enough to $T_c$. Evidently, for the Ising-type compounds $K_2CoF_4$ and $Rb_2CoF_4$, which are strongly anisotropic, the behaviour is Ising-like at all temperatures.

For the weakly anisotropic Heisenberg compounds, in the low-temperature range, where spin-wave theory applies, the temperature dependence is described by 2-d spin-wave theory for the weakly anisotropic Heisenberg antiferromagnet (see the chapter by Arts and de Wijn). So the behaviour appears to be effectively 2-d from $T = 0$ up to $T/T_c \simeq 0.999$ ($t = 1 - T/T_c = 10^{-3}$) for the 2-d antiferromagnets with the $K_2NiF_4$ structure. There is no doubt that this is due to

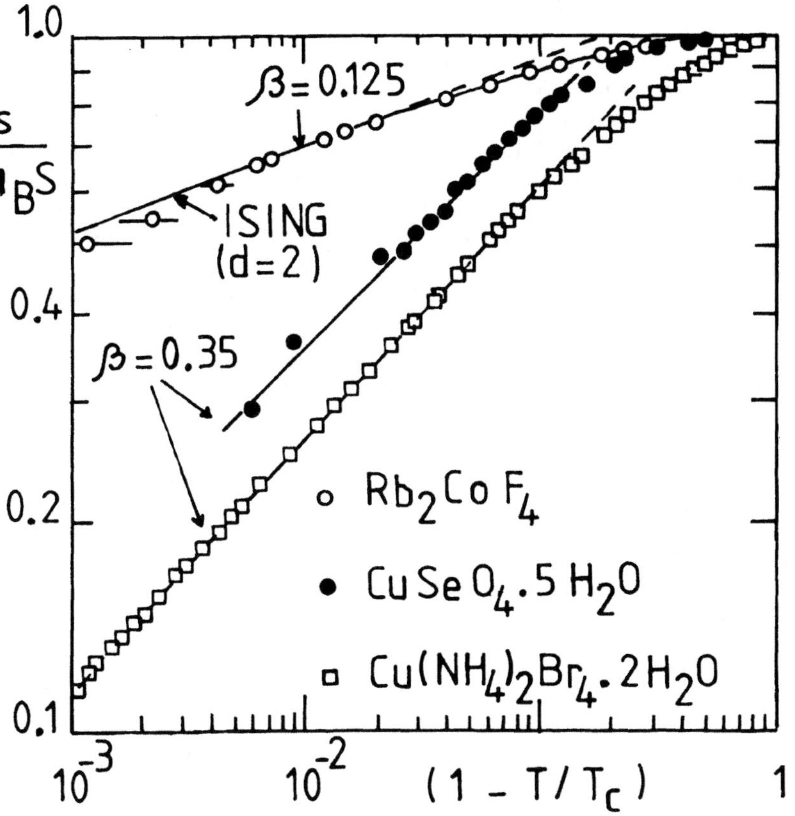


Fig. 15. Critical behaviour of the spontaneous magnetization of the antiferromagnetic Heisenberg, $S = \frac{1}{2}$, linear chain compound $CuSe_4O_4 \cdot 5\ H_2O$, as compared to that of the bcc, $S = \frac{1}{2}$, Heisenberg ferromagnet $Cu(NH_4)_2Br_4 \cdot 2\ H_2O$. Both results were obtained from NMR data [72]. Alos shown are neutron scattering data on the sublattice magnetization for $Rb_2CoF_4$ [73].

the extremely small ratio of $R = |J'/J|$ for these materials, where $R < 10^{-4}$ has been established as an upper limit [70]. (Arguments that $R \simeq 10^{-5}$—$10^{-6}$ can be found in [1]). The 2-d→ 3-d crossover thus probably occurs for $t < 10^{-3}$—$10^{-4}$, i.e. it is inaccessible to the experiments performed so far.

By contrast the 2-d ferromagnet $K_2CuF_4$ has a larger value, $R = 2 \times 10^{-4}$, for the interlayer coupling. This difference can be understood from the following considerations. Whereas with antiferromagnetic order in the layers of the $K_2NiF_4$ structure the interlayer coupling between nearest-neighbour planes is cancelled by symmetry (cf. Figure 3), this cancellation will not occur in the case of ferromagnetically ordered layers, leading to a much stronger interlayer coupling. We think this to be the explanation for the 3-d critical behaviour ($\beta \simeq 0.33$) of the sublattice magnetization observed for $K_2CuF_4$ in the range $t < 0.05$, as seen in Figure 16 [71].

It is of interest to compare the situation in a system of weakly coupled magnetic chains, as found for $CuSeO_4 \cdot 5\ H_2O$ (quasi 1-d Heisenberg-type, $S = \frac{1}{2}$ antiferromagnet) in Figure 15. Since for chain systems such as this the long-range order is brought about by 3-d interchain couplings, the critical behaviour of the sublattice magnetization should be 3-d like, with $\beta \simeq 1/3$ as for $MnF_2$ in Figure 14. Indeed, the slope of the curve for $CuSeO_4 \cdot 5\ H_2O$ is quite similar to that measured for the 3-d, $S = \frac{1}{2}$. Heisenberg copper ferromagnet $Cu(NH_4)_2Br_4 \cdot 2\ H_2O$, plotted in the same figure [72]. The behaviour of these materials is evidently in vivid contrast with the example of $Rb_2CoF_4$ shown in Figure 15, for which the data [73] lie on the prediction [74] for the 2-d Ising model with $S = \frac{1}{2}$ (solid curves; $\beta = 1/8$), and for $K_2MnF_4$, shown in Figure 14, which likewise has $\beta \simeq 1/8$.

We conclude this paragraph by pointing out that a different dimensionality crossover should occur in the magnetic multilayers of the type shown in Figure 3 or, more generally, for very thin magnetic films. In such systems the magnetic behaviour at high temperatures will be 3-d, namely as long as the magnetic correlation length $\xi(T)$ does not exceed the layer thickness $\ell$. As soon as $\xi > \ell$, however, as will happen at low enough temperature, the system will realize its restricted dimensionality and become effectively 2-d. This can be observed in

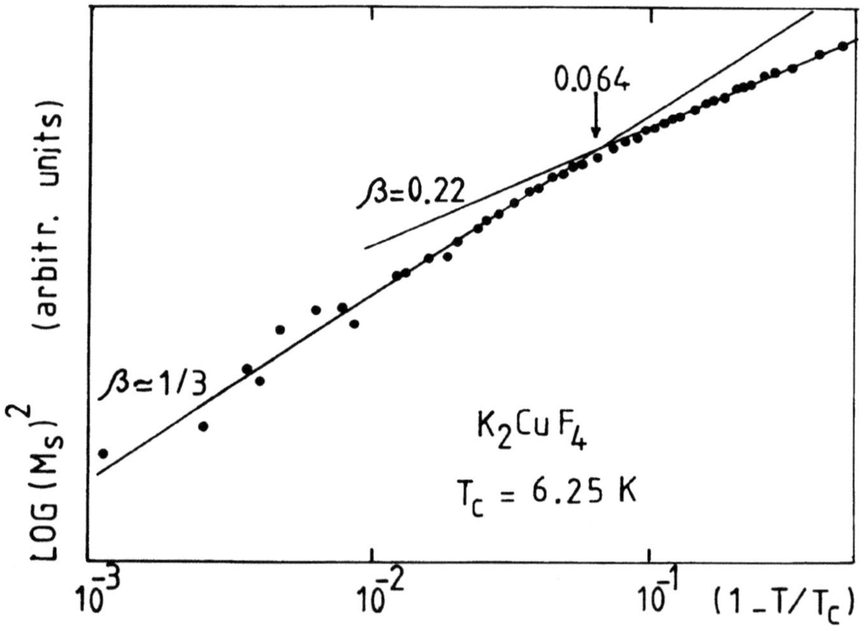


Fig. 16. Example of a 2-d → 3-d lattice dimensionality crossover in the critical behaviour of the spontaneous magnetization of $K_2CuF_4$, a quasi 2-d, $S = \frac{1}{2}$, Heisenberg ferromagnet with weak Ising anisotropy. The order parameter has been measured with elastic neutron scattering [71].

Figure 11 by comparing the $\chi$-curves for $RbMnF_3$, $Rb_3Mn_2F_7$ and $K_2MnF_4$. For the antiferromagnetic double-layer $K_3Mn_2F_7$ the critical behaviour of the sublattice magnetization has been measured by neutron scattering [75], and is shown in Figure 14. The 2-d nature is obvious from this plot, the slope of the curve being again much closer to the 2-d Ising value $\beta = 1/8$, than to the values $\beta = 0.35$ typically found for 3-d systems.

## 5. Magnetic and Nonmagnetic Impurity Doping in Quasi 2-d Magnets

In recent years random or disordered magnetic systems have enjoyed a vivid interest among theoretical and experimental physicists [76—82]. We mention spin glasses, amorphous magnets, diluted magnets, random-field or random-anisotropy magnets, and random systems of two magnetic species with competing spin anisotropies or different exchange interactions. In all these problems low-d systems play an important role, since the dimensionality of the lattice often is a crucial parameter in determining the thermodynamic behaviour. It seems therefore appropriate to review here briefly some of these concepts, more-so since in subsequent chapters these themes will be taken up again.

We first summarize some theoretical and experimental results on random-site diluted magnets, i.e. magnetic systems in which the magnetic ordering temperature $T_c$ is decreased with decreasing concentration $p$ of the magnetic species. For large nonmagnetic impurity concentration the value of $T_c$ is reduced to zero at a critical concentration $p_c$ of magnetic atoms, the value of which depends on the lattice-dimensionality $d$, the magnetic coordination number $z$, and, for 2-d lattices, also on the spin-dimensionality. Here we obviously concentrate on the 2-d aspects of the problem. We shall end with some brief remarks on the magnetic impurity doping and other random problems.

In formulating the percolation problem in terms of a lattice model, one considers a medium defined by lattice sites connected by bonds. In the bond percolation problem these bonds can be either blocked or open, whereas the sites are always open. In the site-percolation problem the reverse is the case. The former medium is e.g. a model for a porous stone, through which water may percolate in case the density of pores (open bonds) is sufficiently high. Calling $p$ the probability that a bond on the lattice is open, it is clear that a certain minimum probability $p_c$ is needed before percolation through the medium is possible ($p$ may be interpreted as the density of open bonds). The value $p_c$ is called the critical probability. For $p < p_c$ only finite clusters of open bonds will be found on the lattice, for $p > p_c$ an infinite cluster of open bonds is formed (and the liquid can flow through the stone). Calling $n(p)$ and $N(p)$ the fraction of open bonds in finite, isolated clusters and in the infinite cluster, respectively, the dependence of $n(p)$ and $N(p)$ is qualitatively as sketched in Figure 17a. A third quantity of interest is the mean size $S(p)$ of a finite, isolated cluster, which diverges for $p \rightarrow p_c$, marking the onset of the formation of the infinite percolation cluster.

Although in the above discussion bond percolation is considered, the site percolation problem is qualitatively the same. The critical probability $p_c$ is however

(a)

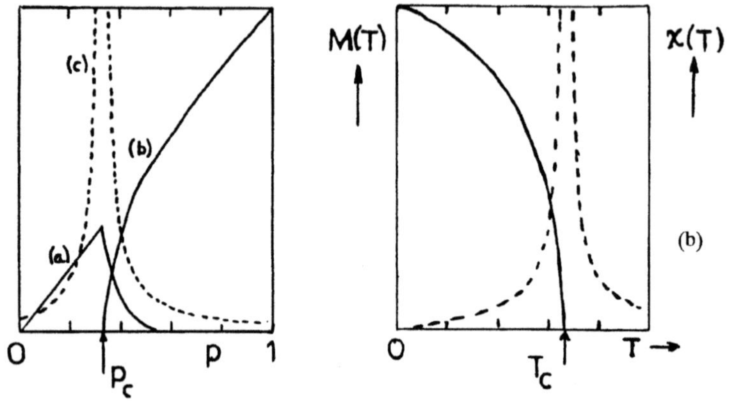


(b)

Fig. 17. **a**: Qualitative behaviour of (a) fraction of open bonds in isolated finite clusters; (b) fraction of open bonds in the infinite cluster, and (c) mean size of the isolated finite clusters. **b**: Qualitative behaviour of the spontaneous magnetization and the susceptibility of the Ising ferromagnet.

numerically different for the two cases. This quantity depends on the number $z$ of interacting magnetic neighbours, and also on the dimensionality of the lattice. Since blocking a single site corresponds with the blocking of $z$ bonds in the lattice, it is not surprising that one always has $p_c(\text{site}) > p_c(\text{bond})$ when the same lattice is considered. Theoretical estimates for these quantities are listed in Table V for the most common 2-d and 3-d lattices [79].

In an important paper [83], Kasteleyn and Fortuin have shown that the percolation problem is but one member of a family of system which includes the ferromagnetic $S = \frac{1}{2}$ Ising model, the Potts model and the linear resistor network. In the Ising ferromagnet a cluster of open bonds corresponds with a correlated spin cluster, the probability $p$ is replaced by the inverse temperature, and the analogues of $N(p)$ and $S(p)$ are, respectively, the spontaneous magnetization and the ferromagnetic susceptibility, which vanishes resp. diverges at the critical temperature $T_c$ (cf. Figure 17b).

Apart from this direct relationship with a model for a pure ferromagnetic system, the percolation problem can also be related to diluted magnetic systems. Although theoretically both bond and site dilution has been studied, the experimental examples so far correspond with the site dilution problem. One may then still distinguish between the annealed-site and the quenched-site case, for which the distribution of nonmagnetic impurities in the magnetic lattice is assumed to be either in thermal equilibrium or completely random, respectively. In the experimental compounds one usually assumes the random site model to apply, which should, however, for each individual case, be checked as carefully as possible.

For the dilute magnet, $p$ corresponds with the concentration of magnetic atoms. The way in which $T_c(p)$ is reduced with decreasing $p$ depends on the lattice symmetry, but also on the symmetry of the interaction Hamiltonian. The variation

TABLE V
Critical bond and site probabilities $p_c(b)$ and $p_c(s)$, for the most common 2-d and 3-d lattices.

| Lattice | $z$ | $p_c(s)$ | $zp_c(s)$ | $p_c(b)$ |
|---|---|---|---|---|
| Honeycomb | 3 | 0.700 | 2.10 | 0.653 |
| Square | 4 | 0.590 | 2.36 | 0.500 |
| Triangular | 6 | 0.500 | 3.00 | 0.347 |
| Diamond | 4 | 0.425 | 1.70 | 0.388 |
| s.c. | 6 | 0.307 | 1.84 | 0.247 |
| bcc | 8 | 0.243 | 1.94 | 0.178 |
| fcc | 12 | 0.195 | 2.34 | 0.119 |
| hcp | 12 | 0.204 | 2.45 | 0.124 |

Critical probabilities with longer-ranged interactions (equivalent neighbour model)

| Lattice | $z$ | $p_c(s)$ | $zp_c(s)$ |
|---|---|---|---|
| Square (1, 2) | 8 | 0.410 | 3.28 |
| Triangular (1, 2) | 12 | 0.295 | 3.54 |
| Honeycomb (1, 2, 3) | 12 | 0.300 | 3.60 |
| Square (1, 2, 3) | 12 | 0.292 | 3.50 |
| Triangular (1, 2, 3) | 18 | 0.225 | 4.06 |
| bcc (1, 2) | 14 | 0.175 | 2.45 |
| s.c. (1, 2) | 18 | 0.137 | 2.47 |
| fcc (1, 2) | 18 | 0.136 | 2.45 |
| s.c. (1, 2, 3) | 26 | 0.097 | 2.52 |
| bcc (1, 2, 3) | 26 | 0.095 | 2.47 |
| fcc (1, 2, 3) | 42 | 0.061 | 2.56 |

$T_c(p)$ as been studied for Ising, XY, and Heisenberg models by many authors, using e.g. series expansion techniques, finite cluster approximations, or numerical (Monte Carlo) calculations [82]. Independently, Elliott and Heap [84], and Morgan and Rushbrooke [85] have proved that for 3-d ferromagnetic lattices the critical concentration $p_c$ at which the transition temperature has been reduced to zero is just equal to the critical percolation probability of the lattice in question. Thus, for 3-d *ferromagnetic* lattices, $p_c$ appears to be a topological property i.e. independent of the symmetry of the hamiltonian or the spinvalue and only dependent on the symmetry of the lattice. The same would hold for the simple 3-d two-sublattice Ising and XY antiferromagnets, since for those models the transition temperatures are independent of the sign of the exchange interaction. However, for 3-d Heisenberg antiferromagnets the transition temperatures differ from the ferromagnetic case, and it is therefore unlikely that the $p_c$ values would be the same as for the ferromagnetic lattices. The same is true for the triangular and fcc Ising or XY antiferromagnets. Lastly, for 2-d lattices there is no longer any relationship possible between the three models, since only the Ising model has a transition to long-range order at a $T_c > 0$. Clearly, the existence of an infinite percolation cluster for $p \geqslant p_c$ is a *necessary* but not a *sufficient* condition for the occurrence of long-range magnetic order.

In Table VI we list a number of magnetic systems for which fairly extensive studies of the effects of dilution have been performed, and which are good approximations of the various 2-d and 3-d magnetic model systems considered in this book The comparison of results for different model systems is worthwhile, since it reveals the effects of lattice-dimensionality and type of interaction on the thermodynamic properties of diluted systems.

TABLE VI
Some experimental examples of diluted systems.

| Compound | Spinvalue | Interaction type | Interaction sign | Lattice |
|---|---|---|---|---|
| $K_2CoF_4$:Mg | $S = 1/2$ | Ising | AF | square |
| $Rb_2CoF_4$:Mg | $S = 1/2$ | Ising | AF | square |
| $K_2MnF_4$:Mg | $S = 5/2$ | Heis. | AF | square |
| $K_2CuF_4$:Zn | $S = 1/2$ | Heis. | F | square |
| $KMnF_3$:Mg | $S = 5/2$ | Heis. | AF | s.c. |
| $CoL_6(ClO_4)_2$:Zn | $S = 1/2$ | XY | AF | s.c. |
| ($L = C_5H_2NO$) | | | | |
| $FeL_6(ClO_4)_2$:Zn | $S = 1/2$ | Ising | AF | s.c. |
| $CoCs_3Cl_5$:Zn | $S = 1/2$ | Ising | AF | b.c.c. |

Starting with the 2-d magnets we show in Figure 18 the the data for $T_c(p)$ obtained by Breed *et al.* [86] from susceptibility experiments on $K_2CoF_4$: Mg and and $K_2MnF_4$ : Mg. It can be seen that the $K_2CoF_4$ : Mg data are very well described by the theoretical prediction for the quadratic layer, $S = \frac{1}{2}$, Ising antiferromagnet, as obtained e.g. by Idogaki and Uryû [87]. Also the results for the isomorphous compounds $Rb_2CoF_4$ : Mg reported by Ikeda *et al.* [88] fall nicely on this same curve, which extrapolates to the critical concentration $p_c \simeq 0.59$ given by percolation theory (cf. Table V). The data on $K_2MnF_4$ : Mg are however considerably below this curve, except close to $p_c$. This reflects no doubt the fact that this compound closely approximates the 2-d Heisenberg model, which ideally has no transition to LRO. The magnetic ordering that occurs is thus rather pathologic, since it is induced by the small Ising component, combined with the very weak interlayer interactions. Similar considerations apply to $K_2CuF_4$, which is to a good approximation an example of the 2-d Heisenberg ferromagnet. The similarity of the $T_c(p)$ curves for the two systems may be coincidental, since $K_2MnF_4$ is an antiferromagnet with spin value $S = 5/2$, whereas $K_2CuF_4$ is ferromagnetic, with $S = \frac{1}{2}$. The extreme sensitivity of the ordering temperature $T_c^{3\text{-}d}$ of the quasi 2-d Heisenberg systems to impurity doping can be understood by the same molecular field agreement that relates $T_c^{3\text{-}d}$ to the correlation length $\xi_{2\text{-}d}$ in the layers, as used above. Putting $kT_c^{3\text{-}d} \simeq \xi_{2\text{-}d}^2 J' S^2$, it is clear that as soon as $\xi_{2\text{-}d}(T)$ becomes limited

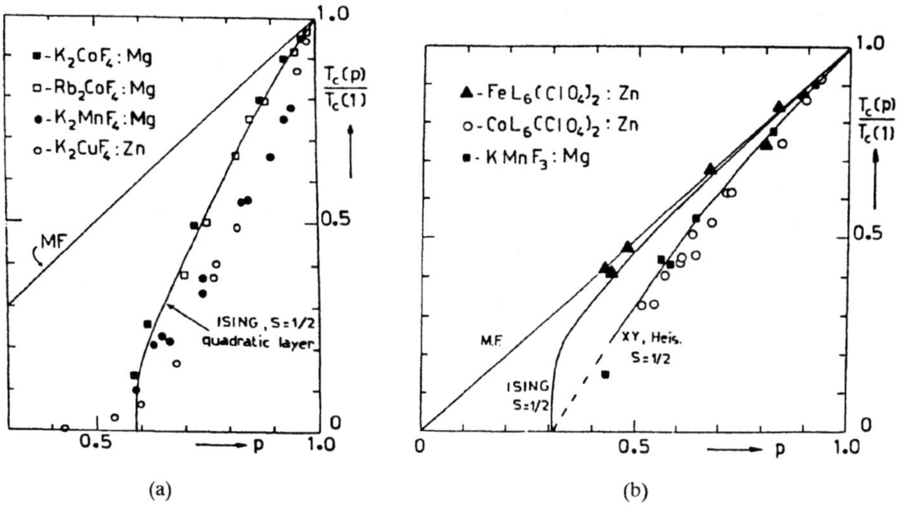


Fig. 18. **a**: Experimental variation of $T_c(p)$ for quasi 2-d magnets. Solid curve is a theoretical prediction for the quadratic layer Ising magnet. **b**: Experimental variation of $T_c(p)$ for simple cubic antiferromagnets compared with theory of the s.c. Ising and s.c. XY, $S = \frac{1}{2}$ models.

by the density of impurities, the $T_c^{3-d}$ will start to decrease with respect to its value in the pure system. In other words, since the impurities are distributed randomly in each of the magnetic layers, the sensitive interlayer order is quite easily broken up through mismatches in the coupling along the direction perpendicular to the sheets. The same argument applies, *a fortiori*, to systems of weakly coupled chains.

Next we show for comparison in Figure 18b the data for three simple cubic antiferromagnets, namely the $S = 5/2$ Heisenberg compound $KMnF_3$ : Mg [86], the $S = \frac{1}{2}$, XY system $CoL_6(ClO_4)_2$ : Zn [89], and the $S = \frac{1}{2}$ Ising system $FeL_6(ClO_4)_2$ : Zn [90]. In the same figure are plotted the theoretical predictions by Reeve and Betts [91] for the s.c., XY magnet, by Stinchcombe [92] for s.c. Heisenberg, and by Stinchcombe [92] and Landau [93] for the s.c. Ising magnet. All predictions are for $S = \frac{1}{2}$. On the scale of the figure, the predictions for s.c. Heisenberg and s.c. XY coincide. It may be seen that the experimental $S = \frac{1}{2}$ Ising and XY compounds show a quite reasonable agreement with theory for these models. The fits are not perfect, but this may be ascribed in part to the fact that the extreme anisotropy inherent to either the Ising or XY model will only be approximately realized in the experimental compounds. One may thus expect data on Ising or XY materials to fall somewhat below the theoretical curves, since the data will be 'shifted' in the direction of the Heisenberg prediction. We note that all the data may be extrapolated to the critical value $pc \simeq 0.31$ expected for the s.c. lattice (cf. Table V). The difference with the 2-d materials is very apparent.

In conclusion we may say that the available data are *grosso modo* well described by the existing predictions. In particular the predicted strong differences in the variation of $T_c$ with $p$ between the 2-d and 3-d magnetic lattices is confirmed by the experiment. As regards the quasi 2-d magnetic materials, we further add that theoretically the $T_c(p)$ curves for the experimental compounds should level off at very small $T_c$ values, because of lattice-dimensionality crossover effects. This arises because in the experimental examples there will always be weak interlayer interactions present, so that the magnetic lattice is basically 3-dimensional. Thus, at very small $T_c(p)$ values the curves should divert from the 2-d prediction, and vanish at the appropriate 3-d value for $p_c$. For the compounds $K_2CoF_4$ and $K_2MnF_4$, however, the ratio of inter- to intralayer couplings is so small ($\simeq 10^{-6}$) that one cannot hope to detect such an effect experimentally. For $K_2CuF_4$; on the other hand, this ratio is only $10^{-4}$, so that the leveling off which is observed in the $T_c(p)$ curve for this material (Figure 18a) might be the effect of the interlayer interactions just described.

Turning next to the effect of dilution on the antiferromagnetic susceptibility, we mention the important experimental finding that the susceptibility of isotropic antiferromagnetic systems increases rather sharply upon dilution. This is illustrated in Figure 19 by data for $K_2MnF_4:Mg$ and $KMnF_3:Mg$ from Breed *et al.* [86]. They found a linear relationship between the value of the perpendicular susceptibility extrapolated to $T = 0$ and the inverse of the transition temperature $T_c(p)$, which would imply that $\chi_\perp(p) \to \infty$ for $p \to p_c$. In explaining this effect, one might think of extra paramagnetic contributions arising from finite isolated clusters containing an odd amount of $Mn^{2+}$ spins. However, the density of spins in isolated clusters decreases very rapidly above $p_c$ (Figure 17a), so that this cannot be the (only) reason for the divergence.

The phenomenon was explained by Harris and Kirkpatrick [94], who have shown that in isotropic antiferromagnets such increases in $\chi_\perp$ are caused by local fluctuations in the total magnetic moment. The latter arise because in the neighbourhood of a nonmagnetic impurity the balance of the two antiferromagnetic sublattices is destroyed [96]. These contributions to $\chi_\perp$ should be suppressed by the presence of crystal-field anisotropy, since this fixes the direction of the moments parallel to the symmetry axis (Ising axis). Indeed, in anisotropic magnets the increases in $\chi_\perp(p)$ are observed to be much less pronounced. For instance in $K_2CoF_4:Mg$ an increase by a factor of 1.7 only is observed in going from $p = 1$ to $p = p_c$ [86]. This should be compared with the factor of 50 that one may estimate for $KMnF_3:Mg$ from the data in Figure 19a when going from $p = 1$ to $p = p_c = 0.31$. Similarly large effects have recently been observed with the susceptibility of Zn-doped $CoL_6(ClO_4)_2$ [95]. On the other hand, in case of Ising-type antiferromagnets with impurities and in a magnetic field, one has to consider the random-field effects that are induced. Basically, such impurities have the effect of producing a coupling between the field-induced uniform magnetization of the antiferromagnet and its staggered (sublattice) magnetization. Accordingly, the application of a uniform field entails simultaneously the presence of random staggered fields that act on the order parameter. Recent susceptibility data on $Rb_2CoF_4:Mg$ by Ikeda [97] were interpreted in this way (see the chapter by

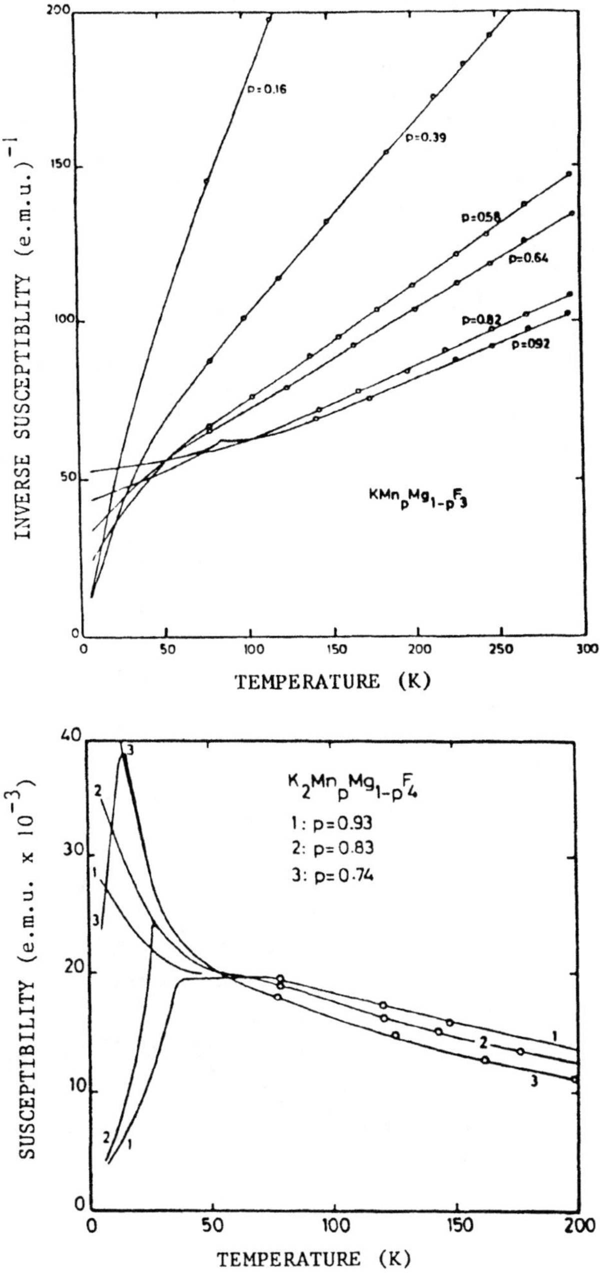


Fig. 19. Experimental susceptibilities of $KMnF_3$ : Mg and $K_2MnF_4$ : Mg. (After [86].)

Hirakawa and Ikeda). We will comment further on the random field problem below.

Quite interesting effects of dilution can also be seen in the magnetic specific heat. As $T_c$ is decreased to zero with increasing nonmagnetic impurity content, a progressively larger fraction of the magnetic entropy will be removed by short-range order processes. Consequently, the size of the $\lambda$-type ordering anomaly will decrease and, as soon as $T_c$ has fallen below the value of $|J/k_B|$, a characteristic broad maximum will appear in the specific heat, corresponding to these short-range order processes. The presence of such a broad maximum would become particularly apparent for $p \to p_c$.

There is in fact a strong similarity between the thermodynamic properties of the infinite magnetic percolation cluster for values of $p$ just above $p_c$, and a system of weakly coupled chains [82, 98]. Decreasing $p$ towards $p_c$ then is analogous to decreasing the interchain interaction. The similarity may be illustrated by the example of a theoretical calculation [99] for the effects of dilution of a 2-d Ising lattice, shown in Figure 20. The strong resemblance with the Onsager calculation

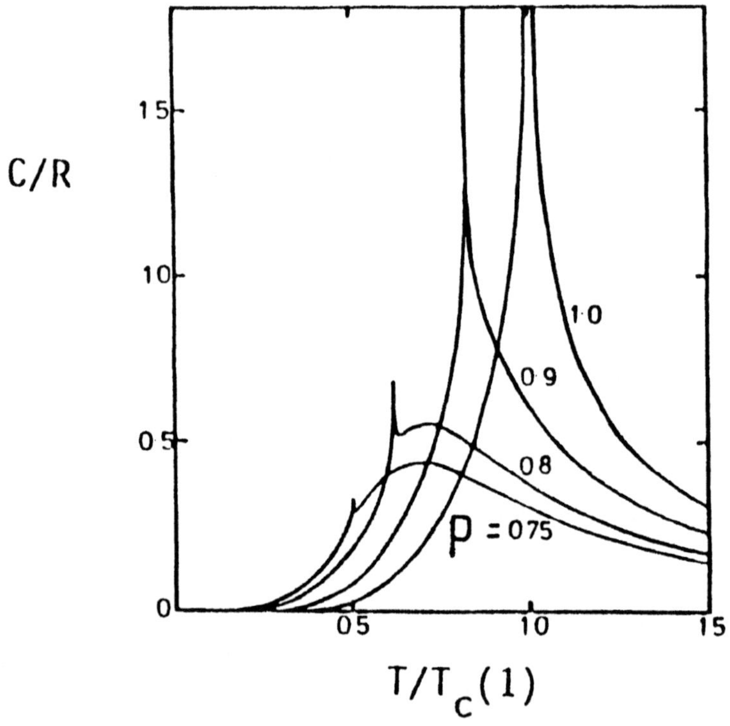


Fig. 20. Calculated specific heat for the quenched random diluted honeycomb Ising lattice. (From [99].)

for the specific heat of a 2-d array of weakly coupled chains (see Figure 1 in the chapter by Navarro) is evident. As pointed out earlier [82], the analogy may be understood on the basis of a simple model, called the 'nodes and links' model, for the description of the infinite percolation cluster. In this model the infinite cluster is represented (for $p \to p_c$) by a network of 1-d macrobonds, which cross one another at modes that are mutually separated by distances of the order of a characteristic length, the connectedness length, $\xi_p$(cf. Figure 21). This length scale $\xi_p$ should be compared to the thermal spin-spin correlation length $\xi_T$. Obviously, at sufficiently low temperatures, one will have $\xi_T \gg \xi_p$, so that the precise structure of the percolation cluster will not be important and a 3-d magnetic long-range order may be established as $\xi_T$ diverges. However, for $\xi_T \leqslant \xi_p$ the fractal structure of the percolation cluster will dominate the thermodynamic behaviour. Most important in this respect is the realization that in this model the extent of spin—spin correlations, i.e. the length $\xi_T$, will be driven by the correlations along the macrobonds. Since these are essentially 1-d objects, $\xi_T$ will be determined by the (quasi) 1-d correlation length $\xi_{1\text{-}d}$.

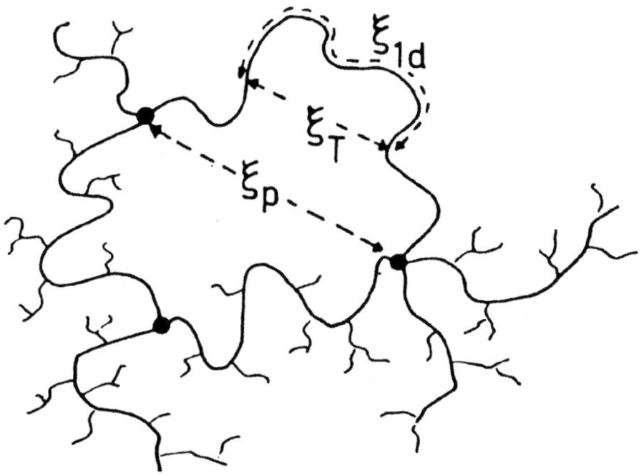


Fig. 21. The structure of the infinite cluster according to the 'nodes and links' model. The nodes, which are at distances of the order $\xi_p$ are connected by macrobonds (bold lines), which form part of the 'backbone' of the infinite cluster. The dangling ends (light lines) are not connected to infinity. The thermal correlation length $\xi_T$ may be correlated with the intrachain correlation length $\xi_{1d}(T)$, which behaves differently for the Ising, XY and Heisenberg type of interaction.

For $p \to p_c$ one has $\xi_p \to \infty$, that is the structure of the percolation cluster becomes increasingly more ramified. Consequently, the 1-d characteristics will become more and more pronounced. The 3-d ordering $T_c$ will move to lower and lower temperature, and the properties above $T_c$ will resemble more closely that of a chain system. For the specific heat this implies a broad, nonanomalous curve as predicted for the linear chain [98]. The presence of such a maximum is indeed apparent in the theoretical predictions for 2-d Ising lattices as shown in Figure 20. Experimentally, although specific heat experiments have been performed on the 2-d Ising compound $Rb_2CoF_4:Mg$ [100], the presence of the Schottky maximum was not resolved in that case, probably because of too low a concentration of Mg atoms in the diluted samples. However, in an earlier experiment on the quasi 2-d Heisenberg compound, $Mn(HCOO)_2 \cdot 2H_2O:Mg$, the maximum was clearly evident [101].

Since 3-d isotropic systems also show strong short-range-order effects, one might expect the just described effects to appear also in the diluted 3-d XY and Heisenberg magnets, in particular for those with low spinvalue (since $T_c \propto S(S+1)$). Indeed, in the specific heat data for the s.c., $S = \frac{1}{2}$, XY antiferromagnet $CoL_6(ClO_4)_2:Zn$ reported by Algra *et al.* [102], the Schottky maximum appears already for $p < 0.7$. As pointed out by these authors, such values of $p$ are still so far above $p_c = 0.31$ that the fraction of magnetic spins in isolated finite clusters is negligible (cf. Figure 17a), so that the broad anomaly has to be associated with the infinite cluster of spins. Interestingly, it was found that, just for $p = p_c$, the experimental Schottky anomaly could be excellently described by the prediction for a $S = \frac{1}{2}$ linear chain XY magnet, with the exchange constant needed for the fit just equal to the value for the pure compound. It was concluded that just at $p_c$ the degree of ramification of the infinite cluster is largest, so that it shows effectively 1-d behaviour as far as the thermodynamic properties are concerned.

Similar conclusions regarding the 1-d character of the infinite cluster just at the percolation limit have been drawn from other types of experiments. Birgeneau *et al.* [103] performed neutron scattering experiments on the 2-d antiferromagnet $Rb_2MnF_4:Mg$ and derived the temperature dependence of the correlation length in samples with $p = 0.54$ and $0.57$, i.e. close to $p_c = 0.59$. They conclude that the data are well explained in terms of a 1-d model of self-avoiding walks. Subsequent neutron scattering experiments by the same group (Cowley *et al.* [104] on the 3-d systems $MnF_2:Zn$ and $KMnF_3:Zn$ could be analysed in similar terms. Finally, we mention the susceptibility experiments of Okuda *et al.* [105] on the 2-d ferromagnetic Heisenberg system $K_2CuF_4:Zn$, in which it was shown that the susceptibility data for a sample with $p = 0.60$, i.e. very close to $p_c = 0.59$, are nicely described by the theoretical prediction for a ferromagnetic Heisenberg chain, whereas the results for higher or lower values of $p$ do not fit the chain prediction.

Turning next to the random-field problem, we recall that in recent years, there has been a vivid interest in *the field-dependent* properties of Ising-type antiferromagnets with nonmagnetic (or magnetic) impurities. The reason is that in such systems the application of a uniform field entails the generation of random staggered fields in the system, with profound consequences for the critical behaviour. As discussed by Fishman and Aharony [106], the uniform field produces a local ferromagnetic magnetization component because the antiferromagnetic sublattices are incompletely balanced in the neighbourhood of the impurity. This uniform component couples with the staggered component of the magnetization, leading to the random staggered fields [107]. Because of space limits we shall only give a very brief outline here, and refer the reader for more information to the many reviews on the theoretical and experimental work that have already appeared [108—112].

For 2-dimensional physics, perhaps the most important result of random-field theory is the prediction that 2-d Ising type systems will no longer have a transition to long-range order as soon as random fields are present. The argument is that the latter cause the formation of large domains, which effectively destroy the long-range order at any temperature $T > 0$ for Ising systems with a dimensionality lower than a critical value $d_c$. In the original domain energy argument of Imry and Ma [113] the value of $d_c = 2$ was obtained. The reasoning is that the formation energy of a domain of size $R$, $E_w \propto J R^{d-1}$, becomes smaller than the gain in random-field energy, $E_r \propto H_0 R^{d/2}$, obtained in forming the domain. Here $H_0$ is the average random field strength, which is supposed to be small compared to $|J|$. For $d > 2$ the net energy is positive, whereas for $d < 2$ it is negative when $R$ is sufficiently large. Thus $d_c = 2$. Although there has been a theoretical debate on this problem for several years, it appears that by now this result has been confirmed by other approaches and should be considered as well established [109].

Even when for $d > 2$ the transition to long-range order remains preserved, the nature of the transition is changed in the presence of weak random fields. For instance critical exponents are predicted to change, and first-order transitions loose their first-order character (become smeared).

As an illustration of the difference in behaviour for $d = 2$ and $d = 3$ random-field systems, we reproduce in Figure 22 the results for the magnetic specific heat of the 3-d system $Fe_{0.46}Zn_{0.54}F_2$ and the 2-d system $Rb_2Co_{0.85}Mg_{0.15}F_4$, both measured in zero and in applied external magnetic fields [114]. As mentioned, in impurity-doped antiferromagnets the application of a uniform external field entails the presence of random staggered fields. Whereas the transition for the 3-d system is seen to be even sharpened by applying the field, for the 2-d system the transition is severely broadened and appears to be destroyed. This has been confirmed in subsequent neutron scattering studies [115].

For small applied fields the data on $Fe_{0.46}Zn_{0.54}F_2$ appear to be independent on the history of the field cycling procedure. However, for fields excluding 0.5 T, differences appear between field cooled and zero-field cooled data. This illustrates at the same time the change in the current interest in this area of research [109]. Whereas initially most of the emphasis was put on the study of changes in critical behaviour and in the determination of the lower critical dimensionality $d_c$, the research is nowadays concentrating on nonequilibrium and history-dependent effects, related to the dynamics of domain walls and metastability. In this respect we refer to the recent extensive neutron scattering studies on these problems [115].

Finally we mention that in antiferromagnets with weak Ising anisotropy, the application of a magnetic field may cause the well-known phenomenon of spin flopping at a critical field $H_{SF}$. At $H_{SF}$ the system has become effectively isotropic, whereas above $H_{SF}$ the symmetry is changed from Ising to XY like. Since random-field effects are only expected for Ising systems, such changes in symmetry are evidently of profound importance for the study of the random-field problem. This

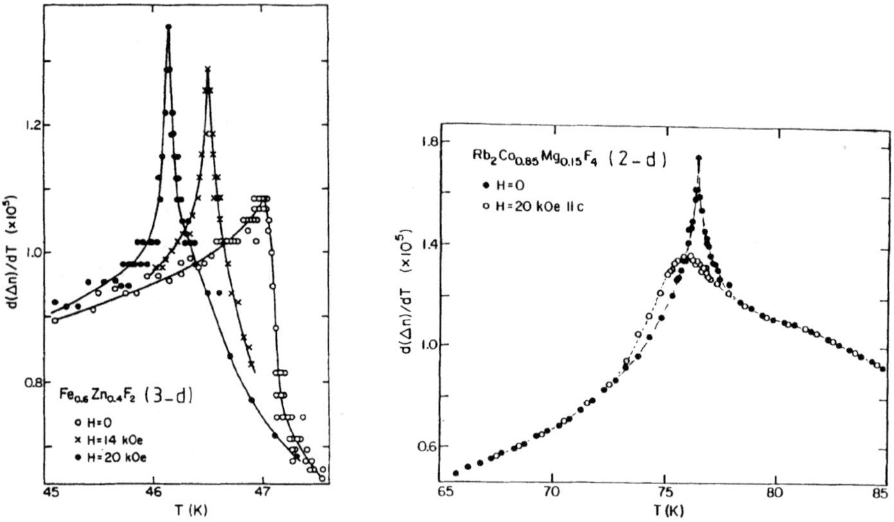


Fig. 22. Comparison of the field-dependence of the specific heat of the Ising-type antiferromagnets $FeF_2$ (3-d) and $Rb_2CoF_4$ (2-d), both doped with nonmagnetic ions (Zn and Mg). The quantity $\Delta n$ is the optical linear birefringence, which is known to be proportional to the magnetic energy, so that its temperature derivative will give the specific heat behaviour. It is seen that the transition becomes sharper and rounded for, respectively, the 3-d and the 2-d Ising-type compound. (After [114].)

subject will be taken up in the chapter by de Groot and de Jongh. It is interesting to mention, however, that some of the earliest studies on random-field behaviour by Rohrer [107], were in fact concerned with the spin-flop problem in a weakly anisotropic antiferromagnetic system.

In concluding this section we want to point to the interesting phenomena that are encountered when, instead of diluting a magnetic system with nonmagnetic ions, a random mixture is considered of two magnetic species with competing interactions and/or competing spin-anisotropies. Let us consider for the sake of the discussion a random mixture of two magnetic ions A and B, described by the Hamiltonian:

$$\mathcal{H} = -2J_{AA} \sum_{\langle ij \rangle} \mathbf{S}_{Ai} \cdot \mathbf{S}_{Aj} - 2J_{BB} \sum_{\langle k,\ell \rangle} \mathbf{S}_{Bk} \cdot \mathbf{S}_{B\ell} - 2J_{AB} \sum_{\langle m,n \rangle} \mathbf{S}_{Am} \cdot \mathbf{S}_{Bn} - \\ - D_A \sum_i (S^z_{Ai})^2 - D_B \sum_k (S^z_{Bk})^2. \tag{15}$$

Here $J_{AA}$, $J_{BB}$ and $J_{AB}$ are the interactions between two neighbouring A moments, B moments, and between an A and a B moment, respectively. We assume only interactions between nearest neighbours to be present. To a good approximation one may take $|J_{AB}| \simeq |J_{AA}J_{BB}|^{\frac{1}{2}}$. The last two terms represent anisotropy terms of the ions A and B, arising e.g. from dipolar interactions or crystal-field effects.

Now, when the exchange interactions are isotropic, the direction of the moments will be determined by the anisotropy, and if $D_A > 0$ and $D_B < 0$, the A moments will tend to align along the $Z$-axis whereas the B moments will order preferentially in the $XY$-plane perpendicular to Z. For neighbouring A and B moments, this clearly leads to a competition between the exchange interaction $J_{AB}$, which will try to align the moments parallel or antiparallel (depending on the sign of $J_{AB}$), and the anisotropy which will tend to keep the A and B moment at right angles. Depending on the ratios of the interaction and anisotropy parameters several situations may then arise. For instance, if the exchange interactions are very small compared to the anisotropy constants, the two competing crystal field terms will keep the A and B moments very nearly perpendicular to one another, leading to an effective decoupling of the two magnetic subsystems. The phase-diagram of $T_c$ versus concentration $p_A$ (or $p_B$) will be as given by the dashed curves in Figure 23, and is just a combination of the $T_c(p)$ curves of an XY and an Ising-type magnet, both diluted with nonmagnetic atoms, since the two spin systems will perceive each other as 'nonmagnetic'. In this case the crossing point of

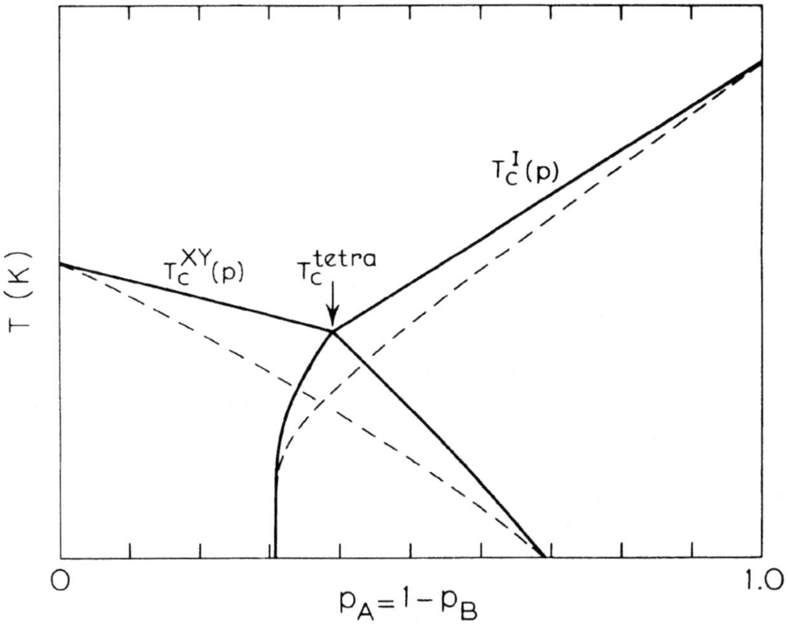


Fig. 23. Qualitative sketch of the phase diagram expected for random mixtures of weakly coupled Ising (I) and XY subsystems. The dashed curves show the fully uncoupled case whereas the full curves represent the weakly coupled case. The crossing point then obtains a special meaning, it is predicted to be a tetracritical point.

the two $T_c$ curves has no special meaning, since the relative position of the two curves is only determined by the ratio of $J_A$ and $J_B$. Such a system was recently studied in [116].

However, as soon as a substantial coupling between neighbouring Ising and XY moments is introduced, the situation becomes different and a diagram will result such as is also sketched schematically in Figure 23. The lower and upper parts of the $T_c^{XY}(p)$ and the $T_c^{I}(p)$ curves will now differ from the diluted case, since each sub-system now experiences the other one as a 'weakly magnetic' impurity. This implies small upward shifts of both $T_c(p)$ curves. In the intermediate concentration range between $p_c$ and $1 - p_c$, there will be two interpenetrating infinite percolation clusters, and, for relatively small coupling between Ising and XY moments, the system may show two transition temperatures associated with the independent magnetic ordering within the two weakly coupled infinite clusters. For not too large a coupling $J_{AB}$, the percolation limits of the two infinite clusters will not be too much different from the $p_c$ value appropriate for magnetic dilution. For larger $J_{AB}$ couplings, evidently, the molecular fields exerted by the two subsystem upon each other will become increasingly important and, in particular because these molecular fields are random, the lower parts (3) and (4) of the $T_c(p)$ curves will be more drastically shifted from the dashed ones.

Phase diagrams of the type shown in Figure 23 for random alloys with competing interactions have been studied *inter alia* within the molecular field approach [117, 118], and by renormalization group techniques [119]. In the latter work an independent ordering of the XY and Ising components is indeed predicted over an intermediate concentration range, the two transition curves $T_c^{XY}(p)$ and $T_c^{I}(p)$ being continuous and crossing each other in a smooth way at what is called a 'decoupled tetracritical point', as indicated in Figure 23. On the other hand, in the molecular field type calculations the two subsystems are not decoupled, and discontinuities are found in the slopes of the $T_c^{XY}(p)$ and $T_c^{I}(p)$ curves at the crossing point.

In the reasoning outlined above, the ordering of the two infinite clusters may also occur independently, provided that the exchange couplings remain small compared to the spin anisotropies. Of crucial importance for the decoupling is the assumption that short-range, nearest neighbour only interactions are present, and this may well be the explanation for the different results obtained within the molecular field treatments. At any rate, the experiments performed so far do show a coupling between the XY and Ising components.

Experimental phase diagrams such as those shown in Figure 23 have been obtained for a number of random mixtures. The first systematic studies on the effects of competing spin-anisotropies were performed by Bevaart *et al.* on the 2-d antiferromagnet $K_2Mn_pFe_{1-p}F_4$. They carried out susceptibility, high-field magnetization and neutron scattering measurements [120—122], the latter both in zero field as well as in applied magnetic fields [122]. Since the Ising anisotropy for $K_2MnF_4$ is of dipolar origin, and thus very small compared to the strong crystal field XY anisotropy in $K_2FeF_4$, the crossing of the two phase-boundaries already occurs at a concentration of $\simeq$ 2.6% of $Fe^{2+}$ impurities in $K_2MnF_4$. In subsequent work by other groups, among others the systems $Fe_pCo_{1-p}Cl_2 \cdot 2H_2O$ [123—125], $Fe_pCo_{1-p}Cl_2$ [126—128] and $Fe_pCo_{1-p}TiO_3$ [129] have been studied by a variety of measurements including specific heat, susceptibility, neutron scattering and Mössbauer effect.

In conclusion, we hope we have called attention in this section to the many interesting aspects of magnetic systems with random magnetic or nonmagnetic impurities. Unfortunately, we could only be very brief; for instance, we have not discussed the problem of random (or competing) interactions that arise when ferro- and antiferromagnetic interactions are introduced by mixing the appropriate magnetic species in the same matrix compound. This concerns the case that $J_{AA} < 0$ and $J_{BB} > 0$ in Equation (15), which may lead to very interesting 2-d spin glass behaviour. At present several groups are working on this problem in such 2-d mixtures as $Rb_2Mn_{1-x}Cr_xCl_4$, $Rb_2Cu_{1-x}Co_xF_4$, and others.

Also, we have only discussed static thermodynamic properties, whereas there are many intriguing dynamical problems involved in these materials (excitations, reentrant behaviour etc.). Clearly, there is still a lot of exciting physics to be explored here, and there will certainly be a need for a more extensive review in the near future.

## References


1. L. J. de Jongh and A. R. Miedema, *Experiments on Simple Magnetic Model Systems*, Taylor and Francis Monographs on Physics, B. R. Coles and N. Mott, eds. (1974). Also published in *Adv. Phys.* **23** (1974) 1—260. This work summarizes experimental and theoretical results up to about 1973. The reader who is new to the field may wish to consult (parts of) this survey as an introductory reading to the present book.
2. See e.g. U. Gradmann, *Appl. Phys.* **3** (1974) 161; *J. Magn. Magn. Mat.* **6** (1977) 173; *Thin Solid Films* **126** (1985).
3. For a review, see e.g. M. Pomerantz in *Phase Transitions in Surface Films*, J. G. Dash and J. Ruvalds, eds. Plenum (1980), pp. 317—346.
4. See e.g. H. Godfrin, R. R. Ruel and D. D. Osheroff, *Phys. Rev. Lett.* **60** (1988) 305; *Proc. Int. Conf. Magn.* (1988), published in *J. Physique, Colloque* **C8**, *Supp. No. 12*, Tome **49** (1988) 2045.
5. See e.g. M. W. Cole, *Rev. Mod. Phys.* **46** (1974) 451.
6. D. W. Hone and P. M. Richards, *Ann. Rev. Mater Sc.* **4** (1974) 337.
7. M. Steiner, J. Villain and C. G. Windsor, *Adv. Phys.* **25** (1976) 87.
8. J. M. Kosterlitz and D. J. Thouless, in *Progr. Low Temp. Phys.* vol. VIIB, D. F. Brewer, ed. North-Holland (1978), Ch. 5, p. 372.
9. V. L. Pokrovsky, *Adv. Phys.* **28** (1979) 595.
10. L. J. de Jongh, *J. Appl. Phys.* **49** (1978) 1305; and in *Recent Developments in Condensed Matter Physics*, J. T. Devreese, ed. vol. I, Plenum Press (1981) p. 343.
11. Proc. NATO-ASI on *Magneto-Structural Correlations in Exchange Coupled Systems*, Italy (1983), R. D. Willett, G. Gatteschi and O. Kahn, eds. Reidel (1985).
12. Proc. NATO-ASI on *Organic and Inorganic Low-Dimensional Crystalline Materials*, Spain (1987), P. Delhaes and M. Drillon, eds. Plenum (1987).
13. J. P. Steadman and R. D. Willett, *Inorg. Chem. Acta* **4** (1970) 367.
14. R. Blinc, M. I. Burgar, V. Rutar, B. Zeks, R. Kind, H. Arend and G. Chapuis, *Phys. Rev. Lett.* **43** (1979) 1679.
15. H. Arend, K. Tichy, K. Baberschke and F. Rys, *Solid State Commun.* **18** (1975) 999.
16. For reviews see e.g. G. A. Gehring and K. A. Gehring, *Rep. Progr. Phys.* **38** (1975) 1—89; D.

Reinen and C. Friebel, *Structure and Bonding* **37** (1979) 1—60; D. Reinen, *J. Solid State Chem.* **27** (1979) 71—85.
17. K. Hirakawa, I. Yamada and Y. Kurogi, *J. Phys.* **32** Suppl. C-1 (1971) 890.
18. M. T. Hutchings, E. J. Samuelsen, G. Shirane and K. Hirakawa, *Phys. Rev.* **188** (1969) 919.
19. K. Hirakawa, *J. Appl. Phys.* **53** (1982) 1893.
20. D. I. Khomskii and K. I. Kugel, *Solid State Commun.* **13** (1973) 763.
21. L. J. de Jongh, *Solid State Commun.* **65** (1988) 963.
22. H. A. Algra, L. J. de Jongh and R. L. Carlin, *Physica* **93B** (1978) 24. Erratum: *Physica* **95 B+C** (1978) 224.
23. L. J. de Jongh and R. L. Carlin, *Chem. Rev.* **86** (1986) 659.
24. H. W. J. Blöte, *J. Appl. Phys.* **50** (1979) 1825.
25. See e.g. J. D. C. McConnell and V. Heine, *J. Phys. C.* **15** (1982) 2387.
26. J. A. Kafalas and J. M. Longo, *J. Solid State Chem.* **4** (1972) 55.
27. A. M. Davies and R. J. D. Tilley, *Nature* **326** (1987) 859; J. B. Torrance, Y. Tokura and A. Nazzal, *Chemtronics* **2** (1987) 120.
28. F. Boersma, W. J. M. de Jonge and K. Kopinga, *Phys. Rev. B* **23** (1981) 186.
29. See e.g. L. J. de Jongh and R. Block, *Physica* **79B** (1975) 568.
30. G. Mennenga, L. J. de Jongh, W. J. Huiskamp and I. Laursen, *J. Magn. Magn. Mater.* **44** (1984) 48 and 59; *ibid.* **53** (1985) 264.
31. L. D. Landau and E. M. Lifshitz, *Statistical Physics*, Pergamon Press, London (1958) p. 482.
32. L. Onsager, *Phys. Rev.* **65** (1942) 117.
33. N. D. Mermin and H. Wagner, *Phys. Rev. Lett.* **17** (1966) 1133.
34. H. E. Stanley and T. A. Kaplan, *Phys. Rev. Lett.* **17** (1966) 913; *J. Appl. Phys.* **38** (1967) 975; H. E. Stanley, *J. Appl. Phys.* **40** (1969) 1546; *Phys. Rev.* **158** (1967) 537 and 546.
35. H. E. Stanley, *Phys. Rev. Lett.* **20** (1968) 150 and 589.
36. V. L. Berezinksii, *Zh. Eksp. Teor. Fiz.* **59** (1970) 907. [(*Sov. Phys. JETP* **32** (1971) 493]; V. L. Berezinksii and Y. Ya. Blank, *Zh. Eksp. Teor. Fiz.* **64** (1973) 725 [(*Sov. Phys. JETP* **37** (1973) 369)].
37. J. M. Kosterlitz and D. J. Thouless, *J. Phys. C.* **6** (1973) 1181; J. M. Kosterlitz, *J. Phys. C.* **7** (1974) 1046.
38. T. Matsubara and H. Matsuda, *Progr. Theor. Phys.* **16** (1956) 569; *ibid* **17** (1957) 19.
39. D. R. Nelson and J. M. Kosterlitz, *Phys. Rev. Lett.* **39** (1977) 1201.
40. D. J. Bishop and J. D. Reppy, *Phys. Rev. Lett.* **40** (1978) 1727; see also *Physics Today*, August 1978, p. 17.
41. H. E. Stanley, *Introduction to Phase Transitions and Critical Phenomena*, Clarendon Press, Oxford (1971).
42. M. E. Fisher, *Rep. Progr. Phys.* **30** (II), (1967) 615—730.
43. Besides [41] and [42], the theory of these models is extensively reviewed in the series of volumes on *Phase Transitions and Critical Phenomena*, edited first by C. Domb and Green (vols. 1—6), and continued by C. Domb and J. L. Lebowitz (vols. 7—), Academic Press (London).
44. The following discussion is based on the paper of M. E. Fisher in *Lectures in Theoretical Physics*, vol. VIIIC (Boulder: Univ. of Colorado Press), 1965, p. 61 (see also [1]).
45. M. E. Fisher, *Phil. Mag.* **7** (1962) 1731.
46. R. F. Wielinga, *Progr. Low Temp. Phys.*, **6**, C. J. Gorter, ed. North-Holland 1971.
47. J. M. Kosterlitz and M. A. Santos, *J. Phys. C.* **11** (1978) 2835.
48. P. Bloembergen, *Physica* **85B** (1977) 51.
49. A. A. Belavin and A. M. Polyakov, *JETP Lett.* **22** (1975) 245; A. M. Kosevich, B. A. Ivanov and A. S. Kovalev, *Physica* **3D** (1981) 363; *Sov. Sci. Rev. A. Phys.* **6** (1985) 161.
50. L. J. de Jongh, *J. Appl. Phys.* **53** (1982) 8018, and references therein.
51. M. E. Fisher, *Rev. Mod. Phys.* **46** (1974) 597; H. J. Maris and L. P. Kadanoff, *Am. J. Phys.* **46** (1978) 652.
52. W. A. C. Erkelens, L. P. Regnault, J. Rossat-Mignod, J. E. Moore, R. A. Butera and L. J. de Jongh, *Europhys. Lett.* **1** (1986) 37.

53. L. P. Kadanoff in *Proc. Enrico Fermi School of Physics*, Varenna 1970, M. S. Green, ed. Academic Press, New York (1971).
54. R. B. Griffiths, *Phys. Rev. Lett.* **24** (1970) 1479.
55. See e.g. D. S. Ritchie and D. D. Betts, *Phys. Rev. B* **11** (1975) 2559, and references therein.
56. R. Navarro and L. J. de Jongh, *Physica* **94B** (1978) 67.
57. J. C. Schouten, H. Hadders, K. Kopinga and W. J. M. de Jonge, *Solid State Commun.* **34** (1980) 667.
58. L. J. de Jongh and H. E. Stanley, *Phys. Rev. Lett.* **36** (1976) 817.
59. C. A. W. Citteur and P. W. Kasteleijn, *Physica* **68** (1973) 491.
60. L. L. Liu and H. E. Stanley, *Phys. Rev. Lett.* **29** (1972) 927; *Phys. Rev. B* **8** (1973) 2279.
61. R. Navarro, J. J. Smit, L. J. de Jongh, W. J. Crama and D. J. W. IJdo, *Physica* **83B** (1976) 97.
62. D. J. Breed, *Phys. Lett.* **23** (1966) 181; *Physica* **37** (1967) 35.
63. L. J. de Jongh and D. J. Breed, *Solid State Commun.* **15** (1974) 1061.
64. R. Navarro and L. J. de Jongh, *Physica* **98b** (1979) 1.
65. R. Navarro and L. J. de Jongh, *Physica* **94B** (1978) 67.
66. L. J. de Jongh, *Physica* **82B** (1976) 247.
67. J. M. Kosterlitz and M. A. Santos, *J. Phys. C* **11** (1978) 2835, and personal communication.
68. R. J. Birgeneau, H. J. Guggenheim, and G. Shirane, *Phys. Rev. Lett.* **22** (1969) 720; *Phys. Rev. B* **1** (1970) 2211.
69. H. Ikeda and K. Hirakawa, *J. Phys. Soc. Japan* **33** (1972) 393; and this volume.
70. Reversing the argument one may conclude from the fact that 2-d behaviour is observed below $T_c$ for $M_s$ that the (staggered) susceptibility in the (anti)ferromagnets, although divergent at $T_c$ itself, does not remain infinite below $T_c$.
71. K. Hirakawa and H. Ikeda, *J. Phys. Soc. Japan* **35** (1973) 1328; *Techn. Report, Inst. for Solid State Physics, Univ. of Tokyo, Ser. A*, no. 584, April 1973; and personal communication. Compare also the behaviour observed in $(CD_3ND_3)_2CuCl_4$, J. J. M. Steijger, E. Frikkee, L. J. de Jongh and W. J. Huiskamp, *Physica* **120B** (1983) 202; *ibid.* **123B** (1984) 271.
72. T. O. Klaassen, A. Gevers and N. J. Poulis, *Physica* **61** (1972) 95; L. S. J. M. Henkens, M. W. van Tol, K. M. Diederix, T. O. Klaassen and N. J. Poulis, *Phys. Rev. Lett.* **36** (1976) 1252.
73. E. J. Samuelsen, *Phys. Rev. Lett.* **31** (1973) 936; *J. Phys. Chem. Solids* **35** (1974) 785; H. Ikeda and K. Hirakawa, *Solid State Commun.* **14** (1974) 529.
74. L. Onsager, Phys. Rev. *65* (1944) 117; C. N. Yang, *Phys. Rev.* **85** (1952) 808.
75. C. M. J. van Uijen, E. Frikkee and H. W. de Wijn, *Phys. Rev. B* **19** (1979) 509.
76. R. B. Stinchcombe, in *Phase Transitions and Critical Phenomena*, C. Domb and J. L. Lebowitz, eds. Academic Press, London, vol. 7 (1987).
77. H. L. Frisch and J. M. Hammersley, *J. Soc. Industr. Appl. Math.* **11** (1963) 894.
78. V. K. S. Shante and S. Kirkpatrick, *Adv. Phys.* **20** (1971) 235.
79. S. Kirkpatrick, *Rev. Mod. Phys.* **45** (1973) 574.
80. D. Stauffer, *Physics Reports* **54** (1979) 1, and *Introduction to Percolation Theory*, Taylor and Francis, London, 1985.
81. H. E. Stanley, *Proc. Int. Conf. on Disordered Systems and Localization*, Springer Lecture Notes in Physics, Heidelberg (1981).
82. A short review of experimental results (far from exhaustive) has been given by L. J. de Jongh in *Magnetic Phase Transitions*, M. Ausloos and R. J. Elliott, eds. Springer Series in Solid State Sciences, vol. 48 (1983) p. 172. A review of the theory by A. Coniglio can be found in the same volume.
83. P. W. Kasteleyn and C. M. Fortuin, *J. Phys. Soc. Japan* **26**, supplement, 11 (1969); see also: *Physica* **57** (1972) 536; *ibid* **58** (1972) 393; *ibid.* **59** (1972) 545.
84. R. J. Elliott and B. R. Heap, *Proc. Roy. Soc. (London)* **A265** (1962) 264.
85. D. J. Morgan and G. S. Rushbrooke, *Mol. Phys.* **4** (1961) 1; *ibid.* **4** (1961) 291; *ibid.* **6** (1963) 477.
86. D. J. Breed, K. Gilijamse, J. W. E. Sterkenburg and A. R. Miedema, *J. Appl. Phys.* **41** (1970) 1267; *Physica* **68** (1973) 303.
87. T. Idogaki and N. Uryû, *J. Phys. Soc. Japan* **43** (1977) 845; *ibid.* **45** (1978) 1498.

88. H. Ikeda, M. T. Hutchings and M. Suzuki, *J. Phys.* **C11** (1978) L359; *J. Phys. Soc. Japan* **46** (1979) 1153.
89. H. A. Algra, L. J. de Jongh, W. J. Huiskamp, and J. Reedijk, *Physica* **86–88B** (1977) 737; *Phys. Rev. Letters* **42** (1979) 606.
90. G. Mennenga, L. J. de Jongh, W. J. Huiskamp and J. Reedijk, *J. Magn. Magn. Mater.* **43** (1984) 3.
91. J. S.Reeve and D. D. Betts, *J. Phys.* **C8** (1975) 2642; *ibid.* **9** (1976) 2575, *ibid.* **10** (1977) 1941.
92. R. B. Stinchcombe, *J. Phys.* **C12** (1979) 4533.
93. D. P. Landau, *Phys. Rev.* **B22** (1980) 2450.
94. A. B. Harris and S. Kirkpatrick, *Phys. Rev.* **B16** (1977) 542.
95. R. Burriel, A. Lambrecht, R. L. Carlin and L. J. de Jongh, *Phys. Rev.* **B36** (1987) 8461.
96. We remark that the increase observed in the susceptibility of the high-$T_c$ superconductors $La_{2-x}Sr_xCuO_4$ upon Sr-doping may be related to this effect, since the doping leads to defects (holes) in the antiferromagnetic Cu-lattice that carry no spin (they are EPR silent).
97. H. Ikeda, *J. Phys.* **C16** (1983) L21.
98. For a more extensive discussion see L. J. de Jongh, G. Mennenga and A. Coniglio, *Physica* **132B** (1985) 100.
99. A. R. McGurn, *J. Phys.* **C13** (1980) 1055.
100. M. Suzuki and H. Ikeda, *J. Phys.* **C11** (1978) 3679.
101. K. Takeda, M. Matsuura and T. Haseda, *J. Phys. Soc. Japan* **28** (1970) 29.
102. H. A. Algra, L. J. de Jongh and J. Reedijk, *Phys. Rev. Letters* **42** (1979) 606; *Physica* **86–88B** (1977) 737.
103. R. J. Birgeneau, R. A. Cowley, G. Shirane and H. J. Guggenheim, *Phys. Rev. Letters* **37** (1976) 940.
104. R. A. Cowley, G. Shirane, R. J. Birgeneau, E. C. Svenson and H. J. Guggenheim, *Phys. Rev.* **B22** (1980) 4412.
105. Y. Okuda, I. Yamada, J. Watanabe and T. Haseda, *J. Phys. Soc. Japan* **49** (1980) 2136.
106. S. Fishman and A. Aharony, *J. Phys.* **C12** (1979) L729.
107. For a discussion of early experiments see e.g. H. Rohrer, *J. Magn. Magn. Mater.* **15–18** (1980) 396; *J. Appl. Phys.* **52** (1981) 1708.
108. Y. Imry, *J. Statistical Physics* **34** (1984) 849.
109. A. Aharony, *J. Magn. Magn. Mater.* **54–57** (1986) 27.
110. G. Grinstein, *J. Appl. Phys.* **55** (1984) 2371.
111. J. Villain, *Magnetic Excitations and Fluctuations*, S. W. Lovesey, U. Balcani, F. Borsa and V. Tognetti, eds Springer Series in Solid State Sciences, **54** (1984) p. 142; *Scaling Phenomena in Disordered Systems*, R. Pynn and A. Skjeltorp, eds. Plenum, (1985), p. 423.
112. T. Nattermann and J. Villain, *Phase Transitions*, vol. 11, Gordon and Breach (1988) p. 5.
113. Y. Imry and S. K. Ma, *Phys. Rev. Lett.* **35** (1975) 1399.
114. D. P. Belanger, A. R. King and V. Jaccarino, *J. Appl. Phys.* **55** (1984) 2383; *J. Appl. Phys.* **53** (1982) 2702; *J. Magn. Magn. Mater.* **54–57** (1986) 19.
115. R. J. Birgeneau, Y. Shapira, G. Shirane, R. A. Cowley and H. Yoshizawa, *Physica* **137B** (1986) 83.
116. G. Mennenga L. J. de Jongh, W. J. Huiskamp and J. Reedijk, *J. Magn. Magn. Mater.* **43** (1984) 13.
117. F. Matsubara and S. Inawashiro, *J. Phys. Soc. Japan* **42** (1977) 1529; *ibid.* **46**, (1979) 1740; *ibid.* **47** (1979) 1102.
118. T. Oguchi and T. Ishikawa, *J. Phys. Soc. Japan* **45** (1978) 1213; *ibid.* **46** (1979) 372.
119. A. Aharony and S. Fishman, *Phys. Rev. Letters* **37** (1976) 1587; S. Fishman and A. Aharony, *Phys. Rev.* **B18** (1978) 3507.
120. L. Bevaart, E. Frikkee, and L. J. de Jongh, *Proc. Conf. Neutron Scatt.*, Gatlinburg, p. 697 (1976) (Nat. Techn. Information Service, U.S. Dept. of Commerce, Springfield, R. M. Moon ed; *Proc. Conf. STATPHYS.* **13**, Haifa (1977) (published in *Ann. Israel. Phys. Soc.* **2**).

121. L. Bevaart, E. Frikkee, J. V. Lebesque and L. J. de Jongh, *Physica* **86—88B** (1977) 729; *Solid State Commun.* **25** (1978) 539; *Phys. Rev.* **B18** (1978) 3376.
122. L. Bevaart, E. Frikkee and L. J. de Jongh, *Solid State Commun.* **25** (1978) 1031; *Phys. Rev.* **B19** (1979) 4741.
123. K. Katsumata, M. Kobayashi, T. Sato and Y. Miyako, *Phys. Rev.* **B19** (1979) 2700; *J. Phys. Soc. Japan* **46** (1979) 1467.
124. K. Katsumata, M. Kobayashi and H. Yoshizawa, *Phys. Rev. Letters* **43** (1979) 960.
125. A. Ito, Y. Someya, and K. Katsumata, *Solid State Commun.* **36** (1980) 681.
126. T. Tawaraya, K. Katsumata and H. Yoshizawa, *J. Phys. Soc. Japan* **49** (1980) 1299.
127. T. Tawaraya and K. Katsumata, *Solid State Commun.* **32** (1979) 337.
128. P. Wong, P. M. Horn, R. J. Birgeneau, C. R. Safinya, and G. Shirane, *Phys. Rev. Letters* **45**, (1980) 1974.
129. A. Ito, S. Morimoto, Y. Someya, H. Ikeda, Y. Syono and H. Takei, *Solid State Commun.* **41** (1982) 507.

# THEORY OF TWO-DIMENSIONAL MAGNETS

V. L. POKROVSKY AND G. V. UIMIN

*Landau Institute for Theoretical Physics, ul. Kosygina 2, Moscow 117940, U.S.S.R.*

## 1. Introduction

Phase transitions in two-dimensional (2-d) magnets are associated with *fluctuational* excitations in these 2-d systems. Excitations in the well-known Ising model (Section 2), whose discrete symmetry differs from the continuous symmetries of the XY (Section 3) and Heisenberg (Section 4) models, are linear defects (domain walls). In continuously degenerate systems (XY and Heisenberg models) the low-energy excitations are spin-waves. The energy of such a fluctuation depends quadratically on its wavenumber, resulting in the absence of long-range order in 2-d systems (the so-called Landau—Peierls theorem). But the low-temperature phase of the XY model corresponds to the quasi-long-range order which denotes an algebraic decay of spin correlation functions. This quasi-order results from spin-waves and vortices, which are topological excitations confined below a certain temperature. The system transforms into a paramagnetic state via the vortex unbinding transition. In the Heisenberg model vortices are prohibited; however, an other type of topological excitation called 'skyrmions' is allowed. Owing to the non-Abelian symmetry group of the order parameter the quasi-long-range order disappears in the Heisenberg model.

The outline of this review is as follows. We start with the 'pure' model theories. Then we investigate their stabilities under the symmetry-breaking fields. The behaviour of layer-type magnets is discussed in Section 5.

The dynamics of 2-d magnets are associated with *dynamical* spin-wave and spin-diffusive modes. The contribution of topological excitations is also relevant.

## 2. Ising Magnets

### 2.1. ISING MODEL. EXCITATIONS AND PHASE TRANSITION

The Ising model represents an ideal lattice with identical magnetic moments positioned at all sites (Figure 1). For each elementary moment (spin) only two states are possible ('up' and 'down'). The lattice site $\mathbf{r}$ is associated with the variable $\sigma_{\mathbf{r}}$, the two values $\pm 1$ corresponding to spin orientation. The energy:

$$\mathscr{H}\{\sigma\} = -\sum_{\mathbf{r},\mathbf{r}'} J(\mathbf{r}-\mathbf{r}')\,\sigma_{\mathbf{r}}\sigma_{\mathbf{r}'} - h\sum_{\mathbf{r}} \sigma_{\mathbf{r}} \tag{1}$$

is attributed to a fixed configuration of spins $\{\sigma\}$. Usually the nearest neighbour interaction is considered, i.e. $J(\mathbf{r}) = 0$ for all $\mathbf{r} \neq \mathbf{a}_i$, where the $\mathbf{a}_i$'s are the basic



*L. J. De Jongh (Ed.), Magnetic Properties of Layered Transition Metal Compounds* 53—103.

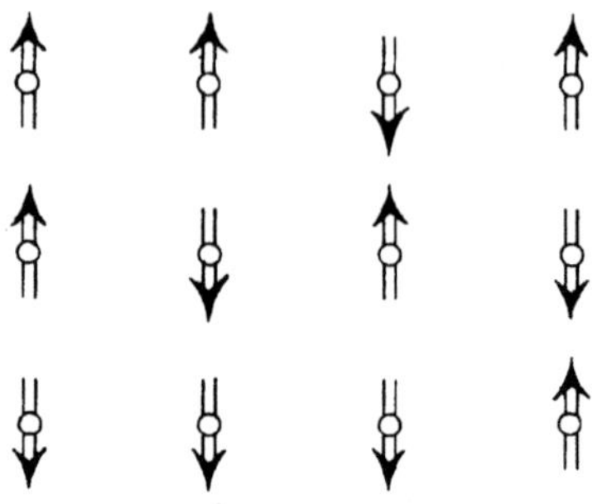

Fig. 1. Square lattice fragment with a configuration $\{\sigma\}$ of uncorrelated spins.

vectors of a lattice. This is the simplest model for a highly anisotropic magnet. The quantities $J(\mathbf{r})$ are the exchange integrals and $h$ is a magnetic field (in units of $g\mu_B$).

We consider first the Ising model in the absence of an external field ($h = 0$). For the sake of simplicity the nonzero quantities $J(\mathbf{r})$ are assumed to be independent of the direction of $\mathbf{r}$. At $h = 0$ the Hamiltonian is invariant under a change in sign of the spins. This transformation together with the identical one forms the symmetry group $Z_2$ for the Ising model.

The ground state (GS) of a system with $J > 0$ corresponds to the ferro-(F-)magnetic order and is doubly degenerate. Its energy is

$$E_0 = -NJz$$

where $N$ is the number of lattice sites, $z$ is the coordination number. Excitations with the lowest energy correspond to a single spin-flip. For a fixed GS (all spins 'up', say) there are $N$ such excited states. The energy of these excitations is

$$\varepsilon_1 = 2Jz$$

More general excitations consist of several groups of reversed spins ('down'). These groups of 'down' spins are separated from the 'sea' of 'up' spins by a 'shore' contour $\Gamma$. It connects the points of the dual lattice, as is shown in Figure 2. Generally a contour $\Gamma$ is disconnected and consists of a set of connected parts. The energy of any excitation is

$$\varepsilon(\Gamma) = 2J\,L(\Gamma) \tag{2}$$

where $L(\Gamma)$ is the total length of a contour $\Gamma$. Equation (2) shows that connected contours can be considered as independent excitations with energies having the form of Equation (2). This statement holds for the hexagonal lattice. For the square lattice some complication arises associated with a possible self-intersection of the contour, while for the hexagonal lattice all contours must be self-avoiding. However, these details are not essential for our qualitative consideration.

Let us introduce the free energy of a connected contour $\Gamma$:

$$\mathscr{F}(\Gamma) = \varepsilon(\Gamma) - TS(\Gamma) \tag{3}$$

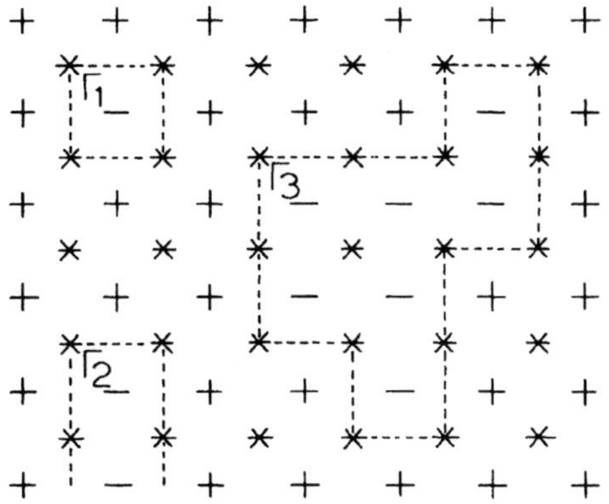


Fig. 2. Spins 'up' ('down') located at sites of the square lattice are marked by '+' ('−'), '∗' denotes site of the dual lattice. Disconnected contour Γ consists of 'connected' parts $\Gamma_1$, $\Gamma_2$, $\Gamma_3$.

where $T$ is the temperature and $S(\Gamma)$ is the entropy of the contour. We consider a connected contour with a fixed length $L$. Then the entropy of a sufficiently long contour Γ is proportional to $L$:

$$S(\Gamma) = \xi L. \tag{4}$$

A rough estimate gives $\xi \approx \ln 3$, corresponding to the three possible directions of a bond on the square lattice, whereas the preceding bond is fixed. Equations (3) and (4) determine the critical temperature $T_c$:

$$T_c = 2J/\xi. \tag{5}$$

At the critical temperature the linear energy of a long contour vanishes. For sufficiently low temperatures a typical state contains small 'islands' of 'down' spins, surrounded by a 'sea' of 'up' spins. With increasing temperature the number of 'islands' increases, as well as their size. At the critical temperature 'islands' join together, forming a 'continent'. There is no difference between the 'sea' and the 'continent' above the phase transition temperature. As a result, the spontaneous magnetization vanishes above $T_c$.

In the vicinity of the critical point the free energy (3) of a long contour can be estimated approximately as follows:

$$\mathcal{F}(\Gamma) = L(2J - \xi T) = \xi_0 L(T_c - T). \tag{6}$$

In this approximation the linear energy of the boundary between the "spin-up" and "spin-down" phases tends to zero linearly with $T - T_c$. Simultaneously, it implies an increase in the correlation length $r_c$ proportional to $|T_c - T|^{-1}$. It follows from the general scaling relations that the critical exponent $\alpha$ of the heat capacity is zero. The exact calculation of these values confirms our simple approach.

An external magnetic field $h$ contributes an additional term to the free energy of a contour:

$$\Delta \mathscr{F} = h A(\Gamma) \bar{\sigma}$$

where $A(\Gamma)$ is the area surrounded by a contour $\Gamma$, and $\bar{\sigma}$ is the average magnetization. For a typical contour $A \sim L^2$. Therefore the phase transition disappears at any finite value of $h$.

The properties of antiferro-(AF-)magnets ($J < 0$) strongly depend on the type of lattice. First we consider those lattices which can be divided into two identical sublattices, so that the nearest-neighbour sites belong to different sublattices. For such Ising AF-magnets the partition function can be transformed into the partition function of the Ising F-magnet. This transformation involves spin inversion in one of the two sublattices with a simultaneous change of sign of the exchange integral $J$. The triangular lattice cannot be divided into two sublattices as has been described above. The GS of a magnet on the triangular lattice with $J < 0$ is highly degenerate and contains a lot of frustrated spins. The phase transition temperature in this case is zero.

In AF-magnets an external magnetic field $h$ does not destroy the long-range order, at least if $h$ is sufficiently small. Indeed, the energy of an 'island' of reversed spins is not proportional to its area as in the case of a F-magnet. A magnetic field favours those segments of a contour which contain spins directed along a field (see Figure 3). This implies that a contour will cause nonzero magnetization per unit length, proportional to $h$. The contour energy contains the following magnetic contribution

$$\Delta E = -\tfrac{1}{2} \chi h^2 L$$

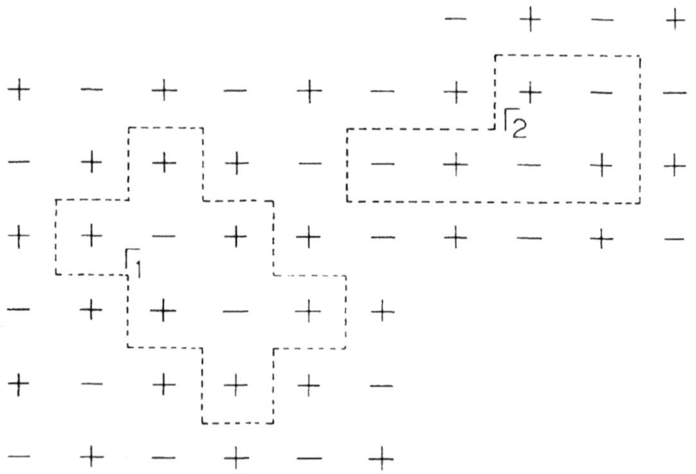


Fig. 3. Contour $\Gamma_1$ consists entirely of elements favoured by a magnetic field; several unfavoured elements are included into contour $\Gamma_2$.

where $\chi$ is the effective susceptibility in the vicinity of a contour. For the free energy of a contour we obtain Equation (6) with the renormalized critical temperature:

$$T_c(h) = T_c(0) - \tfrac{1}{2}\,\chi h^2 \tag{7}$$

Equation (7) determines the line of second-order phase transitions in AF-magnets. The schematic phase diagram is depicted in Figure 4.

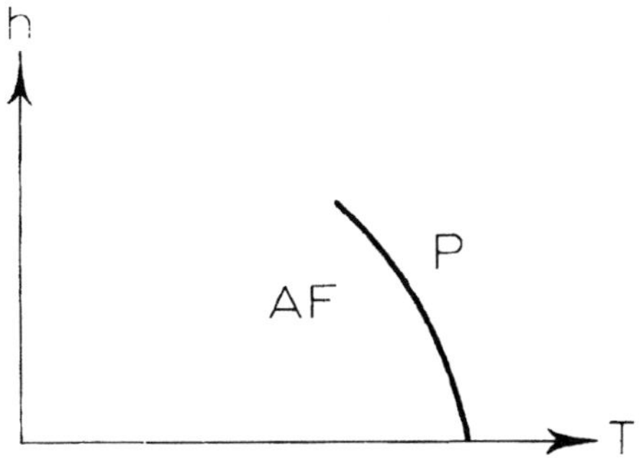


Fig. 4. Phase diagram of the 'antiferro—paramagnetic' transition in an external magnetic field.

HISTORICAL REMARKS. The Ising model was first proposed by Lenz [1]. Ising [2] presented the exact calculation of the partition function for the 1-d case. A great success was achieved by Onsager [3], who calculated the partition function for the 2-d case. He was the first to demonstrate explicitly the existence of the second-order phase transition. Moreover, this has stimulated many theoretical and experimental investigations of critical phenomena. Many authors [4—13] took part in elucidating the exact solution of the 2-d Ising model.

## 2.2. ONSAGER SOLUTION

Onsager applied the transfer-matrix method (TM) proposed originally by Kramers and Wannier [14] in 1941. According to this method, one coordinate (say '$y$') of a square lattice is identified with the discrete time. The TM relates spin configurations in row $y + 1$ with configurations in row $y$. If a row contains $N$ spins, the total number of different states is $2^N$. Hence, the TM is a square $2^N \times 2^N$ matrix. The partition function can be expressed in terms of TM:

$$Z = \mathrm{Tr}\,(\hat{T}^M)$$

where $\hat{T}$ is the TM and $M$ is the total number of rows. In the thermodynamic limit ($M, N \to \infty$) the partition function and free energy are defined by the maximum eigenvalue $\lambda_0$ of $\hat{T}$ only. For instance, the free energy per site is given by

$$f = -\frac{1}{N}\,T \ln \lambda_0. \tag{8}$$

The problem is reduced to the calculation of $\lambda_0$. This problem has been solved by Onsager for the case of the nearest-neighbour Ising model on the rectangular lattice, i.e. a quadratic lattice with different exchange integrals $J_1$ an $J_2$ in the two directions. The Hamiltonian is:

$$\mathcal{H} = -\sum_{m=1}^{M} \sum_{n=1}^{N} \{J_1 \sigma_{m,n} \sigma_{m,n+1} + J_2 \sigma_{m,n} \sigma_{m+1,n}\}. \tag{9}$$

The cyclic boundary conditions imposed are:

$$\sigma_{m+M,n} = \sigma_{m,n+N} = \sigma_{m,n}.$$

The TM corresponding to Hamiltonian (9) has the form:

$$\hat{T} = \hat{T}_1 \times \hat{T}_2$$

where

$$\hat{T}_1 = \exp\left\{ K_1 \sum_n \hat{\sigma}^z_n \hat{\sigma}^z_{n+1} \right\} \tag{10}$$

$$\hat{T}_2 = \exp\left\{ \tilde{K}_2 \sum_n \hat{\sigma}^x_n \right\}. \tag{11}$$

Here $\hat{\sigma}^{x,y,z}_n$ are Pauli matrices with the usual commutation relations, and

$$K_1 = J_1/T, \quad K_2 = J_2/T, \quad \tilde{K}_2 = -\tfrac{1}{2} \ln \tanh K_2.$$

The partition function is invariant under the transformation $\hat{\sigma}^x_n \rightleftarrows \hat{\sigma}^z_n$. Performing this transformation and subsequently the Jordan—Wigner transformation defined by

$$\sigma^z_n = 1 - 2a^+_n a_n$$

$$s^+_n = \frac{1}{2}(\sigma^x_n + i\sigma^y_n) = a^+_n \prod_{l=0}^{n-1} (1 - 2a^+_l a_l)$$

one can express the TM in terms of fermion operators $a_n$, $a^+_n$ (Schultz *et al.* [12])

$$\hat{T}_1 = \exp\left\{ K_1 \sum_n (a^+_{n+1} a_n + a^+_n a_{n+1} + a^+_n a^+_{n+1} + a_{n+1} a_n) \right\} \tag{12}$$

$$\hat{T}_2 = \exp\left\{ -2\tilde{K}_2 \sum_n a^+_n a_n \right\}. \tag{13}$$

The Bogolyubov diagonalization of the quadratic form in the exponents of

Equations (12) and (13) leads to the following expression for the value of $\lambda_0$:

$$\ln \lambda_0 = \int_{-\pi}^{\pi} \frac{dp}{4\pi}\, \varepsilon_p \tag{14}$$

where

$$\cosh \varepsilon_p = \cosh 2K_1 \cosh 2\tilde{K}_2 - \cos p \sinh 2K_1 \sinh 2\tilde{K}_2. \tag{15}$$

The critical point is determined by the condition

$$K_1 = \tilde{K}_2. \tag{16}$$

At this point $\varepsilon_0 = 0$.

The free energy defined by Equations (8) and (14) has the following singularity:

$$f \sim (T - T_c)^2 \ln |T - T_c|.$$

As a consequence the specific heat diverges at the critical point as $\ln |T_c - T|$ (see Figure 5).

The fermion description can be easily understood within the framework of the qualitative picture developed in Section 2.1. If the $y$-coordinate is considered as the time, the contours described in Section 2.1 can be treated as world lines of fermions. They can be created spontaneously at any finite temperature, which plays the role of Planck's constant in the quantum-mechanical analogy. Fermions can be created only in pairs to satisfy the $Z_2$-symmetry constraints. The lifetime of such a pair is proportional to the inverse mass $m^{-1}$ of a fermion according to the uncertainty principle. The mass in the quantum language is equivalent to the free

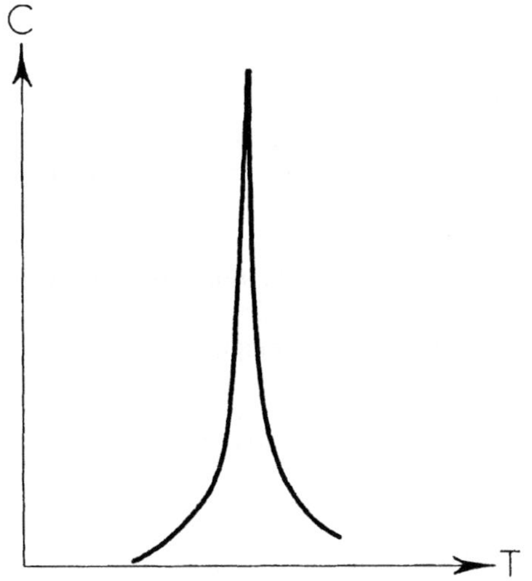


Fig. 5. Heat capacity signularity in the 2-d Ising model.

energy per unit length of a contour in a statistical description. At low temperature the mass is sufficiently large and typical contours are small. At the critical point the mass of fermions vanishes and the length of some contours becomes infinite (see Section 2.1). The exact energy of free fermions is given by Equation (15), while contours or domain walls can be treated as original fermions.

The behavior of the spontaneous magnetization was also solved by Onsager [4] in 1948. It has the form:

$$M(T) = [1 - (\sinh 2K_1 \sinh 2K_2)^{-2}]. \tag{17}$$

However, this result was only mentioned by him at several conferences without proof or derivation. The first publication of the proof of Equation (17) was by C. N. Yang [5] in 1952.

The two-spin correlation function has been first calculated by Kaufman and Onsager [7]. A detailed analysis of the correlation functions given by McCoy and Wu [15] should also be mentioned.

## 2.3. CRITICAL EXPONENTS AND SCALING

Four critical exponents are known from exact results, namely:

$$\alpha = 0, \quad \beta = 1/8, \quad \nu = 1, \quad \eta = 1/4$$

(we use the conventional notation). They satisfy the scaling relations:

$$d\nu = 2 - \alpha \quad \text{and} \quad \beta = \tfrac{1}{2}(\mathrm{d} - 2 + \eta)\nu \qquad (d = 2).$$

Two other scaling relations define the critical exponents $\gamma$ and $\delta$:

$$\gamma = \nu(2 - \eta) = 7/4 \tag{18}$$

$$\delta = (d + 2 - \eta)/(d - 2 + \eta) = 15. \tag{19}$$

Equations (18) and (19) have been checked numerically (Domb and Sykes [16]).

It is worthwhile noting that the correlation function $\langle \sigma^z(0)\sigma^z(x)\rangle$ can be transformed into the following average over fermion variables:

$$\langle \exp(i\,\pi\,N(x))\rangle$$

where $N(x)$ is the number of original fermions (domain walls) in the length interval $(0, x)$. For massless fermions (at the critical temperature) this average is proportional to $x^{-1/4}$, confirming the value of $\eta$ mentioned above.

## 2.4. DUAL TRANSFORMATION. ORDER AND DISORDER

Equations (10) and (11) can be transformed:

$$\mu^x_{n-1/2} = \sigma^z_n \sigma^z_{n-1} \tag{20}$$

$$\sigma^x_n = \mu^z_{n-1/2}\mu^z_{n+1/2}. \tag{21}$$

The matrices $\mu_{\tilde{n}}^{x}$, $\mu_{\tilde{n}}^{z}$ introduced in this way satisfy the same permutation relations as the original Pauli matrices $\sigma_n^x$ and $\sigma_n^z$. Obviously the partition function remains unchanged, if one simultaneously substitutes $K_1 \rightleftarrows \tilde{K}_2$. It implies the following property of dual symmetry

$$Z(K_1, \tilde{K}_2) = Z(\tilde{K}_2, K_1) \tag{22}$$

Kramers and Wannier [14] were the first to establish the phase transition point applying Equation (22) (see also Equation (16)).

The matrices $\hat{\sigma}_n^z$ and $\hat{\mu}_{\tilde{n}}^z$ can be interpreted as order and disorder operators, respectively. This terminology is justified by appearance of the nonzero average below (above) $T_c$. It is clear from Equations (20) and (21) that the order (disorder) operator $\sigma^z(\mu^z)$ can be expressed as a nonlocal product of disorder (order) operators $\mu^x(\sigma^x)$. The partition function and correlation functions are invariant with respect to the dual transformation.

## 3. Planar Magnets

### 3.1. XY MODEL

Magnets with easy plane anisotropy can be described by the so-called XY model. This is a lattice model of classical spins rotating in a plane and interacting with each other via the Hamiltonian:

$$\mathcal{H} = -\sum_{(\mathbf{r},\mathbf{r}')} JS^2 \cos(\varphi(\mathbf{r}) - \varphi(\mathbf{r}')). \tag{23}$$

The spins are fixed by their angles of rotation $\varphi(\mathbf{r})$. Their length $S$ is assumed to be constant. Summation in Equation (23) proceeds over the nearest-neighbour sites on a lattice; $J$ is the exchange constant. The GS is F-magnetic for $J > 0$ and arbitrary lattice. In the opposite case $J < 0$, the GS is AF-magnetic if the lattice can be divided into two sublattices in such a way that the nearest-neighbour sites belong to different sublattices. Otherwise, the situation is more complicated. In the important case of a triangular lattice with $J < 0$ the GS consists of three sublattices with the spins in different sublattices tilted at angles of $\pm$ 120° (Figure 6).

An external magnetic field fixes the orientation of spins in F-magnets and in AF-magnets with two sublattices. In the last case spins are directed approximately perpendicular to the field direction. In the case of a three-sublattice AF-magnet an external magnetic field does not interfere with the freedom of spin rotation in the GS [17].

The quantum effects can in principle change the GS of nontrivial AF-magnets, especially for the spin value $S = 1/2$. One can expect the GS to be nondegenerate if the quantum fluctuations are sufficiently large [18].

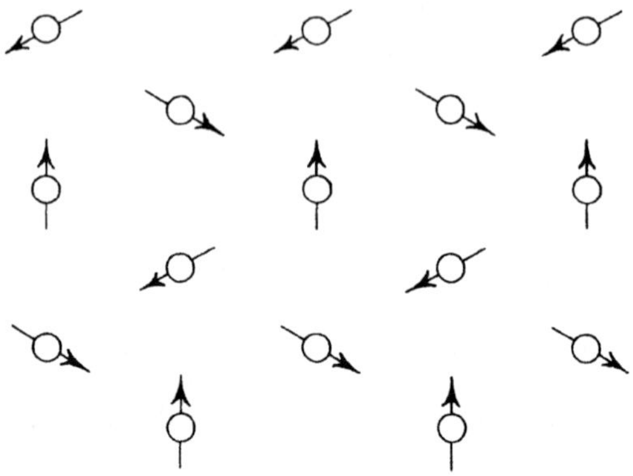

Fig. 6. The 3-sublattice GS of the AF—XY model on the triangular lattice.

### 3.2. EXCITATIONS

The most simple excitations of a planar magnet are spinwaves with a gapless spectrum, according to the Goldstone theorem. They will be considered later. Here we emphasize the static aspect of the Goldstone theorem: the enhancement of long-wave spin fluctuations. To confirm this we represent the energy of spin interactions $U$ in the long-wave approximation:

$$U = \frac{1}{2}\bar{J}\int d^2x(\nabla\varphi)^2. \tag{24}$$

$\bar{J}$ differs from $J$ (cf. Equation (23)) by the factor $z\,S^2/2a^2$ ($z$ is the coordination number). The quadratic form (24) can be diagonalized in terms of Fourier transforms $\varphi_{\mathbf{q}}$ of the field $\varphi(\mathbf{r})$:

$$U = \frac{1}{2}\bar{J}\sum_{\mathbf{q}} q^2\,|\varphi_{\mathbf{q}}|^2\,.$$

In the classical limit we obtain

$$\langle|\varphi_{\mathbf{q}}|^2\rangle = T/\bar{J}\,q^2. \tag{25}$$

The enhancement of long-wave fluctuations characterized by Equation (25) is a starting point in the proof of the Landau—Peierls theorem [19, 20], which states that there is no long-range order in 2-d systems with a continuous symmetry. Indeed, the mean value of the magnetization at a nonzero temperature

$$\langle|\mathbf{S}|\rangle = S\langle\exp(i\varphi(\mathbf{r}))\rangle$$

can be expressed in terms of $\langle\varphi^2\rangle$:

$$\langle e^{i\varphi(\mathbf{r})}\rangle = \exp(-\tfrac{1}{2}\langle\varphi^2(r)\rangle)$$

as a consequence of the Gaussian distribution of fluctuations. The mean square of the field $\varphi$ can be easily evaluated with a logarithmic accuracy, using Equation (25):

$$\langle \varphi^2(\mathbf{r}) \rangle = 2\Delta \ln L/a; \qquad \Delta = T/4\pi \tilde{J} \tag{26}$$

where $L$ is the linear size of the system. Finally, one obtains

$$\langle |\mathbf{S}| \rangle \sim S(L/a)^{-\Delta}. \tag{27}$$

Equation (27) shows that magnetization vanishes in the thermodynamic limit ($L \to \infty$ at fixed $T$).

HISTORICAL REMARKS. This theorem was first formulated by Landau [20] and Peierls [19]. A rigorous proof was given by Hohenberg [21] for a superfluid liquid, which has the same symmetry as a planar magnet, and by Mermin and Wagner [22] for Heisenberg magnets. Their proofs are based on thermodynamic inequalities due to Bogolyubov [23].

A new type of excitations in a planar magnet has been discovered theoretically by Berezinskii [24]. Spins in such a type of localized excitation have a nonzero circulation along any contour $\Gamma$ surrounding a singular point $O$ — the center of a vortex. The analogy with a vortex in a quantum liquid is obvious. A schematic picture of spin vortices is shown in Figure 7. Magnetic vortices are a very simple example of topological excitations.

The order parameter $\mathbf{S}$ takes values on the circle $C$. Generally, for a continuous symmetry group there exists a manifold of order parameter values, the free energy being minimal. This manifold is usually called a degeneracy space. Any continuous distribution of the order parameter produces a mapping of the contour $\Gamma$ onto a contour $\Gamma'$ in the degeneracy space $C$. All the mappings can be classified according to their topological properties. If $\Gamma'$ can be restricted to a point by a continuous transformation, the mapping is called trivial. For example, any spin-wave excitation produces a trivial mapping. Generally, a mapping must be characterized by the mapping degree. This is an integer showing how many times the circumference $C$ is closed when a spatial point passess contour $\Gamma$. A detailed consideration of topological methods in the physics of condensed matter is given in the review by Mineev [25].

The nontrivial equilibrium configuration satisfies the Laplace equation:

$$\Delta\varphi = 0 \tag{28}$$

with the constraint

$$\oint \nabla\varphi \cdot d\mathbf{l} = 2\pi\, Q. \tag{29}$$

The solution of this problem is $\varphi = Q\phi + \text{const}$, where $\phi$ is the azimuthal angle in the plane. The energy of a single vortex diverges logarithmically with the size of the system:

$$E = 2\pi\, \tilde{J}Q^2 \ln L/a. \tag{30}$$

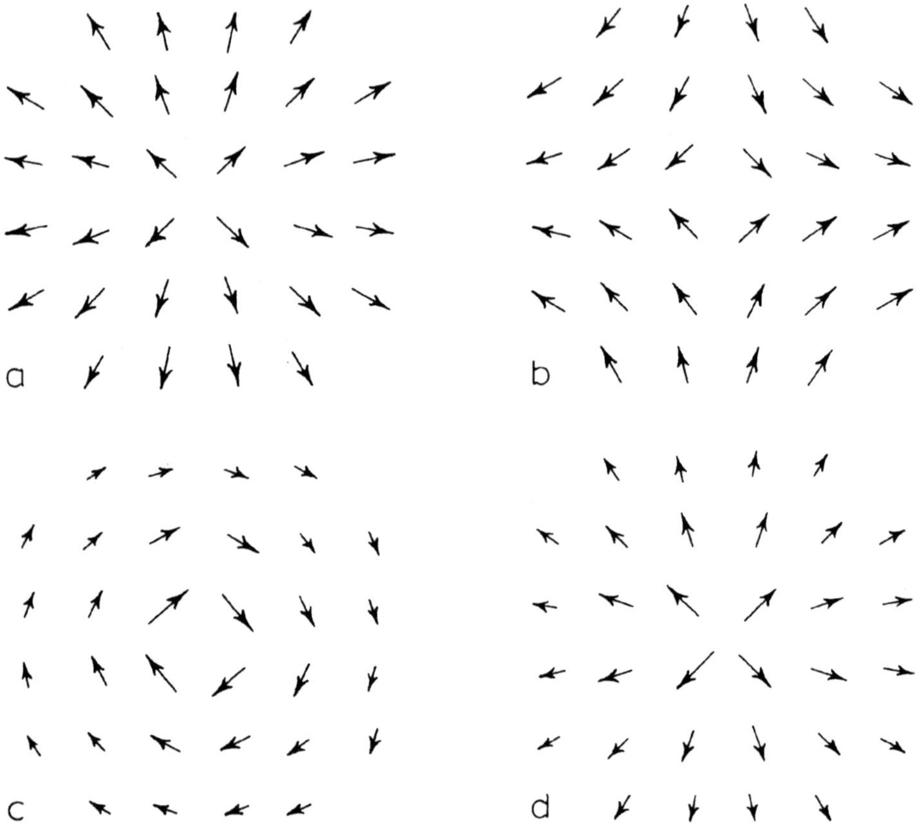


Fig. 7. Spin orientations in vortices with (a) $Q = 1$ and (b) $Q = -1$; (c) 'tension' field of a vortex; (d) the same after the dual transformation (Eq. (36)).

The topological invariant $Q$ in the particular case of a vortex is usually called the vorticity.

To be rigorous, the above considerations are justified for F-magnets only. The excitations in AF-magnets will be discussed later.

### 3.3. SCALING AND CORRELATIONS

The XY model exhibits critical behaviour in the temperature range $0 < T < T_c$. The correlation functions of the spins can be found by complete analogy with our calculation of the magnetization (see Equation (27)), yielding the following formula:

$$\langle \mathbf{S}(0) \cdot \mathbf{S}(\mathbf{r}) \rangle = S^2 (r/a)^{-2\Delta}.$$

Therefore, $\Delta$ has a meaning of a scaling dimensionality of the magnetization. The dimensionality $\Delta$ varies continuously with temperature from zero at $T = 0$ to 1/8 at the critical temperature (the latter fact will be demonstrated in the following section).

The scaling dimensionality of an external magnetic field, which is the quantity conjugated thermodynamically with the magnetization, is determined by the scaling relation [26]

$$\Delta_h + \Delta = 2. \tag{31}$$

Generally, the r.h.s. of Equation (31) is equal to a space dimensionality $d$. In an external magnetic field $h$ the nonzero magnetization $m$ can be given by:

$$m = S\langle \cos \varphi \rangle \sim h^{\Delta/(2-\Delta)}. \tag{32}$$

Equation (32) is the consequence of the scaling invariance. The relationship between $m$ and $h$ is always nonlinear since the exponent is less than unity. In particular, this implies that the magnetic susceptibility at zero field is infinite at any finite temperature below $T_c$.

The anisotropy can be taken into account by adding the following term to the free energy:

$$\delta \mathscr{F} = - h_p \int \cos p\varphi \, d^2 x$$

where $p(= 2, 3, 4, 6)$ is a parameter giving the symmetry of the anisotropy. For $p = 2$ the symmetry group of the model is $Z_2$ (Ising model). For $p = 3$ the model is equivalent in its symmetry to the 3-state Potts model. Generally, the model has the symmetry group $Z_p$, which describes the $p$-fold axis symmetry, or the invariance relative to the modulo $p$ addition.

The scaling dimensionality of cos $p\varphi$ is

$$\Delta_p = p^2 \Delta$$

in accordance with the well-known rule of averaging over the Gaussian distribution

$$\langle \cos p\varphi \rangle = \exp(-\tfrac{1}{2} p^2 \langle \varphi^2 \rangle) = (\langle \cos \varphi \rangle)^{p^2}.$$

The anisotropy constant $h_p$ is thermodynamically conjugated with cos $p\varphi$. The field $h_p$ induces spontaneous magnetization $m$. The analogue of Equation (32) reads:

$$m \sim h^{\Delta/(2-p^2\Delta)}. \tag{33}$$

It should be noted that Equation (32) is a particular case of the more general Equation (33) for $p = 1$.

HISTORICAL REMARKS. Stanley and Kaplan [27] have put forward a hypothesis (1966) that there exists a low-temperature phase with infinite susceptibility but without long-range order. Berezinskii [24] and Berezinskii and Blank [28] have obtained formulae (32) and (33) in the framework of the spin-wave approximation. The results of this section are based on the work by the authors of [29].

### 3.4. PHASE TRANSITION

Berezinskii [24] has established that vortices with opposite circulations form bound states ('molecules') at sufficiently low temperatures. The distortion of the magnetization field decreases rapidly away from the molecule. Hence, they do not destroy the algebraic decay of the correlations between spins. On the other hand a high-temperature phase is characterized by an exponential decay of correlations. The transition from an algebraic to an exponential decay cannot be continuous. Therefore a phase transition should exist. The exponential decay of the correlations can be treated as a screening effect caused by a free plasma of vortices. In contrast with the 3-d case, in this case molecules dissociate at a certain temperature. Kosterlitz and Thouless [30] have suggested defining the phase transition point by equating the free energy of a vortex to zero. The energy of a vortex is given by Equation (30), its entropy $S$ is equal to $\ln(L/a)^2$. Thus the transition temperature obeys the condition:

$$T_c = \tfrac{1}{2}\pi\bar{J}. \tag{34}$$

The quantity $\bar{J}$ is the 'effective' exchange constant at large distances. It vanishes above the phase transition temperature. Below $T_c$ the quantity $\bar{J}$ depends on temperature due to the contribution of anharmonicities and vortices. Equation (34) states the universality of the ratio $T/\bar{J}$ at the critical point, and can be rewritten in the invariant form:

$$\Delta(T_c) = 1/8. \tag{35}$$

The quantity $\bar{J}$ acquires a finite value at $T_c$. This conclusion is consistent with the arguments propounded by Josephson [31], who has found the scaling dimensionality of $\bar{J}$ to be equal to $d$-2, i.e., to zero in the 2-d case.

### 3.5. MAGNETIC VORTICES AS A COULOMB GAS

In Section 3.2 we have introduced a magnetic vortex as an elementary topological excitation, which is defined by Equations (28) and (29). The vector field $\nabla\varphi$ of a vortex with $Q = \pm 1$ is shown in Figure 7c. Let us define the dual transformation $\tilde{\varphi}$ of the field $\varphi$ by the following equation

$$\frac{\partial \varphi}{\partial x_\mu} = e_{\mu\nu}\frac{\partial \tilde{\varphi}}{\partial x_\nu} \qquad (\mu, \nu = 1 \text{ or } 2) \tag{36}$$

where $e_{\mu\nu}$ is the antisymmetric unit tensor: $e_{12} = -e_{21} = 1$, $e_{11} = e_{22} = 0$. At any point of the plane the vectors $\nabla\varphi$ and $\nabla\tilde{\varphi}$ are mutually orthogonal. In particular, the vector field of $\nabla\tilde{\varphi}(Q = \pm 1)$ is depicted in Figure 7d. It looks like an electric field generated by a unit charge placed in the center of the vortex. After the dual transformation an arbitrary system of vortices with vorticities $Q_1, Q_2, \ldots, Q_n$

placed at points $\mathbf{r}_1, \mathbf{r}_2, \ldots, \mathbf{r}_n$ becomes a system of Coulomb charges $Q_1, Q_2, \ldots, Q_n$ placed at the same points. The energies of these two systems coincide. This follows from Equation (36).

A continuum approximation as used in the preceding considerations does not completely define the energy of a vortex. There remains a question about the core energy, which can be solved only by the microscopic theory. In the framework of the continuum theory this energy strongly depends on the cut-off procedure. We assume the core energy to be very large for all $Q$ except for the case $Q = \pm 1$. Further on, we shall consider only unit charges. This assumption, which is not essential, simplifies the previous consideration by reducing the problem to a neutral plasma consisting of unit charges.

At low temperatures only neutral clusters of charges have nonzero probabilities. The most simple example is a molecule consisting of two charges of opposite sign. The energy of such a molecule is:

$$E(r) = \varepsilon^{-1} \ln r/a \qquad (Q = \pm 1) \tag{37}$$

where $r$ is the spacing between charges; $\varepsilon$ is the dielectric constant, which is a slowly varying function of the spacing. Comparing Equations (37) and (30), one obtains the following relationship between $\varepsilon$ and the exchange constant $\bar{J}$:

$$\varepsilon^{-1} = \pi \bar{J}.$$

Vortex molecules are polarized by an external field and contribute to $\varepsilon$. The field caused by molecules of a fixed spacing $R$ plays the role of an external field for smaller molecules. This is the main reason for the $R$-dependence of the dielectric permeability $\varepsilon(R)$, which has the form:

$$\varepsilon(R) = \varepsilon_0 + 2\pi \int_{R_0}^{R} W(r)\, \alpha(r) \frac{r\, \mathrm{d}r}{R_0^2}. \tag{38}$$

Here $\alpha(r)$ is the polarizability of a molecule with spacing $r$ between charges; $W(r)$ is the Boltzmann factor of such a molecule; $R_0$ is a conventional cut-off limit at small distances. The polarizability of a rigid dipole is proportional to $r^2$. The statistical weight $W(r)$ can be written as:

$$W(r) = z \exp(-E(r)/T)$$

where $z$ is the activity of the molecule. It is more convenient to consider the scale dimensionality $\Delta$

$$\Delta = T/4\pi J = T/2\varepsilon \tag{39}$$

instead of the dielectric constant $\varepsilon$. The equation for $\Delta$ can be derived from Equations (38) and (39):

$$\Delta(R) = \Delta(R_0) + 2\pi \int_{R_0}^{R} z(R_0) \left( \frac{r}{R_0} \right)^{3 - 1/2\Delta} \frac{\mathrm{d}r}{R_0}. \tag{40}$$

All the extra constant factors are absorbed by the redefined quantity of $z$. Following José *et al.*[32], we divide the integration range in the r.h.s. of Equation (40) into two parts: (1) from $R_0$ to $R_0(1 + \xi)$ and (2) from $R_0(1 + \xi)$ to $R$. After rescaling one obtains:

$$\Delta(R) = \Delta(R_0) + 2\pi\, z(R_0)\, \xi +$$

$$+ 2\pi(1+\xi)^{4-1/2\Delta} \int_{R_0(1+\xi)}^{R} z(R_0) \left( \frac{r}{R_0(1+\xi)} \right)^{3-1/2\Delta} \frac{\mathrm{d}r}{R_0(1+\xi)}. \tag{41}$$

Equation (41) has the same form as Equation (40) with only the substitution of $R_0$ by $R_0(1 + \xi)$, if we accept:

$$\Delta(R_0(1+\xi)) = \Delta(R_0) + 2\pi\, z(R_0)\xi$$

$$z(R_0(1+\xi)) = (1+\xi)^{4-1/2\Delta}\, z(R_0)$$

or, in differential form:

$$\frac{\mathrm{d}\Delta}{\mathrm{d}\xi} = 2\pi\, z, \qquad \frac{\mathrm{d}z}{\mathrm{d}\xi} = \left( 4 - \frac{1}{2\Delta} \right) z. \tag{42}$$

These are the Gell-Mann–Low equations for a 2-d plasma. The equations of this type were first obtained by Anderson *et al.* [33] for the Kondo problem. The direct application of these equations to a planar magnet has been done by Kosterlitz [34]. Their form is not unambiguous. However, the most important characteristics, namely, the existence and location of a line of fixed points and the general topology of the phase plane do not depend on a special choice of the renormalization procedure. Equation (42) are valid at $z \ll 1$, because we neglect contributions of higher charges and complex molecules.

Obviously, $z = 0$ is the fixed line of Equation (42). At sufficiently small $z$ the renormalization trajectories flow in and out of this line at $\Delta < 1/8$ and $\Delta > 1/8$, respectively (see Figure 8). In the last case a renormalized value $z$ increases and exits the region of validity of Equation (42). Hence, the point ($\Delta = 1/8$, $z = 0$) marks the phase transition. As has been discussed earlier, this transition is associated with the dissociation of vortex molecules. At temperatures below the phase transition point an average distance between molecules of a fixed spacing $R$ can be estimated as

$$L(R) \sim (z(R)/R^2)^{-1/2} \sim R^{1/4\Delta - 1} \gg R.$$

This inequality is consistent with the consideration of molecules as an ideal gas. Turning again to the renormalization trajectories shown in Figure 8 we note that region I can be identified with the manifold of states below the phase transition, while regions II and III correspond to the states above the phase transition. To obtain the critical temperature we must know $z_0$ and $\Delta_0$ as a function of temperature; or, in other words, the locus of the initial Hamiltonian shown by the dashed curve in Figure 8. The intersection of these parametrically represented curves $z_0(T)$ and $\Delta_0(T)$ and the separatrix $S$ determines $T_c$. At the transition point

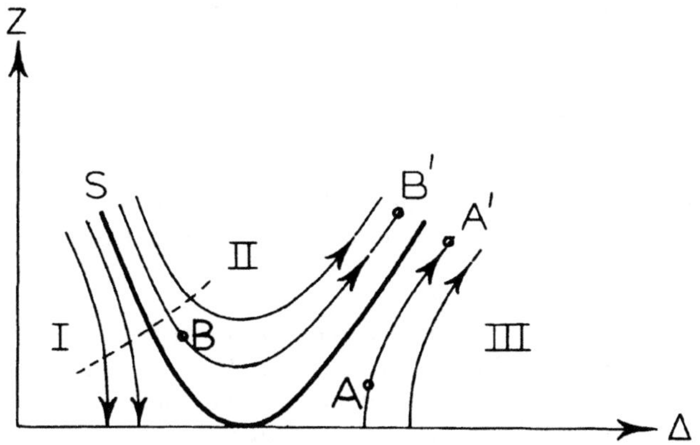


Fig. 8. Renormalization-group flow diagram for the Coulomb gas model.

$T_c$ and below it, the quantity $z$ tends asymptotically to zero. The scale dimensionality $\Delta$ goes asymptotically to the universal value 1/8 at $T_c$.

The renormalization group approach permits an estimate of the correlation length $R_c$ above the phase transition. From the initial point $z_0$ (point $A$ in Figure 8) and along the renormalization trajectory, the quantity $z$ increases up to $z \sim 1$ (point $A'$ in Figure 8). One might speculate that the further increase of $z$ stops somewhere near this point. An approximate solution of the second Equation (42) gives

$$z = z_0 \exp\left[\left(4 - \frac{1}{2\Delta}\right)\xi\right].$$

Putting $z \sim 1$ we obtain the value of $\xi$ at the vicinity of point $A'$. The correlation length satisfies

$$R_c \approx ae^{\xi} \approx az_0^{(4-1/2\Delta)^{-1}}.$$

This estimate is valid provided that the denominator of the exponent is not too small. In the close vicinity of the phase transition the best choice of variables is $x = \Delta - 1/8$ and $z' = z$. After rescaling, Equation (42) are reduced to the following form:

$$\frac{\mathrm{d}x}{\mathrm{d}\xi} = z; \qquad \frac{\mathrm{d}z}{\mathrm{d}\xi} = xz. \tag{43}$$

The renormalization trajectories of Equation (43) satisfy the parabolic equation:

$$\frac{x^2}{2} - z = -\tau$$

where $\tau$ has the physical meaning of the temperature deviation from the transition point. Starting from the point $B$ placed near the separatrix $S$ the trajectory line reaches point $B'$, where $z$ becomes of the order of unity. Almost everywhere on this line $z \approx x^2/2$. Hence, the value $\xi$ can be found as follows

$$\xi = \int \frac{dx}{z} \approx \int \frac{dx}{x^2/2 + \tau}.$$

The contribution to this integral is due to the close vicinity of the point ($x = 0$, $z = \tau$). Obviously, $\xi \sim \tau^{-1/2}$, resulting in the correlation length (Kosterlitz [34]):

$$R_c \sim a \exp(\pi \sqrt{2}\, \tau^{-1/2}). \tag{44}$$

Equation (44) defines the screening length of the vortex-free plasma formed above the phase transition point. The inverse square of $R_c$ is proportional to the density of dissociated molecules. An alternative interpretation of $R_c$ may be related to the maximal spacing of molecules, which are not yet decayed.

The heat capacity and the magnetic susceptibility can be expressed in terms of the correlation length $R_c$:

$$C \sim 1/R_c^2 \tau^2 \tag{45}$$

$$\chi \sim R_c^2. \tag{46}$$

### 3.6. RELATIONSHIP WITH OTHER MODELS

We have already shown the equivalence of the XY model and a 2-d vortex plasma with logarithmic Coulomb interaction. In this section its equivalence to some other statistical models will be established.

We start with the modification of the XY model proposed by Berezinskii [24] and Villain [35]. The statistical weight $W\{\varphi(\mathbf{r})\}$ of the configuration $\{\varphi(\mathbf{r})\}$ in the Berezinskii–Villain (BV) model is written as follows:

$$W\{\varphi(\mathbf{r})\} = \sum_{\{m_{\mathbf{r},\mathbf{r}'}\}} \exp\left[-\frac{1}{2T} \sum_{(\mathbf{r},\mathbf{r}')} (\varphi_{\mathbf{r}} - \varphi_{\mathbf{r}'} - 2\pi\, m_{\mathbf{r},\mathbf{r}'})^2\right] \tag{47}$$

where the exchange integral is absorbed into the reduced temperature $T$, $m_{\mathbf{r},\mathbf{r}'}$ is an integer associated with the bond $(\mathbf{r}, \mathbf{r}')$. The summation proceeds over all $m_{\mathbf{r},\mathbf{r}'}$ from $-\infty$ to $+\infty$. Strictly speaking the BV model cannot describe a statistical system with a temperature-independent Hamiltonian. Nevertheless, it has the same symmetry as the XY model and coincides to a high accuracy with the last one at low temperatures. Moreover, these two models demonstrate the phase transition to be of the same nature. So, whereas its properties are similar to those of the XY model, the BV model is mathematically more simple, because a statistical weight of any configuration is represented as a sum over Gaussian distributions.

The statistical weight $W\{\varphi(\mathbf{r})\}$ is a periodic function of any variables $\varphi(\mathbf{r})$ with the period $2\pi$. So, it can be expanded into a Fourier series. After integration over $\varphi(\mathbf{r})$ and summation over $m_{\mathbf{r},\mathbf{r}'}$, the partition function depends on new integer-valued variables $n_{\boldsymbol{\rho}}$ associated with the sites of a dual lattice. A statistical weight of the bond is the Fourier-transform of the corresponding statistical weight on a dual lattice:

$$W^* = \sum_{\{n_{\mathbf{r}}^*\}} \exp\left(-\frac{1}{2T^*}\sum_{(\mathbf{r}^*,\mathbf{r}'^*)}(n_{\mathbf{r}}^* - n_{\mathbf{r}'}^*)^2\right) \tag{48}$$

where the dual temperature $T^*$ is connected with the initial temperature $T$ by the relation: $T\,T^* = 1$. Model (48) is usually called the discrete Gaussian (DG) model or, alternatively, the solid-on-solid (SOS) model. This last name is associated with the description of a solid—gas or solid—liquid interface introduced by Cabrera *et al.* [36]. The DG (SOS) model may be shown to be the dual transformation of the BV model. The DG model is invariant under the discrete translations $n_{\mathbf{r}}^* \to n_{\mathbf{r}}^* + a$, forming group $Z$ of all the integers with respect to their addition. Spontaneous breaking of this symmetry is conventionally called a roughening transition. At a roughening transition temperature a solid interface changes its form from macroscopically smooth to a macroscopically rough shape (step formation).

The statistical weight (48) can be formally written as a function of continuous variables $u_{\mathbf{r}}^*$, if the necessary $\delta$-functions are introduced. Instead, we include an additional term into the energy, so the Hamiltonian takes the form:

$$\mathcal{H} = \frac{1}{2}\sum_{(\mathbf{r},\mathbf{r}')}(u_{\mathbf{r}} - u_{\mathbf{r}'})^2 - V\sum_{\mathbf{r}}\cos 2\pi\, u_{\mathbf{r}}. \tag{49}$$

When $V \to \infty$, the model coincides with the SOS model. However, the Hamiltonian (49) has a meaning for arbitrary value of the coupling $V$. In fact it is a discrete version of the so-called Sine—Gordon model. Its renormalization has been presented by Wiegmann [37] and José *et al.* [32]. The phase plane has the same topological features as that of the XY model (Figure 9, cf. Figure 8). At a sufficiently high temperature the renormalized coupling $V_R$ goes to zero. Below the phase transition temperature it reaches a constant value.

The reader may conclude that all 2-d models with the Abelian infinite group of symmetry are very closely interrelated, and are in fact various appearances of the same basic model.

### 3.7. PLANAR ANTIFERROMAGNETS

Here we restrict ourselves to a consideration of classical planar AF-magnets ($J < 0$ in Hamiltonian (23)). Any configuration of spins $\{\mathbf{m}_{\mathbf{r}}\}$ in an AF-magnet can be transformed into a configuration $\{\mathbf{m}_{\mathbf{r}}\}$, which differs from the initial one by a 180°-rotation of spins of the same sublattice. Simultaneously changing the sign of the exchange coupling $J$ in Hamiltonian (23) we reduce the partition function of an AF-magnet to that of an F-magnet. This transformation also establishes an unambiguous relationship between the partition functions of an AF-magnet in a homogeneous magnetic field $h$ and an F-magnet in a staggered field.

The GS of an AF-magnet on a square lattice is shown in Figure 10. It is continuously (two-fold) degenerate in a zero (nonzero) field. These peculiarities of

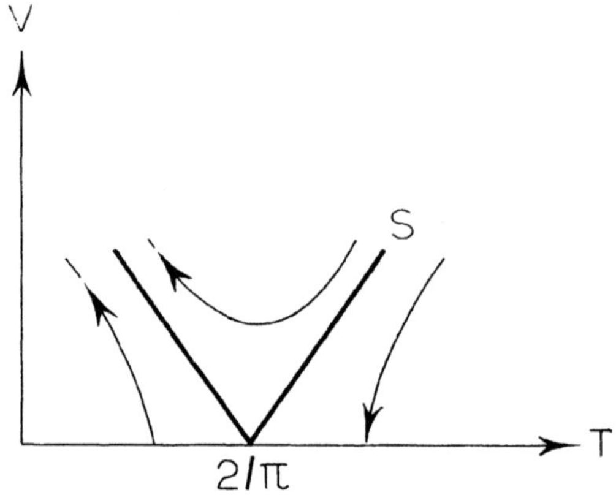


Fig. 9. Renormalization group flow diagram for the Sine—Gordon model.

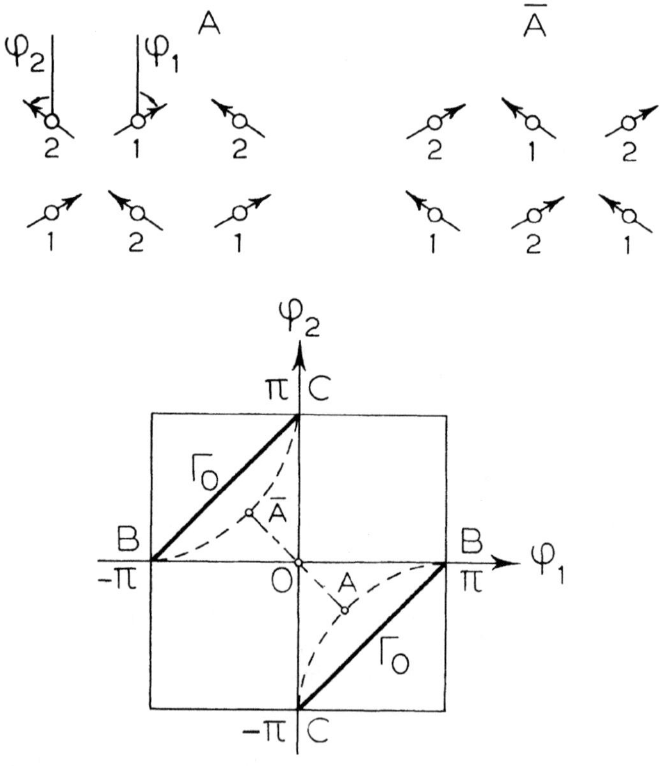


Fig. 10. Fragments ($A$ and $\bar{A}$) of the two-fold degenerate GS of a planar AF on the square lattice.

the GS are illustrated in Figure 10: the degeneracy space shown in the plane ($\varphi_1$, $\varphi_2$) is represented by line $\Gamma_0$ at $h = 0$ and by two points $A$ and $\bar{A}$ at $h \neq 0$. In the last case the GS has the Ising-type symmetry. Relevant excitations (walls) separate 'vacuum' states $A$ and $\bar{A}$, as in a conventional Ising AF-magnet. In weak fields ($h \ll J$) these walls are 'easy' (energetically most favourable), and their inner spatial distribution of spins can be mapped onto the contours: either $AB\bar{A}$ or $AC\bar{A}$ (Figure 10). The 'easy' wall structure in a real lattice whose mapping onto the plane ($\varphi_1$, $\varphi_2$) corresponds to contour $\bar{A}CA$ is shown in Figure 11a. Its 'twin' ($AB\bar{A}$) can be represented by the same drawing after renumbering the sublattices. To obtain the phase transition temperature the 'easy' wall energy per unit length has to be estimated. The whole energy can be divided into two parts, the 'barrier' and 'gradient' energies, whose contributions are of the order of $lh^2/J$ and $J/l$, respectively, where $l$ is the width of the wall. The minimal energy $\varepsilon$ is achieved at

$$l \sim J/h \tag{50}$$

so

$$\varepsilon \sim h. \tag{51}$$

It is worthwhile discussing the 'easy' wall structure initially in more detail. The abovementioned domain walls, which are topologically equivalent to contours $AB\bar{A}$ and $AC\bar{A}$, should be the constituents of their unification and could be named 'white' and 'black'. They can be transformed into one another by a Bloch wall (see Figure 12), which plays the role of a point defect. This results in a 'grey' colour of the unified wall, since point defects are not ordered inside the domain wall. In other words, the 'easy' walls should be isotropic with respect to their constituents, and their free energy is renormalized slightly due to the presence of Bloch walls. The temperature of the 'easy' wall's 'melting' is given by $T_{is} \propto \varepsilon l$, since $\varepsilon l$ is the wall energy per new characteristic length $l$ (cf. estimates of Section 2.1).

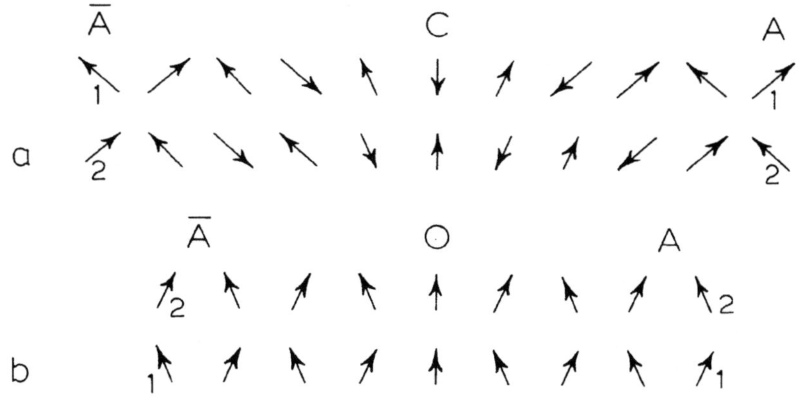


Fig. 11. Domain walls corresponding to contour $AC\bar{A}$ (a) and $AO\bar{A}$ (b).

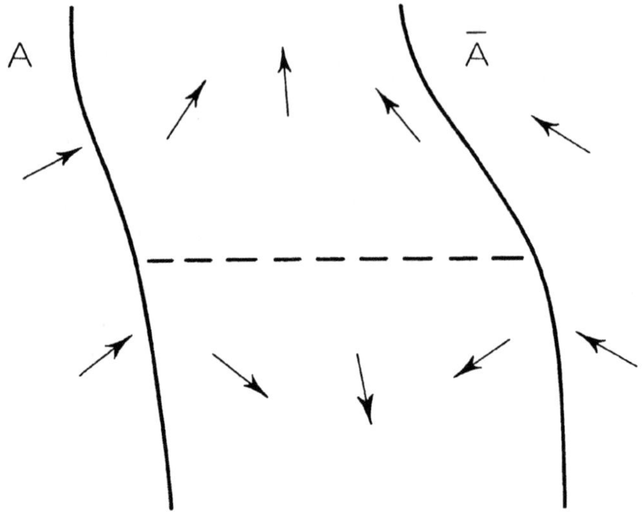


Fig. 12. Bloch-wall as a 'point' defect.

Symmetry breaking under such a transition has the following meaning: at low temperatures ($T < T_{is}$) an AF-state corresponds to '$A$-sea' with '$\bar{A}$-islands' or *vice versa*. At $T > T_{is}$ the symmetry is restored to an isotropic one with an '$A$-sea' coexisting with an '$\bar{A}$-continent'.

The points $A$ and $\bar{A}$ draw closer and closer to the origin with increasing field. The line $AO\bar{A}$ (Figure 10) corresponds to the walls which will have to be considered. Their spatial spin distribution is shown in Figure 11b. Near the critical field $h_c = 8J$ these walls can be called 'easy', unlike those represented by contours $AB\bar{A}$ and $AC\bar{A}$. The energy and the width of these high field 'easy' walls can be estimated in the same simple manner, as has been done for Equations (50) and (51),

$$l \sim J^{1/2}(h_c - h)^{-1/2}, \qquad \varepsilon \sim J^{-1/2}(h_c - h)^{3/2}.$$

The critical field $h_c$ marks the AF—P (paramagnetic) transition at zero temperature. Below $h_c$ the critical temperature is proportional to $h_c - h$.

This Ising-type phase transition is illustrated by the phase diagram in the $T$—$h$ plane (Figure 13). The phase transition curve has the endpoint $T = T_{BKT}$, $h = 0$.

A most fascinating critical behaviour is associated with the case of an AF-magnet on a triangular lattice in an external field. The energy minimization leads to the following relation for sublattice magnetic moments (the GS appears to have

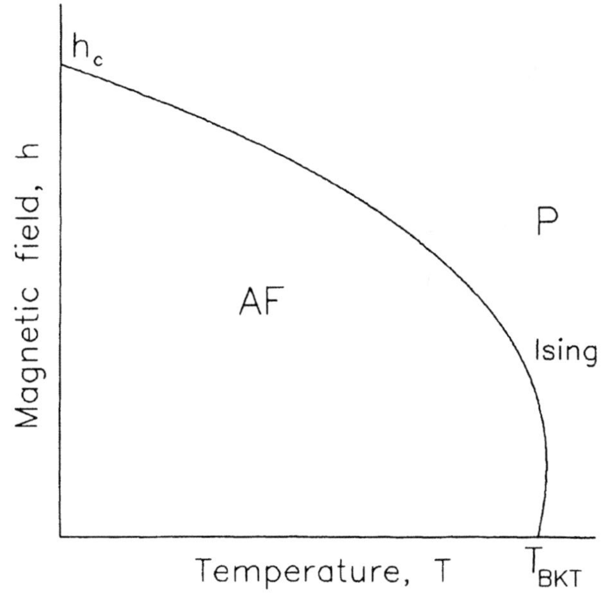


Fig. 13. Phase diagram of the planar AF on the square lattice: AF — ordered AF-magnet; P — paramagnetic phase.

three sublattices):

$$\cos\varphi_1 + \cos\varphi_2 + \cos\varphi_3 = 0, \sin\varphi_1 + \sin\varphi_2 + \sin\varphi_3 = \frac{h}{3J} \tag{52}$$

The Equation (52) displays the continuous degeneracy of the GS at $0 < h < h_{c2} = 9J$. This degeneracy is occasional and is lifted by zero oscillations of spin waves. The free energy of spin waves is proportional to

$$S(\varphi_1, \varphi_2, \varphi_3) = -(\cos^2(\varphi_1 - \varphi_2) + \cos^2(\varphi_2 - \varphi_3) + \cos^2(\varphi_3 - \varphi_1))$$

(the angles $\varphi_j$ are supposed to satisfy the constraint (52)). Different types of the ordering are shown in Figure 14. Their positions on the phase diagram in the $T - h$ plane are indicated in Figure 15.

The ordered states *a* and *b* of Figure 14 are six-fold degenerated states while *c* is three-fold degenerated. Almost all of the phase transition lines in Figure 15 correspond to a nonuniversal critical behaviour excluding the line $c - p$ which corresponds to the 3-state Potts model universality class [17]. In the region *d* of Figure 15 three sublattices are equivalent. Still, the staggered ordering of vortices is maintained [38]. There is a disagreement between Korshunov [38] and Lee *et al.* [38] on the properties of AF in the vicinity of BKT transition.

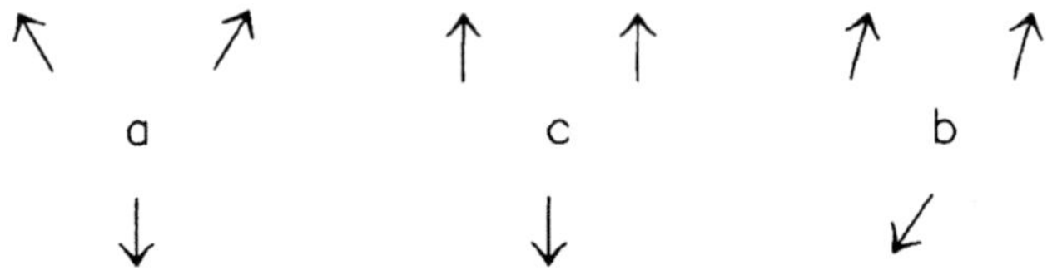


Fig. 14. In *a*- and *b*-cases degeneracy spaces include six states, while in *c*-case it consists of three states. All the representatives can be constructed by permutations of sublattice spins and by changing $\varphi \to -\varphi$.

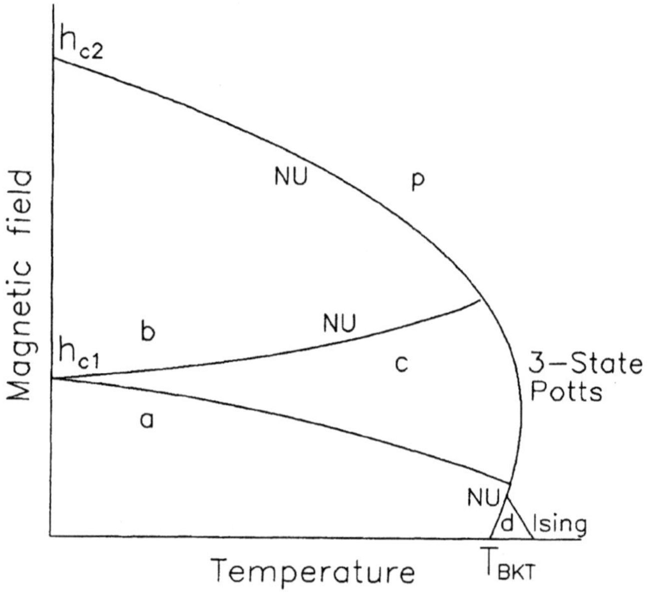


Fig. 15. Phase diagram of the planar AF on the triangular lattice. Low-temperature phases *a, b, c* reflects the GS topology shown in Fig. 14a, b, c, respectively. *d*-phase corresponds to the staggered vorticity ordering.

## 4. Heisenberg Magnets

### 4.1. HEISENBERG MODEL AND REAL MAGNETS

The Heisenberg model was proposed as an idealization of purely exchange magnets. The energy of such a magnet is invariant with respect to homogeneous rotation of all spins (group $SO_3$). The spins are three-dimensional quantum vectors with the components satisfying the usual permutation relations:

$$[S_\alpha, S_\beta] = i\, e_{\alpha\beta\gamma} S_\gamma.$$

A most frequently used model is one with nearest-neighbour interactions only, described by the Hamiltonian:

$$\mathcal{H} = -J \sum_{(\mathbf{r},\mathbf{r}')} \mathbf{S}_\mathbf{r} \cdot \mathbf{S}_{\mathbf{r}'}$$

where summation proceeds over the pairs of nearest-neighbour sites ($\mathbf{r}$, $\mathbf{r}'$). This is the so-called Heisenberg model. It appears to be the general form of the $SO_3$-invariant Hamiltonian for $S = 1/2$. It can easily be generalized to higher values of $S$ by adding the terms $(\mathbf{S}_\mathbf{r} \cdot \mathbf{S}_{\mathbf{r}'})^n$ with integer exponents $n$. Another type of generalization, which has not been widely used so far, implies the addition of further-neighbor interactions.

All the spins in the GS of a Heisenberg F-magnet are along the same direction, aligned and acquire their maximal values. The GS of a Heisenberg AF-magnet has not yet been found analytically. Approximate calculations for the lattices which can be divided into two sublattices demonstrate magnetizations of opposite signs for different sublattices. In both cases the GS's continuously degenerate. However, the degree of degeneracy does not depend on the number of sites $N$. Therefore, the entropy per site due to this degeneracy is zero.

In real magnets there always exists an anisotropy due either to spin-orbit coupling or to dipole-dipole interactions. Sometimes these effects can be rather small, and an isotropic model may be used with a great accuracy. For rare-earth elements the inter-atomic spin-orbit coupling is strong, resulting in the formation of a sufficiently rigid total momentum $\mathbf{J}$ of an atom. Some rare-earth compounds can be described by the Heisenberg model, or its generalization with the vector $\mathbf{S}$ playing the role of the total atomic moment of the rare-earth element.

### 4.2. RENORMALIZATION OF THE TEMPERATURE

The Landau—Peierls—Hohenberg—Mermin—Wagner theorem on the absence of a long-range order is quite applicable to Heisenberg magnets: spin waves destroy the magnetization (staggered magnetization at nonzero temperature) in 2 dimensions, similar to the XY model. However, the phase diagram of a Heisenberg magnet is quite different: there are no phase transitions at any finite temperature. Quite recently this fact has been proved rigorously by Wiegmann and Polyakov [39]. Their proof, based on the Bethe-ansatz is rather cumbersome, so we prefer a nonrigorous and more qualitative treatment developed by Polyakov [40] and Berezinskii and Blank [28].

Since the phase transition is a purely classical phenomenon, we concentrate on a classical continuum model of an isotropic magnet. The local magnetization can be described by a vector $\mathbf{S}(\mathbf{r})$ of fixed length. For definiteness we put $|\mathbf{S}| = 1$. The simplest Hamiltonian describing such an isotropic magnet is defined as follows:

$$\mathscr{H} = \frac{1}{2T} \int (\nabla \mathbf{S})^2 \, d^2x. \tag{53}$$

The terms proportional to $(\nabla \mathbf{S})^{2n}$ with $n > 1$ can be neglected in a continuum version. Fluctuations of the magnetization can be divided into rapidly-varying ($\mathbf{S}_1$) and slowly-varying parts ($\mathbf{S}_0$):

$$\mathbf{S} = \mathbf{S}_0 \sqrt{1 - S_1^2} + \mathbf{S}_1. \tag{54}$$

Vectors $\mathbf{S}_0$ and $\mathbf{S}_1$ can be chosen to be mutually orthogonal, $S_0^2 = 1$ together with $S^2 = 1$, while $\mathbf{S}_1$ is assumed to be a small vector. Substituting $\mathbf{S}$ from Equation (54) into Equation (53), and integrating $w = \exp(-\mathscr{H})$ over $S_1$ we find the resulting distribution reduced to the form $w^* = \exp(-\mathscr{H}^*)$, where $\mathscr{H}^*$ differs from $\mathscr{H}$ by changing the effective temperature $T \to T^*$ (besides the substitution of the variable $\mathbf{S}_0$ for $\mathbf{S}$). In other to calculate the renormalized temperature $T^*$ we have to specify the rapidly-varying variable $\mathbf{S}_1$. Starting from this point we consider a more general problem in which $\mathbf{S}$ is an $n$-component vector. Let $\mathbf{e}_\alpha(\alpha = 1, \ldots, n-1)$ be a set of unit vectors, orthogonal to $\mathbf{S}_0$. Then:

$$\mathbf{S}_1 = \sum_\alpha S_{1\alpha} \mathbf{e}_\alpha.$$

The vector field $\mathbf{S}_1$ is approximately a free one, i.e., it obeys the Gaussian distribution, and so does each $S_{1\alpha}$. So, integrating Fourier components of $S_{1\alpha}$ from $q$ to $q + \delta q$ we obtain

$$\langle \mathbf{S}_1^2 \rangle = \frac{n-1}{2\pi} T \ln \frac{q + \delta q}{q} \equiv \frac{n-1}{2\pi} T \xi. \tag{55}$$

Since the derivatives $\partial \mathbf{S}_0 / \partial x_\alpha$ are orthogonal to $\mathbf{S}_0$ itself, they can be written in the form:

$$\partial \mathbf{S}_0 / \partial x_\alpha = \sum_\beta C_{\alpha\beta} \mathbf{e}_\beta. \tag{56}$$

Obviously, the antisymmetric matrix $\| C_{\alpha\beta} \|$ satisfies:

$$\sum_{\alpha\beta} | C_{\alpha\beta} |^2 = (\nabla \mathbf{S}_0)^2. \tag{57}$$

A proper choice of vectors $\mathbf{e}_\beta$ should be made, otherwise their dependence on spatial coordinates will be 'fast'. When they rotate slowly the derivatives $\partial \mathbf{e}_\alpha / \partial x_\beta$ can be approximately equated to

$$\partial \mathbf{e}_\alpha / \partial x_\beta = -C_{\alpha\beta} \mathbf{S}_0. \tag{58}$$

Substituting **S** from Equation (54) into Equation (53) and using Equations (55)–(58) we get:

$$T^* = T + \frac{T^2(n-2)}{2\pi}\,\xi. \tag{59}$$

Equation (59) holds for a sufficiently small $\xi$. It has to be replaced by the differential equation:

$$\mathrm{d}T/\mathrm{d}\xi = T^2(n-2)/2\pi$$

which is correct as far as $T$ is small. It has the solution;

$$T(\xi) = T_0 \Bigg/ \left(1 - \frac{T_0(n-2)}{2\pi}\,\xi\right) \tag{60}$$

where $\xi$ is the logarithm of a spatial scale, $T_0$ is the temperature of the bath (at $\xi = 0$). For $n = 2$ the temperature $T$ does not depend on the scale, as we have already observed in Section 3.5. For $n > 2$ the temperature increases with the wavelength of the fluctuations, reaching a value of the order of unity in the vicinity of the logarithmic pole. The approximation $T \ll 1$ is inconsistent in this region. However, the effective temperature of the long-range fluctuations appears to be of the same order of magnitude, whatever the temperature of the thermal bath. Hence we can conclude that no phase transition takes place at any finite temperature.

The correlation length $l_c$ has to be identified with the location of the logarithmic pole in Equation (60):

$$l_c \sim \exp\left(\frac{2\pi}{T(n-2)}\right).$$

A slightly more complicated calculation must be performed to obtain a spin correlation function. We present it here without proof:

$$\langle \mathbf{S}(\mathbf{r})\mathbf{S}(0)\rangle = \left(1 - \frac{T_0(n-2)}{2\pi}\ln r\right)^{(n-1)/2(n-2)}. \tag{61}$$

For $n = 2$ this results in an algebraic decay of correlations. For $n = 3$ the exponent in the r.h.s. of Equation (61) equals unity. This form of the correlator holds for $r < l_c$. At $r > l_c$ correlations decay exponentially as $\exp(-r/l_c)$. This last result can be obtained correctly only in the framework of an exact solution.

## 4.3. HEISENBERG FERROMAGNETS IN AN EXTERNAL MAGNETIC FIELD

The external magnetic field $h$ can be regarded as strong, if $hl_c^2 \gg T$. In this case the decay of correlations stops at $r \approx r_h$, defined by the relation $hr_h^2 \sim T_0$. Substituting this value of $r$ into Equation (61) we find the induced magnetization

($n = 3$) (Khokhlachev [41]):

$$\langle S\rangle \sim \left(1 - \frac{1}{4\pi} T_0 \ln T_0/h\right). \tag{62}$$

The logarithmic dependence of the average magnetization on the magnetic field has been found in an experiment by Karimov [42] on $NiCl_2$ intercalated into graphite.

In the opposite case, $hl_c^2 \ll T_0$, the magnetization depends linearly on the magnetic field with the following magnetic susceptibility:

$$\chi \sim l_c^2 \sim \exp(4\pi/T_0). \tag{63}$$

### 4.4. EXCITATIONS OF THE 2-D HEISENBERG MODEL

The simplest excitations of the 2-d Heisenberg model are the spin waves. When discussing an analogous question in Section 3.2, we have emphasized that the spectrum of spin waves must be gapless due to the Goldstone theorem. In the long-wave limit the energy spectrum of F-(AF-)magnets satisfies $\omega \sim k^2(\omega \sim k)$. This difference is a consequence of the different symmetries of these states with respect to the time reversal. The dynamical aspects of spin waves will be considered in detail in Section 6.2.

In a Heisenberg F-magnet a classical spin wave state can be continuously transformed into the GS. Now we decribe those states which are not topologically equivalent to the GS. They were first introduced by Skyrme [43]. Spins are assumed to be parallel far away from the excitation core. Each point of the plane or its spherical projection has its mapping on the sphere, on which the ends of the vectors $\mathbf{S}(\mathbf{r})$ are positioned. There exists a topological invariant $D$ called the mapping degree. This is an integer that shows how many times the sphere in the spin space is covered in the course of mapping. For the GS and spin wave states, $D = 0$. The simplest case of excitations with non-zero mapping degree is shown in Figure 16. This excitation corresponds to a conditional minimum of energy (53), with the constraint $S^2(\bar{r}) = 1$. The corresponding Euler variational equation is:

$$\Delta\mathbf{S} + (\nabla\mathbf{S})^2\mathbf{S} = 0, \tag{64}$$

The vector $\mathbf{S}$ can be parametrized by two angles, $\theta$ and $\varphi$, on the sphere. Instead of those, Belavin and Polyakov [44] have suggested consideration of the complex quantity $w = \tan\theta/2 \exp(i\varphi)$, which is a function of a complex variable $z = x + iy$, describing the point of a plane with coordinates $(x, y)$. In these terms a most general solution $w$ of Equation (64) satisfying the requirements of unambiguity and $w \to \infty$ at $z \to \infty$, is a ratio of polynomials in $z$. As a result of the second condition ($w \to \infty$) the degree $n$ of the nominator in the polynominal exceeds the degree $m$ of the denominator. In this case $D = n - m$. The general solution has been found in [44]. The most simple solution due to Skyrm is $w = z/R$.

Its section is schematically depicted in Figure 16. The quantity $R$ has the

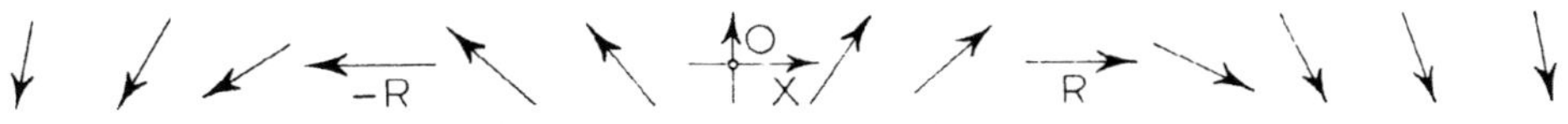


Fig. 16. Cross-section of the Skyrmion, the topological excitation in the 2-d Heisenberg magnet.

meaning of an excitation radius. The energy of such an excitation is finite and equals $4\pi$ (the exchange coupling is equated to unity). The energy of an arbitrary solution with the mapping degree $D$ is equal to $4\pi D$. Such a solution can be interpreted as $D$ noninteracting excitations (skyrmions).

We can introduce the concept of an antiskyrmion in the same way as that of a skyrmion. In these excitations the vector $\mathbf{S}(\mathbf{r})$ rotates in opposite directions in an azimuthal plane. Mathematically we have to consider the complex conjugate function $\bar{w} = \tan \theta/2 \exp(-i\varphi)$, depending on $\bar{z} = x - iy$.

The role of skyrmions and antiskyrmions in the statics and dynamics remains unclear to present. In contrast to magnetic vortices there may, in principle, arise single skyrmions which suppress the long-range order. This point of view, however, meets strong objection, since there are no phase transitions and no topological excitations for $n > 3$. Fateev *et al.* [60] have proved that skyrmion fluctuations only destroy the long-range order.

### 4.5. DIPOLAR INTERACTION

In 2-d Heisenberg F-magnets, the dipolar interaction results in an easy-plane anisotropy (Maleev [45], Pokrovsky and Feigelman [46]). This effect can be explained qualitatively as follows. In the GS the dipolar and exchange energies can be minimized separately. Therefore, the average magnetization cannot have a component normal to the plane because the dipolar energy is minimized when all the spins are aligned in the plane. The spin alignment remains unchanged with increasing temperature, until the thermal energy exceeds the characteristic dipolar energy $\mu_B^2/a^3$. For $T \gg \mu_B^2/a^3$ the dipolar interaction can be neglected. Although it is small compared with the exchange interaction, the former cannot be neglected at $T \lesssim \mu_B^2/a^3$ due to its long-range nature. The magnetic dipolar forces change fundamentally the state at low enough temperatures and induce in fact a long-range order. Indeed, let us suppose a long-range order existing initially. In the absence of the weak dipolar interaction fluctuations would destroy the order, since thermal fluctuations of the magnetization are proportional to the integral of mean squares $\langle |\delta \mathbf{S}_\mathbf{q}|^2\rangle \sim T/q^2$ and diverge. The non-zero dipolar interaction changes the spectrum of the fluctuations at small wave-numbers. The dipolar part of the Hamiltonian can be written as follows

$$\mathcal{H}_d = \frac{1}{2}(g\mu_B)^2 \sum_{\mathbf{r} \neq \mathbf{r}'} \frac{\delta\mathbf{S}(\mathbf{r})\delta\mathbf{S}(\mathbf{r}) - 3(\delta\mathbf{S}(\mathbf{r})\boldsymbol{\nu})(\delta\mathbf{S}(\mathbf{r}')\boldsymbol{\nu})}{|\mathbf{r} - \mathbf{r}|^3} \tag{65}$$

where $\boldsymbol{\nu}$ is the unit vector parallel to $\mathbf{r}' - \mathbf{r}$. The Fourier transform of Hamiltonian

(65) reads:

$$\mathscr{H}_{\rm d} = \pi(g\mu_{\rm B})^2 \sum_{\mathbf{q}} \frac{q_a q_b}{q} \delta S_\alpha(\mathbf{q}) \delta S_\beta(-\mathbf{q}). \tag{66}$$

Adding the dipolar Hamiltonian (66) to the exchange Hamiltonian (53), we obtain the mean square of the fluctuations:

$$\langle |\delta \mathbf{S}(\mathbf{q})|^2\rangle = T/(q^2 + 2\pi(g\mu_{\rm B})^2 q \sin^2 \gamma_{\mathbf{q}})$$

where $\gamma_{\mathbf{q}}$ is the angle between the vector $\mathbf{q}$ and the total magnetization. The integral of $\langle |\delta \mathbf{S}(\mathbf{q})|^2\rangle$ now no longer diverges, and the Landau—Peierls theorem becomes invalid. Hence, the dipolar interaction stabilizes the long-range order at distances exceeding $(g\mu_{\rm B})^{-2}$. At smaller distances the magnet is effectively isotropic.

The average magnetization can be calculated like a Debye—Waller factor:

$$\langle S\rangle = \exp\left[-\frac{1}{2}\int \frac{\mathrm{d}^2 q}{(2\pi)^2} \langle |\delta \mathbf{S}(\mathbf{q})|^2\rangle\right]. \tag{67}$$

The explicit calculation in the exponent of Equation (67) gives the following equation of state:

$$y^{1-2\Delta} = \exp(f(z))^{\Delta/2} \tag{68}$$

where

$$y = \langle S\rangle/(\pi(g\mu_{\rm B})^2)^{\Delta/(1-2\Delta)}; \quad z = h(2\pi(g\mu_{\rm B})^2)^{-2}\langle S\rangle^{-3}; \quad \Delta = T/4\pi$$

and

$$f(z) = \int_0^\infty s\mathrm{d}s\left[\frac{1}{s\sqrt{s^2+s}} - \frac{1}{\sqrt{s^2+z}\,\sqrt{s^2+s+z}}\right].$$

The function $f$ has the characteristic scaling form. In a weak field region ($z \ll 1$) the solution of Equation (68) is approximately as follows:

$$\langle S\rangle = (\pi(g\mu_{\rm B})^2)^{\Delta/(1-2\Delta)}\left(1 + \frac{\Delta}{1-2\Delta}\frac{4}{\sqrt{\pi}}\Gamma\left(\frac{3}{4}\right)z^{1/4}\right). \tag{69}$$

As a consequence of Equation (69), the magnetic susceptibility $\chi = \partial S/\partial h$ tends to infinity as $h^{-3/4}$ when $h$ goes to zero. In the range of moderately strong fields $(g\mu_{\rm B})^{2(2-\Delta)/(1-2\Delta)} \ll h \ll \exp(-2\pi/T)$ i.e. at large $z$, the solution of Equation (68) is approximately $S \sim h^{\Delta/(2-\Delta)}$ in accordance with the scale dimensionality consideration of Section 2.3. The dependence of the dimensionless magnetization $y$ on the dimensionless field $x = h(g\mu_{\rm B})^{-2(2-\Delta)/(1-2\Delta)}$ is demonstrated in Figure 17.

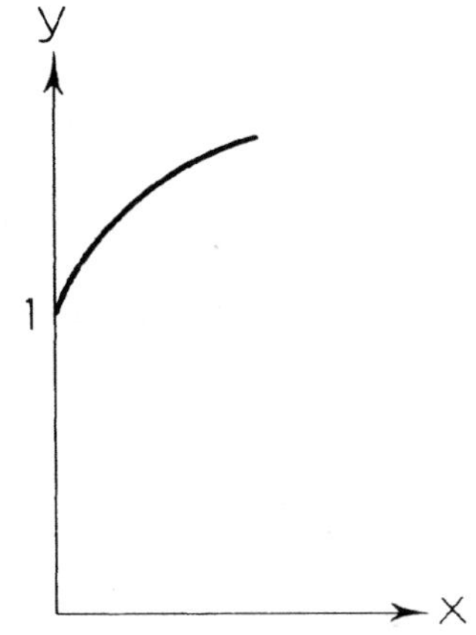


Fig. 17. Dependence of the dimensionless magnetization ($y$) on the dimensionless field ($x$).

## 5. Experimental Layered Magnets

### 5.1. ISING LAYERED MAGNETS. ANNNI MODEL: APPLICATION TO CeSb AND CeBi

The interaction in an experimental layered magnet can be conveniently divided into two parts:

$$\mathcal{H} = \mathcal{H}_0 + \mathcal{H}_1 \tag{70}$$

where the first term in the r.h.s. of Equation (70)

$$\mathcal{H}_0 = -\sum_j \sum_{(\mathbf{r},\mathbf{r}')} J_0 S_j(\mathbf{r}) S_j(\mathbf{r}') \tag{71}$$

describes the in-plane interaction of Ising spins, while the second term

$$\mathcal{H}_1 = \sum_j \sum_{\mathbf{r}} J_1 S_j(\mathbf{r}) S_{j+1}(\mathbf{r}) \tag{72}$$

represents the weak interaction ($J_1 \ll J_0$) of spins, belonging to nearest-neighbour layers. In Equations (71) and (72) the index $j$ marks the planes, whereas $\mathbf{r}$ and $\mathbf{r}'$ denote a spin coordinate in a plane. Hereafter we set $|S| = 1$ for convenience.

To solve this problem of the layer-type magnet the method of high-temperature expansion in the small constant $J_1$ appears to be the most appropriate. Indeed, the partition function of the layered system

$$Z = \prod_j \mathrm{Tr}_j \exp\left(-\frac{1}{T}\mathcal{H}_0(j)\right) \exp\left(-\frac{1}{T}\mathcal{H}_1\right)$$

can be expanded in a series in $\mathscr{H}_1$:

$$Z = Z_0^{N_e}\left(1 - \frac{1}{T}\overline{\mathscr{H}}_1 + \frac{1}{2T^2}\overline{\mathscr{H}_1^2} - \ldots\right)$$

where $Z_0(T)$ is the partition function of the 2-d Ising model; $N_l$ is the total number of layers. Bars denote a 2-d statistical average in the spin layer, neglecting the interaction with the others. The free energy expansion is:

$$\mathscr{F} = N_l\mathscr{F}_0 + \overline{\mathscr{H}}_1 - \frac{1}{2T}\langle\mathscr{H}_1^2\rangle + \ldots \tag{73}$$

where $\mathscr{F}_0$ is the free energy of the 2-d Ising model; angular brackets denote irreducible averages over the Gibbs ensemble of decoupled Ising planes. In the zero order of the expansion all the planes are decoupled. At temperatures below the critical temperature $T_{is}$ of the 2-d Ising model described by Hamiltonian $\mathscr{H}_0(j)$ each magnetic plane is ordered in the state with either magnetization $+\overline{S}(T)$ or $-\overline{S}(T)$, where $\overline{S}(T)$ is the absolute value of the spontaneous magnetization of a decoupled plane. The stacking sequence of magnetization signs is chaotic. The term $\overline{\mathscr{H}}_1$ (Equation (73)) corresponds to the mean-field approximation and has the 1-d Hamiltonian form:

$$\overline{\mathscr{H}}_1 = N_2 \sum_j J_1 \overline{S}_j \overline{S}_{j+1} \tag{74}$$

where $N_2$ is the number of magnetic sites in a layer. Obviously, Hamiltonian (74) leads to F (or AF) stacking depending on the sign of $J_1$.

Let us find an expansion parameter for the free energy expansion (73). In Equation (73) the second-order correction includes the following contribution alongside the others:

$$\Delta F^{(2)} = -N_2 J_1^2\left(\frac{1}{T}\sum_{\mathbf{r}}\langle S(0)S(\mathbf{r})\rangle\right)\sum_j \overline{S}_j\overline{S}_{j+2}\,. \tag{75}$$

So, in addition to the effective 1-d Hamiltonian (74) we obtain the effective 1-d interaction (75) between the magnetizations of next-nearest neighbour layers. An expansion parameter appears to be of the order of $\chi J_1$, where

$$\chi = \frac{1}{T}\sum_{\mathbf{r}}\langle S(0)S(\mathbf{r})\rangle$$

is the magnetic susceptibility of the 2-d Ising model. In the vicinity of $T_{is}$ the parameter $\chi J_1$ becomes of the order of unity, and the high-temperature expansion ceases to be convergent. This narrow temperature range near $T_{is}$ can be estimated as:

$$\Delta T \sim T_c(J_1/J_0)^{4/7}$$

(see formula (18) for the susceptibility exponent).

Therefore, in a layer-type system at $T < T_{is}$ the high-temperature expansion method allows one to construct an effective 1-d Hamiltonian, where the sign of the average magnetization of a plane plays the role of an Ising variable.

Systems with competing interactions are more suitable for applying the high-temperature expansion. To introduce a competition into Hamiltonian (72) the interaction between spins situated in next-nearest-neighbour layers must be taken into account, i.e. we may generalize the Hamiltonian $\mathscr{H}_1$ to the following form:

$$\mathscr{H}_1 = \sum_{\mathbf{r}} \sum_j \sum_{l=1}^{2} J_l S_j(\mathbf{r}) S_{j+l}(\mathbf{r}). \tag{72'}$$

This generalization is known as the so-called ANNNI model. Its phase diagram at zero temperature is shown in Figure 18. It is worthwhile paying attention to the existence of the degeneracy lines: at a certain ratio of the couplings $J_1$ and $J_2$ the GS of the ANNI model is degenerate. For instance, in the case

$$J_1 < 0, \qquad J_2 = \tfrac{1}{2} |J_1| \tag{76}$$

any stacking is allowed excluding the elements of the AF-structure: $\cdots + - + \cdots$ and $\cdots - + - \cdots$. This manifold of degenerate states has two limiting configurations: the first one is F-magnetic, and the second is the so-called 'up-up-down-down' stacking (its conventional designation is $\langle 2 \rangle$).

In the vicinity of the degeneracy lines, contributions of the form given in Equation (75) can be derived:

$$\Delta F^{(2)} = -N_2 \chi J_2 \sum_j (2J_1 \bar{S}_j \bar{S}_{j+3} + J_2 \bar{S}_j \bar{S}_{j+4}). \tag{75'}$$

Due to an increase of the interaction radius, the structure $\langle 3 \rangle$ with the following periodical stacking $+ + + - - -$ is stabilized in the narrow vicinity of the

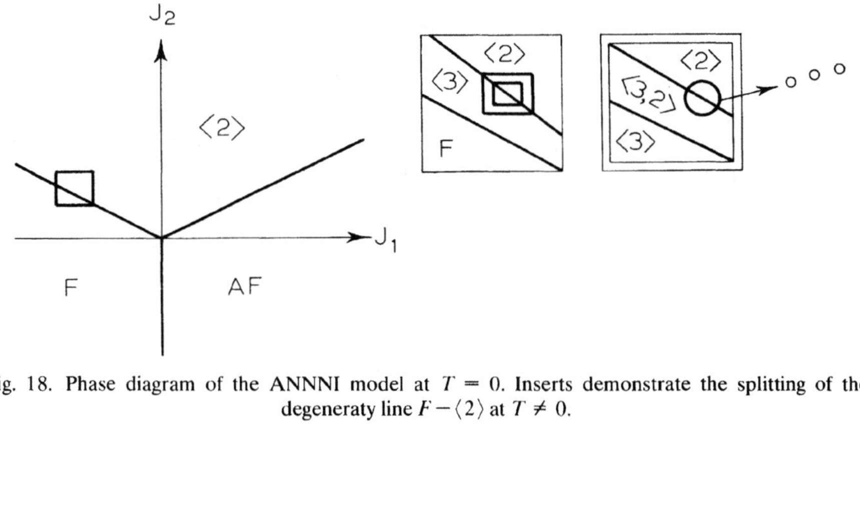


Fig. 18. Phase diagram of the ANNNI model at $T = 0$. Inserts demonstrate the splitting of the degeneraty line $F - \langle 2 \rangle$ at $T \neq 0$.

degeneracy line defined by Equation (76). It indicates that two degeneracy lines F $-\langle 3\rangle$ and $\langle 3\rangle - \langle 2\rangle$ appear instead of the initial line F $-\langle 2\rangle$. Further the stability of the two new lines must be elucidated. Fisher and Selke [47] have been the first to analyse this stability problem by the method of low-temperature expansion of the free energy. It appears that the line F $-\langle 3\rangle$ is stable against thermal fluctuations of higher orders, while the line $\langle 3\rangle - \langle 2\rangle$ splits, generating a narrow strip of the stable phase $\langle 3.2\rangle$ with a periodical stacking, either $+++--$ or $---++$. The ratio of the widths of phase $\langle 3.2\rangle$ and $\langle 3\rangle$ demonstrated in the phase diagram of Figure 18 is proportional to the small parameter $\chi J_1$. Furthermore, this analytical procedure results in a cascade of first-order phase transitions:

$$\mathrm{F}-\langle 3\rangle-\langle 3.2\rangle - \langle 3.2^2\rangle \equiv \langle 3.2.2\rangle-\langle 3.2^3\rangle - \cdots - \langle 3.2^\infty\rangle \equiv \langle 2\rangle.$$

In the vicinity of the zero-temperature degeneracy line AF $-\langle 2\rangle$ an analogous cascade is:

$$\mathrm{AF}-\langle 2.1\rangle-\langle 2^2.1\rangle-\langle 2^3.1\rangle-\cdots-\langle 2^\infty.1\rangle \equiv \langle 2\rangle.$$

Some of these transitions can be realized by changing the temperature.

An analytic investigation of phase transitions in the ANNNI model has been done in the framework of the low-temperature expansion method by Fisher and Selke [47], in the mean-field approximation by Bak and von Boehm [48] and by the high-temperature expansion by Pokrovsky and Uimin [49]. The last method appeared to be convenient to analyse the ANNNI model behaviour in an external magnetic field. This problem was particularly relevant in connection with the experimental data on magnetic behaviour of GeSb reviewed by Rossat-Mignod *et al.* [50]. The phase diagram for CeSb is qualitatively depicted in the $T$—$h$ plane in Figure 19. The main features of the experimental phase diagram [50] have been subject to the high-temperature analysis based on the ANNNI model [49].

A similar problem exists for the material CeBi. The theoretical interpretation of the phase diagram has been done by Uimin [51]. These two interpretations (cf. [49] and [51]) differ in their modifications of the conventional ANNNI model. For example, in the case of CeBi an additional competing coupling $J_3$ had to be taken into consideration. In the case of CeSb at moderately low temperatures a competition of some other type appears. It is connected with arising separated paramagnetic layers. The reason for this can be explained as follows. The multiplet $J = 5/2$ of the $Ce^{2+}$ ion in a cubic crystal field splits into the quartet and doublet levels. The doublet level of an isolated $Ce^{2+}$ ion lies lower than the quartet one. But, due to the exchange interaction of Ce ions, these levels change their relative position at low temperatures. The magnitude of the exchange energy decreases with increasing temperature. A possible intersection of the 'magnetically ordered quartet' — 'paramagnetic doublet' levels is illustrated in Figure 20. For this case beginning at a definite temperature a competition of magnetic and paramagnetic levels should be allowed.

An alternative explanation of the CeSb phase diagram is entirely based on the ANNNI model [49].

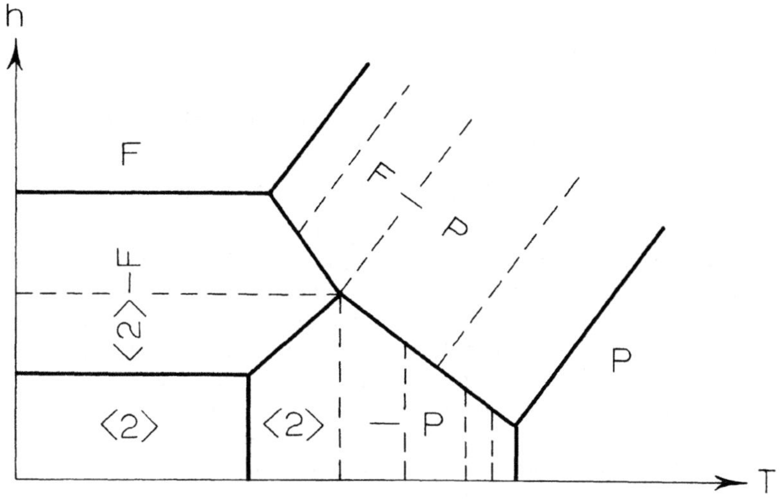


Fig. 19. Quantitative picture of the experimental phase diagram in CeSb [50]. Different magnetic structures are separated by solid lines.

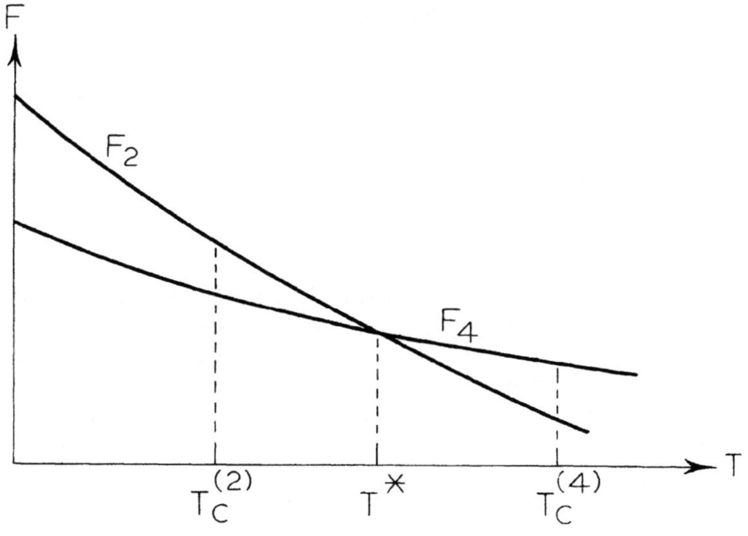


Fig. 20. Possible explanation of paramagnetic layer arising near $T^*$. $F_2$, $F_4$ are the free energies of magnetic 'doublet' and 'quartet' systems, respectively, with corresponding ordered temperatures ($T_c^{(2)}$ and $T_c^{(4)}$).

## 5.2. LAYERED PLANAR MAGNETS

In a layer-type magnet with planar anisotropy even a small interlayer exchange interaction $J_1$ results in the stabilization of the long-range order (magnetization or staggered magnetization). The value of the magnetization depends on the ratio $J_1/J_0$ and the temperature according to the law:

$$m = |\langle \mathbf{S} \rangle| = \text{const}\,(J_1/J_0)^{\Delta/(2-2\Delta)}, \qquad \Delta = T/8T_c \tag{77}$$

as follows from the scale invariance [28, 29]. The magnetization given by Equation (77), which can be attributed to spin waves, remains finite at the phase transition point, $\Delta_c = 1/8$. This result contradicts the obvious 3-d character of the phase transition in the close vicinity of the transition point. Therefore, the crucial role of vortex excitations becomes clear. It has been elucidated by Hikami and Tsuneto [52]. They have found out that independent vortex molecules in each plane form 3-d vortex rings with the characteristic size $R_1 \sim a\sqrt{J_0/J_1}$. At these distances the gain in energy due to a coherent arrangement of vortices in different planes can be estimated as $J_1 R_1^2$, which has to be compared with the loss of the free energy ($\sim J_0$) of an in-plane molecule (we omit the slowly changing factor $\ln R_1$). So a layer system undergoes a transition from a 2-d one to the crossover behaviour at $R_1 \sim R_c$, or at $\tau \sim (\ln R_1)^{-2}$ (cf. Equation (44)).

Let us consider the same problem from another point of view. The exchange interaction for a layered planar magnet can be written as follows:

$$\mathcal{H}_{cx} = J_0 \int d^2x\, dz \left[ \frac{1}{2} (\nabla \mathbf{m})^2 + \frac{1}{2} \frac{J_1}{J_0} \left( \frac{\partial \mathbf{m}}{\partial z} \right)^2 \right]$$

where $\nabla$ denotes the in-plane gradient. The corresponding propagator $G(\mathbf{r}, z)$ has a typical Ornstein–Zernike form:

$$G(\mathbf{r}, z) = \text{const}\,\exp(-R\kappa)/R$$

where

$$R = \sqrt{r^2 + z^2 J_0/J_1}/a.$$

The correlation length is anisotropic and is defined by the usual relation $R_{3c} \sim \tau^{-\nu}$ (the exponent $\nu \approx 2/3$ for 3-d planar systems). Putting $r = 0$ and $z = a$ we obtain the estimate $\tau \sim (J_1/J_0)^{3/4}$ for a boundary of the 3-d fluctuational region. For a sufficiently small ratio $J_1/J_0$, there appears a temperature range in which vortex molecules are glued together forming vortex rings, but 3-d fluctuations have not yet developed:

$$(J_1/J_0)^{3/4} \ll \tau \ll (\ln J_1/J_0)^{-2}.$$

However, for the finite ratio $J_1/J_0 \gtrsim 10^{-2}$ this range does not exist. Numerically, $\tau$ is not very small in the 3-d region, even for $J_1/J_0 \sim 10^{-2}$. This means that the observation of a pure 2-d behaviour will be a difficult experimental problem. Nevertheless, the developement of a theory for the 2-d region is of a certain interest, since those two regions can be distinguished only by comparing the theory and the experimental data.

A most interesting one is the case of F-magnetic planes interacting via weak AF-bonds. Introducing a weak in-plane anisotropy one can distinguish the in-plane longitudinal and transverse susceptibilities. Their ratio in the quasi 2-d region has been derived by Pokrovsky and Uimin [29]:

$$\chi_{\parallel}/\chi_{\perp} = \Delta. \tag{78}$$

In the quasi 2-d region $\Delta$ should not exceed its critical value 1/8. Moreover, the real 3-d phase transition point, where $\chi_{\parallel}/\chi_{\perp} \to 1$ should be separated from the point $\Delta = 1/8$ by the value of the order of $(\ln J_0/J_1)^{-2}$ on the $\tau$-scale, provided that the dipolar interaction or some other factors are not essential.

If a magnetic field is applied along the 'easy' magnetization axis, the well-known spin-flop transition occurs at a certain critical magnitude of a field $H_{sf}(T)$, which is proportional to the spontaneous magnetization $m_s(T)$. Thus:

$$H_{sf}(T)/H_{sf}(0) = m_s(T)/m_s(0). \tag{79}$$

The second critical field $H_2(T)$ is associated with the second-order phase transition from AF-magnetic to a paramagnetic state. It appears to be proportional to the spontaneous magnetization as well:

$$H_2(T)/H_2(0) = H_{sf}(T)/H_{sf}(0). \tag{80}$$

Equations (78)—(80) can be checked experimentally. In the 2-d region the logarithm of the ratio $H_{sf}(T)/H_{sf}(0)$ can be equated to $\Delta/(2 - 2\Delta)$ and compared with the independent experimental definition of $\Delta$ given by Equation (78).

As an example of a layer magnet with in-plane F- and interplane AF-bonds we may mention the AF-magnetic copper compound $(C_2H_5NH_3)_2CuCl_4$, investigated by de Jongh *et al.* [53]. In Figure 21 the experimental values of the ratio $\chi_{\parallel}/\chi_{\perp}$ versus the argument $T/T_c$ are plotted. One can see that the value $\chi_{\parallel}/\chi_{\perp} = 1/8$ is achieved rather far from the phase transition point, approximately at $T/T_c \sim 0.7$, while our estimate results in $T/T_c \sim 1 - (\ln J_0/J_1)^{-2} \sim 0.9$.

The case of a layer F-magnet has been analysed in the work by Hirakawa and Ikeda [54] (see also the chapter by these authors).

### 5.3. LAYERED HEISENBERG MAGNETS

Stabilization of the long-range order due to the weak inter-layer coupling $J_1$ also occurs in Heisenberg layer-type magnets. The characteristic length $R_1 = a(J_0/J_1)^{1/2}$ has to be compared with the correlation length $R_c = a \exp(2\pi J_0/T)$ in a Heisenberg plane magnet. When $R_1 < R_c$, the interplane coupling favours magnetization. So the transition temperature can be estimated as:

$$T_c = 4\pi J_0/\ln (J_0/J_1). \tag{81}$$

The spontaneous magnetization at $T < T_c$ can be found from Equation (61),

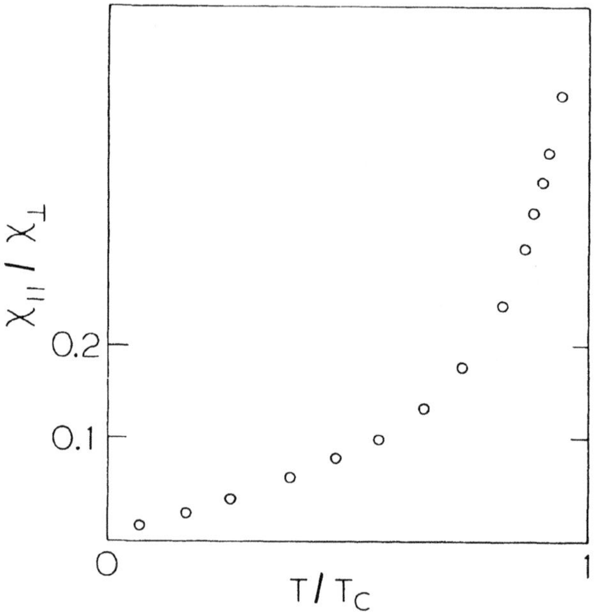


Fig. 21. Experimental ratio $\chi_{\parallel}/\chi_{\perp}$ in the layer-type compound $(C_2H_5NH_3)_2CuCl_4$ (data from [53]).

where the special values $n = 3$ and $R = R_1$ should be substituted:

$$m_s = 1 - \frac{T}{4\pi J_0} \ln J_0/J_1. \tag{82}$$

The region in which Equation (82) holds is restricted by the smallness of the renormalized temperature:

$$T/4\pi J_0(1 - (T \ln J_0/J_1)/4\pi J_0) \ll 1.$$

In the close vicinity of $T_c$ the 3-d critical fluctuations play a crucial role resulting in the usual 3-d critical behaviour.

## 6. Dynamics of 2-d Magnets

### 6.1. EQUATIONS OF MOTION

So far we have considered an ensemble of static excitations, i.e. an instantaneous picture of a 2-d magnet has been averaged over a large area. Owing to ergodicity the same result could be obtained by averaging over a long time interval. Hereafter, dynamical excitations and responses will be discussed. In addition, a dynamical behaviour of magnets depends on dissipation, which in turn is controlled by the interaction with non-magnetic subsystems, such as phonons, electrons, etc. In this case the energy $\varepsilon$ of a magnetic excitation and its decay time $\tau_d$ satisfy

$$\tau_d \sim \varepsilon^{-1}$$

as an order of magnitude estimate. In particular, in the case of large dissipation spin waves get transformed into a spin-diffusion mode.

For a macroscopic description of spin motions we use the Landau–Lifshitz equation in the following form:

$$\frac{d\mathbf{M}}{dt} = -g\mathbf{M} \times \mathbf{H}_{eff} - \frac{\Gamma}{M} \mathbf{M} \times (\mathbf{M} \times \mathbf{H}_{eff}). \tag{83}$$

This equation is tantamount to the condition that the magnitude of $\mathbf{M}$ does not change. This approximation is reasonable for a sufficiently weak anisotropy. The effective field $\mathbf{H}_{eff}$ is defined as a variational derivative of the free energy over the local magnetization:

$$\mathbf{H}_{eff} = -\delta\mathcal{F}/\delta\mathbf{M}(\mathbf{r}).$$

The first term in the r.h.s. of Equation (83) corresponds to a non-dissipative motion. The second one is responsible for dissipation. The different types of elementary excitations are principally defined by the ratio $\Gamma/g$.

## 6.2. SPIN WAVE DYNAMICS

We start with the planar case, adding a strong anisotropy term to the exchange Hamiltonian:

$$\mathcal{H} = \frac{1}{2}\int J(\nabla\mathbf{M})^2\, d^2r + \frac{\beta}{2}\int M_z^2\, d^2r. \tag{84}$$

Equation (83) holds for degenerate magnetic systems, so it remains unchanged under the easy-plane anisotropy. Now the effective field satisfies:

$$\mathbf{H}_{eff} = J\Delta\mathbf{M} - \beta M_z\hat{z}$$

where $\hat{z}$ is a unit vector in the $Z$ direction. For definiteness the magnetization vector $\mathbf{M}$ is assumed to be directed along the $X$-axis in the GS. In the linearized version of the Landau–Lifshitz equation only the transverse deviations of the magnetization, that is the $Y$- and $Z$-components, have to be taken into account:

$$\begin{aligned} \dot{M}_y &= gM(J\Delta M_z - \beta M_z) \\ \dot{M}_z &= -gMJ\Delta M_y. \end{aligned} \tag{85}$$

Thus, a spin wave frequency $\omega$ depends on its wave vector $k$ as

$$\omega_k = sk\sqrt{1 + Jk^2/\beta}.$$

The spin wave velocity $s$ is expressed in terms of the parameters of the

Hamiltonian (84) as:

$$s = gM\sqrt{J\beta}$$

Using Equation (85) we generalize the static correlation function (see Equation (25)) to the following time-dependent form:

$$D_q(t) = \langle \varphi_q(t)\varphi_{-q}(0)\rangle = \frac{T}{JM^2q^2}\cos\omega_q t$$

where $\varphi_q(t)$ is a spatial Fourier transform of the azimuthal angle of the magnetization vector.

In neutron scattering experiments as well as in measurements of the electromagnetic response the Fourier transform of the time-dependent correlation function

$$K_{ij}(\mathbf{r}, t; \mathbf{r}', t') = \langle M_i(\mathbf{r}, t) M_j(\mathbf{r}', t')\rangle$$

can be found. For a planar magnet $i, j = 1, 2$ and $K_{ij}$ is proportional to $\delta_{ij}$. The invariant correlator can be written as:

$$K(\mathbf{r}-\mathbf{r}', t-t') = \tfrac{1}{2}\,K_{ii}(\mathbf{r}, t; \mathbf{r}', t; \mathbf{r}, t') = \langle\cos(\varphi(\mathbf{r}, t) - \varphi(\mathbf{r}', t'))\rangle. \quad (86)$$

Using the Gaussian distribution of the variable $\varphi$, we find (cf. the equations of Section 3.2, describing static correlations):

$$K(\mathbf{r}-\mathbf{r}', t-t') = \exp[-\langle\varphi^2(\mathbf{r}, t)\rangle + \langle\varphi(\mathbf{r}, t)\,\varphi(\mathbf{r}', t')\rangle]. \quad (87)$$

The exponent in Equation (87) is

$$D(0,0) - D(\mathbf{r}-\mathbf{r}', t-t') = \int \frac{d^2q}{(2\pi)^2}\,[\langle\varphi_q(t)\varphi_{-q}(t)\rangle - -\langle\varphi_q(t)\varphi_{-q}(t')\rangle\, e^{i\mathbf{q}(\mathbf{r}-\mathbf{r}')}]. \quad (88)$$

According to the fluctuation—dissipation theorem the Fourier transforms of the correlator $K(\mathbf{r}, t)$ and of the imaginary part of the causal Green function $\mathcal{K}_c(\mathbf{r}, t)$ are equal to one another. By definition the latter is a statistical average of the time-ordered product of quantum spin operators. The real part of $\mathcal{K}_c(\mathbf{r}, t)$ can be restored by means of the analyticity conditions. Physical measurements are associated with the causal and retarded Green functions, $\mathcal{K}_c$ and $\mathcal{K}_r$, respectively. Their Fourier transforms are related by:

$$\mathcal{K}_r(\mathbf{q}, \omega) = \operatorname{Re}\mathcal{K}_c(\mathbf{q}, \omega) + i\tanh\frac{\omega}{2T}\operatorname{Im}\mathcal{K}_c(\mathbf{q}, \omega).$$

Below we list those observable quantities which may allow a direct comparison of experimental data with theoretical predictions.

The electromagnetic response function has the form:

$$\chi(\mathbf{q}, \omega) = -\mu_0^2\mathcal{K}_r(\mathbf{q}, \omega). \quad (89)$$

The differential cross-section for magnetic neutron scattering is given by the expression:

$$d\sigma/d\Omega d\omega = (g_n r_0)^2 \frac{k}{k_0}\frac{K(\mathbf{q}, \omega)}{4\pi}, \qquad \mathbf{q} = \mathbf{k} - \mathbf{k}_0 \tag{90}$$

where $r_0$ is the classical radius of the electron, $g_n$ is the neutron gyromagnetic ratio, and $\mathbf{q}$ and $\hbar\omega$ are the momentum and energy transfers, respectively.

Finally, the linewidth of the nuclear magnetic resonance is proportional to the integral:

$$\delta\omega \sim \mathrm{Re} \int \frac{d^2 q}{(2\pi)^2} K(\mathbf{q}, \omega) \tag{91}$$

where $\omega_0$ is the nuclear magnetic resonance frequency.

In a neutron scattering study the most interesting pieces of information are associated with the intensity and shape of the resonance peak at $\omega \approx \omega_q$. The main difference from the well-known resonant scattering is found in the non-Lorenzian shape of the resonance curve. The approximate expression for the correlator $K(\mathbf{q}, \omega)$ has a nontrivial resonance-like behaviour:

$$K(\mathbf{q}, \omega) = \frac{2\pi\Delta \sin(\pi\Delta)\,\Gamma(1-2\Delta)}{\hbar\omega_q q^{2-2\Delta}} \left| 1 - \frac{\omega_q}{\omega} \right|^{-1+2\Delta}, \qquad \left| \frac{\omega - \omega_q}{\omega_q} \right| \ll 1. \tag{92}$$

The observation of spin wave resonances in inelastic neutron scattering at small momentum transfer would be one of the best demonstrations of the BKT phase transition. The theory predicts a special temperature-dependent form of the resonance peak. At $T = 0$ it coincides with the usual Lorenzian form, at any temperature within the interval $0 < T \leqslant T_c$ the resonance curve decays algebraically (Equation (92)) with the exponent varying between $-1$ and $-3/4$.

High-frequency susceptibility ($\omega \gg \omega_q$) can be found from Equation (89) where $q$ should be set to zero:

$$\mathcal{K}_r(0, \omega) = A(\Delta)\, \omega^{-2+2\Delta}.$$

The explicit form of the coefficient $A(\Delta)$ can be taken from the work by Blank *et al.* [55] This asymptotic formula is valid in the frequency interval:

$$\hbar\omega_q \ll \hbar\omega \ll T.$$

The critical slowing down in the nuclear magnetic resonance curve is defined by the relation:

$$\delta\omega \sim h^{2\Delta/(2-\Delta)}$$

where $h$ is an external magnetic field.

The results of this section are based on the work described in [55].

### 6.3. SPIN-DIFFUSION DYNAMICS

Let us consider pure planar magnets ($\beta \to \infty$) in the high dissipation limit. Neglecting the first term in the r.h.s. of Equation (83) we obtain the equation of motion for the angular variable $\varphi$:

$$\frac{\partial \varphi}{\partial t} = J\Gamma M^2 \Delta\varphi. \tag{93}$$

The corresponding correlation function in the $q, \omega$ representation can be related to the linear response function as

$$G^0_{\mathbf{q},\omega} = \frac{\Gamma}{i\omega - \Gamma J M^2 q^2} \tag{94}$$

by means of the fluctuation—dissipation theorem, which reads:

$$\langle \varphi_{\mathbf{q},\omega}\varphi_{-\mathbf{q},-\omega}\rangle = -\frac{2T}{\omega}\,\mathrm{Im}\,\mathbf{G}^0_{\mathbf{q},\omega} = \frac{2T\Gamma}{\omega^2 + (\Gamma\, JM^2q^2)^2}\,. \tag{95}$$

Using the general relations (87) and (88), the time-dependent correlation function (86) can be derived:

$$K(\mathbf{r}, t) = (r/a)^{-2\Delta} \exp\left(-\Delta E_1\left(\frac{r^2}{4\Gamma\, JM^2|t|}\right)\right) \tag{96}$$

where

$$E_1(x) = \int_x^\infty \frac{e^{-S}}{S}\,\mathrm{d}S$$

is the integral exponential function. The Fourier transform of the spin—spin correlation function (96) has a typical scaling form:

$$K(\mathbf{q}, \omega) = a^{2\Delta}\omega^{-1}q^{-2+2\Delta} F\left(\frac{|\omega|}{4\Gamma\, JM^2q^2}\right) \tag{97}$$

where the function $F(x)$ is defined by

$$F(x) = 4\pi\Delta \int_0^\infty \frac{\sin y}{y}\,\mathrm{d}y \int_0^\infty z^{1-2\Delta} J_0(z)\, e^{-\Delta E_1(z^2x/y)}\, e^{-z^2x/y}\,\mathrm{d}z.$$

Here $J_0(z)$ is the Bessel function.

The observable quantity $K(\mathbf{q}, \omega)$ has the following asymptotic behaviour:

$$K(\mathbf{q}, \omega) \sim q^{-4+2\Delta}, \qquad \omega \ll \Gamma JM^2q^2$$

and

$$K(\mathbf{q}, \omega) \sim \omega^{-2+\Delta}, \qquad \omega \gg \Gamma JM^2q^2.$$

So, the differential cross-section of inelastic neutron scattering has a non-resonant maximum at zero energy transfer, with a characteristic width $\delta\omega \sim \Gamma J q^2$. Experimental tests of various aspects of the theory are more appropriate for the spin-diffusion region than for the nondissipative case, where high resolution methods must be applied.

## 6.4. DYNAMICS OF LOCALIZED EXCITATIONS

Let us consider a moving localized excitation with the magnetization field $\mathbf{M}$, depending on the variable $\mathbf{r}' = \mathbf{r} - \mathbf{v}t$. This implies that $\mathbf{M}$ does not change in a moving frame of reference. Our purpose is to derive the equation of motion for such localized excitations provided their velocities $\mathbf{v}_j$ are sufficiently low. Analogous equations were first derived by Thiele [56] for magnetic bubbles.

Let $\mathbf{M}_0(\mathbf{r})$ be a static localized solution of the nondissipative Landau—Lifshitz equation that satisfies the equation:

$$\mathbf{H}_{\text{eff}}\{\mathbf{M}_0(\mathbf{r})\} = \lambda(\mathbf{r})\mathbf{M}_0(\mathbf{r})$$

where $\lambda(\mathbf{r})$ is the Lagrange factor. It is worthwhile finding a moving solution in the following form

$$\mathbf{M}(\mathbf{r}') = \mathbf{M}_0(\mathbf{r}') + \mathbf{M}_1(\mathbf{r}') \qquad (\mathbf{r}' = \mathbf{r} - \mathbf{v}t)$$

where $|\mathbf{M}_1|$ is much smaller than $|\mathbf{M}_0|$. The linearized Landau—Lifshitz equation for $\mathbf{M}_1$ reads:

$$(\mathbf{v}\nabla)\mathbf{M}_0 = g(\mathbf{M}_0 \times \delta\mathbf{H}_{\text{eff}} + \mathbf{M}_1 \times \mathbf{H}_{\text{eff}}) + \\ + \Gamma(gM_0)^{-1}\mathbf{M}_0 \times (\mathbf{v}\nabla)\mathbf{M}_0. \tag{98}$$

Owing to the translational invariance of the $\mathbf{M}_0$ solution, we should take into consideration as the $\mathbf{M}_1$ candidates those vectors which are perpendicular to the vector fields $\nabla\mathbf{M}_0$. In order to obtain an equation for an averaged velocity field, we form the vector product of both sides of Equation (98) with $\mathbf{M}_0$, and then the scalar product with $\nabla\mathbf{M}_0$. Subsequent integration of these products over the planar coordinates $X$ and $Y$ results in the equation:

$$\mathbf{G} \times \mathbf{v} + D_0\mathbf{v} + \mathbf{F} = 0 \tag{99}$$

where

$$\mathbf{G} = \frac{1}{2}\int \mathrm{d}^2r\, e_{ijk}\mathbf{M}_{0i}\, \nabla M_{0j} \times \nabla M_{0k} \tag{100}$$

$$D_0 = -\frac{1}{2}\frac{\Gamma M_0}{g}\int \nabla M_{0j}\, \nabla M_{0j}\, \mathrm{d}^2r \tag{101}$$

and

$$\mathbf{F} = -gM_0^2 \int H_j \nabla M_j \, \mathrm{d}^2 r. \tag{102}$$

In Equations (100)—(102) $\nabla$ is a two-dimensional gradient vector and the summation runs over the pairs of coinciding indices which mark three Cartesian vector components.

The gyration vector **G** is perpendicular to the plane. Its absolute value coincides with the mapping degree $D$ for a localized excitation in a 2-d Heisenberg magnet if $|M| = 1$ (see Section 4.4). For the case of a vortex-like excitation in the XY model the gyrovector has been calculated by Huber [57]. For a weak easy-plane anisotropy the normal component of the magnetization appears inside the core of the vortex-like excitation. Symmetry considerations imply that the magnetization direction in the centre of the vortex is normal to the plane. From the topological point of view this means that a vortex excitation is mapped onto a half of the sphere. So, to point out the mapped half of the sphere the additional topological number $\sigma$ is necessary which takes the values $+1$ or $-1$ depending on the direction of the central spin: 'up' or 'down', respectively. The gyrovector **G** is associated with topological numbers of a vortex $q$ and $\sigma$ by the relation

$$\mathbf{G} = -2\pi q \sigma \hat{z} \tag{103}$$

where $\hat{z}$ is a unit vector normal to the plane.

$D_0$ defined by Equation (101) is a trace of the dissipation dyadic $D_{\alpha\beta}$, which is diagonal in the absence of in-plane anisotropy. The dominant contribution to the integral (101) comes from large $r$, resulting in the equation

$$D_0 = -\frac{\Gamma}{g} \pi \ln R/l \tag{104}$$

where $R$ is the 'outer radius' of a vortex and $l$ is the core size.

To determine the motion of vortices bound in a molecule we should specify the 'force' **F** (see Equation (102)). We consider the effective field $\mathbf{H}_{\text{eff}}$ induced by one of the vortices as an external field for the other vortex. Using Equation (102) we obtain

$$\mathbf{F} = -g \frac{\mathbf{R}}{R^2} 2\pi J M^2 q^2. \tag{105}$$

The same result can be found by differentiation of the vortex pair energy.

There are two types of vortex molecules. The first type is characterized by different $\sigma$: $\sigma_1 = -\sigma_2$. If dissipation could be neglected, these molecules would rotate around their nonmoving centers with the angular velocities

$$\omega = \frac{\gamma}{R^2}, \qquad \gamma = 4gJM^2.$$

Molecules of the second type have $\sigma_1 = \sigma_2$. Vortices bound in these molecules move in parallel with one an other, and perpendicular to a line connecting them, with a linear velocity:

$$\mathbf{v} = gJM^2 \frac{\hat{z} \times \mathbf{R}}{R^2} = \gamma \frac{\hat{z} \times \mathbf{R}}{4R^2}.$$

This picture alludes to vortex pair motion in hydrodynamics (see, e.g. [58]). The essential difference between these two problems lies in the absence of the second topological number $\sigma$ from the usual hydrodynamic motion.

Dissipation results in decreasing vortex spacing according to the law:

$$R^2 = R_0^2 - \frac{2\pi |\gamma D_0|}{G^2 + D^2} t.$$

Obviously this equation is valid for those spacings which are large as compared to the vortex core radius.

In the continuum approximation all spin interactions except the dipolar interaction result in the dependence of the energy and the total magnetic moment of a vortex pair on the pair size only. This follows from the invariance of the energy $E$ and the total magnetization $\mathbf{M}$ with respect to rotation in the plane without any change of magnetization (see Figure 22). It is noteworthy that this invariance is not violated by uniaxial, cubic and exchange anisotropies. However, the dipolar interaction destroys the abovementioned invariance and produces a weak angular

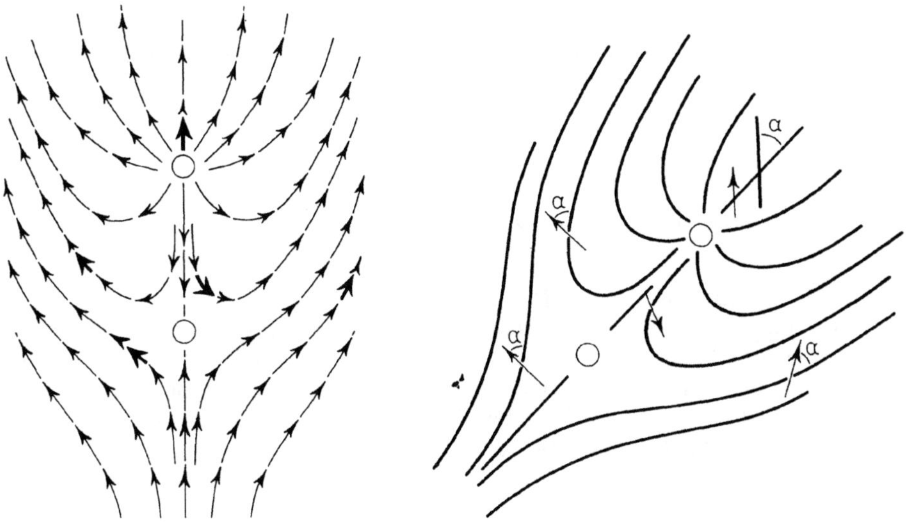


Fig. 22. Rotation of the molecular axis with unchanged direction of spins. The energies of the depicted spin configurations differ in the dipolar interaction terms only.

dependence of the energy and the total magnetic moment [59]:

$$E = E_0(R) + E_1(R)\cos 2\phi + \dots \tag{106}$$

$$M_x = M_0(R) + M_1(R)\cos 2\phi + \dots \tag{107}$$

$$M_y = M_1(R)\sin 2\phi + \dots \tag{108}$$

where $\phi$ is the angle between the preferred magnetization direction at large distances, treated as the $X$-axis, and the axis of a vortex molecule. The quantity $M_1$ calculated by the authors [59] is

$$M_1 = \frac{Sfg}{4\gamma} R \ln R$$

where $f = \mu_B^2/a^4$ is the coupling constant of the dipolar interaction; $S \sim J/\beta$ is the vortex core area.

## 6.5. RESONANT PARAMETRIC EXCITATION OF VORTEX PAIRS

A minor asymmetry of a vortex pair orbit caused by the dipolar interaction can give rise to a resonant enhancement of the rotation energy by an alternating external magnetic field. This pump wave can be included in the scheme of Equation (99) as an additional term in the external 'force':

$$\mathbf{F}_1 = -g \frac{\partial}{\partial \mathbf{R}} (\mathbf{HM}(\mathbf{R})) \frac{\mu_B}{a^2}. \tag{109}$$

Confining ourselves to the second harmonics in Equations (107) and (108) we obtain the following equations of motion for a vortex pair:

$$\begin{aligned} -\dot{y} + \bar{D}\dot{x} &= -\gamma \frac{x}{R^2} - hM_1 \frac{4xy^2}{R^4} \\ \dot{x} + \bar{D}\dot{y} &= -\gamma \frac{y}{R^2} + hM_1 \frac{4x^2y}{R^4} \end{aligned} \tag{110}$$

where $\bar{D} = D_0/2\pi$ and $h = g\mu_B H(t)/2\pi a^2$.

Since the products $xy^2$ and $x^2y$ contain the first and third rotational harmonics, the resonance should be expected at an external field frequency $\omega_0$ close to $2\omega(R)$, i.e., the double frequency of pair rotation. In the vicinity of the resonance Equation (110) can be solved by the well-known method of averaging over the fast variable $\phi$, defined in the following way:

$$x = R\cos(\phi + \alpha), \quad y = R\sin(\phi + \alpha), \quad \phi = \int_0^t \omega(R)\,\mathrm{d}t.$$

Slow variables $R$ and $\alpha$ satisfy the equations

$$\dot{\alpha} = \bar{D}\,\dot{R}/R \tag{111}$$

$$(1+\bar{D}^2)\,R\dot{R} + \bar{D}\gamma = \frac{Sfg}{2\gamma}\,R \ln R\langle h \sin\phi\rangle. \tag{112}$$

Here the angular brackets denote the averaging over the period of unperturbed rotation. In the resonant approximation Equation (112) can be written as:

$$A\ddot{\psi} + \cos\psi - B = 0 \tag{113}$$

where

$$\psi = \int_0^t (2\omega(R) - \omega_0)\,\mathrm{d}t + \bar{D}\ln R_0$$

and

$$A = 2R_0^3(g\,Sfh_0 \ln R_0)^{-1} \tag{114}$$

$$B = 4\bar{D}\gamma^2(g\,Sfh_0\,R \ln R_0)^{-1}. \tag{115}$$

The quantity $R_0$ in Equations (114)—(115) is determined by the resonance point:

$$\omega_0 = 2\omega(R_0).$$

Equation (113) has the first integral

$$\tfrac{1}{2}A\dot{\psi}^2 - B\psi + \sin\psi = \text{const.} \tag{116}$$

The first term in the l.h.s. of Equation (116) can be treated as the kinetic energy of an imaginary particle moving in the multi-valley potential (see Figure 23):

$$V(\psi) = -B\psi + \sin\psi.$$

The bound states in this potential arise when $B < 1$. They correspond to the quasistationary regime of a rotating molecule. Hence the threshold $H_c$ is determined by the equation:

$$B = 1 \quad \text{or} \quad H_c = \frac{4D_0\gamma^2}{SfR_0 \ln R_0}\,\frac{a^2}{g^2\mu_B}.$$

In the case of $(C_2H_5NH_3)_2CuCl_4$ the couplings $J$ and $\beta a^2$ are of the order of 10 K and 0.1 K, respectively (see [53]). So upper estimates of $H_c$ and $\nu = \omega/2\pi$ are

$$H_c \sim 10\,\text{Oe} \quad \text{and} \quad \nu \sim 10^8 \sim 10^9\,\text{Hz}$$

The vortex spacing range $\Delta R$ of the molecules which are resonantly trapped by an external r.f. field that satisfies the equation:

$$\frac{\Delta R}{R_0} = \frac{1}{2\omega_0}\sqrt{2A^{-1}\,\Delta V} \tag{117}$$

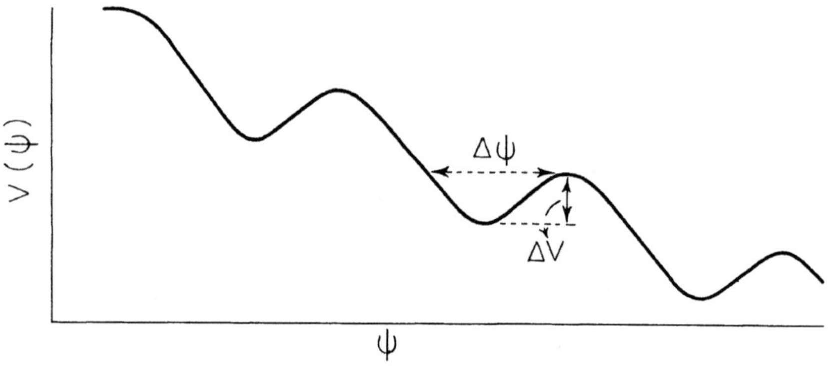


Fig. 23. Schematic plot of $V(\psi)$ as a function of $\psi$ (Eq. 116).

where $\Delta V$ is the height of the potential barrier shown in Figure 23. Near the threshold

$$\Delta V = \frac{4\sqrt{2}}{3}\left(1 - \frac{H_c}{H}\right)^{3/2}.$$

The resonant absorption is determined by the density of vortex molecules trapped in the resonant rotation by the r.f. field:

$$\Delta N_{tr} = \tfrac{1}{2}\, W(P)\, R \Delta R \Delta \psi. \tag{118}$$

Here $W(R)$ is the Boltzmann factor of vortex molecules (see Equation (38)); $\Delta R$ is defined by Equation (117); $\delta\psi$ is the interval of phases which corresponds to bound states (see Figure 23). The factor 1/2 implies that half of all the molecules rotate.

Each molecule that is trapped in the resonant rotation absorbs the power

$$Q(R) = \frac{1}{g}\, DV^2 = \frac{D}{g}\,\frac{\gamma^2}{R^2}.$$

The total r.f. field attenuation per unit area is

$$q(\omega_0) = \Delta N_{tr} Q(R_0)$$

$$\sim \omega^{\lambda}\left(1 - \frac{H_c}{H}\right)^{5/4}, \qquad \lambda = \frac{1}{4\Delta} - 1. \tag{119}$$

The most interesting features of Equation (119) are the strong frequency and temperature dependences. The exponent $\lambda$ varies from $+\infty$ to 1 when the temperature increases from zero to $T_c$. Near the threshold the attenuation behaves as $(1 - H_c/H)^{5/4}$.

The uniform r.f. external field directed along the $Y$ axis does not excite spin waves, since the $Y$ component of the total magnetic moment in a spin wave is zero.* So, the attenuation of r.f. pump waves is determined by resonant rotation of molecules and can be used in principal for investigating their dynamical properties.

Our argument is restricted by the condition of sufficiently rare collisions creating or destroying vortex pairs. The lifetime $\tau(R)$ with respect to those collisions can be estimated from the simplified kinetic equation for the distribution function $W(R, t)$ of vortex pairs

$$\frac{\partial W}{\partial t} + \frac{\partial}{\partial R}(W\dot{R}) = -\frac{W}{\tau(R)}. \tag{120}$$

This equation has a static solution given by the relations in Section 3.5. Inserting this value of $W$ and $\dot{R} = \bar{D}\gamma/2R$ we find

$$\tau(R) = 2R^2/\lambda\bar{D}\gamma. \tag{121}$$

The characteristic frequency $\Omega$ of a 'trapped' molecule can be evaluated as

$$\Omega(R, B) = \frac{2\bar{D}(1-B)^{1/4}}{B^{1/2}R^2} \qquad B = \frac{H_c}{H}. \tag{122}$$

So collisions are rare, if the following inequality holds:

$$\Omega\tau = \frac{2T}{D^{1/2}T_c}\,\frac{(1-B)^{1/4}}{B} \gg 1.$$

This can be succesfully fulfilled for a sufficiently small value of $D(\sim 0.01 \sim 0.1)$ and not too low a temperature.

## 6.6. SUMMARY

The study of dynamics of 2-d magnets is not yet complete. Even for a linear response the contribution of localized defects has not been calculated. Huber [57] has calculated the longitudinal autocorrelation spin function above the phase transition point. Unfortunately, there is no direct agreement between his theory and experiment. For the description of neutron scattering or of the r.f. field attenuation the data of correlations of different spins is necessary. We do not know anything yet about the skyrmion dynamics; we have analysed above the dynamics of an individual vortex pair. However, the kinetics of the molecular ensemble has not yet been investigated.

* If a magnetic field is directed along the $X$ axis, parametric excitation of spin waves appears. It might probably be a rather difficult experimental problem to separate the different contributions to the r.f. field attenuation.

**Acknowledgement**

We would like to thank M. V. Feigelman for useful remarks.

## References

1. Lenz W., *Zs. Phys.* **21** (1920) 613.
2. Ising E., *Zs. Phys.* **31** (1925) 253.
3. Onsager L., *Phys. Rev.* **65** (1944) 117.
4. Onsager L., *Nuovo Cim. Suppl.* **6** (1949) 261.
5. Yang C. N., *Phys. Rev.* **85** (1952) 808.
6. Kaufman B., *Phys. Rev.* **76** (1949) 1232.
7. Kaufman B., Onsager L., *Phys. Rev.* **76** (1949) 1244.
8. Kac M., Ward J. C., *Phys. Rev.* **88** (1952) 1332.
9. Newell G. F., Montroll E. W., *Rev. Mod. Phys.* **25** (1953) 253.
10. Potts R. B., Ward J. C., *Prog. Theor. Phys.* **13** (1955) 38.
11. Vdovitchenko N. V., *ZhETF* **48** (1965) 526.
12. Schultz T., Mattis D., Lieb E., *Rev. Mod. Phys.* **36** (1964) 856.
13. Hurst C. A., Green H. S., *J. Chem. Phys.* **33** (1960) 1059.
14. Kramers H. A., Wannier G. H., *Phys. Rev.* **60** (1941) 252.
15. Mc Coy B. M., Wu T. T., *The Two-Dimensional Ising Model*, Harvard University Press, Cambridge MA (1973).
16. Domb C., Sykes M. F., *Proc. Roy. Soc.* **A240** (1957) 214.
17. Lee D. H., Joannopoulos J. D., Negele J. W., Landau D. P., *Phys. Rev. Lett.* **52** (1984) 433.
18. Anderson P. W., *Mater. Res. Bull.* **8** (1973) 153.
19. Peierls R., *Helv. Phys. Acta, VII, Suppl. 2,* (1936) 81.
20. Landau L. D. *ZhETF* **7** (1937) 627.
21. Hohenberg P. C., *Phys. Rev.* **158** (1967) 383.
22. Mermin N. D., Wagner H., *Phys. Rev. Lett.* **17** (1966) 1133.
23. Bogolyubov N. N., *Preprint JINR*, D-781, Dubna (1961).
24. Berezinskii V. L., *ZhETF* **59** (1970) 907; *ZhETF* **61** (1971) 1144.
25. Mineev V. P., *Sov. Sci. Rev.* **IIA** (1981) 173.
26. Patashinskii A. Z., Pokrovskii V. L. *Fluctuational Theory of Phase Transitions* p. 66, Pergamon Press (1979).
27. Stanley H. E., Kaplan T. A., *Phys. Rev. Lett.* **17** (1966) 913.
28. Berezinskii V. L., Blank A. Ya., *ZhETF* **64** (1973) 723.
29. Pokrovskii V. L., Uimin G. V., *ZhETF* **65** (1973) 1691.
30. Kosterlitz J. M., Thouless D., *J. Phys.* **C6** (1973) 1186.
31. Josephson B. D., *Phys. Lett.* **21** (1966) 608.
32. Jose J. V., Kadanoff L. P., Kirkpatrick S., Nelson D. R., *Phys. Rev.* **16** (1977) 1217.
33. Anderson P. W., Yuval J., Hamman P. R., *Phys. Rev.* **B1** (1970) 4664.
34. Kosterlitz J. M., *J. Phys.* **C7** (1974) 1046.
35. Villain J., *J. Physique* **36** (1975) 581.
36. Cabrera N., Burton W. K., Frank F. C., *Phil. Trans. Roy. Soc.* **A243** (1951) 299.
37. Wiegmann P. B., *J. Phys.* **C11** (1978) 1583.
38. Korshunov S. E., *Pis'ma v ZhETF* **41** (1985) 525.
39. Polyakov A. M., Wiegmann P. B., *Phys. Lett.* **B131** (1983) 121.
40. Polyakov A. M., *ZhETF* **68** (1975) 1975.
41. Khokhlachev S. B., *ZhETF* **71** (1976) 812.
42. Karimov Yu. S, *ZhETF* **65** (1973) 263.
43. Skyrme T. H. R., *Proc. Roy. Soc.* **A247** (1958) 260.
44. Belavin A. A., Polyakov A. M., *Pis'ma v ZhETF* **28** (1975) 503.
45. Maleev S. V., *ZhETF* **76** (1976) 2375.

46. Pokrovskii V. L., Feigelman M. V., *ZhETF* **76** (1979) 812.
47. Fisher M. E., Selke W., *Phys. Rev. Lett.* **44** (1980) 1502.
48. Bak P., von Boehm J., *Phys. Rev.* **B21** (1980) 5297.
49. Pokrovskii V. L. Uimin G. V. *ZhETF* **82** (1982) 1640.
50. Rossat-Mignot J., Burlet P., Quezel S., Effantin J. H., Delacote D., Bartholin H., Vogt O., Ravot D., *J. Magn. Magn. Mater.* **31–34** (1983) 398.
51. Uimin G., *J. Physique Letters* **43** (1982) 1665.
52. Hikami S., Tsuneto T., *Prog. Theor. Phys.* **63** (1980) 387.
53. de Jongh L. J., van Amstel W. D., Miedema A. R., *Physica* **58** (1972) 277.
54. Hirakawa K., Ikeda H., this volume.
55. Blank A. Y., Pokrovsky V. L., Uimin G. V., *J. Low Temp. Phys.* **14** (1974) 459.
56. Thiele A. A., *Phys. Rev. Lett.* **30** (1973) 230.
57. Huber D. L., *Phys. Rev.* **B26** (1982) 3758.
58. Sommerfeld A., *Mechanics of Deformable Bodies*, p. 156, Academic Press, NY (1964).
59. Pokrovsky V. L., Uimin G. V., *Pis'ma v ZhETF* **41** (1985) 105.
60. Fateev V. A., Frolov I. V., Schwarz A. S., *Nucl. Phys.* **154B** (1979) 1.

# APPLICATION OF HIGH- AND LOW-TEMPERATURE SERIES EXPANSIONS TO TWO-DIMENSIONAL MAGNETIC SYSTEMS

R. NAVARRO

*Departamento de Ciencia y Tecnología de Materiales y Fluídos, Instituto de Ciencia de Materiales de Aragón, C.S.I.C. — Universidad de Zaragoza, 50015 Zaragoza, Spain*

## 1. Introduction

Despite the success of molecular field theory in describing qualitatively the behaviour of ferro- and antiferromagnetic materials [1—4], its predictions fail completely in the critical region. The importance of the fluctuations that are not accounted for in this theory was pointed out by Néel [3] and has inspired the search for other, more realistic models, such as the Ising [5] and Heisenberg [6] models. To apply statistical mechanical methods to the problem, one may start e.g. from the Hamiltonian with uniaxial anisotropy:

$$\mathscr{H} = -2 \sum_{\langle i,j \rangle} J_{ij}(aS_i^z S_j^z + b(S_i^x S_j^x + S_i^y S_j^y)) - g\mu_B \mathbf{H} \cdot \sum_i \mathbf{S}_i \tag{1}$$

where $J_{ij}$ stands for the exchange interaction between ions at sites $i$ and $j$, $a$ and $b$ are constants, and the last term accounts for the presence of a uniform magnetic field.

For $a = 1$, $b = 0$ the Ising model (I) is obtained, while $a = b = 1$ gives the isotropic Heisenberg model (H) and $a = 0$, $b = 1$ the XY model. These three limiting cases have been most commonly used in magnetism. It has been pointed out that the Hamiltonian (1) is particularly applicable to magnetic insulators [7, 8], for which the electronic magnetic moments are well localized.

Unfortunately, even for the simple bilinear form (1) exact solutions have been obtained in a few cases only [9], and further restrictions or simplifications have been used in theoretical work.

It should also be remarked that the interest of the Hamiltonian (1) is not restricted to magnetic phenomena. In fact, it has played a very important role in the development of the extensive field of phase transitions and critical phenomena in general [10, 11].

### 1.1. SERIES EXPANSIONS

The above mentioned difficulties in solving the Hamiltonian (1) have led to a great number of approximate methods, often with a quite different physical basis and with a varying degree of reliability and mathematical complexity. A review of these methods has been given by Domb [10].



*L. J. De Jongh (Ed.), Magnetic Properties of Layered Transition Metal Compounds* 105—190.

In 1936 Kramers [12] laid down the first foundations for one of these approximate methods, namely the derivation of high-temperature series (h.t.s.) expansion of the partition function, e.g. in the parameter $K = J/k_B T$. Opechowsky [13] exploited Kramers' method and applied it to the discussion of the ferromagnetic interaction in the spin 1/2 Heisenberg model, H(1/2). Later, Kramers and Wannier [14] applied the method to obtain the h.t.s. for the $S = 1/2$ Ising model, I(1/2), on the square planar (s.p.) lattice. They also derived a low-temperature series (l.t.s.) expansion as a function of $\exp(-J/k_B T)$ and found that the partition function at low temperatures can be related to the corresponding one at high temperatures by means of a transformation. The existence of such a transformation allows the determination of the critical temperature, $T_c$, at which the ordering occurs.

Following this work, Onsager [15] succeeded in obtaining the exact solution of the partition function at zero field for the I(1/2) s.p. lattice. The existence of an exact solution was encouraging, since it offered the possibility of testing the approximate methods. Early comparisons by Kramers and Wannier [14] of the most reliable known approximate methods with h.t.s. had pointed to substantial differences with the latter. Therefore, the agreement of the series expansion results with the exact Onsager's solution, in the appropriate temperature range, proved the validity of h.t.s. methods and, at the same time, the limitation of the other approximate methods [10].

Another interesting aspect of Onsager's work was the critical behaviour obtained for the specific heat, $C(T)$, in the neighbourhood of $T_c$. The magnetic specific heat was found to diverge logarithmically in a symmetric way for temperatures above the below $T_c$:

$$C(T) \approx C\,|\ln(|1 - T_c/T|)| \tag{2}$$

where $\approx$ stands for asymptotic behaviour for $T \rightarrow T_c$. A singular behaviour may also be expected for other thermodynamic functions, such as the magnetic susceptibility at zero field, $\chi$, for which a simple relation in $(1 - T_c/T)$ is thought to describe the critical behaviour for $T \rightarrow T_c^+$:

$$\chi \approx A(1 - T_c/T)^{-\gamma} = A(1 - K/K_c)^{-\gamma} \tag{3}$$

Indeed, it was later found that Equation (3) holds for the two-dimensional (2-d) Ising models with the critical exponent $\gamma = 1.75$ [10]. The agreement between the exact solution and the series inspired Domb [16] to suggest that if a sufficient number of coefficients of the series could be calculated, they might provide a direct determination of the critical behaviour, and to this end he derived more coefficients of the h.t.s. for I(1/2) s.p. lattice. The success of the first analysis and the absence of exact solutions for three dimensional (3-d) lattices led to the application of series calculations to the simple cubic (s.c.) lattice by Wakefield [17] and to a great variety of other 2-d and 3-d lattices by Domb and Potts [18], Domb [19] and Domb and Sykes [20, 21]. An extensive review of the work on the Ising model with $S = 1/2$ and arbitrary $S$ may be found in the reviews by Domb [10, 22].

For the Heisenberg model, the pioneering work of Kramers-Opechowsky was extended by Zehler [23] to $H(1/2)$, by Brown and Luttinger [24] and Brown [25] to general spin, $H(S)$, and later by Rushbrooke and Wood to $H(1/2)$ and $H(S)$ [26]. A review of the literature on the isotropic Heisenherg model up to 1973 has been reported by Rushbrooke, Baker and Wood [27], where the methods used in the derivation of h.t.s. and the results are described in detail.

Undoubtedly, the earlier molecular field predictions for ferro- and antiferromagnetic systems have been very much improved upon by the results from the analyses of the series for the Ising and Heisenberg models. Moreover, the information obtained on critical temperatures, exponents, amplitudes, etc., has inspired many further theoretical investigations on the nature of phase transitions and critical phenomena [10, 11]. In order to study the way in which the critical behaviour depends on quantities such as the lattice dimensionality d, the spin-dimensionality n, the number ($z$) of interacting magnetic neighbours, and the symmetry of the lattice, the spin value, an externally applied magnetic field, etc., a great effort has been made to increase the number of available coefficients in the series for the various 2-d and 3-d models. Series with next-nearest and even further neighbour interactions have been derived. Series for general spin value were calculated, the number of terms being less, however, than for the lower quantum limit of $S = 1/2$ or for the classical limit of $S = \infty$. Series have been used to obtain magnetic phase diagrams and critical isotherms. In order to study the influence of spin anisotropy, series have also been derived for models related to the general Hamiltonian (1), like the Ising—Heisenberg (IH) and XYZ [28] models or new ones like the $n$-vector model [29].

Until 1971 the use of series expansions was virtually the only available method to obtain reliable information on the critical behaviour for those models for which no (or only scarce) exact results were available. Thereafter, the application of renormalization group techniques to critical phenomena by Wilson [30] and the subsequent developments of Monte Carlo simulations for classical and quantum statistical mechanics [31] have given alternative and independent ways to obtain successive approximations to critical parameters. The renormalization group (R.G.) techniques also allows the derivation of series expansions for the critical parameters, the expansion variables being $\varepsilon = 4 - d$ and/or $1/n$. Indeed, the R.G. results have greatly increased our knowledge and understanding of phase transitions and critical phenomena, and have also been widely applied to the analysis of other quantum fields.

However, even after the R.G. developements, the 'classical' series expansions still play an important role in the theoretical understanding of critical phenomena, and are particularly useful in their application to real magnetic systems. Several arguments may be mentioned to justify this statement.

The series expansions often provide very precise estimates of the critical temperatures, exponents, amplitudes and correction terms. The only assumption required is the asymptotic form of the thermodynamic function in the neighbourhood of $T_c$, in common with the R.G. methods, which in fact start from the same hypothesis. However, it should be noted that although the R.G. methods may derive relations between the critical exponents, they need at least quantitative estimate of one of them derived in an independent way.

The comparison of the results from high- and low-temperature series with those from R.G. formalism is quite useful to asses the degree to which a given set of estimates approximates the true behaviour. The comparison for the 3-d Ising model, for which no exact results exist, may illustrate the validity and mutual consistency of both types of analysis. In Table I, the predictions obtained by the R.G. formalism [32] have been listed for such characteristic critical parameters as the susceptibility exponent, $\gamma$, cf. Equation (3), the magnetization exponent, $\beta$, defined as:

TABLE I
Critical exponents for the 3-d Ising model deduced from high-temperature series and renormalization group methods. The exponent $\beta$ has been averaged from the values reported in [22] and [29].

| | h.t.s. [33] | h.t.s. [34] | h.t.s. [35] | R.G. [32] |
|---|---|---|---|---|
| $\gamma$ | 1.250 ± 0.003 | 1.2385 ± 0.0015 | 1.2406 ± 0.0017 | 1.241 ± 0.004 |
| $\beta$ | 0.312 ± 0.008 | — | — | 0.324 ± 0.006 |
| $\alpha$ | 0.125 ± 0.015 | — | — | 0.110 ± 0.008 |
| $\nu$ | 0.638 + 0.002 − 0.001 | 0.6305 ± 0.0015 | 0.632 ± 0.002 | 0.630 ± 0.0025 |

$$M(T) \approx B(1 - T/T_c)^{\beta} \qquad T \to T_c^- \tag{4}$$

the specific heat exponent, $\alpha$,:

$$C(T) \approx C(1 - T_c/T)^{-\alpha} \qquad T \to T_c^+ \tag{5}$$

and the correlation length exponent, $\nu$,:

$$\xi(T) \approx D(1 - T_c/T)^{-\nu} \qquad T \to T_c^+ \tag{6}$$

where the $+(-)$ sign refers to the region $T > T_c$ ($T < T_c$). The results are compared with those obtained from h.t.s. analyses by Gaunt [33] by Zinn-Justin *et al.* [34] and also more recently by Guttmann *et al.* [35].

Another basic assumption in series expansion analysis is that they represent a convergent series of sufficient length to derive the critical behaviour. In many cases the number of available coefficients has been found to be large enough to allow extrapolation with a useful degree of confidence. Also in the R.G. methods, the series obtained in $\varepsilon$ and $1/n$ are probably asymptotic, but only very few terms are known. Since $\varepsilon = 4 - d$ this leads to unreliable results, in particular for 2-d lattices.

As far as the comparison with experimental data is concerned the following points should be noted. In case of h.t.s. expansions one may construct closed-form expressions for the thermodynamic quantities valid for any temperature range above and/or below $T_c$, based on the available coefficients and the estimated critical parameters. These results may be very easily fitted to the experimental data. On the other hand the R.G. results are valid in the 'critical region' only, and the experimental dificulties in obtaining reliable data in this region are often very large. In fact, for each thermodynamic function the extent of the critical region will be different, so that any comparison of experimental data with the predicted critical behaviour will only be meaningful if this extent is known. At this moment only h.t.s. analyses and Monte Carlo simulations may provide us with such estimates.

In some cases the thermodynamic function in question is nonsingular at $T_c$, as for instance the antiferromagnetic susceptibility or the specific heat for the 2-d Heisenberg and XY models. In these cases expansions still allow to make estimates of the thermodynamic behaviour. Lastly [36] the modern R.G. formalism has been combined with the h.t.s. techniques that have already been in use for decades.

Obviously, the success of series expansion is closely related to the mathematical methods of analysis of the series coefficients, and in the remainder of this section a brief survey of the various methods used is given. In Section 2 the series expansions for the Ising model on 2-d lattices are reviewed and the main conclusions regarding the critical behaviour summarized. Sections 3 and 4 deal with the series analyses for the Heisenberg, IH and XY models on the 2-d lattices. Finally, in Section 5 some examples are given which illustrate the power of the method to describe the observed behaviour in some well-known quasi 2-d magnetic systems.

## 1.2. METHODS APPLIED IN SERIES ANALYSIS

Rather than referring to a general thermodynamic function, the suceptibility will be specifically considered for convenience, since for the Ising and Heisenberg models the series for $\chi$, at zero-field, provide [22, 27] the smoothest, longest and most regular pattern of coefficients.

### 1.2.1. *Ratio Methods*

The ratio method is based on the assumption that a given thermodynamic function has a singular behaviour at $T_c$. If an appropriate low- or high-temperature variable ($K = k_B T/J$ or $K = J/k_B T$) is defined and if it is assumed that the singularity at $T_c$ is the dominant one and closest to the origin $K = 0$ of the expansion, then it may be expected that for sufficiently high order the coefficients of the h.t.s. expansion should closely approximate those derived from the asymptotic critical behaviour. The binomial expansion of the form (3) is easily obtained as:

$$(1 - K/K_c)^{-\gamma} = 1 + \sum_{n=1}^{\infty} a_n K^n, \qquad a_n = \gamma(\gamma - 1) \ldots (\gamma + n - 1)/K_c^n \, n! \tag{7}$$

It follows that the ratios of successive coefficients $r_n = a_n/a_{n-1} = (\gamma + n - 1)/nK_c$ plotted against $1/n$ will tend towards $1/K_c$, when $n \to \infty$. At the same time the limiting slope of the plot gives $(\gamma - 1)/K_c$. Thus, if the above assumptions of the ratio methods are valid, it should be possible to deduce $K_c$ and $\gamma$ from the finite number of known coefficients

$$\chi = \sum_{n=0}^{\mathcal{N}} c_n K^n + O(K^{\mathcal{N}}) \tag{8}$$

where $O(K^{\mathcal{N}})$ stands for the unknown coefficients of degree higher than $\mathcal{N}$. In some cases, e.g. for the loosely-packed lattices, the series coefficients show marked differences between odd and even terms [21]. Then, a combination of ratios called 'linear extrapolant' may avoid these dificulties:

$$l_n = ((n+1)r_n - (n-1)r_{n-1})/2 \to 1/K_c \quad \text{when} \quad n \to \infty. \tag{9}$$

On the basis of the form (7) for the expansion, other choices of combinations may be made in order to obtain unbiased estimations of $K_c$, when $\gamma$ is known [37]. If the ratios $nr_n/(\gamma + n - 1)$ are plotted against $1/n$ as above, the value $1/K_c$ will be reached with zero slope in the limit $1/n \to 0$. If the same manipulation is performed with $c_n$ one obtains:

$$nr_n/(\gamma + n - 1) \to 1/K_c \quad \text{when} \quad n \to \infty. \tag{10}$$

Moreover, if accurate values for $K_c$ are known, unbiased estimations of $\gamma$ may be obtained from $n(K_c r_n - 1) = \gamma - 1$, the $(\gamma - 1)$ value being reached with zero slope in the limit $n \to \infty$.

When accurate values for both $\gamma$ and $K_c$ are known, quantitative predictions for $\chi(K)$ in the critical region may be obtained from the estimates of the critical amplitude, $A$, since the ratio $c_n/a_n$ should tend to $A$ in the limit $n \to \infty$.

Many different manipulations of the ratios have been applied, which are reviewed elsewhere [37]. Another important point already mentioned is the construction of mimic functions from the series expansions. If all critical parameters $K_c$, $\gamma$ and $A$ of (3) are known, one may assume that the susceptibility should be given, to a good approximation, by an expression of the form:

$$A(1 - K/K_c)^{-\gamma} + \Psi(K) \tag{11}$$

where the first term represents the singular part and $\Psi(K)$ is a smooth function of $K$ that is continuous at $K_c$, and is presumably well convergent. The combination of Equation (7) and (11) allows the calculation of the polynomial $\Psi(K)$ as:

$$\Psi(K) = \sum_{n=0}^{\mathcal{N}} (c_n - Aa_n)K^n. \tag{12}$$

If the singular behaviour is properly accounted for by expression (3), the use of the closed form (11) should give a much more accurate estimate of $\chi(T)$ than those obtained directly from the finite series.

The ratio method may be safely applied when the coefficients have a smooth behaviour, but the uncertainties involved prevent the estimation of absolute lower and upper bounds. Indeed, if the sequence of series coefficients is not very smooth, it is always possible to improve the situation by suitable grouping of the coefficients, conformal transformations, or changes of variable [38]. However, a systematic approach to the choice of the most suitable transformation for a given series does not exist.

### 1.2.2. *Padé Approximant Methods*

A method that has yielded a significant improvement of the series analysis was first applied by Baker [39] to the I(1/2) model and uses the Padé approximant (P.A.). The method starts from a different mathematical basis, and not only reproduces the ratio results in the case of smooth coefficients, but can also be applied succesfully to series with a more irregular character.

A Pade approximant is defined mathematicaly [40, 41] as the ratio of two polynomials, namely a numerator $P(K)$ of degree $N$ and a denominator $Q(K)$ of degree $D$. It can be denoted by:

$$(N/D)(K) \equiv P(K)/Q(K). \tag{13}$$

Equating the P.A. to the susceptibility expansion (8), the coefficients of the polynomials $P$ and $Q$ are obtained from the original series, using the system of $N + D + 1$ linear equations

$$\chi(K)Q(K) - P(K) = O(K^{N+D+1}) \quad \text{and} \quad Q(K=0) = 1. \tag{14}$$

From the definition it immediately follows that the maximum degree of $P$ and $Q$ is limited by the number of series coefficients available, $N + D + 1 \leqslant \mathcal{N}$.

The main advantage of the P.A. method compared to the ratio method stems from the mathematical form of the P.A., which incorporates the possibility of more than one simple singularity (pole), namely those corresponding to the roots of the polynomial $Q$.

In the P.A. analysis one quite often uses the so-called Padé conjecture [40, 41], that states that P.A.'s enlarge the convergency radius of the series and thus may give information on the function in a larger region.

The way in which the P.A. methods proceed is as follows. For a given series, a table is built with all possible P.A.'s. Then the higher-order diagonal and near-diagonal elements ($N$; $N$), ($N - 1$; $N$), ($N$; $N - 1$), . . . are considered to give the most reliable estimates. The sequence of successive approximations, quite explicit in the ratio method, is much less obvious here, because the entries in the P.A. table do not allow extrapolation.

Assuming the critical behaviour of the susceptibility to be given by Equation (3), the best transformation using P.A.'s consists in reducing the essential singularity to a simple pole. An appropriate way [42] is to take the logarithmic derivative of the series $\mathrm{d}/\mathrm{d}K(\ln((1 - K/K_c)^{-\gamma})) = -\gamma/(K_c - K)$, abreviated by Dlog. As a consequence, a simple way to derive the $K_c$ value by P.A. techniques is via:

$$\text{P.A.}\left(\text{Dlog}\left(\sum_{n=0}^{I} c_n K^n\right)\right) \approx -\gamma/(K_c - K). \tag{15}$$

Thus, the roots of the different denominators of the P.A.s close to the origin should correspond with the $K_c$ values, and the residues of the P.A.s obtained by the product $(K_c - K)(N/D)$ evaluated at $K_c$ yields the value of the exponent. In some P.A.'s spurious poles may appear, which would affect its value if they were closer to the origin than $K_c$. These weak singularities, called 'defects', are easily detected, because the associated residues are some orders of magnitude different from those which are physically meaningful. Although the use of Equation (15) can determine $\gamma$ and $K_c$, the method gives biased estimations, because any spread in the values of $K_c$ is amplified in the spread of the corresponding residues [42].

If $K_c$ is known, unbiased estimates of $\gamma$ may be obtained, by means of the P.A. table

$$\text{P.A.}\left((K - K_c)\,\text{Dlog}\left(\sum_{n=0}^{\mathcal{N}} c_n K^n\right)\right) \approx \gamma. \tag{16}$$

If the above method is still sensitive to the value chosen for $K_c$, the exponent $\gamma$ may also be determined in a less sensitive way, namely by evaluating the P.A. tables at $K_c$:

$$\text{P.A.}\left(\text{Dlog}\left(\partial/\partial K\left(\sum_{n=0}^{\mathcal{N}} c_n \text{K}^n\right)\right)\Big/\,\text{Dlog}\left(\sum_{n=0}^{\mathcal{N}} c_n K^n\right)\right)\Bigg|_{K=K_c} \approx (\gamma + 1)/K_c. \tag{17}$$

When $\gamma$ is known, the transformation $((1 - K/K_c)^{-\gamma})^{1/\gamma} = K_c/(K_c - K)$ reduces (3) to a simple pole. If the P.A. table

$$\text{P.A.}\left(\left(\sum_{n=0}^{\mathcal{N}} c_n K^n\right)^{1/\gamma}\right) \approx (A)^{1/\gamma} K_c/(K_c - K) \tag{18}$$

is formed and similar procedures as above are used, the values for $K_c$ and the amplitudes $A$ may then be calculated. Another related method to obtain the amplitude is by using the $K_c$ and $\gamma$ values in the transformation

$$\text{P.A.}\left((1 - K/K_c)^{\gamma}\left(\sum_{n=0}^{\mathcal{N}} c_n K^n\right)\right) \approx A. \tag{19}$$

Generally speaking, it has been observed that the use of only one method for the determination of $\gamma$ and $K_c$ is not conclusive, because there are many factors that may interfere with the convergency of the P.A.s. A combination of methods is therefore used in most cases.

The P.A.'s also facilitate the construction of mimic functions, as in (12), if the function $\Psi(K)$ is evaluated using the P.A. table of $(\Sigma_{n=0}^{\mathcal{N}} c_n K^n - A(1 - K/K_c)^{-\gamma})$ for any $K$. Following the Padé conjecture [40, 41] the direct application of P.A.s would increase the convergency radius and, in fact, in most cases the use of P.A.s yields better results than just the series alone.

In some problems there is additional numerical information available regarding the value of a thermodynamic function at some $K$ and, if there is no physical singularity between the origin of the expansion and that particular $K$ value, it is possible to combine the additional information with the h.t.s. by means of the so-called 'two-point Padé-approximant'. This method, also introduced by Baker *et al.* [43], allows the inclusion of new conditions, for example: $(N; N)$ $(K = \infty) =$ const. Often the usual oscillations of the P.A. then diminish, so that it may be used as a mimic function over the whole range of $K$. However, the convergency of these approximants is not known and few applications have appeared in the literature to date.

The basic idea of the reduction of an essential singularity (3) to a simple pole may be applied when another functional form is assumed. For example, in case of the asymptotic behaviour of $C(T)$ in the 2-d I(1/2) model the logarithmic divergence (Equation (2)) is reduced to a simple pole just by taking the derivative. The critical amplitude and temperature may then be derived from the P.A. table of

$$\text{P.A.}\left(\partial/\partial K\left(\sum_{n=0}^{N} c_n K^n\right)\right) \approx C/(K_c - K). \tag{20}$$

In other 2-d models an exponential singularity of $\chi$ of the form

$$\chi(K) \approx \chi_0 \exp(A(1 - K/K_c)^{-\nu}) \tag{21}$$

has been suggested [44, 45]. If the logarithm of (21) is taken the expression is reduced to an essential singularity, like (3), and the above methods may be applied, e.g. by forming a P.A. table

$$\text{P.A.}\left(\text{Dlog}\left(\ln\left(\sum_{n=0}^{N} c_c K^n\right)\right)\right) \approx -\nu/(K_c - K). \tag{22}$$

### 1.2.3. *Other Methods of Series Analysis*

The increasing number of available series coefficients, the minor but persistent differences found for the critical exponents, together with the verification of such hypotheses as universality, scaling, etc., have lead to a further evolution of the ratio and P.A. methods. Nowadays the asymptotic behaviour of Equation (3) is considered to be too simplified. One rather believes it to be the dominant contribution to a singularity of the form [22, 42, 46]

$$\chi(K) \approx \chi_0|1 - K/K_c|^{-\gamma} + \chi_1|1 - K/K_c|^{1-\gamma} + \cdots \tag{23}$$

i.e. at $K_c$ there is a confluent singular behaviour. Generalizations of the above methods have been proposed [40] to analyse forms such as (23), or functions with more than one essential singularity. One of the latest successful and comprehensive studies has been reported by Hunter and Baker [47], who discuss the approximations of a function by means of the solution of certain differential equations derived from the series coefficients. These 'differential approximants' of a function

$f(K)$ are defined as:

$$\sum_{\nu=0}^{M} Q_\nu(K)\, \partial^\nu/\partial K^\nu f(K) + R(K) = 0 \tag{24}$$

where $Q_\nu(K)$ ($\nu = 0$ to $M$) and $R(K)$ are polynomials. For $M = 0$, and identifying $R(K) = P(K)$ and $Q_0(K) = Q(K)$, Equations (14) are fulfilled and the direct P.A.'s of $f(K)$ are obtained. Moreover, setting $M = 1$ and $R(K) = 0$ and calling $Q_0(K) = P(K)$ and $Q_1(K) = Q(K)$, Equation (24) is reduced to $Q(K)\mathrm{Dlog}(f(K)) + P(K) = 0$ and the P.A. tables of Equation (15) are obtained.

The ratio method, and other ratio—based extrapolations [47] are also special cases of differential approximants. Thus, if $M = 0$, $Q_0(K) = (1 + a_0K)$, and $R_L(K)$ an arbitrary polynomial of degree $L$, Equation (24) reduces to

$$(1 + a_0K)\left(\sum_{n=0}^{\mathcal{N}} c_nK^n\right) - R_L(K) = O(K^{L+2}), \qquad \nu = 0.$$

The solution of this system of $L + 1$ linear equations allows the determination of the coefficients of $R_L$ and $a_0$. The value of $a_0$ for different $L$ are the ratios of the coefficients $c_L/c_{L-1}$ Moreover, this procedure may be generalized for $M \neq 0$, considering that

$$(1 + a_0K)\, \partial/\partial K(K\partial/\partial K)^\nu \cdot f(K) + \sum_{j=1}^{\nu} a_j(K\partial/\partial K)^j f(K) + R_L(K) =$$
$$O(K^{L+\nu+2}) \quad \text{for} \quad \nu > 0$$

which reduces to (24) if appropriate $Q_\nu(K)$ polynomials are taken. The solution of the above equation for $\nu = 1$ amounts to the linear extrapolant (9), and for $\nu \geqslant 2$ gives other ratio extrapolations. Other methods of approximation are also included in the general form of the differential approximant, which, in fact, is a synthesis of the most common methods used in the analysis of series expansion in one variable.

Still further generalizations of the above methods have been introduced for series expansions in two variables. They have been applied e.g. to the analysis of bicritical behaviour. The interested reader is refered to the work by Fisher and Chen [48] for more information.

## 2. Series Expansion and Predictions for the 2-d Ising Model

The comparison between theory and experiment has been mainly restricted to the series for the static thermodynamic functions like the susceptibility, specific heat and magnetization. Therefore, these functions will be the main theme of the present review.

For the Ising model there are three types of series expansions [22]:

(i) High-temperature series expansions of the 'moment' or 'cumulant' type.
(ii) Density expansions of the 'cluster' type.
(iii) Low temperature expansions of the 'excitation' type.

(i) H.t.s. start from the point of view that, at very high temperatures, the spin configurations are completely random and the coefficients of the expansions reflect the deviations from randomness. At zero field, it is easy to prove that the partition function is the same for ferro- and antiferromagnets [10], whereas at finite field the properties are different and so are the series. In Ising antiferromagnets there exists a finite $T_c$ in some field range $H \neq 0$. This is not the case for ferromagnets, for which the phase diagram in the $T$—$H$ plane is just the segment of the $H = 0$ axis up to $T_c$ [11, 22].

In the analysis of antiferromagnets it is useful to introduce the 'staggered' susceptibility, $\chi^{st}$, which corresponds to the response of the system to a staggered field, $H^{st}$, that takes opposite directions on each of the two sublattices of the (loosely-packed) lattices. The behaviour of $\chi^{st}$ in the neighbourhood of the Néel temperature, $T_N$, should be the same as that of the ferromagnetic susceptibility, $\chi^F$, at $T_c$. Moreover, it is easily proved [22] that the series for both $\chi^{st}$ and $\chi^F$ are identical, and that $|J/k_B T_N| = |J/k_B T_c|$.

When a uniform external magnetic field is applied to antiferromagnets the response is given by the direct susceptibility, $\chi^{AF}$, and in the zero-field limit the h.t.s. of $\chi^{AF}$ can be obtained from the h.t.s. of $\chi^F$ by substituting $+J$ for $-J (J > 0)$. This amounts to an alternation in sign, i.e. a factor $(-1)^n$, of the h.t.s. coefficients. At nonzero field $\chi^F$, $\chi^{AF}$ and $\chi^{st}$ are different, and only $\chi^{st}$ remains singular at a certain temperature, $T_c(H)$, so that it may be used to derive the antiferromagnetic phase diagram. Furthermore, as regards the direction of the applied field, one should distinguish between the parallel ($\chi_{\parallel}$) and the perpendicular ($\chi_{\perp}$) component of the susceptibility, according to whether the field is applied parallel or perpendicular to the quantization axis.

(ii) In the density expansions the presence of a strong magnetic field, which aligns all the spins, is assumed. If the field decreases at any given temperature groups of spins overturn. A double series in the parameters $u = \exp(-2J/k_B T)$ and $y = \exp(-mH/k_B T)$, where $m$ is the magnetic moment, is obtained [22], which can be analysed using similar methods as for the l.t.s. It should be noted that in the literature for the Ising model the $J$ used is most often defined as half the of $J$ used here.

(iii) The l.t.s. start from $T = 0$, where all the spins should be aligned, even at zero field. The excited states arise from groups of overturned spins and the natural variable is $u$. In this case the l.t.s. expansion of antiferromagnets can not be obtained from the ferromagnetic one as happens in the h.t.s., because if $J$ is replaced by $-J$, the transformed variable is $u' = \exp(2J/k_B T)$ and the range of variation of $u$, (0, 1) is changed to ($\infty$, 1). The change from ferro- to antiferromagnetism is only possible at zero field, by substituting $u$ by $(u')^{-1}$ [22].

### 2.1. SPIN 1/2 MODEL WITH NEAREST NEIGHBOURS ONLY (SIMPLE 2-d LATTICES)

There exist exact expressions for the partition function and the magnetization at zero field for the common 2-d lattices, i.e. s.p., honeycomb (h.c.), and planar triangular (p.t.), lattices [11, 22].

Onsager's solution of the partition function for the s.p. lattice with two different exchange interactions $J_x$ and $J_y$ in the $x$, and the $y$ directions, has the form:

$$-F/Nk_B T = 1/2 \int_{-\pi}^{+\pi} \mathrm{d}\theta_1/2\pi \int_{-\pi}^{+\pi} \mathrm{d}\theta_2/2\pi \ln(\cosh K_x \cosh K_y - \\ - \sinh K_x \cos\theta_1 - \sinh K_y \cos\theta_2) \qquad (25)$$

where $K_{x,y} = J_{x,y}/k_B T$. The integrand becomes infinite at a transition temperature given by the relation:

$$\sinh(J_x/k_B T_c)\sinh(J_y/k_B T_c) = 1. \qquad (26)$$

For any nonzero value of $J_x$ and $J_y$ the specific heat has a symmetric logarithmic divergence of the form (2), as may be seen in Figure 1 for different $J_x/J_y$ ratios. There is also an analytic expression for the magnetization obtained by Onsager [49] and Yang [50] of the form:

$$M(T)/mN = (1 - (\sinh(J_x/k_B T)\sinh(J_y/k_B T))^{-2})^{1/8} \qquad (27)$$

where $m$ is the magnetic moment per site ($m = g\mu_\beta S$). Close enough to $T_c$ Equation (27) reduces to the asymptotic expression (4), with a critical exponent $\beta = 1/8$.

Fisher [51] has reported exact results for $\chi_\perp$ in the s.p. ($J_x = J_y = J$) and h.c. lattices. $\chi_\perp$ is continuous at any temperature, but shows a singularity in $\partial\chi_\perp/\partial T$ at $T_c$, with asymptotic behaviour

$$\chi_\perp(T) \approx 2Nm^2/z|J|(\bar{\chi}^0_\perp + b(1 - T/T_c)\ln|1 - T/T_c|). \qquad (28)$$

Moreover, as $T \to 0$ the constant value $\chi_\perp(0) = 2Nm^2/z|J|$ is reached. The susceptibility shows a maximum of height $\chi_\perp^{max}$ at a temperature $T_\perp^{max} > T_c$. Fisher also proved that in the s.p. and h.c. lattices the expression (28) is valid for ferro- and antiferromagnetic interactions.

For other 2-d lattices similar expressions may be found, which have been reviewed elsewhere and have lastly been extended to the Kagomé and decorated Kagomé lattices [53]. In Table II some significant parameters for equivalent nearest neighbour interactions have been listed.

In addition to the exact results, both high- and low-temperature series are available for $\chi$, $C(T)$ and $M$, with and without field, which all together lead to a very comprehensive understanding of the model.

#### 2.1.1. *High-Temperature Series*

H.t.s. expansions for the thermodynamic functions of the $S = 1/2$ Ising model are

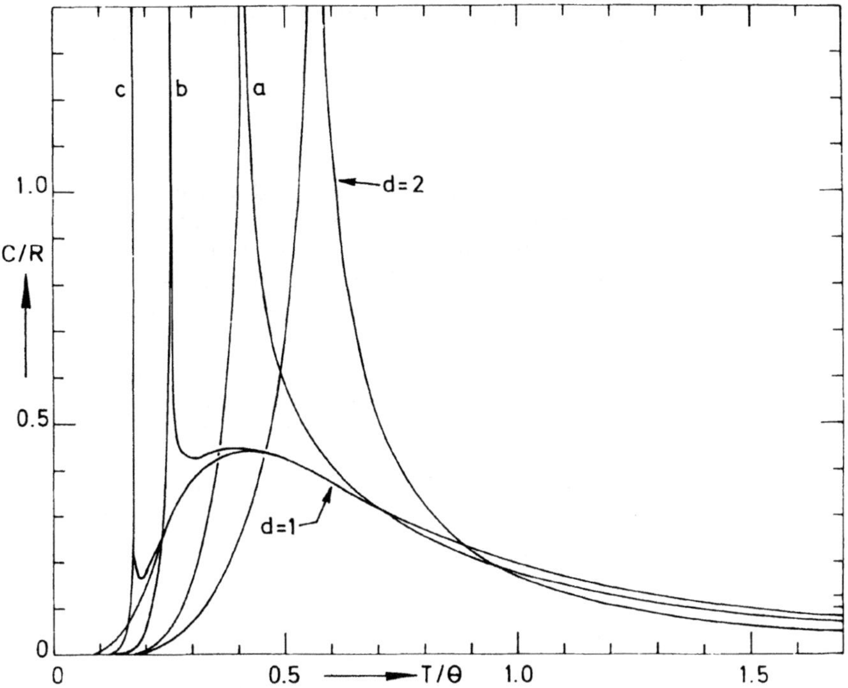


Fig. 1. Onsager's exact specific heat predictions for the 2-d $S = 1/2$ Ising model on the s.p. lattice for different exchange interactions $J_x$ and $J_y$ in the $x$ and $y$ directions. The curves labeled a, b and c, correspond to $|J_x/J_y| = 0.1$, 0.01 and 0.001, respectively ($\Theta = (J_x + J_y)/k_B$).

TABLE II

A summary of critical temperatures, $w_c = \tanh(J/2k_BT_c)$, energy and entropy parameters below $T_c$, critical amplitudes for the specific heat (Eq. (2)), magnetization (Eq. (4)) and perpendicular susceptibility (Eq. (28)) for 2-d Ising models. The relative values of $\chi_\perp^{max} = \chi_\perp^m$ and $T(\chi_\perp^{max}) = T_\perp^m$ are also included. Data taken from Fisher [11], and Barry and Khatun [53].

| lattice ($z$) | h.c. (3) | s.p. (4) | p.t. (6) | Kagome (4) |
|---|---|---|---|---|
| $1/w_c$ | $\sqrt{3}$ | $1+\sqrt{2}$ | $2+\sqrt{3}$ | 2.29663 |
| $k_BT_c/zJ$ | 1.0124343 | 1.1345977 | 1.2136523 | 1.0716595 |
| $-U_c/k_BT_c$ | 0.760346 | 0.623225 | 0.549306 | 0.694266 |
| $S_c/R$ | 0.26471 | 0.306470 | 0.33028 | 0.28052 |
| $C$ | 0.478106 | 0.494538 | 0.499070 | 0.480062 |
| $B$ | 1.264904 | 1.222410 | 1.203270 | — |
| $\bar{\chi}_\perp^0$ | 1.154701 | 1.136951 | 1.180253 | 1.1426 |
| $b$ | 0.4840 | 0.4701715 | — | — |
| $\chi_\perp^m/\chi_\perp^{(0)}$ | 1.2126 | 1.18314 | 1.15976 | 1.1945 |
| $T_\perp^m/T_c$ | 1.104 | 1.08872 | 1.077 | 1.0994 |

usually developments in the dimensionless variable $w = \tanh(J/2k_B T)$. For $\chi^{F,AF}$ the usual h.t.s. takes the form:

$$\bar{\chi} = k_B T\chi_{\parallel}/Nm^2 = 1 + \sum_{n=1}^{\infty} a_n w^n. \tag{29}$$

Susceptibility series of sufficient length to be analyzed were calculated firstly by Brooks and Domb [54], who reported nine terms of the s.p. lattice. Up to now the successive contributions by Domb and Sykes [21], Sykes and collaborators [55] and finally Nickel [56] have provided 16, 34 and 32 coefficients for the p.t., s.p., and h.c. lattices, respectively, and in Table A-I (Appendix A) these terms have been collected.

The initial analysis of $\chi^F$ by Domb and Sykes [21] gave $K_c$ values close to the exact ones and also pointed to strong similarities between the 2-d lattices. They appeared to have a common critical exponent $\gamma = 7/4$, fully comfirmed later by Domb and Sykes [57]. Finally, a confluent singularity behaviour (Equation (23)) has been worked out by Gartenhaus and Scott McCullough [58] considering for $T > T_c (T \to T_c^+)$ a form such as:

$$C_0^+|t|^{-7/4} + C_1^+|t|^{-3/4} + C_2^+|t|^{1/4} + C_3^+|t|^{5/4} + \\ + D_0^+ + E_0^+\, t \ln|t| + D_1^+\, t + O(t^2 \ln|t|, t^2) \tag{30}$$

where $t = (1 - T_c/T)$ and

$$G_0(|1 + T_c/T|) \ln(|1 + T_c/T|) + \text{higher order terms} \tag{31}$$

when $T \to -T_c^+$. According the analytic structure [58] the terms with $C_i^{\pm}$ $(i = 0, 1, 2)$ represent the ferromagnetic singularity, those with $D_i^{\pm}$, and additive function expanded around $T_c$, the one with $E_0^+$ a logarithmic singularity and $G_0$ for the loosely packed lattices the amplitude of the antiferromagnetic slope singularity at $T = -T_c$.

Barouch *et al.* [59] proved the validity of the expressions (30) and (31), and were able to obtain exact results for $C_0^{\pm}$ and $C_1^{\pm}$ for the s.p. lattice. Guttmann [60] extended the calculation to other planar lattices, using only the 'generalized law of corresponding states', and obtained good agreement with results by Sykes *et al* [55]. Combining all information available [58, 60] together with the last exact results of $D_0^+$ and $E_0^+$ of Kong *et al.* [61] we derived the amplitudes listed in Table III.

Sykes and Fisher [55, 62, 63] analysed $\chi_{\parallel}^{AF}$ and found that it is continuous through the transition, but with an infinite temperature gradient at $T_N$. Close to $T_N$ it may be written appronimately as:

$$\chi^{AF}(T) \approx 1 - q(T)U(T) \tag{32}$$

where $U(T)$ is the reduced energy per spin and $q(T)$ is a slowly varying function. Using the exact result for $U(T)$, the behaviour of $\chi^{AF}$ for $T \to T_N$ takes the form

$$\bar{\chi}^{AF} \approx \bar{\chi}_0^{AF} - a(1 - T_N/T) \ln(|1 - T_N/T|) \tag{33}$$

TABLE III

Critical amplitudes above and below $T_c$ for 2-d, $S = 1/2$ Ising models. The parameters for the antiferromagnetic susceptibility maximum, $T(\chi^{max}) = T^m$, $\chi_{\parallel}'^m = z|J|\chi^{max}/2Nm^2$ and the values at $T_N$, $\chi'_{\parallel c} = \chi^{AF}(T_N)$, are also listed. (a) Domb and Sykes [57]; (*) exact results.

| | h.c. | s.p. | p.t. |
|---|---|---|---|
| $C_0^+$ | 1.0464170761 | 0.9625817322(*) | 0.9242069582 |
| $C_1^+$ | 0.0994548793 | 0.0749881538(*) | 0.0634590701 |
| $C_2^+$ | 0.0150 ± 0.004 | 0.1470189801(*) | 0.0690 ± 0.002 |
| $C_3^+$ | — | 0.1399419352(*) | — |
| $D_0^+$ | −0.24 | −0.10413324511(*) | −0.0496 ± 0.002 |
| $D_1^+$ | — | −0.14869 ± 0.00001 | — |
| $E_0^+$ | — | −0.0403255003 | — |
| $G_0$ | 0.182 ± 0.001(a) | 0.196 ± 0.002 | Non existent |
| $\chi'_{\parallel c}$ | 0.2398(a) | 0.2768(a) | — |
| $T^m_{\parallel}/T_c$ | 1.6877(a) | 1.5371(a) | — |
| $\chi'^m$ | 0.416828(a) | 0.42957(a) | — |
| $C_0^-$ | 0.0277610956 | 0.0255369719(*) | 0.0245189020 |
| $C_1^-$ | −0.0026385047 | −0.0019894107(*) | −0.0016835479 |
| $C_2^-$ | $0.02^{+0.02}_{-0.01}$ | $0.0095^{+0.0024}_{-0.0013}$ | $0.007^{+0.002}_{-0.001}$ |
| $D_0^-$ | $-0.24^{+0.02}_{-0.03}$ | −0.10413324511(*) | −0.0487 ± 0.001 |

where $\bar{\chi}_0^{AF}$ and $a$ are characteristic constants for each lattice. Moreover, the presence of a broad maximum in the susceptibility above $T_N$ was revealed [55] and estimations of $\chi^{AF}$ over the whole temperature range were obtained and are shown in Figure 2. In addition, in Table III the values of $\chi_{\parallel}^{AF}$ at $T_N$ and at the maximum are collected.

Although in two dimensions there are exact results for the zero-field partition function, Equation (25), it is still useful to compare with the h.t.s. expansions, in order to check on the extrapolation procedures or to verify the relations between series. In Table A-II (Appendix A) the coefficients of the internal energy given by Domb [22] are listed:

$$-2U/NJ = 1/2zw + \sum_{n=2}^{\infty} u_n w^n. \tag{34}$$

From this series the partition function, $\ln Z(\beta, H)$ ($\beta = 1/k_B T$), or the specific heat may be easily deduced from the relations $U = -\partial/\partial\beta(\ln Z(\beta, H))$ and $C(T) = -k_B\beta\partial U/\partial\beta$.

### 2.1.2. *Low-Temperature Series*

Low-temperature series at zero field have been calculated for $\chi^F$ and $M$ for the 2-d lattices, and for $\chi^{AF}$ for the loosely-packed ones. In addition, the configura-

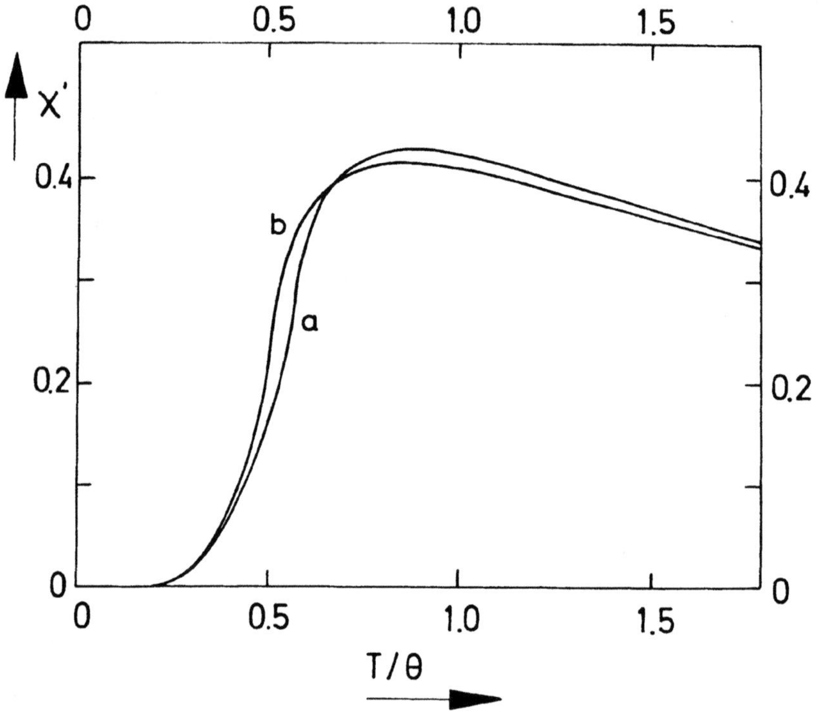


Fig. 2. Reduced antiferromagnetic parallel susceptibilities $\chi'_{\|} = zJ\chi_{\|}/2Nm^2$ versus $T/\Theta$, $\Theta = 2J/k_B$ of the $S = 1/2$ Ising model for the square planar (curve a) and honeycomb (curve b) lattices, taken from Sykes and Fisher [55]. Reprinted by permission of Elsevier Science Publishers.

tional partition function may be expandend as a double series in the variables $u$ and $y$.

$$(\ln Z)/N = -(\ln y)/2 - (z \ln u)/8 + \ln \Lambda(y, u), \qquad \ln \Lambda(y, u) = \sum_{n=0}^{\infty} y^n g_n(u) \quad (35)$$

where $g_n(u)$ are polynomials in $u$. Indeed, all these series are not independent and the series for $\chi^F$, $C(T)$ and $M$ at zero and at finite field may in principle be derived from (35), using well-known thermodynamic relations. The spontaneous magnetization is obtained as

$$M/Nm = 1 - 2y\partial/\partial y\left(\sum_{n=0}^{\infty} y^n g_n(u)\right) \quad (36)$$

for the limit $y = 1$, and the zero-field magnetic susceptibility is given by

$$\bar{\chi}_{\parallel}^{F} = (2y\partial/\partial y)^2 \left( \sum_{n=0}^{\infty} y^n g_n(u) \right) \tag{37}$$

in the same limit. However, in practice it is much simpler to obtain zero-field expansions for $\chi^F$, $C(T)$ and $M$ than for the $g_n(u)$ polynomials, since the lengths of the series are larger than the order of the known polynomials.

From ln $\Lambda(y, u)$ one may also derive the corresponding high-temperature expansion of the form

$$\ln \Lambda(y, u) = \sum \phi_r(y)(1 - u)^r/(1 + y)^{2r-1} \tag{38}$$

where $\phi_r(y)$ are symmetric polynomials in $y$ obtained by manipulation of Equation (35) [10]. Expanding the parameter $(1 - u)$ as a function of $w$, an h.t.s. analogous to the ones given above is obtained.

The first l.t.s. expansion for the magnetization was derived for the s.p. lattice by Ashkin and Lamb [64]. Thereafter successive contributions by Domb [16], Brooks and Domb [54], Domb and Sykes [20], Sykes and collaborators [55, 65—70], Baxter and Enting [71], and Sykes and Hunter [72] have yielded a number of l.t.s. expansions of varying lengths.

For $\chi^F$ there are 18, 21 and 18 coefficients for the h.c., s.p. and p.t. lattices, These may be normalized into the form:

$$\bar{\chi}^{F,AF} = 4u^{z/2} \sum_{n=1}^{\infty} b_n^{F,AF} u^n \tag{39}$$

and have been listed in Table A-III (Appendix A). The analysis of such $\chi^F$ series by Essam and Fisher [73] was not as satisfactory as the corresponding one for the h.t.s., but nevertheless a critical index $\gamma' = 1.75 \pm 0.01$ was found. The agreement with the h.t.s. value of $\gamma = 7/4$ was sufficiently close to suggest that the critical exponents above and below $T_c$ should be equal.

The derivation of an l.t.s. expansion for $\chi^{AF}$ for the h.c. lattice has been improved by the relation given by Fisher [74]

$$\bar{\chi}^F(\text{p.t.}; u(\text{p.t.})) = 1/2(\bar{\chi}^F(\text{h.c.}; u(\text{h.c.})) + \bar{\chi}^{AF}(\text{h.c.}; 1/u(\text{h.c.})) \tag{40}$$

$$u(\text{p.t.}) = u(\text{h.c.})/(1 - u(\text{h.c.}) + u^2(\text{h.c.})) \tag{41}$$

where $u$(h.c.) is the usual l.t.s. variable of the h.c. lattice related to $u$ by $u(\text{h.c.}) = z = \exp(-J/k_B T) = u^{1/2}$. Therefore, there are 18 coefficients available for the h.c. lattice and 9 for the s.p. one, which are also listed in Table A-III (Appendix A).

Ferromagnetic $g_n(u)$ polynomials have also been obtained up to $g_{23}(u)$, $g_{12}(u)$ and $g_{25}(u)$ for the s.p., p.t. and h.c. lattices, respectively, from which the spontaneous magnetization may be derived. Moreover, using specific methods the l.t.s.

expansions for $M$

$$M/Nm = 1 - 2u^{z/2} - u^{z/1} \sum_{n=1}^{\infty} b_n u^n \tag{42}$$

with 14, 15 and 16 terms for the s.p., p.t. and h.c. lattices, respectively, have been obtained. In Table A-IV (Appendix A) the coefficients $b_n$ have been listed. Furthermore, for comparative purposes, the spontaneous magnetization for $S = 1/2$, $S = 1$ and $S = 3/2$ in the p.t. lattice, together with estimations for the b.c.c. lattice, are displayed in Figure 3.

In conclusion, it may be said that the collection of exact, long and well-convergent series for the 2-d, $S = 1/2$ Ising model with nearest-neighbour interactions, provides quite accurate estimates of the critical behaviour, as well as of the termodynamic behaviour at any temperature. In Table IV the critical exponents characteristic for the 2-d lattices, the scaling relations and the proper references have been summarized.

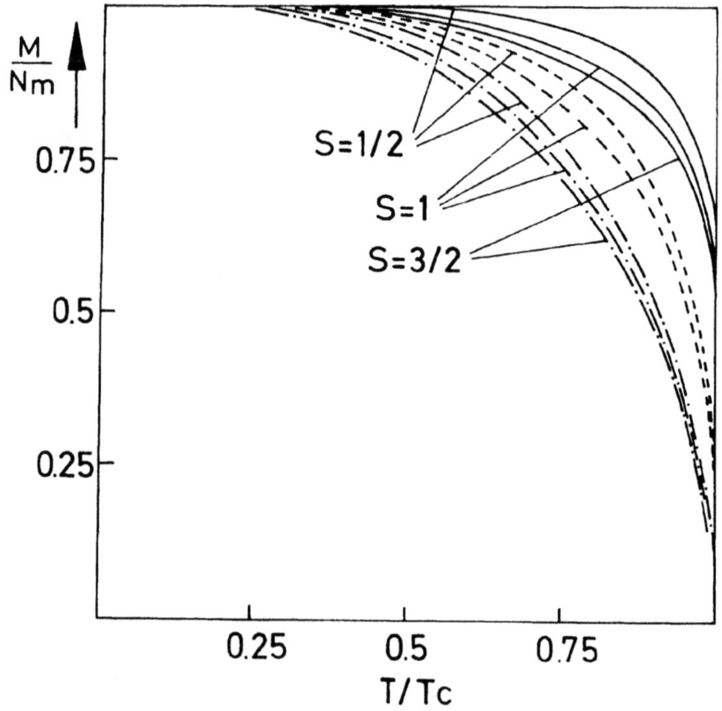


Fig. 3. Spontaneous magnetization of the Ising model with $S = 1/2$, 1 and 3/2, for the plane triangular (full line) and body center cubic (dashed line) lattices. Also the molecular field predictions (point dashed line) for the same spin values are included. Graphs taken from Fox and Guttmann [98]. Reprinted by permission of IOP Publishing Ltd.

TABLE IV
Critical exponent and scaling relations for the 2-d, $S = 1/2$ Ising model. The dashed values correspond to exponents below $T_c$. (a) Exact result [59], (b) series expansion [75], (c) exact result [15], (d) exact result [49, 50], (e) series expansion [75, 76]. Reprinted by permission of the American Physical Society.

| Thermodynamic quantity | C. Exponent | $S = 1/2$ Ising model |
|---|---|---|
| High temperature $\chi$ | $\gamma$ | 7/4 (a) |
| High temperature gap | $2\Delta$ | 15/4 (b) |
| High temperature specific heat | $\alpha = 2 + \gamma - 2\Delta$ | 0 (c) |
| Low temperature specific heat | $\alpha = \alpha'$ | 0(c) |
| Spontaneous Magnetization | $\beta = \delta - \gamma$ | 1/8 (d) |
| $M$ vs. field at $T_c$ | $\delta = \Delta/(\Delta - \gamma)$ | 15 (e) |
| Low temperature $\chi$ | $\gamma = \gamma'$ | 7/4 (a) |
| Low temperature gap | $\Delta = \Delta'$ | 15/8 (b) |

### 2.1.3. *Properties in Nonzero Parallel Field*

For finite parallel fields there are no exact results available, but the analysis of the series and high-field polynomials, together with the use of relation between critical exponents, provide quite a lot of information for both ferro- and antiferromagnetic lattices, which will be discussed separately.

For a ferromagnetic interaction one may insert the critical temperature, $u_c$ in expression (35) for ln $Z$, and then a series of $M$ versus $y$ may be obtained, which represents the critical isotherm. Gaunt *et al.* [76], following the analogy with the critical isotherm near the gas—liquid critical point, have proposed a critical behaviour of the form:

$$M(H) = D|H|^{1/\delta} \qquad (T = T_c, H \to 0). \tag{43}$$

The analyses for the s.p. and p.t. lattices yielded a value of $\delta = 15.00 \pm 0.08$, quite different from the value $\delta = 5$ expected for 3-d lattices. Subsequently, Domb and Hunter [77] have proposed an equation of state in the critical region of the form

$$mH/k_BT = (1 - T_c/T)^{\Delta}F(1/M(1 - T_c/T)^{\Delta-\gamma}) \tag{44}$$

where $F(x)$ is a continuous function that characterizes the critical behaviour, implying that all the critical indexes may be derived from the two indices $\gamma$ and $\Delta$.

Successive analyses by Gaunt [78], Domb [79], Gaunt and Domb [80] and Gaunt and Sykes [81], using expressions like (43) and (44), and the longest l.t.s. [65—70], have fielded the value $\delta = 15$ with an accuracy of 0.5%, and confirmed the critical exponent relations of Table IV. In addition, Betts and Filipow [82] have analysed the first six temperature derivatives of the magnetization with respect to the field in the 2-d lattices, obtaining amplitudes and exponents. The combination of all these analyses characterizes fairly well the behaviour around $T_c$.

The nature of the possible singularity of the $S = 1/2$ Ising ferromagnet along the first-order transition line $H = 0$, $T \leqslant T_c$, see Figure 4, has also been the subject of considerable theoretical interest. Because of space limits, only the latest contributions by Baker and Kim [83], and Enting and Baxter [84], who used the longest series [83] (35 terms for $M(T, H)$) will be summarized. Baker and Kim's analysis rejects the possibility of a 'spinodal' line [85] and points to a divergent character of $M(T, H)$ for $H = 0$ in the range $0 \leqslant T \leqslant T_c$. This result would imply the existence of a line of sigularities exactly at $H = 0$, for $0 \leqslant T \leqslant T_c$. Enting and Baxter's analysis also rejects the spinodal possibility, and claims the series to be consistent with an infinitely differentiable singularity of the type predicted for the droplet model [85]. Nevertheless, more complicated forms could not be precluded and no convincing evidence for a displacement of the line or singularities to $H = 0$ has been found as yet.

The polynomials of ln $Z$ for the antiferromagnetic case have also been analysed, although the existence of a finite $T_c$ at nonzero fields changes qualitatively the physics involved. The problem to be solved in antiferromagnets is how to determine the variation of the critical temperature, $T_c(H)$, with the external magnetic field. The action of the field is to lower the critical temperature up to some critical field, which in the Ising model is given by $H_c = zJ/2m$, for which the value of $T_c(H \geqslant H_c)$ has been reduced to zero.

Bienenstock and Lewis [86, 87] undertook the calculation of the phase diagram,

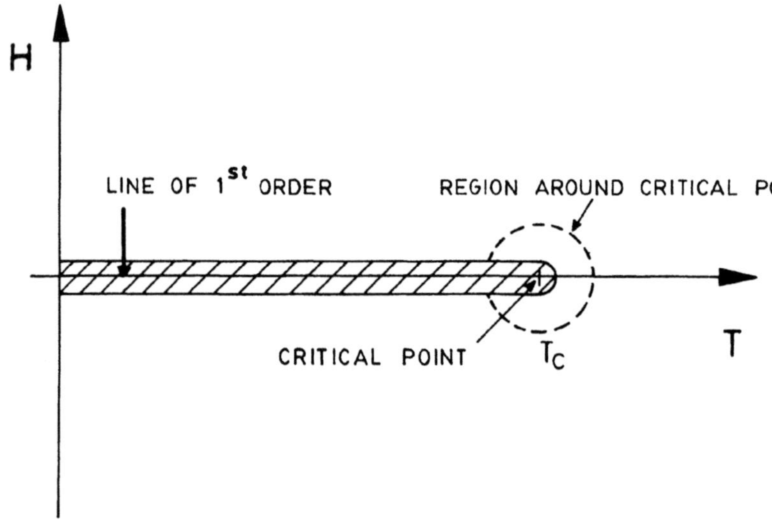


Fig. 4. Schematic representation of the main features of the ferromagnetic Ising phase diagram for fields parallel to the quantization axis.

$T_c(H)$, determining the antiferromagnetic ln $Z$ in the presence of a constant field, $H$, in addition to a small staggered one, $H^{st}$. From such calculations a triple series in the temperature, in $H$ and in $H^{st}$ emerges, and straightforward manipulations allow the estimation of $\chi^{st}(H)$, at $H^{st} = 0$, in finite field. $\chi^{st}(H)$ should have a singularity at $T_c(H)$ with the usual critical exponent $\gamma = 7/4$ for 2-d Ising lattices, and, by applying the Dlog transformation (15) for given $H$ values, the P.A. tables should determine $T_c(H)$. The values obtained for the s.p. and the 3-d s.c. and b.c.c. lattices are represented in Figure 5.

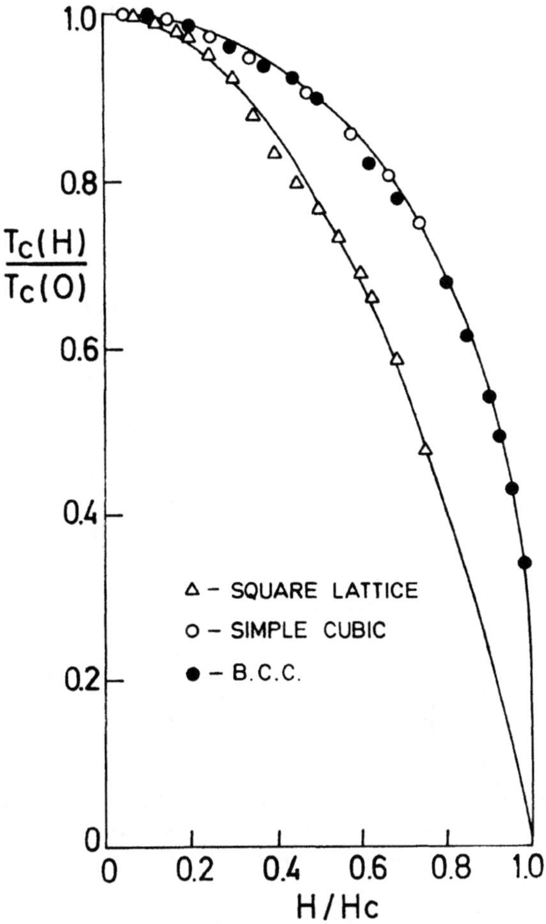


Fig. 5. Phase diagram of the antiferromagnetic $S = 1/2$ Ising model for the square planar (triangles), simple cubic (circles), and body center cubic (full circles) lattices, for fields parallel to the quantization axis. The continuous curves were obtained from Eq. (45) with $\xi = 0.87$ for the square planar and $\xi = 0.36$ for the cubic lattices (Bienenstock [86]). Reprinted by permission of the American Institute of Physics.

From the coincidence of the curves for the s.c. and b.c.c. lattice in the $T_c(H)/T_c(0)$ ($T_c(0) = T_N$) versus $H/H_c$ plot, Bienenstock [86] concluded that its shape is primarily determined by the dimensionality, similar as for other thermodynamic functions in the Ising model. The curves obtained may be fitted by a simple expression of the form

$$T_c(H)/T_c(0) = (1 - (H/H_c)^2)^{\xi} \tag{45}$$

where $\xi = 0.87$ for the s.p. lattice and $\xi = 0.36$ for the cubic ones (the range of application is restricted to $H/H_c \leqslant 0.7$ for the 2-d lattices [22]).

Rapaport and Domb [88] reexamined the $T_c(H)$ phase diagram using longer series as well as the results available for the specific heat and the antiferromagnetic susceptibility, with special attention to the region around $H = 0$. Thus, they performed a careful examination of the free energy, using Griffiths' [89] smoothness postulate and the relation of critical amplitudes, and arrived at an expression similar to the following:

$$T_c(H)/T_c(0) = 1 - 0.0373(2m/J)^2H^2 + O(H^4) \tag{46}$$

for the s.p. lattice. In reduced units the same form should also be valid for the h.c. lattice.

It is well known that at zero field the triangular lattice with antiferromagnetic interaction between nearest-neighbours only cannot order at any finite $T_c$. However, in the presence of a field it may have a nonzero $T_c(H)$, because the lattice can be divided in three interpenetrating sublattices, $A$, $B$ and $C$, as shown in Figure 6. Each moment $S_A$ on the sublattice $A$ has three neighbours ($S_B$) on

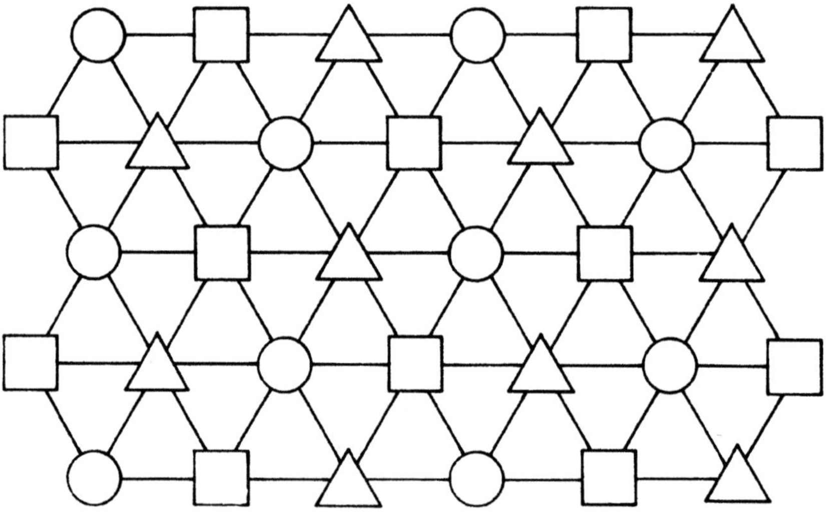

Fig. 6. Schematic representation of the three equivalent magnetic sublattices which appear in the triangular net for antiferromagnetic interaction between nearest-neighbours, in the presence of a field parallel to the quantization axis.

sublattice $B$ and three on sublattice $C(S_C)$. The energy for the ferromagnetic ground state $S_A = S_B = S_C = S$ at $T = 0$ would be $6NJS^2 - NHm$ and $-2NJS^2 - NHm/3$ for the triply degenerate ground state $-S_A = S_B = S_C$. Then, for $0 \leqslant mH \leqslant 12JS^2$ the system orders in one of the lower antiferromagnetic states. Ferer [90] followed a procedure analogous to the one applied for the p.t. lattice, obtaining high-field polynomials and deduced the staggered susceptibility at constant field. With these results the phase diagram in the $M$–$T$ plane, could be derived by the usual analysis.

### 2.1.4. *Properties in Nonzero Perpendicular Field*

In the case of an external magnetic field applied in the direction perpendicular to the magnetization axis $z$, there exists an exact solution for $\chi_\perp$ in the zero field limit for the s.p. and h.c. lattices, as well as series expansions, first derived by Fisher [51]. The exact results are represented in Figure 7. Again, for nonzero field there are only series expansion results available. The effect of the transverse field $H_\perp$ is to opposed the cooperative spin ordering and, as $H_\perp$ increases, the critical temperature decreases smoothly from $T_c(H_\perp = 0)$ to $T_c(H_\perp \geqslant H_{\perp c}) = 0$ when the cooperative ordering can no longer occur.

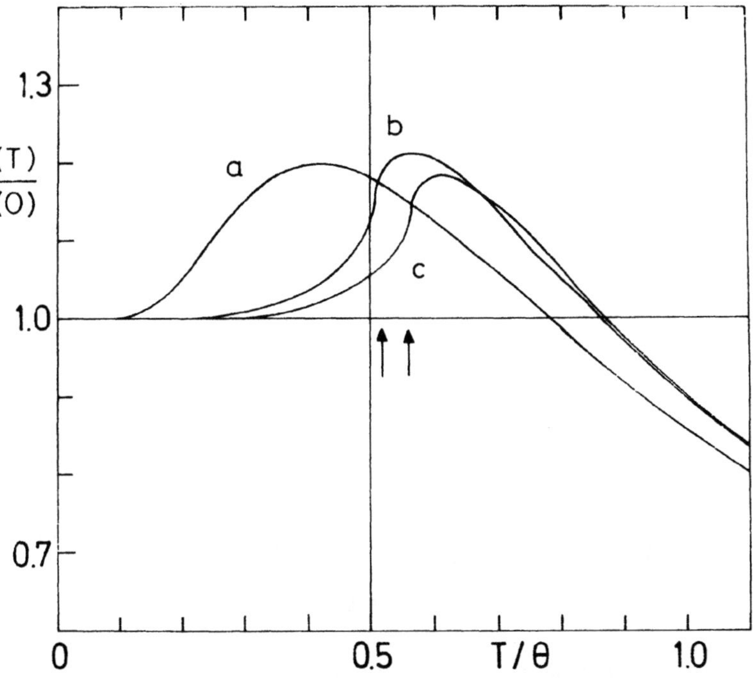


Fig. 7. Exact zero-field perpendicular susceptibilities of the Ising $S = 1/2$ model for the linear chain (curve a), honeycomb (curve b) and plane square (curve c) lattices, taken from Fisher [51] ($\chi_\perp(0) = Ng^2\mu_B^2/2q|J|$, $\Theta = 2J/k_B$). Reprinted by permission of the American Institute of Physics.

Calculations for the $S = 1/2$ Ising model in a perpendicular field were first performed by Elliott and Wood [91], and later by Yanase *et al.* [92] and by Oitmaa and Plischke [93]. With the aim to derive the phase diagram $T_c(H_\perp)$, the h.t.s. expansion for the $\chi_{zz}$ susceptibility was derived [91, 92]. $\chi_{zz}$ is defined as the susceptibility in the $z$ direction at the limit of vanishing field, $H_\parallel$, when a perpendicular constant field $H_\perp$ is acting, and therefore it should diverge at some $T_c(H_\perp)$ at which the system orders. The studies of Yanase *et al.* [92] with seven coefficients in the $\chi_{zz}$ series for the s.p. and p.t. lattices, established the universality of the critical exponent, $\gamma = 7/4$ for $0 \leqslant H_\perp \leqslant H_{\perp c}$. They also derived the values of the critical field $2mH_{\perp c}/zJ = 0.761$ and 0.794 for the s.p. and p.t. lattices, respectively. In Figure 8 the $T_c(H_\perp)$ diagram obtained by Yanase *et al.* [92] has been represented.

Oitmaa and Plischke [93] have improved the analysis of Elliott and Wood [91] for 3-d lattices, considering instead of $\chi_{zz}$ the square of the order parameter

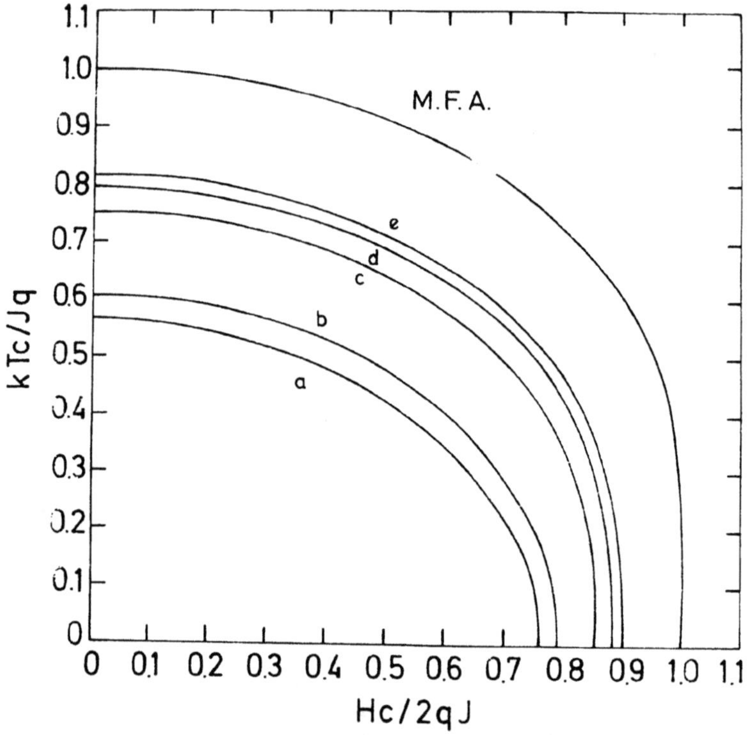


Fig. 8. Phase diagram of the ferromagnetic $S = 1/2$ Ising model in the presence of a field perpendicular to the quantization axis. Curves a and b correspond to the square and triangular planar lattices, whereas curves c, d, and e, correspond to the 3-d cubic lattices, s.c. b.c.c. and f.c.c., respectively. The results has been taken from Yanase *et al.* [92] ($q = z$). Reprinted by permission of the Physical Society of Japan.

fluctuation, $y_{\|} = \langle (M^z)^2 \rangle$, which has the same asymptotic beahviour in the critical region. The authors derived the universality of the critical exponent for 3-d lattices and furthermore, using the reduced scales $T_c(H_\perp)/T_c(0)$ versus $H_\perp/H_{\perp c}$, they found that the phase diagrams of the 3-d lattices coincide, as it was the case for antiferromagnets in a parallel field. The same scaling law is expected to hold for the 2-d lattices.

We note that the Ising model in a transverse field has quite number of different applications. Amongst other things, it may serve as a model for singlet ground state systems, as are found in rare earth intermetallic compounds.

### 2.2. ISING MODEL WITH GENERAL $S$

The fundamental question as to whether the critical indices depend on the spin value has stimulated the derivation of series for $S > 1/2$. The first h.t.s. for the magnetic susceptibility and the partition function at zero-field were derived by Domb and Sykes [94] for the f.c.c. lattice. Their analysis in terms of the ratio method gave $\gamma$-values independent of $S$, pointing to universality of the critical exponent. With the same aim, h.t.s. and l.t.s. expansions were also derived for different $S$ values on the most common 2-d and 3-d lattices [95—100].

The h.t.s. derivation for the Ising model with general $S$ is more difficult than for $S = 1/2$, and therefore the series are much shorter. Van Dyke and Camp [99] derived h.t.s. for the susceptibility on the p.t. and s.p. lattices up to order ten. The expansion of $\chi$ takes the form:

$$3\chi k_B T/Ng^2\mu_B^2 X = \sum_{n=0}^{\infty} h_n(X)K^n, \qquad h_n(X) = 1/D_n \sum_{i=1}^{n} b_{n,i}X^i \tag{47}$$

where $X = S(S+1)$, $K = 2J/k_B T$, $h_n(X)$ is a polynomial of degree $n$ in $X$, and the coefficients $D_n$ and $b_{n,i}$ have been listed in Table A-V (Appendix A). Camp and Van Dyke [99] have analysed the series for the p.t. net, which apparently does not exhibit good convergence for any value of $S$, whereas for $S = 1/2$ the series for the same lattice converges well [55]. The $\gamma$-values obtained show a marked dependence on $S$, varying between $\gamma = 1.75$ for $S = 1/2$ to $\gamma = 1.89$ for $S \geqslant 3$. Other analyses performed by the same authors and based on a confluent singularity behaviour, as in Equation (23), also proved to be inconclusive, probably due to the shortness of the series. Assuming $\gamma = 7/4$ and using P.A.s of Equation (18), and for some selected $S$ values, the critical temperatures have been derived and are listed in Table V. The positions of the antiferromagnetic maxima are also included.

The l.t.s. expansions for $\chi$, $M$ and $\ln \Lambda$ were first generated for the $S = 1$ Ising model by Fox and Gaunt [95] for various 2-d and 3-d lattices, and later extended to $S = 3/2$. The authors analysed the critical isotherm for $S = 1$ in the s.p. lattices and concluded that $14.5 \leqslant \delta \leqslant 15.5$, in agreement with the value for $S = 1/2$. Guttmann *et al.* [96], working with the s.p. and p.t. lattices, derived the values $\gamma' =$

TABLE V
Critical temperature $K_c = J/k_B T_c$ for the 2-d Ising model with different spin values $S$ for the p.t. and s.p. lattices. The position of the antiferromagnetic maximum $K_m = K(\chi_{max})$ and $\chi'_m = \chi_{max}|J|/Ng^2\mu_B^2$ for the s.p. lattice are included. The estimated uncertainties are indicated.

| | | $S=1/2$ | $S=1$ | $S=3/2$ | $S=2$ | $S=5/2$ | $S=\infty$ |
|---|---|---|---|---|---|---|---|
| p.t. | $K_c$ | 0.549306 | 0.1873 | 0.09851 | 0.0596 ± 2 | 0.0406 ± 1 | 0.698 |
| s.p. | $K_c$ | 0.881374 | 0.2959 | 0.1527 | 0.0940 | 0.0640 | 1.100 |
| | $K_m$ | 0.574 | 0.216 | 0.116 | 0.072 | 0.050 | 0.434 |
| | $\chi'_m$ | 0.0536 | 0.0556 | 0.0563 | 0.0566 | 0.0567 | 0.0570 |
| h.c. | $K_c$ | 1.31698 | 0.4317 | — | — | — | — |

1.72 ± 0.05 and $\beta = 0.125 \pm 0.004$, also in agreement with the corresponding $S = 1/2$ values.

Fox and Gaunt [95], using the basic principles involved in the derivation of the l.t.s. for $S = 1/2$ [65], derived high-field expansions for ln $\Lambda(u, y)$ in the usual form, up to $g_{12}$(h.c.), $g_{10}$(s.p.) and $g_7$(p.t.) for $S = 1$ and $g_7$(p.t.) for $S = 3/2$. Moreover, using Equations (36) and (37) the values of $M$ and $\chi$ were derived and analysed. The $T_c$ values found for $S = 1$ and $S = 3/2$ agree with those of Table V. Furthermore, De'Bell and Betts [97], using a high-temperature R.G. series transformation, have derived $T_c$ values for any $S$ on the s.p. and p.t. lattices, which agree, within the estimated error, with those of Table V and clearly support the smoothness postulate [89]. With these series critical amplitudes for $\chi$ and $M$ were also derived [98], and found to be in agreement with previous estimations.

There is no h.t.s. for ln $Z$ nor for $\chi$ available for the honeycomb lattice. However, short series may easily be calculated from the known h.t.s. of the full Hamiltonian (1). Seven series coefficients for ln $Z$ and six for $\chi$ have been reported by Wood and Dalton [101] for any lattice, spin value and parameters $a$ and $b$. For $a = 1$, $b = 0$, the series of the $S$-spin Ising model may be obtained.

All the h.t.s. and l.t.s. expansion analyses for general $S$, although not fully conclusive in themselves, appear to indicate the same values for the critical exponents and the same relations between them as for $S = 1/2$.

Series expansions for mixed-spin Ising systems in 2-d have recently been derived by Bowers and co-workers [102]. This model is based on a two-sublattice description (only valid for loosely-packed lattices) in which the sublattice $A$ is occupied by spins $S_A$ that are different from the spins in the $B$ sublattice ($S_B$). The interest of such a model is two-fold, firstly it allows the description of very anisotropic, simple ferrimagnets, and secondly it has less translational symmetry, which may affect the critical behaviour. The analyses performed so far on $\chi$, $C(T)$ and on the critical isotherm [102] for $S_A = 1/2$, $S_B = 1$, agree with the universality hypothesis, however.

We finally mention that recently the critical behaviour of Ising spin-glass with $\pm J$ has been studied [103], using h.t.s. of the Edwards—Anderson susceptibility and of the free energy. For the s.p. lattice (i.e. 2-d) the analysis does not show any convergent singularity.

### 2.3. OTHER SERIES FOR $I(1/2)$

#### 2.3.1. *Restricted Dimensionality Systems*

The knowledge of the strong dependence of the critical behaviour upon the dimensionality of the lattice has opened many other questions about the effects of finite size on critical exponents and amplitudes. Thus, systems of finite length in some directions and infinite in the others, which may be called 'restricted dimensionality systems', were studied. These systems are interesting for several reasons. First, they provide a means of studying crossover phenomena in thin films from 3-d behaviour (at high $T$) to low-d behaviour as the temperature is lowered. Second, they have increasing importance for the field of surface physics. Third, they fill in the region between the 'pure', i.e. the monolayer 2-d magnets, and the 'thin films', on which a large amount of literature exits, but which are not precisely thin in the sense defined below, because the film thicknesses considered are of the order of $10^3$—$10^5$ layers.

The simplest lattice with restricted dimensionality which may have a finite $T_c$ is built up of two infinite layers stacked one upon the other, i.e. a double layer, DL, which has been represented in Figure 9a. The DL is 2-d in the sense of Kasteleyn [104], but it is not planar, each spin having a coordination number of five; if it is mapped in a mono layer, a complicated structure appears, with interactions between first and second neighbours. The DL is, obviously, a member of a series of systems with dimensionality restricted to 2-d, that may be formed by piling up $n$ layers, obtaining a multilayered system. For more than two layers a difference of coordination number appears between the surface ($z = 5$) and the bulk ($z = 6$) spins. The latter have the same coordination and surrounding as in the s.c. lattice, if the inter- and intra-layer distances are equal. In Figure 9b the triple layer, TL, has been plotted for the case of s.c. elementary cubes.

Following the methods described by Domb [10] and assuming equal inter- and intralayer exchange constants, Ballentine [105], Allan [106] and Capehart and Fisher [107] have derived series for multilayers (or films) with 2 to 10 monolayers and free boundary conditions, and with 2 to 14 monolayers and periodic boundary conditions. Also Oitmaa and Enting [108] have considered the DL case with different exchange constants $J_1$ and $J_2$ for layers 1 and 2 and interlayer interaction $J_{12}$, in an obvious notation.

The $\chi$ analysis of the DL series [105, 106] leads to a critical exponent, $\gamma = 7/4$, equal to the one obtained for the normal monolayer 2-d lattices. Ballentine [105] also analysed the distribution of entropy below $T_c$, which proved to be higher than that obtained in the s.p. and p.t. lattices, but lower than for the s.c. net. For $J_{12} \neq 0$ Oitmaa and Enting [108] derived the same $\gamma$ and only one critical temperature, in agreement with scaling results.

From the analysis of the susceptibility $\chi(n)$ as a function of the number $n$ of monolayers, Allan [106] and later Capehart and Fisher [107] anticipated a critical behaviour of the form:

$$\bar{\chi}(n, T) = A(n)(1 - w/w_c(n))^{-7/4} \tag{48}$$

with $w_c(n) = \tanh(J/2k_B T_c(n))$. Moreover, the authors analysed the crossover from

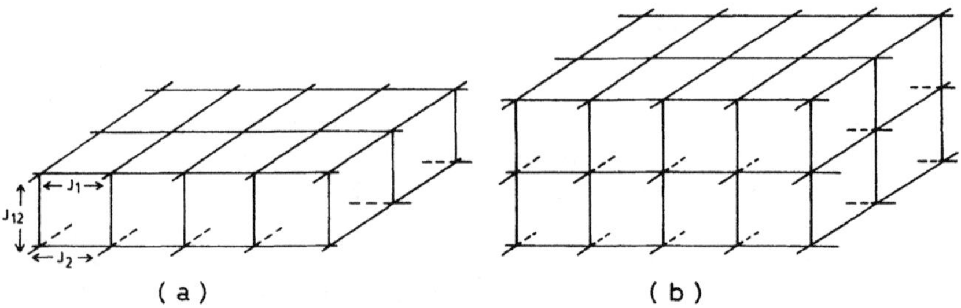


Fig. 9. Schematic representation of a thin film formed by two quadratic layers (double-layer, DL), figure a, and three layers, (triple-layer, TL), figure b. The exchange constant interlayer, $J_{12}$, and intralayer, $J_1$ and $J_2$, have also been shown.

this 2-d behaviour to the 3-d one, $\bar{\chi}(\infty, T) \approx A(\infty)(1 - w/w_c(\infty))^{-5/4}$. The standard ratio analysis of the $(\bar{\chi}(n))^{1/\gamma}$ series yields the critical temperatures summarized in Table VI. The $w_c(n)$ results smoothly join the $w_c(\infty)$ limit according to the approximate representation:

$$w_c(n) = w_c(\infty) + 0.350(n + 1/2)^{-1/\nu} \tag{49}$$

where $1/\nu = \phi = 1.56$ for free boundary conditions and:

$$w_c(n) = w_c(\infty) + 0.177(n - 1/4)^{-2} \tag{50}$$

for periodic boundary conditions. The use of the ratio method (10) with $\gamma = 7/4$ and the above $w_c(n)$ yields the amplitude values $A(n)$ listed in Table VI. Also, an approximate representation for $A(n)$ can be derived, which shows an asymptotic behaviour like

$$A(n) = X_0(n + m)^{-7/9} \tag{51}$$

where $X_0 = 1.27 \pm 0.02$ and $m = -2/3$ for free boundary conditions and $X_0 = 0.98 \pm 0.02$ and $m = -3/2$ for periodic multilayers. The apparent bad convergence of the amplitudes is due to the 3-d $\leftrightarrow$ 2-d crossover behaviour, because the imposed 2-d exponent is larger than the 3-d one.

Capehart and Fisher [107], paid special attention to the effective reduced surface susceptibility of a $n$ layer system, $\bar{\chi}^x(n)$, through the definition

$$-\sigma^x \bar{\chi}^x(n, T) = 1/2n(\bar{\chi}(\infty, T) - \bar{\chi}(n, T)) \tag{52}$$

where $\sigma^x = -1$ for the s.c. net. In the limit $n \to \infty$, $\bar{\chi}^x(n, T)$ tends towards the surface susceptibility of a semiinfinite system, $\bar{\chi}^x$, previously considered by Binder and Hohenberg [109].

Semiinfinite 2-d and 3-d lattices are a natural extension of multichain and multilayered systems, and have been analysed in order to understand the effect of finite systems with real surfaces. Whereas the bulk behaviour has the same exponent as above, new exponents should be defined for the surface layer or limit chain ($\gamma^x$), for the $n$th layer or chain ($\gamma_n$) and for the local ($\gamma_{nn}$) susceptibilities. The layer susceptibility ($\chi_n$) corresponds to the $n$th layer contribution to the bulk

TABLE VI

Critical temperatures, $w_c(n) = \tanh(J/2k_B T_c(n))$, and critical susceptibility amplitudes $A(n)$ for thin films consisting of $n = 1, 2, \ldots$ Ising ($S = 1/2$) s.c. layers, with free and periodic boundary conditions, as taken from Capehart and Fisher [107]. The estimated uncertainty in the last digit is indicated. Reprinted by permission of the American Physical Society.

| | | $n=2$ | $n=3$ | $n=4$ | $n=5$ | $n=6$ | $n=7$ | $n=8$ | $n=\infty$ |
|---|---|---|---|---|---|---|---|---|---|
| Free | $w_c(n)$ | 0.3020 ± 6 | 0.2675 ± 4 | 0.2520 ± 3 | 0.2430 ± 3 | 0.2371 ± 4 | 0.2334 ± 4 | — | 0.21813 ± 1 |
| | $A(n)$ | 0.6863 ± 10 | 0.5200 ± 15 | 0.449 ± 4 | 0.387 ± 5 | 0.323 ± 6 | 0.290 ± 7 | — | 1.0585 ± 10 |
| Per. | $w_c(n)$ | 0.2692 ± 11 | 0.2401 ± 1 | 0.2305 ± 1 | 0.2261 ± 15 | 0.2235 ± 3 | 0.2220 ± 3 | 0.2211 ± 5 | 0.21813 ± 1 |
| | $A(n)$ | 0.836 | 0.590 ± 1 | 0.452 ± 1 | 0.366 ± 1 | 0.299 ± 2 | 0.249 ± 3 | 0.220 ± 4 | 1.0585 ± 10 |

susceptibility when a uniform field is applied, and the local susceptibility ($\chi_{nn}$) corresponds to the susceptibility when a field is present only in the $n$th layer. Different R.G. relations hold exactly for such partial 2-d or 3-d systems namely

$$\text{(i)}\quad \gamma_{11} = \nu - 1, \qquad \text{(ii)}\quad \gamma^{x} = \gamma + \nu, \qquad \text{(iii)}\quad 2\gamma_{1} - \gamma_{11} = \gamma + \nu \tag{53}$$

where the exponents without indexes correspond to those for the conventional 2-d or 3-d lattices and all the susceptibilities diverge at the same critical temperature $w_c(\infty)$ as the bulk $\bar{\chi}$.

Successive enlargements of the lengths of the series and their corresponding analyses [110] are consistent with $\gamma_1 = 1.375 \pm 0.005$ and $\gamma_{11} = 0.00 \pm 0.01$ for the 2-d s.p. semiinfinite lattice, which verifies the Equations (i) and (iii) of (53). Moreover, for the s.c. semiinfinite lattice $\gamma_1 = 0.78 \pm 0.02$ and $\gamma_{11} = -0.33 \pm 0.04$, the confidence limit being sufficiently broad to include the R.G. scaling results [111].

### 2.3.2. *Further-Neighbour Interactions*

The inclusion of further-neighbour interactions in the 2-d, $S = 1/2$ Ising model also renders the problem insoluble in an exact way by the existing techniques. However, such a generalization has physical interest for the following reasons. First, to study the dependence of the critical indices on the number of interacting neighbours in 2-d lattices. Second, to analyse the convergence of the values for the indices to the molecular field predictions, which are expected to hold in the limit of an infinite number of neighbours. Third, in many magnets further-neighbour interactions are present, and it is worthwhile investigating the effects of these. Finally, the presence of different interactions may lead to competing effects, as it may entail the presence of degenerate ground states, modulated ordering arrangements and a variety of commensurate or incommensurate phases.

The first analysis of the effects of first ($J_1$) and second ($J_2$) neighbour exchange interactions (NNN) was performed by Domb and Potts [18], followed by work by Domb and Dalton [112], Dalton and Wood [113], Plischke and Oitmaa [114], and Oitmaa [115].

The difficulty of obtaining long series for the general further-neighbour model, was avoided by Domb and Dalton [112] by the introduction of the 'equivalent-neighbour model', in which the interactions within the first and second (1, 2), or the first, second and third (1, 2, 3), shells of neighbouring spins were taken as equivalent. In spite of the obvious losses of generality, the series for such models depend only on one parameter and the authors succeeded in generalizing the methods developed for nearest-neighbours. Their analyses by the ratio methods (Equations (7—10)) gave the estimates of $K_c$ listed in Table VII, and moreover a value $\gamma = 7/4$ equal to that for the nearest-neighbours-only model.

The use of the $K_c$ values and the h.t.s. for the partition function allows the estimation of the entropy content $\Delta S_c$ above $T_c$, and the values for the s.p. and p.t. lattices are listed in Table VII. The decrease of $\Delta S_c$ with the number of neighbours agrees with the idea that it should be zero for a model with infinite number of equivalent neighbours, since this corresponds to the molecular field model.

TABLE VII
Number of neighbouring spins, critical temperature $K_c = J/k_B T_c$ and entropy $\Delta S_c = (S(\infty) - S(T_c))/R$ parameters for the $S = 1/2$ Ising equivalent neighbours model. Only nearest (1), first and second, (1, 2), and first, second and third, (1, 2, 3), neighbouring shells have been considered. Data taken from Domb and Dalton [112]. Reprinted by permission of IOP Publishing Ltd.

| | s.p. | | | p.t. | | |
|---|---|---|---|---|---|---|
| | (1) | (1, 2) | (1, 2, 3) | (1) | (1, 2) | (1, 2, 3) |
| $z$ | 4 | 8 | 12 | 6 | 12 | 18 |
| $K_c$ | 0.881374 | 0.3802 | 0.2270 | 0.549306 | 0.2272 | 0.1424 |
| $\Delta S_c$ | 0.38668 | 0.332 | 0.240 | 0.36287 | 0.250 | 0.182 |

However, just in this limit the critical exponents should change abruptly to $\gamma = 1$, $\alpha = 0$ (discontinuous), $\beta = 1/2$. Also the analyses by means of P.A. methods of the l.t.s. for $M$ in the (1, 2) model gave an exponent $0.122 \leqslant \beta \leqslant 0.134$, consistent with the prediction for the nearest-neighbour model, although with a large uncertainty. Dalton and Wood [113] also derived series for arbitrary values of $J_1$ and $J_2$ for the NNN-Ising model, and found that the variation of the critical temperature is well represented by

$$T_c(J_2/J_1) = T_c(0)(1 + m_1 J_2/J_1) \tag{54}$$

where $0 \leqslant J_2/J_1 \leqslant 1$ and $m_1$ takes the values $m_1(\text{s.p.}) = 1.45$ and $m_1(\text{p.t.}) = 1.35$. Dalton and Wood's main conclusion was that the critical exponent seems to be independent of the number of interacting shells considered.

Additional studies on the NNN-Ising model have been performed as the interest in knowing the variation of the critical exponents with the dimension of the order parameter arose. Namely, the existence of degenerate ground states for some competing values of $J_1$ and $J_2$ gives the possibility to have order parameters of higher dimension than for the Ising model, and consequently a different critical behaviour should be expected in such cases. With this idea, h.t.s. for the s.p. and p.t. lattices [114] and l.t.s. for the s.p. one [115] have been reported. In the s.p. lattice three ordered states, ferromagnetic, F, antiferromagnetic, AF, and super-antiferromagnetic, SAF (see Figure 10) are possible. The SAF state has parallel chains of spins up and down and a twofold degeneracy appears, the order parameter being bidimensional. In addition, on the separation line of the F and AF phases another type of degeneracy is possible and a 3-d order parameter is appropriate for $J_2 = -1/2|J_1|$.

Oitmaa's analysis [114] of the F phase reproduces the critical exponent $\gamma = 7/4$. For the SAF state the analysis is rather complicated by the presence of an unphysical singularity, but after appropriate mapping the result gives a variation of $\gamma$ with the ratio $J_2/J_1$, contradicting the universality hypothesis. The analysis of the lines F-SAF and AF-SAF by series expansion has not yet been performed.

For the p.t. lattice four ground states are possible; ferromagnetic, F, Potts state, P, alternating up—down spin lines, $A1$, and alternating double up—down spin

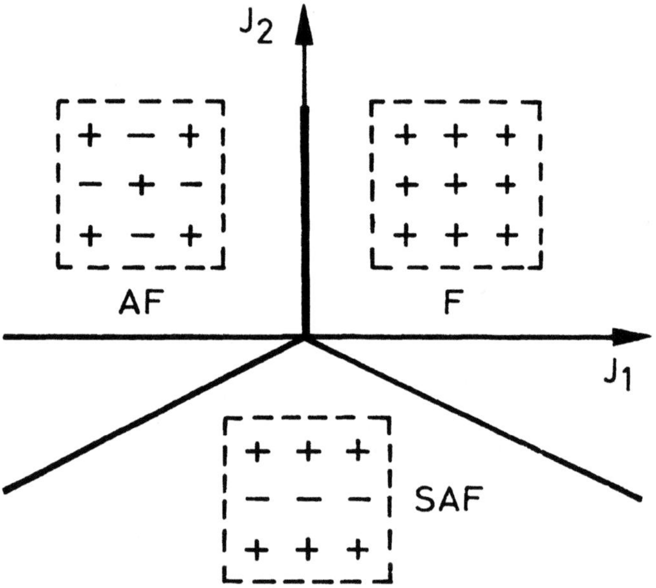


Fig. 10. Ground state arrangement regions of the Ising $S = 1/2$ model for the s.p. lattice with different interactions between nearest, $J_1$, and next-nearest, $J_2$, neighbours, as a function of the $J_2/J_1$ ratio. The region labeled F corresponds to ferromagnetic order; AF corresponds to the ususal two sublattice antiferromagnetic order; and SAF to a degenerate antiferromagnetic arrangement with alternating chain sublattices.

lines, $A2$. Along the separation lines $P$—$A1$, $A1$—$A2$ and $F$—$A2$ there are an infinite number of ground states, and a nonzero ground state entropy. The analysis of the $F$ phase ($J_2 > 0$, $J_2 > -1/2J_1$) gave results consistent with $\gamma = 7/4$, the series being more regular for $J_2/J_1 > 0$ than for $J_2/J_1 < 0$. The $P$ state ($J_1 < 0$, $J_2 > 0$) has very irregular series, and an unexpected high value, $\gamma = 2.4 \pm 0.02$, is obtained. Probably, longer series will be necessary to clarify all these aspects, as well as the properties of the phases $A1$ and $A2$.

### 2.3.3. *Crossover from 2-d to 3-d Behaviour*

The Onsager solution (25) of the partition function of the $S = 1/2$ Ising model on the rectangular lattice ($J_x \neq J_y$) has clearly shown that for $J_y \neq 0$ there is a finite $T_c$ at which the system orders. Such lattices with $J_x \gg J_y$ may be described as a set of linear chains in the $x$ direction, weakly coupled in the $y$ one. At high enough temperature linear chain characteristics are to be expected, and upon lowering the temperature a crossover should occur from the 1-d behaviour (no long range order) to the 2-d behaviour with ordering at a finite $T_c$. Similar crossover effects should appear when a set of layers, with exchange constant $J_{xy}$, are weakly coupled in the third direction, $J_z(|J_z| \ll |J_{xy}|)$, the behaviour changing from 2-d to 3-d. In real magnetic system this 3-d interaction may be very small, but it is always present and it is of great interest to know its effects, because if they are relevant at any temperature lower-dimensional behaviour could not be observable. In addition, the anisotropy of the interaction is in fact a particular case of further-neighbour interactions, and competing situations, multicritical points and modulated structures may also be obtained.

Griffiths [89] included the phenomenology of further-neighbours and of lattice crossovers in a general study of lattice models for phase transition and critical phenomena, which can be related by variation of linear parameters appearing in the Hamiltonian. He suggested that "the critical indices remain constant as function of these parameters, except at points where the nature of the associated phase transition itself is altered by varying a parameter". The verification of this 'universality hypothesis' was another interesting point in the analysis of dimensionality crossover-effects.

Abe [116] reached the same conclusion independently for the crossover from 2-d → 3-d in the Ising model, and predicted for small $R = J_z/J_{xy}$ a shift of the transition temperature according to

$$T_c(R) = T_c(0)(1 + BR^{1/\phi}) \qquad \phi = 7/4 \tag{55}$$

with a critical exponent corresponding to the 3-d Ising model. An h.t.s. for the $S = 1/2$ Ising model with spatial anisotropy was first derived by Oitmaa and Enting [117] for $\chi$ in the s.p. → s.c. crossover, and later extended by Stanley and coworkers [118—120]. The extensive analyses of these series yield results that are consistent with the universality hypothesis [118], and are also in agreement with Abe's Equation (55) [119]. Moreover, several relations between the succesive derivatives of $\chi$ with $R$ have been obtained [120]. Considering the expansion

$$\bar{\chi}(J_{xy}, R) = \bar{\chi}_0^{(0)}(J_{xy}) + R\bar{\chi}_0^{(1)}(J_{xy}) + R^2/2!\bar{\chi}_0^{(2)}(J_{xy}) + \cdots$$

where $\bar{\chi}_0^{(n)} = (\partial^n \bar{\chi}/\partial R^n)|_0$ behaves asymptotically as

$$\bar{\chi}_0^{(n)} \approx (1 - T_c/T)^{-\gamma^{(n)}} \qquad \gamma^{(n)} = \gamma_0 - n\phi, \tag{56}$$

and $\gamma_0$ is the critical $\chi$ exponent for 2-d lattices, the 'crossover exponent' $\phi = 7/4$ was found and it is now believed to be equal to $\gamma_0$. Indeed, $\bar{\chi}_0^{(0)}(J_{xy})$ coincides with the $\bar{\chi}$ value for the 2-d system. Moreover, it was proved [120] that

$$\bar{\chi}_0^{(1)} = 2J_{xy}/k_B T(\bar{\chi}_0^{(0)2} + M_2 \partial \bar{\chi}_0^{(0)}/\partial H) \tag{57}$$

where $M_2$ is the reduced magnetization of a 2-d lattice and $\partial \bar{\chi}_0^{(0)}/\partial H$ is evaluated at zero-field. Lower and upper bounds were also established for the higher-order $R$ derivatives [120]. Furhtermore, from scaling with the parameter $R$ predictions for the critical amplitude could be also obtained, and a double power law of the form:

$$\bar{\chi}(J_{xy}, R) \approx R^{(5/4 - 7/4)\phi} \cdot (T - T_c(R))^{-5/4} \tag{58}$$

has been proposed [120].

Expansions as a function of $R$ for the other thermodynamic functions may be very useful in analysing the nature of the crossover phenomena. At first sight the above results would indicate that below some crossover temperature ($T_{cr}(R)$) all systems should have a 3-d character. However, this is not always apparent since, depending on the sign of the interaction [121], some thermodynamic quantities are nondivergent and the pure 2-d character may be dominant at all temperatures. Taking as an example the susceptibility for ferromagnetic intralayer interactions $J_{xy}$, a ferromagnetic interlayer coupling increases the susceptibility whereas for $R < 0$ it decreases. This has been depicted in Figure 11, which clearly shows the crossover effects [122]. However, for an antiferromagnetic interaction $J_{xy}$ none of the $\bar{\chi}_0^{(n)}$ derivatives are divergent, so that only minor crossover effects will be present and for small $R$ almost pure 2-d behaviour will be found over the whole temperature range.

The above method of analytic expansions in terms of the ratio between the weakest and dominant interactions has been also applied in the study of the 1-d → 2-d crossover [123] or 1-d → 3-d crossover [118], and relations similar to (55–

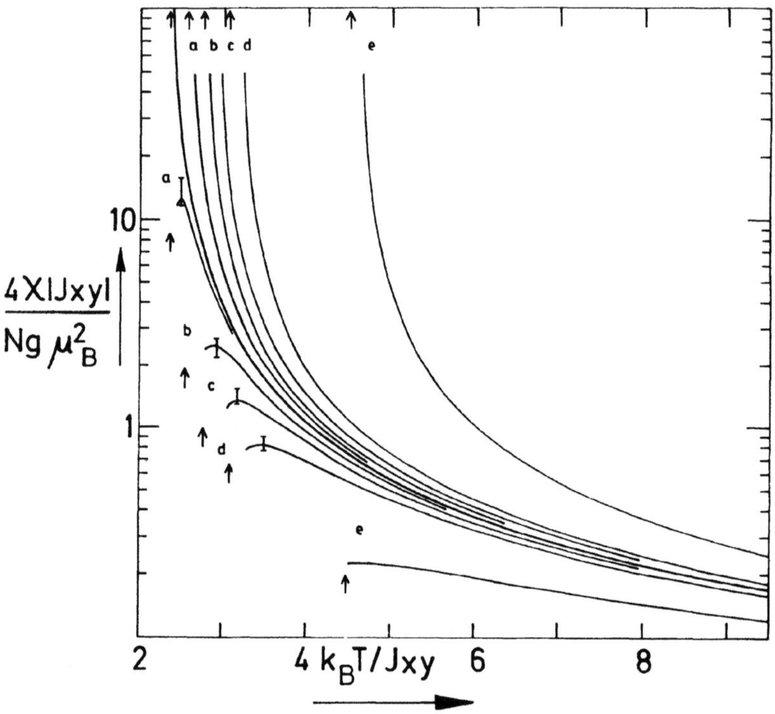


Fig. 11. Lattice dimensionality crossover effects in the susceptibility of the ferromagnetic Ising $S = 1/2$ model for the s.p. lattice, with ferro- and antiferromagnetic 3-d coupling in an s.c. net. The curves labeled a, b, c, d and e, refer to the ratio $|J_z/J_{xy}| = 0.01, 0.05, 0.1, 0.2$ and 1, respectively taken from Navarro and de Jongh [122]. Reprinted by permission of Elsevier Science Publishers.

58) may be obtained, with appropriate critical exponents. A summary of the $T_c(R)$ values and the positions of the susceptibility maxima in the 1-d → 3-d, 2-d → 3-d crossover was reported by Navarro and de Jongh [122] and is given in Table VIII.

TABLE VIII

Critical temperatures, position and height of susceptibility maximum $\chi'_m = 2\chi_{max}|J_z|/Ng^2\mu_B^2$ for the 1-d → 3-d and $\chi'_m = 2\chi_{max}|J_{xy}|/Ng^2\mu_B^2$ for the 2-d → 3-d lattice-dimensionality crossover for the $S = 1/2$ Ising model. Data taken from Navarro and de Jongh [122]. Reproduced by permission of Elsevier Science Publishers.

| 1-d → 3-d | | | $R = J_{xy}/J_z(J_z > 0)$ | | | |
|---|---|---|---|---|---|---|
| R | −1.0 | −0.2 | −0.1 | −0.05 | −0.01 | 0.0 |
| $k_B T_c/J_z$ | 2.2553 | 0.907 | 0.672 | 0.521 | 0.33 | 0.0 |
| $k_B T_m/J_z$ | 2.40 ± 0.03 | 0.98 ± 0.03 | 0.70 ± 0.03 | 0.60 ± 0.1 | 0.39 ± 0.05 | — |
| $\chi'_m$ | 0.116 | 0.57 ± 0.02 | 1.15 ± 0.05 | 2.05 ± 0.05 | 8.8 ± 1.5 | — |
| **2-d → 3-d** | | | $R = J_z/J_{xy}(J_{xy} > 0)$ | | | |
| $k_B T_c/J_{xy}$ | 2.2553 | 1.5286 | 1.3808 | 1.301 | 1.180 | 1.1346 |
| $k_B T_m/J_{xy}$ | 2.33 | 1.75 ± 0.03 | 1.6 ± 0.1 | 1.5 ± 0.1 | 1.3 ± 0.1 | — |
| $\chi'_m$ | 0.226 | 0.82 ± 0.07 | 1.35 ± 0.15 | 2.5 ± 0.3 | 12.5 ± 2 | — |

Another point of interest in systems with lattice anisotropy showing 2-d → 3-d crossover, is the existence of a tricritical point for $J_{xy} > 0$, $J_z < 0$ [124]. The term 'tricritical point', TCP, was proposed by Griffiths [89], and corresponds to a special symmetry point, where a changeover from a second-order to a first-order phase transition occurs. Harbus and Stanley's [124] analysis of the h.t.s. for $\chi^F$ and $\chi^{st}$ in an external field (for $J_{xy} > 0$, $R = -1$), located the critical line in the $T$—$H$ plane and verified that the $\chi^{st}$ exponent is $\gamma = 5/4$, as expected from the universality hypothesis. However, at the TCP a different exponent, $\gamma = 1/2$, was obtained.

A further extension of the model with lattice dimensionality crossover (2-d → 3-d) consisted of the inclusion of an interaction, $J'_z$, between the next-nearest neighbour spin pairs along the $z$-axis only [124, 125] (Anisotropic Next-Nearest Neighbour Interaction model ANNNI). The interest of such a model arises because it is the simplest system with a complex spacially modulated behaviour, leading to a transition between ferromagnetic and helical order at a Lifshitz point.

The helical phase is characterized by the direction of the moments of each layer, which vary sinusoidally in space with an associate wave vector $\mathbf{q}_0$ that is a continuous function of the exchange interaction. The Lifshitz point is the locus of the phase diagram at which $\mathbf{q}_0 \to \mathbf{0}$.

The helical phase appears as a consequence of the competition between $J_z$ and $J'_z$, for $J_z/J'_z < 0$. For certain values of $J_z/J'_z$ the helical phase is energetically favoured. Redner and Stanley [124] reported h.t.s. expansions for $\chi$ and the reduced second and fourth moment in the form of a triple series in $J_{xy}$, $J_z$, $J'_z$. The molecular field theory predicts a helical phase for $J'_z/J_{xy} < -1/4|R|$ with $|\mathbf{q}_0| = \cos^{-1}(-|R|J_{xy}/4J_z)$. The analysis of the h.t.s. locates the Lifshitz point at $J'_z/J_{xy}|R| = -0.281 \pm 0.002$ and enables the determination of $\mathbf{q}_0$ [124]. The complex phase diagram [125] of the ANNNI model has been studied using h.t.s. expansions of the wavevector dependent susceptibility [126] and for 2-d the paramagnetic phase extends to zero temperature. Systematic studies [127] have shown that in 3-d an infinite sequence of spatially modulated commensurate phases with wave vectors $\mathbf{q}_j = \pi j/(2j+1)\mathbf{a}$, ($j = 1, 2, \ldots$ and $a$ = the interlayer distance) occurs for $J_{xy}, J_z > 0$ and $J'_z = -J_z < 0$.

Furthermore, an Ising model with competing first- and third-nearest neighbour interactions (Biaxial Next-Nearest-Neighbour Ising or BNNNI model) has been investigated with h.t.s. and l.t.s. expansions [128].

## 3. Series Expansions and Predictions for the Heisenberg Model

The actual theoretical knowledge on the Heisenberg model in two dimensions is considerably less than for the Ising model. In the Heisenberg case there are hardly any exact results available and the analyses are performed only with h.t.s. expansions and R.G. methods, complemented in the low-temperature region by spin wave theory.

The differences in the development of the Heisenberg and Ising models may be understood by considering two main properties of the Heisenberg Hamiltonian namely:

(i) The individual quantum operator, $S_i^x$, $S_i^y$, $S_i^z$ of any lattice site do not commute.
(ii) The exchange term has spherical symmetry.

As will be seen below, both properties lead to drastic changes in the methods of analysis and in the calculation of the h.t.s.

Perhaps the most important characteristic of the 2-d-Heisenberg magnet is the absence of spontaneous magnetization at any finite temperature. Mermin and Wagner [129] proved rigorously that: "The 1-d and 2-d isotropic Heisenberg model, with interactions of finite range, can be neither ferromagnetic nor antiferromagnetic at non-zero temperature." They showed that for isotropic 2-d systems the spontaneous magnetization, for sufficiently small field $H_T$, is bounded by

$$M(H, T) < \text{const.}/(T|\ln|H_T||)^{1/2}. \tag{59}$$

In the zero-field limit the magnetization will thus be zero, except when $T = 0$. If a small anisotropic term, like $D\,\Sigma_i(S_i^z)^2$ or $H_A\,\Sigma_i\,S_i^z$, is added to the Hamiltonian (1) (as will actually always be present in experimental systems), it may be interpreted as a total field $H_T = H + H_A$ acting upon the systems. According to Equation (59) the existence of an ordered state is then allowed at a finite temperature.

Other important differences between the 2-d Ising and Heisenberg models are found for the loose-packed lattices with antiferromagnetic exchange interactions.

Whereas the staggered and ferromagnetic h.t.s. susceptibility expansions coincide for the Ising model, implying that $|J/k_B T_c| = |J/k_B T_N|$, this it is not so for the Heisenberg model, except in the limit $S \to \infty$. These differences between the critical temperatures for ferro- and antiferromagnetic interactions may be easily understood from elementary arguments of quantum mechanics. The magnetization operator in the ferromagnet, $M^{F,z} = \Sigma_j S_j^z$, commutes with $\Sigma_{\langle i,j \rangle} \mathbf{S}_i \cdot \mathbf{S}_j$, whereas the sum of the sublattice magnetizations in the antiferromagnet, $M^{AF,z} = \Sigma_j^a S_j^z - \Sigma_j^b S_j^z$, (where the superscripts $a$ and $b$ denote the two interpenetrating sublattices) does not. This implies that for $J > 0$ the complete ferromagnetic alignment (which is an eigenstate of the magnetization) is also an eigenstate of the Hamiltonian and corresponds to the ground state. However, for $J < 0$ the Néel state (two complete antiparallel sublattices), which is an eigenstate of $M^{AF}$, can not be an eigenstate of the Hamiltonian, and in contrast to the Ising model it is not the ground state. In fact the problem of obtaining the ground state for the Heisenberg antiferromagnet has proven to be a notoriously difficult one, and is at present still unresolved.

In addition, as a consequence of the spherical symmetry it is not possible to obtain l.t.s. expansions of the excitation type, as for the Ising model. Even for ferromagnetic interaction, in which case the ground state is known exactly, this is not possible because collective excitations in the form of 'spin waves' are more favoured energetically than the local inversion of a single spin. Although descriptions of the excitations in the ordered phase do exist, as there are the spin wave theory [130], or the Green function methods [131], these theories are applicable at low temperatures only ($T < 0.5\ T_c$), and can not provide information about the critical region.

### 3.1. SERIES FOR $S = 1/2$, ARBITRARY $S$ AND $S = \infty$

High-temperature series expansions for the Heisenberg model have been derived using the following three basic procedures; the moment, the cumulant and the finite cluster methods [27].

The moment method (Kramers [12] and Opechowski [13]) was in fact the pioneering work on the h.t.s. for the H(1/2) model, and nowadays is still the method applied to obtain the h.t.s. for the H($S$) model, with minor modifications introduced by Rushbrooke and Wood [26].

The cumulant and the finite cluster methods, initially proposed for the I(1/2) model [132], have also been applied to the Heisenberg case, but due to computational limitations the derivation of h.t.s. expansions has been restricted to H(1/2), for which it has produced the longest series available [133].

The classical limit, H($\infty$), has received special attention because, for this case, the Hamiltonian operators commute and therefore some generalizations of the methods used in the I(1/2) model with commuting operators are applicable [134]. These similarities between H($\infty$) and I(1/2) have allowed the calculation of series much longer than for the H(1/2) case.

Most of the analyses and studies have focused on the extreme quantum limit

H(1/2) and on H(∞), because those series appear to be long enough to derive the critical behaviour. However, an important diference between the H(∞) and the other H($S$) models should be noted here, namely the coincidence of the ferro- and antiferromagnetic reduced critical temperature for H(∞) [27].

Rushbrooke and Wood performed the first systematic h.t.s. derivation for 2-d lattices in the H($S$) model, reporting series for $\chi^F$, $C(T)$ [26] and for $\chi^{st}$ [135], which were later extended by Stephenson *et al.* [136], and Yamaji and Kondo [137] to eight coefficients in $\chi^F$ and seven in $C(T)$.

Rushbrooke and Wood [26] have derived critical temperatures for 2-d lattices. Later analyses of the same series by Gammel *et al.* [138] with P.A. methods, and by Stanley and Kaplan [139] with the ratio method, pointed to the existence of a phase transition for 2-d lattices at a finite temperature with $\gamma \approx 8/3$. This result disagreed with Mermin and Wagner's proof [129] of the absence of long-range order at a finite $T_c$, and, as a consequence, a new type of phase transition to a state with infinite $\chi$ but with zero magnetization was proposed [139].

The analysis of $\chi^{st}$ for H($S$) on the 3-d loose-packed lattices, showed a complete analogy of $\chi^{st}$ and $\chi^F$ at $T_N$ and $T_c$, respectively, notwithstanding that $T_N < T_c$ ($K_N > K_c$). The analyses and arguments about the ferromagnetic ordering in 2-d lattices may therefore be extended straightforwardly to the antiferromagnetic one.

The controversy between the h.t.s. predictions and the rigorous proof stimulated the derivation of longer series, in particular for $S = 1/2$ [133, 140] and $S = \infty$ [141—143], where the difficulties are greatly reduced compared with those for general spin.

Stanley [141] performed the first extension of the h.t.s. for $S = \infty$. Later, Stanley [141] considered the isotropic Hamiltonian with classical spins of dimensions $n = 1$, 2 and 3 ($n$-vector model) in 2-d lattices and found a similar behaviour for $n = 2$ (planar rotor, PR) and $n = 3$. Moore [143] extended the h.t.s. for the susceptibility and the spin—spin correlation function on the p.t. lattice for $n = 1$, 2 and 3 up to order nine, and his analysis supported the existence of a phase transition in 2-d lattices.

On the other hand, for $S = 1/2$, Baker *et al.* [133] extended the h.t.s. of $\chi$ up to order ten for the s.p. and nine for the p.t. lattice and noted that the series were very irregular, containing both positive and negative coefficients, so that the analysis was quite inconclusive. Starting from the anisotropic Ising—Heisenberg Hamiltonian (1), Obokata *et al.* [140] and Dalton and Wood [144] have derived seven terms of the h.t.s. for $\chi$, $S = 1/2$ for the s.p. lattice. The P.A. methods located the position of $T_c$, except in the proximity of the isotropic case, where the convergence was worse for every lattice.

The experimental evidence that systems which approximate very closely the ideal 2-d antiferromagnets do exist in nature [145], stimulated Lines's studies [146] on the paramagnetic susceptibility curves, similar to those previously performed for 3-d lattices by Rushbrooke and Wood [26] and Danielian and Stevens [147]. Lines derived extrapolations from the series of different lengths for $1/\chi$, and deduced the positions and heights of the characteristic $\chi^{AF}$ maxima. This allowed estimates of the exchange constants that are considerably more accurate than those deduced at much higher temperatures by the method of Danielian and Stevens [147]. For the 2-d antiferromagnets the $\chi$ maximum is located at temperatures $T_{max}$ much higher than $T_N$, and it may be determined without further assumptions about the existence or nonexistence of a $T_N$. In the experimental systems the instability of the 2-d ordering is removed by the presence of weak anisotropy and/or interlayer coupling. The value of $T_N$ is strongly dependent on such parameters, whereas $T_{max}$ and $\chi_{max}$ are not so sensitive. The heights and positions of these maxima for loosely packed 2-d lattices have been listed in Table IX [148]. For comparative purposes Table IX also includes the values of $T_c$, $T_N$, $T_{max}$ and $\chi_{max}$ for the s.c. and b.c.c. lattices at the same $S$ values, which illustrates the main features of the Heisenberg model predictions as well as the differences between 2-d and 3-d systems. In Figure 12 the antiferromagnetic susceptibilities for the h.c. lattice for some selected $S$ values have been plotted.

TABLE IX

Position, $\tau_m = k_B T_{max}/|J|S(S+1)$, and height, $\chi'_m = \chi_{max}|J|/Ng^2\mu_B^2$ of the antiferromagnetic $\chi$ maximum, for different $S$ values for the 2-d and 3-d Heisenberg models and loosely packed lattices, taken from Navarro [148]. The bold digit indicates an error ± 1 in the last digit.

| | Spin | 1/2 | 1 | 3/2 | 2 | 5/2 | ∞ |
|---|---|---|---|---|---|---|---|
| s.p. | $\tau_m$ | 2.495 ± 5 | 2.18 ± 2 | 2.10 ± 2 | 2.07 ± 3 | 2.06 ± 3 | 2.00 ± 2 |
| | $\chi'_m$ | 0.046**9** | 0.052**1** | 0.053**9** | 0.054**7** | 0.055**1** | 0.056**1** |
| h.c. | $\tau_m$ | 1.87 ± 5 | 1.60 ± 5 | 1.53 ± 3 | 1.50 ± 3 | 1.45 ± 5 | 1.42 ± 2 |
| | $\chi'_m$ | 0.058**4** | 0.066**5** | 0.069**5** | 0.070**8** | 0.071**7** | 0.073**3** |
| | $\tau_m$ | 3.22 ± 1 | 3.15 ± 6 | 3.11 ± 5 | 3.08 ± 6 | 3.07 ± 8 | 3.03 ± 1 |
| s.c. | $\chi'_m$ | 0.0352**4** | 0.037**8** | 0.038**7** | 0.039**2** | 0.039**4** | 0.0398**4** |
| | $\tau_c$ | 2.497 ± 5 | 2.740 ± 2 | 2.798 ± 3 | 2.825 ± 5 | 2.843 ± 6 | 2.882 ± 8 |
| | $T_N/T_c$ | 1.1**2** | 1.05**3** | 1.03**0** | 1.01**9** | 1.01**3** | 1.000 |
| b.c.c. | $\tau_m$ | 4.30 ± | 4.22 ± 4 | 4.20 ± 5 | 4.19 ± 8 | 4.18 ± 9 | 4.20 ± 1 |
| | $\chi'_m$ | 0.027**5** | 0.029**3** | 0.029**7** | 0.030**0** | 0.030**1** | 0.030**2** |
| | $\tau_c$ | 3.663 ± 10 | 3.91 ± 3 | 3.992 ± 6 | 4.026 ± 10 | 4.044 ± 10 | 4.105 ± 2 |
| | $T_N/T_c$ | 1.09**0** | 1.03**9** | 1.02**1** | 1.0133 | 1.009**2** | 1.000 |

The analogies found in the h.t.s. for the 2-d, H(∞) and PR models [142, 143] were extended also to the 2-d spherical model [149] (i.e. classical spins with infinite dimensions $n = \infty$), to the 2-d classical XY model [150], XY(∞), (i.e. classical spins with $n = 3$, but without a $z$-term in the Hamiltonian (1)), and to the 2-d, $S = 1/2$ XY model, XY(1/2), [151].

On the other hand, the nature of the magnetic ordering in 2-d isotropic magnets has been the subject of several other theoretical studies, namely by Lines [146], Mubayi and Lange [152] and Oguchi [153], with the aim of revealing the nature of the phase transition from a microscopic basis. The first 'picture' of a phase transition with a divergent $\chi$ and without magnetization, for some years called the Stanley—Kaplan type transition (S.K.), is due to Berezinskii [154] and to Kosterlitz

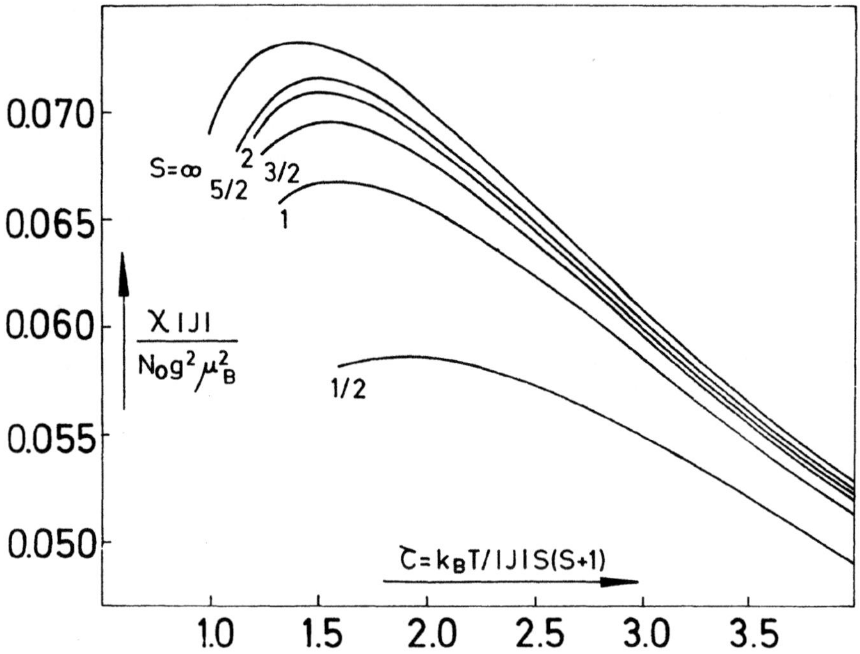


Fig. 12. Reduced antiferromagnetic susceptibility versus temperature for the isotropic Heisenberg model on the honeycomb lattice. The spin values are indicated on the curves, taken from Navarro [148].

and Thouless [44], who proposed a new type of order, namely a topological order, for the PR model, in which the spins form 'vortex' and 'antivortex' structures in the easy plane wherein they are free to rotate. It should be noted that with the definition used here, the PR model corresponds with the so called XY model in [44]. This microscopic picture fulfills the condition for a S.K. phase transition, and further gives an exponential critical behaviour for $\chi$ (Equation (21)). However, it should be noted that the explanation of the S.K. phase transition on basis of a vortex and antivortex pair structure can not be applied to the H($\infty$) model because the singularity at the core of the vortex may be avoided by allowing the core spins to point outside the plane.

Camp and Van Dyke [45] reanalysed the h.t.s. of $\chi$ for classical spins in the s.p. and p.t. lattices and concluded that there are nonzero critical temperatures for the PR and XY($\infty$) models and, with less confidence, also in the H($\infty$) case. Moreover, the usual P.A. and ratio methods indicated that $\chi$ displays a stronger divergence than given by the power law (3).

Navarro and De Jongh [155] performed a similar analysis on 2-d $H(1/2)$ model and concluded that the power law form of $\chi$ may be put in doubt. In addition, suitable fits on the experimental $\chi$ data were found to agree quite well with an exponential form. Further analyses of the 2-d isotropic models by Guttmann [156] and by Ferer and Velgakis [157] yielded the same conclusions about the form of the singularities. Moreover, for the PR and XY($\infty$) models they found strong indications for a phase transition at finite temperatures, whereas in the H($\infty$) case the evidence was weak. Although the exponential form (21) was only clearly supported in the study of the PR case, values of $\nu$(PR) = 0.4 and $\nu$(XY) = 0.5 were indicated.

Praveczki [158] has recently rewritten the h.t.s. in the H($\infty$) and H(1/2) models for the s.p. lattice, expanding $K$ in powers of $(1 - 1/\chi)$, and for the limit of $1/\chi \to 0$ finite $K_c$ values were deduced. From his plots the values $K_c(\infty) = 0.9$ and $K_c(1/2) = 0.7 \pm 0.1$ have been estimated.

At present the analyses of the available h.t.s. for the 2-d Heisenberg model are still not conclusive, and the presence of a phase transition of the S.K. type at a finite temperature can not be excluded. However, the results of the h.t.s. analyses and of other theoretical studies [159—161] point towards the existence of finite $T_c$ for the isotropic Hamiltonian with $n = 2$, and the absence of $T_c$ for $n = 3$. Thus, for the 2-d Heisenberg model the specific heat will in all probability show a rounded maximum, represented in Figure 13 for ferro- and antiferromagnetic interactions [162, 163].

In Tables B-I, B-II and B-III (Appendix B) the susceptibility h.t.s. coefficients are collected for the H(1/2), H($S$) and H($\infty$) models, respectively. The following definition has been used

$$\chi = Ng^2\mu_B^2 S(S+1)/k_B T \left(1 + \sum_{n=1}^{\infty} a_n K^n\right). \tag{60}$$

In addition, in Table B-IV, B-V and B-VI the corresponding specific heat h.t.s. are listed for the following expansion

$$C(T) = 2/3 N k_B z S^2 (S+1)^2 K^2 \left(1 + \sum_{n=1}^{\infty} c_n K^n\right). \tag{61}$$

### 3.1.1. *Properties at Nonzero Field*

The predictions for the 2-d Heisenberg model at nonzero field are even more scarce, and there are h.t.s. expansions of the partition function only for the $S$ = 1/2 and $S = \infty$ ferromagnetic Heisenberg models in the form of a double series in $K$ and $y = mH/k_B T$, of the form:

$$\ln Z/N = \ln(2\cosh y) + \sum_j \sum_n K^n f(n,j)\,(\tanh y)^{2j}/2^n n! \tag{62}$$

where the coefficients $f(n, j) = 0$ for $n < j$.

For $S = 1/2$, Baker *et al.* [133] have given the first coefficients for $n \leqslant 8$ in the p.t. and s.p. lattices, which are explicitly listed in [27]. The analogous 3-d series analysis was directed to the determinations of the gap exponent, $2\Delta$, or to deduce

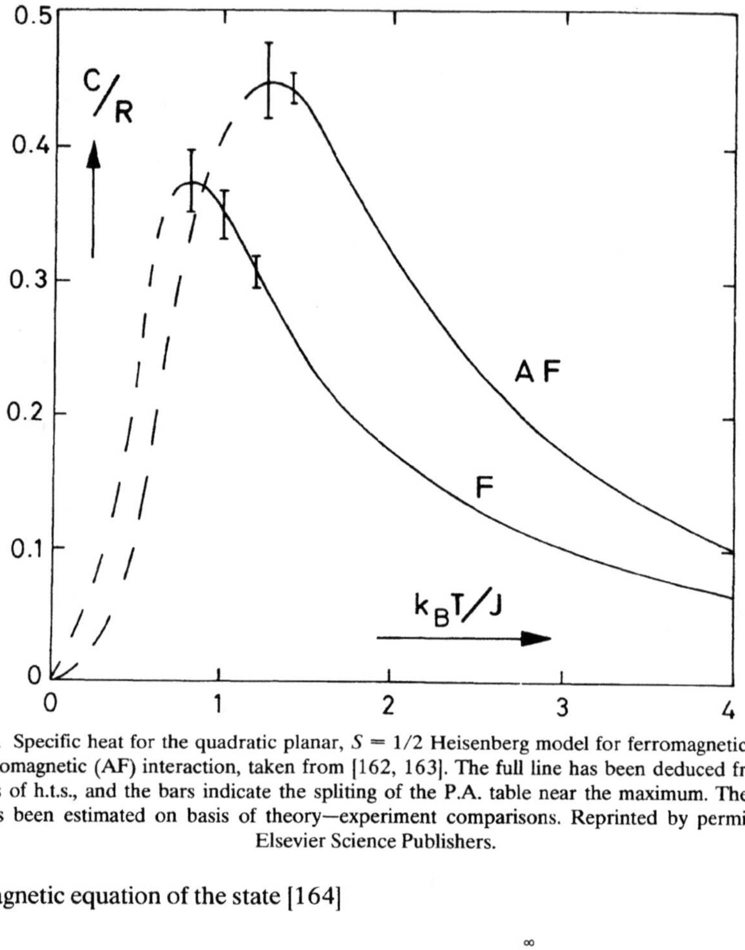


Fig. 13. Specific heat for the quadratic planar, $S = 1/2$ Heisenberg model for ferromagnetic (F) and antiferromagnetic (AF) interaction, taken from [162, 163]. The full line has been deduced from P.A. analysis of h.t.s., and the bars indicate the spliting of the P.A. table near the maximum. The dashed line has been estimated on basis of theory—experiment comparisons. Reprinted by permission of Elsevier Science Publishers.

the magnetic equation of the state [164]

$$\tanh y = M\,g(K, M^2), \qquad g(K, M^2) = 1 + \sum_{n=1}^{\infty} K^n/2^n n!\; p_n(M^2)$$

where $p_n(M^2)$ are polynomials in $M^2$, which allow the estimation of the exponents $\beta$ and $\delta$.

According to Mermin and Wagner's work on 2-d lattices, one should have $M \to 0$ for $H \to 0$ and $T \neq 0$ (Equation (59)). However, Baker *et al.* [164], following Stanley—Kaplan's [139] suggestion, analysed the equation of state for the p.t. lattice, and found that $M = 1$ at $T = 0$. At $M \neq 0$, they tried to find the possibility of a pseudocritical point, but the P.A. method gave no evidence of its existence. Moreover, a qualitative difference with the 3-d h.t.s. behaviour was also observed.

Stephenson and Wood [165] derived the series analogous to Equation (62) for $S = \infty$, up to seventh order in $K$ and the tenth in $y$, for the p.t. and s.p. lattices.

They also followed the analysis performed by Stanley [139, 141] for 2-d systems, using both the P.A. and the ratio methods, and concluded that more terms are needed to obtain reliable information about the behaviour of 2-d lattices.

## 3.2. OTHER SERIES FOR THE HEISENBERG MODEL

### 3.2.1. *Restricted Dimensionality*

The restriction of the size in one spatial direction for a 3-d system, as in thin films, should affect the position and nature of the critical singularity. For the Ising model (Section 2.3) the essential 2-d character of the critical behaviour of these restricted 2-d systems has been discussed. Nevertheless, it must be stressed that for the Ising model there is a long-range order phase transition for both 2-d and 3-d lattices, whereas for the Heisenberg model the Mermin and Wagner argument [129] excludes such type of transition in the 2-d case. Indeed, Fisher and Jasnow [166] rigorously proved the absence of spontaneous magnetization in Heisenberg-like films. This difference between the Ising and Heisenberg model points towards more drastic effects in the case of restricted 2-d Heisenberg films.

The properties of Heisenberg ferromagnetic films (see Figure 9) have been studied by Ritchie and Fisher [167], who derived h.t.s. for $\chi(n)$ and $C(n, T)$ for free ($n = 1$ to 7 layers) and periodic ($n = 2$ to 5 layers) boundary conditions, both for varying $S$. Moreover, h.t.s. of the surface contribution to the susceptibility, $\chi^x$, defined as in Equation (52) were obtained. Ritchie and Fisher's analysis of the (3-d → 2-d) crossover in thin films was performed in the critical region, assuming the existence of a finite $T_c$ and, in the light of present knowledge, was not conclusive.

However, the analysis performed by Navarro *et al.* [168] on the h.t.s. for $\chi(n)$ with antiferromagnetic interaction using direct P.A. tables (see Figure 14), proved clearly that a crossover from 3-d behaviour at high temperatures to 2-d at low temperatures was present. The broad $\chi$ maxima characteristic of the short-range interactions for 2-d systems were clearly observed. These 2-d properties for thin films up to 7 layers agree with the spin wave results [169] for ferro- and antiferromagnetic interactions, in which differences with the bulk s.c. behaviour are detected even for $n = 20$. In Table X, the position and height of the $\chi$ maxima have been listed for different values of $S$.

The $\chi^x$ series has essentially a bulk 3-d character, and Ritchie and Fisher [167], using in Equation (16) the s.c. critical temperature, obtain $\gamma^x = 2.16 \pm 0.02$ for all $S$. This value, for free boundary conditions, ranges between $\gamma_3 + \nu_3 = 2.08 \pm 0.03 < \gamma^x < \gamma_3 + 1 = 2.38 \pm 0.03$, where the subindex 3 refers to the 3-d lattices, $\gamma^x = \gamma_3 + \nu_3$ is the relation that holds for the Ising model [107] (see Equations (53)) and $\gamma^x = \gamma_3 + 1$ are the corresponding relation for the spherical model and the Bose gas films [170]. Clearly, the results of $\gamma^x$ are close to the lower bound.

Semiinfinite 3-d lattices have also been analysed, using h.t.s. for the H($\infty$) model. For general $n$ values, Ohno and Okabe [171] have derived nine coefficients of the surface-layer susceptibility, $\chi_1$, in a semi-infinite s.c. lattice. The aim was to

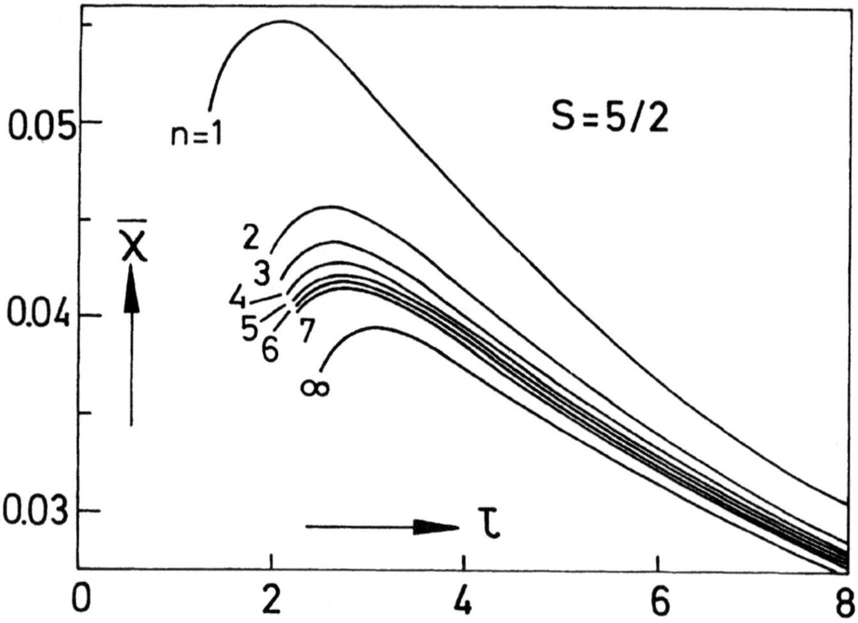


Fig. 14. High temperature series predictions for the antiferromagnetic susceptibility $\bar{\chi} = \chi(n)|J|/Ng^2\mu_B^2$, $\tau = k_BT/|J|S(S+1)$ of the $S = 5/2$ Heisenberg model for $n$-square layers and free boundary conditions, taken from Navarro *et al.* [168]. Reprinted by permission of Elsevier Science Publishers.

TABLE X

Values of the susceptibility maximum $\chi'_m = \chi_{max}|J|/Ng^2\mu_B^2$ and $\tau_m = k_BT_{max}/|J|S(S+1)$ for antiferromagnetic double layer ($n = 2$) Heisenberg systems with varying spin value. The estimated uncertainties are indicated. Data taken from Navarro *et al.* [168]. Reproduced by permission of Elsevier Science Publishers.

| $S$ | 1/2 | 1 | 3/2 | 2 | 5/2 | ∞ |
|---|---|---|---|---|---|---|
| $\chi'_m$ | 0.0396 ± 2 | 0.0433 ± 2 | 0.0445 ± 2 | 0.0450 ± 2 | 0.0453 ± 2 | 0.0460 ± 2 |
| $\tau_m$ | 3.05 ± 5 | 2.81 ± 2 | 2.74 ± 6 | 2.71 ± 6 | 2.70 ± 5 | 2.64 ± 5 |

test the scaling Equations (53), and their results for $n = 1$, $\gamma_1 = 0.79 + 0.05$, reproduced the surface-layer exponent for the Ising model [110]. They also derived the corresponding value for $n = 2$ (PR) $\gamma_1 = 0.83 + 0.05$, and for $n = 3$ (H(∞)) $\gamma_1 = 0.91 + 0.05$. Furthermore, using these results they verified the additional R.G. scaling law obtained by Bray and Moore [172], which could be straighforwardly derived from Equations (53) (i) and (iii), namely $\gamma_1 = \nu_3 + (\gamma_3 - 1)/2$.

### 3.2.2. *Further-Neighbour Interactions*

The first derivation of H($S$) h.t.s. expansions with different interactions between nearest, $J_1$, and next-nearest neighbours, $J_2$, was performed [173, 174] with the aim of increasing the utility of the $\chi$ and $C(T)$ expansions derived by Rushbrooke and Wood [26] for the comparison of theory with experiment. Wojtowicz and Joseph [173] derived four coefficients for $\chi$, and Wojtowicz [174] three coefficients for $C(T)$ for any lattice, spin value and ratio $J_1/J_2$.

Later, Domb and Dalton [112] studied the effect of higher-order neighbours using the equivalent neighbour model, already discussed in Section 2.3. Following Wojtowicz's work, Dalton [175] added a new term to the h.t.s. expansion of $\chi$ and $C(T)$ for any spin value and lattice and up to our knowledge there are no longer h.t.s. for 2-d lattices.

Indeed the shortness of the series and the poor convergence of the coefficients already observed for the nearest neighbour 2-d Heisenberg models, renders any conclusive analysis about $T_c$ impossible. Furthermore, the length of the h.t.s. is not sufficient to perform detailed comparisons with the available experimental data [155].

### 3.2.3. *Crossover from 2-d to 3-d behaviour*

The interest in materials with lattice-dimensionality crossover has incited the extension of the Ising crossover studies [116—124] to other models, which has been achieved in the framework of the $n$-vector model. For $n = 1$, the situation has been described in Section 2.3 and for $n = \infty$, there are exact results [29]. The results for $n = 2$ and $n = 3$ may be tested by comparison with these limits. Following this line, Lambeth and Stanley [176] derived h.t.s. expansion for $\chi$, $C(T)$ and the two-spin correlation function in the $n = 2$ and $n = 3$ model with lattice anisotropy namely s.p. $\leftrightarrow$ s.c. and s.p. $\leftrightarrow$ f.c.c. lattices.

A double h.t.s. expansion was derived as a function of the intralayer $J_{xy}$, and interlayer $J_z$, interactions. Therefore, it may be used with $|J_{xy}| \gg |J_z|$ to analyse the 2-d $\rightarrow$ 3-d crossover, but also with $|J_z| \gg |J_{xy}|$ to consider the 1-d $\rightarrow$ 3-d crossover or the b.c.c. $\leftrightarrow$ f.c.c. one. Lambeth and Stanley paid attention to $n \leqslant 3$ cases, which corresponds with the range of experimental interest, though the main contributions of their analysis are the theoretical aspects of scaling and crossover, when a system with lattice anisotropy changes its universality class.

Lambeth and Stanley [176] performed a parallel analysis for the $n = 1, 2, 3$ and in some cases $n = \infty$ models by means of the ratio and P.A. methods. In the direct ratio sequence, they found similarities, with minor differences, in the sets of $\bar{\chi}_0^{(n)}$ series ratios of the $n = 1$, 2 and 3 cases, consistent with the scaling prediction Equations (55) and (56). However, the direct ratio alone is not sufficiently accurate to predict values for the critical temperatures. The Dlog P.A. analysis gives both $T_c$ and $\gamma^{(n)}$, but the increase of $\gamma^{(n)}$ with $n$ is not regular enough to derive the crossover exponent $\phi$. Moreover, the P.A. tables corresponding to $n =$ 2 and $n = 3$ differ clearly from the $n = \infty$ case, which has $T_c = 0$. This seems to be an indication of the different behaviour of the series. Appropriate sequence or combination of ratios, tested in the $n = 1$ case, may improve the convergence of the h.t.s. [176] and allow to estimate the exponent $\phi = \gamma_0 = 2.53 \pm 0.30$ for the PR model and, with less confidence, $\phi = 3.5$ for the Heisenberg one. These last conclusions imply a finite $T_c$ in both models with a power-law behaviour for the $\chi$ divergence, which disagrees with the results stated for the 2-d isotropic lattices.

Redner and Stanley [124] have considered the extension of the ANNNI model for $n = 2$ and $n = 3$, deriving h.t.s. for these cases. The results are similar to the ones derived for $n = 1$ [124, 127] with the Lifshitz point located at $J_z'/J_{xy}|R| = -0.263 \pm 0.002$ and $-0.259 \pm 0.002$ for $n = 2$ and $n = 3$, respectively. However, some differences were noted and no evidence for the predicted first-order phase transition from the paramagnetic to helical phases could be detected for $n = 3$ [126].

The existence of many well-known isotropic magnetic systems with spin 5/2 [145], offers the possibility of suitable comparisons with the crossover predictions of the classical Heisenberg model. The results of these fits, however, have physical meaning only in the paramagnetic phase [177] where the h.t.s. predictions hold. Puertolas *et al.* [177], have obtained predictions for $\chi$ and $C(T)$ over a wide temperature range using the Lambeth and Stanley h.t.s. expansions and direct P.A. In Table XI the results for the susceptibility maximum in the s.q. ↔ s.c. crossover are listed for various $R$ values. Moreover, using the 3-d $S = \infty$ critical exponent $\gamma = 1.405$, the P.A. table of Equation (18) gives biased $T_c$ values as a function of $R$, which are also listed in Table XI.

TABLE XI

Estimations of $K_c = 2J_{xy}/3k_BT_c$, $K_m = 2J_{xy}/3k_BT(\chi_{max})$ and $\chi_{max} = 2\chi_{max}|J_{xy}|/Nm^2$ for some $R = J_z/J_{xy}$ values for the $S = \infty$ Heisenberg model with s.p. → s.c. lattice-dimensionality crossover. Estimated uncertainties are in the last printed digit. Data taken from Puertolas *et al.* [177].

| | | $J_{xy} > 0$ | $J_z < 0$ | $J_{xy} < 0$ | $J_z > 0$ | $J_{xy} < 0$ | $J_z < 0$ |
|---|---|---|---|---|---|---|---|
| $\lvert R\rvert$ | $K_c$ | $K_m$ | $\chi_{max}$ | $K_m$ | $\chi_{max}$ | $K_m$ | $\chi_{max}$ |
| 1 | 0.2313 | 0.235 | 0.237 | 0.227 | 0.119 | 0.217 | 0.0796 |
| 0.5 | 0.284 | 0.289 | 0.476 | 0.27 | 0.1175 | 0.267 | 0.0947 |
| 0.1 | 0.392 | 0.44 | 3.13 | 0.326 | 0.1143 | 0.316 | 0.1093 |
| 0.05 | 0.431 | 0.48 | 5.70 | 0.331 | 0.1133 | 0.330 | 0.1109 |
| 0.0 | — | — | $\infty$ | 0.333 | 0.1123 | 0.333 | 0.1123 |

## 4. Series Expansion in the XY and Ising–Heisenberg Models

The analysis of the general Ising–Heisenberg Hamiltonian, Equation (1) may be simplified if the order parameter properties are considered. Two associated independent magnetization operators (or sublattice magnetizations) may be considered

$$M_\parallel = m_\parallel \sum_i S_i^z, \qquad M_\perp = m_\perp \sum_i S_i^x.$$

For $|a| \gg |b|$, the Ising limit is attained and, at sufficiently low temperature, the magnetic moments should be completely aligned in the $z$-direction. Then $M_{\parallel}$ is the order parameter, which commutes with the Hamiltonian. For $|b| \gg |a|$, the XY limit holds, and the moments should lie in the XY plane, $M_{\perp}$ being the order parameter. It is simple to see that $M_{\perp}$ does not commute with the XY Hamiltonian neither for ferro- nor for antiferromagnetic interaction. According to the arguments advanced for Heisenberg antiferromagnetis this implies that in both cases the ground state is unknown. For $b = a$, the Heisenberg limit is reached, which means that there is no preferred direction for alignment, and $M_{\parallel}$ is taken as the order parameter. This quantity commutes with the Hamiltonian for ferromagnetic interaction, but not for the antiferromagnetic one ($M^{\mathrm{AF}}$). All these characteristics point to the difficulty of obtaining the l.t.s. expansion in general cases and up till now no progress on this topic has been reported. However, the spin wave formalism has been applied [178] to derive the low-temperature thermodynamic properties of systems with ferro- and antiferromagnetic interactions.

Most of the theoretical studies on the Ising—Heisenberg model have focussed on the analysis of critical properties, and therefore the h.t.s. expansion calculations have concentrated on such thermodynamic functions as the partition function and its first derivatives, the specific heat and the parallel susceptibility $\chi_{\parallel} = \partial M_{\parallel}/\partial H_{\parallel}$. For this model, however, also $\chi_{\perp} = \partial M_{\perp}/\partial H_{\perp}$ should be considered. For ferromagnetic interaction and positive $a$ and $b$, two situations are possible. First, if $a > b$, the zero field $\chi_{\parallel}$ diverges at the critical temperature whereas $\chi_{\perp}$ remains finite, as in the pure Ising model. On the other hand, if $a < b$, $\chi_{\parallel}$ and $\chi_{\perp}$ interchange role, and this is the situation for the XY model. Up to now the possibility of a singular behaviour of $\chi_{\perp}$ has not been considered and, before proceding further with the properties of the complete Hamiltonian, it is worthwhile considering the properties in the XY limit.

## 4.1. SERIES FOR THE 2-d XY MODEL

The XY model avoids some of the difficulties mentioned for the general Ising—Heisenberg model, and has some special properties that should be discussed before we continue further. One of the simplifications is the presence of a parallelism for the XY ferro- and antiferromagnets, because the Hamiltonians for both interactions are isomorphic [179]. This implies that on the loosely-packed lattices the static and dynamic properties for both type of interactions must be exactly the same, after the appropriate change of variables. In particular, $\chi^{\mathrm{F}}$ and $\chi^{\mathrm{st}}$ coincide and the critical properties are equal, i.e. $k_{\mathrm{B}}T_{\mathrm{N}}/|J| = k_{\mathrm{B}}T_{\mathrm{c}}/|J|$, similar to the Ising model.

However, the derivation of the h.t.s. expansion for $\chi_{\perp}$ is somewhat more complicated than for $\chi_{\parallel}$. Since $M_{\parallel}$ and $\mathscr{H}$ commute, the parallel susceptibility and the fluctuation of the parallel magnetization, $Y$, are related by [28]

$$k_{\mathrm{B}}T\chi_{\parallel} = \langle M_{\parallel}^2 \rangle - \langle M_{\parallel} \rangle^2 = Y_{\parallel} \tag{63}$$

and the h.t.s. expansion may be derived by similar procedures as above. The commutation properties of $M_\perp$ imply differences between $\chi_\perp$ and the fluctuation of the perpendicular magnetization, $Y_\perp$, for which long h.t.s. are easier to achieve. However, h.t.s. analyses of $Y_\perp$ are of interest because it has the same critical behaviour as $\chi_\perp$. This may be easily proved, since for the XY model there is an upper bound for $\chi_\perp$ [180], namely

$$k_B T\chi_\perp \leqslant \langle M_\perp^2\rangle - \langle M_\perp\rangle^2 = Y_\perp \tag{64}$$

and also a lower bound

$$k_B T\chi_\perp > \langle M_\perp^2\rangle \cdot A \tag{65}$$

where $A$ is a function of the temperature that may be found elsewhere [28, 180]. In the critical region $A$ tends to unity, and moreover for $T > T_c$, $\langle M_\perp\rangle = 0$ at zero field. Therefore, the upper and the lower bounds become identical and $k_B T\chi_\perp$ and $Y_\perp$ coincide in the critical region above $T_c$.

The derivation of h.t.s. expansions for the zero-field partition function is performed by the usual perturbation methods. The series of increasing powers of $1/k_B T$, and $\mathscr{H}$ is reexpressed in terms of creation and annihilation operators; thereafter a correspondence between graphs and terms is performed [181] and later the graphs are reduced and calculated. The same methods apply to the h.t.s. expansions of $\chi_\perp$ and $Y_\perp$ with small modifications. However, in view of the above stated relations among $\chi_\perp$ and $Y_\perp$ in the critical region, and the greater difficulty to derive the $\chi_\perp$ series, most of the expansions have been performed for $Y_\perp$.

As mentioned in Section 3, the initial interest in obtaining series for the XY model was the possible existence of an S.K. phase transition in isotropic models, as studied by Jasnow and Wortis [150] ($S = \infty$) and by Betts *et al.* [151] ($S = 1/2$). Also, other isotropic and related models, such as the planar rotator (PR) [141], the step model [142] and the $O(2)$ symmetric spin model with discretely valued interactions [182], have been considered in the same framework. All these models likewise have a 2-d order parameter, so that in the light of the universality hypothesis they may be experted to have the same critical exponents as the XY model.

For the XY(1/2) model, Betts *et al.* [151] have reported the first nine terms for $Y_\perp$, and found the critical temperatures to be $k_B T_c/J = 1.500 \pm 0.006$ and $0.900 \pm 0.006$ for the p.t. and s.p. lattices, both with a susceptibility exponent $\gamma \simeq 3/2$. Later, Betts *et al.* [181] derived the h.t.s. of the partition function up to $K^{10}$ in the p.t., s.p. and h.c. lattices, and deduced a critical exponent $\alpha \simeq -1$, which means that $C(T)$ is not singular at $T_c$. The existing series for $\chi_\perp$ are rather short, and, for arbitrary lattice, only the first four coefficients as a function of graphs [28] are known. Moreover, in a comparison with experimental data six terms of $\chi_\perp$ have been reported for the s.p. lattice [183]. In view of Equations (64) and (65), the higher-order coefficients of $Y_\perp$ were used instead of the unknown coefficients for $\chi_\perp$ [183].

Rogiers *et al.* [184] have derived the longest h.t.s. for the $S = 1/2$ XY model on the 2-d and 3-d lattices. In particular for the s.p. and p.t. lattices, the previous series of $Y_\perp$ and ln $\Lambda$ were extended up to the twelfth power and in addition series of the fourth-order fluctuations of the transverse magnetization ($Q$) were reported. The analysis of these h.t.s. [184] pointed to finite critical temperatures $k_B T_c/J$ = 1.315 ± 0.014 and 0.79 ± 0.01 for the p.t. and s.p. lattices, with a conventional power-law critical behaviour for $\chi_\perp$ and indices $\gamma = 2.50 \pm 0.3$, $\Delta = 2.38 \pm 0.2$; $\nu = 1.43 \pm 0.1$ and $\alpha = -1$. The coefficients of these h.t.s. have been collected in Tables C-I and C-III (Appendix C).

For $S = \infty$, Camp and Van Dyke [45] have derived the h.t.s. for $\chi_\perp$, on the s.p. and p.t. lattices up to the tenth power for the XY($\infty$) and PR models. Their analyses, as well as Guttmann's [156] work, support the existence of a finite critical temperature, but with an exponential behaviour of $\chi_\perp$. Ferer and Velgakis's [157] analysis, with longer series for the p.t. lattice, also supports the exponential form with $\nu = 0.5 \pm 0.1$ and $k_B T_c/J = 3.32 \pm 0.03$ in the XY($\infty$) model. Moreover, longer h.t.s. series for the planar rotor on the s.p. lattice for $\chi$ and the two-point correlation function [185] likewise agree with the Kosterlitz—Thouless theory. Also the l.t.s. for the energy density in the PR model s.p. lattice has been recently reported [186] to reproduce the Monte Carlo simulations. In Table C-II (Appendix C) the coefficients of $\chi_\perp$ for both the XY($\infty$) and the PR models have been collected.

Series for the nondivergent $\chi_\parallel$, in the $S = 1/2$ XY model may be deduced from the h.t.s. of the general IH model. In this way Obokata *et al.* [140] derived seven coefficients for the s.p. lattice, Dalton and Wood five [144] and six [101] coefficients for arbitrary spin and lattice, and Jou and Chen [187] seven for arbitrary lattice with $S = 1/2$. Moreover, Lee [188] has performed direct calculations up to seventh order. The h.t.s. of $\chi_\parallel$ are related to the zero-field energy, which in the critical region has a dependence $(T - T_c)^{1-\alpha}$, where $\alpha = -1$ is the specific heat exponent [188]. Furthermore, $\chi_\parallel$ has only even powers of $K$, and is the same for ferro- and antiferromagnetic interactions. These coefficients have been included in Table C-I (Appendix C).

In contrast to the $n = 3$ case, the h.t.s. analysis for 2-d, $n = 2$ models are now generally believed to indicate a phase transition at a finite temperature of the S.K. type, with infinite $\chi$, non-divergent $C(T)$ and zero magnetization.

## 4.2. SERIES FOR THE 2-d ISING—HEISENBERG MODEL

Besides the study of the pure models the analysis of the general Ising—Heisenberg is of interest since, from the experimental point of view, the presence of some degree of anisotropy in the interaction is to be expected in nearly all cases. Therefore, it is important to know the changes induced by this anisotropy on the thermodynamic functions and on the critical behaviour. Theoretically, one deals with a Hamiltonian that depends on a single parameter, and the universality hypothesis should be applicable.

Most of the h.t.s. derivations for the IH model have been performed for the thermodynamic functions that are easiest to expand, namely ln $Z$ and $\chi_\parallel$ (see Section 4.1). Consequently, the studies on the 2-d critical behaviour have been mainly restricted to the $(a, b)$ region in which $\chi_\parallel$ is divergent, i.e. for $a > b$. The intermediate model going from the Ising limit $b/a = 0$ to the Heisenberg limit $b/a = 1$ has been considered in some detail elsewhere [101, 140, 143, 187]. Jou and Chen [187] have analysed the susceptibility series using different P.A. techniques, and derived the critical temperatures collected in Table XII. For any value of $b/a \neq 1$, a finite $T_c$ appears which is almost constant between $0 \leqslant b/a \leqslant 0.5$. For higher $b/a$ values, $T_c$ decreases rapidly, and at the same time the error bounds increase, in such a way that for $b/a > 0.8$ the methods of series analysis no longer yield reliable $T_c$ values. Presumably the phase transition will occur at a finite $T_c$ for all values of $b/a$ except for $b/a = 1$, and from this point of view the Heisenberg limit is singular. Indeed, the critical exponents found correspond to the 2-d Ising value $\gamma = 7/4$. The singular character of the Heisenberg limit is also met for the 3-d lattices, where for $S = 1/2$ [187, 189] and $S = \infty$ [150], similar h.t.s. analyses have been performed.

TABLE XII
Critical temperatures derived for the $S = 1/2$ Ising-Heisenberg model by Jou and Chen [187] in some 2-d lattices. The error in the last digits are indicated. Reprinted by permission of IOP Publishing Ltd.

| | Ising—Heisenberg spin crossover ($n = 1 \leftrightarrow n = 3$) | | | | | | |
|---|---|---|---|---|---|---|---|
| $b/a$ | 0 | 0.2 | 0.4 | 0.6 | 0.7 | 0.8 | 1.0 |
| $k_B T_c/J$(p.t.) | 1.82 ± 1 | 1.81 ± 1 | 1.75 ± 2 | 1.62 ± 4 | 1.51 ± 6 | 1.35 ± 15 | — |
| $k_B T_c/J$(s.p.) | 1.13 ± 1 | 1.12 ± 1 | 1.09 ± 1 | 0.99 ± 3 | 0.91 ± 3 | 0.75 ± 15 | — |

For $b > a$, the intermediate model interpolates between the limits $a/b = 0$ (XY) to $a/b = 1$ (Heisenberg). The order parameter now is $M_\perp$, and $\chi_\perp$ shows a singular behaviour at $T_c$. However, as stated in Section 4.1, it is easier to derive the series for $Y_\perp$. For $S = 1/2$ on the 3-d lattices, Jou and Chen [189] derived seven coefficients for the s.c. and b.c.c., and six for the f.c.c. lattices. Their analyses lead to XY critical exponents, except at $a/b = 1$. Similar analyses for $S = \infty$ on the 3-d lattices by Jasnow and Wortis [150] support this conclusion. Although there are no h.t.s. studies on 2-d lattices, a similar behaviour is expected.

A model that is intermediate between Ising and XY has also been considered by Ishikawa and Oguchi [190] for the s.p. lattice and for arbitrary spin. The Hamiltonian takes the form

$$\mathscr{H} = -2J \sum_{\langle i,j \rangle} (\xi S_i^x S_j^x + S_i^z S_j^z) - g\mu_B H \sum_i S_i^z. \tag{66}$$

For $\xi = 0$ and $\xi = 1$ the Ising and XY limits are reached. The authors derived five coefficients for $\chi$, and their analysis by the ratio method gave finite critical temperatures and $\gamma$ values varying with $S$. However, the behaviour is similar to that of the Ising—Heisenberg model, so that values of $\gamma = 7/4$ are to be expected. The fact that these are not indicated is probably due to the brevity of the series.

From the abovementioned analyses the following conclusions may be drawn. The presence of uniaxial anisotropy in the interaction ($a > b$) should produce Ising-like critical behaviour at temperatures close enough to $T_c$. On the other hand, if there is a planar anisotropy in the interaction ($b > a$) the critical behaviour should be XY-like. These points are important experimentally, since they indicate the possibility of spin-dimensionality crossover effects in nonisotropic magnets. However, as regards the experiments there are usually still other contributions to the Hamiltonian (1) to consider, namely crystal field and dipolar anisotropy terms which could both lead to an orthorhombic (biaxial) type of anisotropy, instead of the uniaxial cases considered here. Furthermore in 2-d magnets the interlayer coupling will lead to a 2-d/3-d dimensionality crossover. All these additional complications may easily mask the expected spin-dimensionality crossovers in the experimental systems.

## 5. Applications to Magnetic Systems

After reviewing the predictions derived mainly from series expansion analyses of the 2-d Ising, Heisenberg and XY models, it is illustrative to consider a number of magnetic insulators which appear to be closely described by these models. The aim is not to present an exhaustive summary of all the examples known, but to call attention to the high degree of approximation of these models that may be reached in the experimental systems found in the literature.

### 5.1. ISING MODEL

In some compounds which contain transition metal or rare earth ions, the crystal field effects may split the electronic levels in such a way that a well-isolated ground doublet with highly anisotropic uniaxial properties results [191]. If this doublet is the only populated level at temperatures where the magnetic ordering takes place, all the conditions are fulfilled to describe the system by an effective $S = 1/2$ Hamiltonian and to have an Ising- or XY-type magnet.

In Figure 15, the magnetic specific heats of some cobalt salts with strong Ising-type anisotropy are collected, together with available exact and series expansion results for the $S = 1/2$ Ising model [192]. The compound $Rb_3CoCl_5$ [193] orders antiferromagnetically at $T_N = 1.14$ K and its magnetic specific heat approximates the predictions derived from low- and high-temperature series for the s.c. lattice (curve a) with $J/k_B = -0.51$ K. The result for this 3-d system may be compared with that for the 2-d antiferromagnet $Cs_3CoBr_5$ ($T_N = 0.282$ K, $J/k_B = -0.22$ K) [194, 195] which nicely follows the exact Onsager result (curve b) [15]. This last compound, together with $K_2CoF_4$ and $Rb_2CoF_4$ [196], are up till now the best physical realizations of 2-d Ising magnets. In addition, it should be noted that also non-cobalt compounds with dominant 2-d Ising behaviour have been reported, such as Fe(triazole)$_2$(NCS)$_2$ [197], or the rare earth oxisulphates $Re_2O_2SO_4$ (Re = Dy, Tb, Tm) [198]. However, it should be remarked that the Co

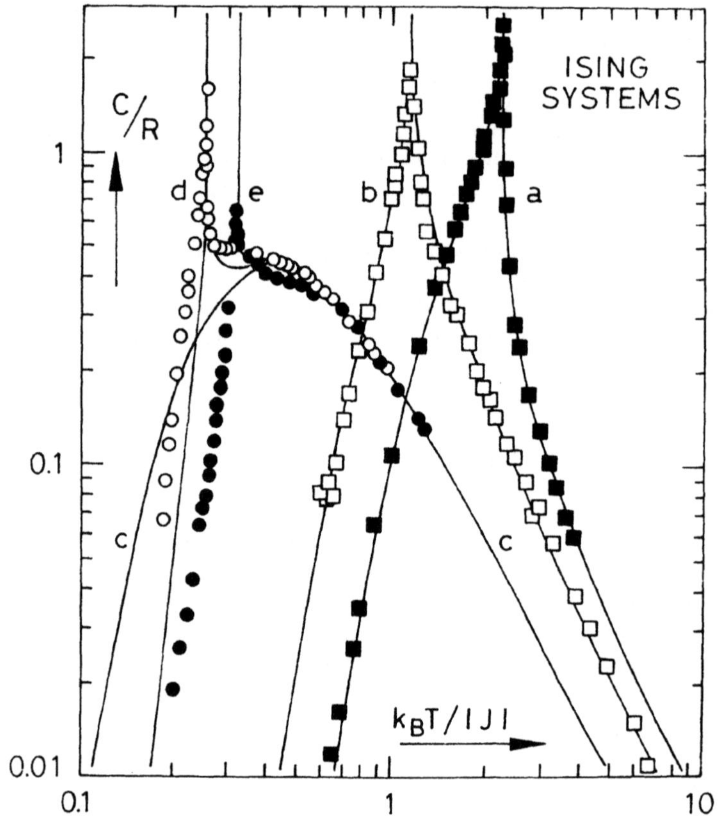


Fig. 15. Magnetic specific heat ($C(T)/R$) of some cobalt salts with high uniaxial (Ising-type) anisotropy; (■) $Rb_3CoCl_5$, (□) $Cs_3CoBr_5$, (●), $CoCl_2 \cdot 2\ NC_5H_5$ and (○) $((CH_3)_3NH)CoCl_3 \cdot 2\ H_2O$ scaled with the exchange constants given in the text. The continuous curves represent the predictions for the $S = 1/2$, Ising model on various lattices; a- (s.c.), b- (s.p.) c- (l.c.) d- (crossover 1-d → 2-d in the s.p. lattice with $R = 0.01$) e- (crossover 1-d → 3-d in the s.c. lattice with $R = 0.01$). Taken from de Jongh [192]. Reproduced by permission of the American Institute of Physics.

salts, together with rare-earth compounds of Dy, Ho or Tb are known to show the highest anisotropy values [145].

Although not systematically studied, one may find in the literature different low-dimensional magnetic compounds in which 1-d → 2-d, 1-d → 3-d [122] (Table VIII) or 2-d → 3-d [199] crossover effects are present. Thus, $CoCl_2 \cdot 2\ NC_5H_5$, is composed of Ising-type linear chains, with a ferromagnetic intrachain interaction $J_z$, which are weakly coupled by an antiferromagnetic interaction $J'_{xy}$ in an s.c. arrangement ($T_N = 3.17$ K, $J_z/k_B = 9.6$ K, $J_{xy}/k_B = -0.1$ K [122]). The specific heat data for this material are also plotted in Figure 15, and clearly show a crossover from the 1-d behaviour at high temperatures to the 3-d ordered state at low temperatures. The theoretical h.t.s. prediction for such a crossover (curve e) [122] also coincides at high temperatures with the exact linear chain predictions (curve c) [5] and reproduces the main features of the experimental results. The compound $((CH_3)_3NH)CoCl_3 \cdot 2\ H_2O$ is also composed of anisotropic ferromagnetic linear chains (interaction) $J_x$, which are now coupled ferromagnetically into magnetic layers by a weaker interaction $J_y$, and by an even weaker coupling in the third direction. The scaled data ($T_N = 4.13$ K, $J_x/k_B = 17.5$ K and $J_y/k_B = 0.2$ K) [122] show the 1-d → 2-d crossover effect, and nicely fit Onsager's predictions for $J_x/J_y \simeq 0.01$ (curve d).

A similar agreement between theory and experiments as found for the specific heat is also obtained for other quantities, such as the susceptibility, the magnetization, etc., as has been described elsewhere [122, 145]. However, it should be noted that the description of the lowest doublet with an effective Hamiltonian is only valid if the thermal population of the higher lying levels is negligible. This implies that the paramagnetic temperature region in which the fit is meaningfull is limited. The same argument also holds when very high fields are applied, since the energy gaps between the levels may then be reduced. In fact, these deviations due to the thermal population of excited levels have been detected in the comparisons of the data for $K_2CoF_4$, $Rb_2CoF_4$ [122, 196] and Fe(triazole)$_2$(NCS)$_2$ [197] with predictions based on effective spin $S = 1/2$.

Another interesting realization deals with the phase diagram for Ising antiferromagnets. In Figure 16, the $T_c(H)$ data of the $Cs_3CoBr_5$, deduced from the dependence of the position of the specific heat lambda peak on an applied field ($H < 880$ Oe) (squares), are plotted. For $H > H_c/2$ ($H_c = 2100$ Oe) the position of $T_c(H)$ has been assumed to be coincident with the sharp maximum of $\chi$ found from temperature scans at constant fields (circles) [95]. It should be noted that no corrections for demagnetizing effects have been applied, which may slightly affect the data in highest fields. Nevertheless, a comparison with Bienenstock's predictions [86] for the 2-d Ising $S = 1/2$ antiferromagnet yield a good agreement with the experimental data.

Much of the theoretical effort on simple magnetic models has been focused upon the derivation of critical exponents, amplitudes and functional forms in the critical region. Such predictions have been extensively compared with many experimental results for different thermodynamic quantities. Most of these studies are hampered by the fact that the extent of the critical region is not known, as well as, by experimental uncertainties. The best magnetic functions to analyse in the critical region are probably the susceptibility and magnetization [145].

An example of the critical behaviour of the staggered susceptibility of the antiferromagnetic $K_2CoF_4$, measured with neutron differaction, may be found in a companion chapter. Moreover the parallel and perpendicular susceptibilities also show singularities at $T_N$ in $\partial\chi/\partial T$, although they are weaker. These data on the $K_2CoF_4$ and $Rb_2CoF_4$ compounds [196] are represented in Figure 17 with an appropriate scale to linearize the Equations (28) and (33). The compounds $K_2CoF_4$ ($T_N = 107$ K, $J/k_B = -(19.3 \pm 2.5)$K) and $Rb_2CoF_4$ ($T_N = 101$ K, $J/k_B = -(15.1 \pm 1.8)$K) have the well-known 2-d magnetic structure of $K_2NiF_4$, and are

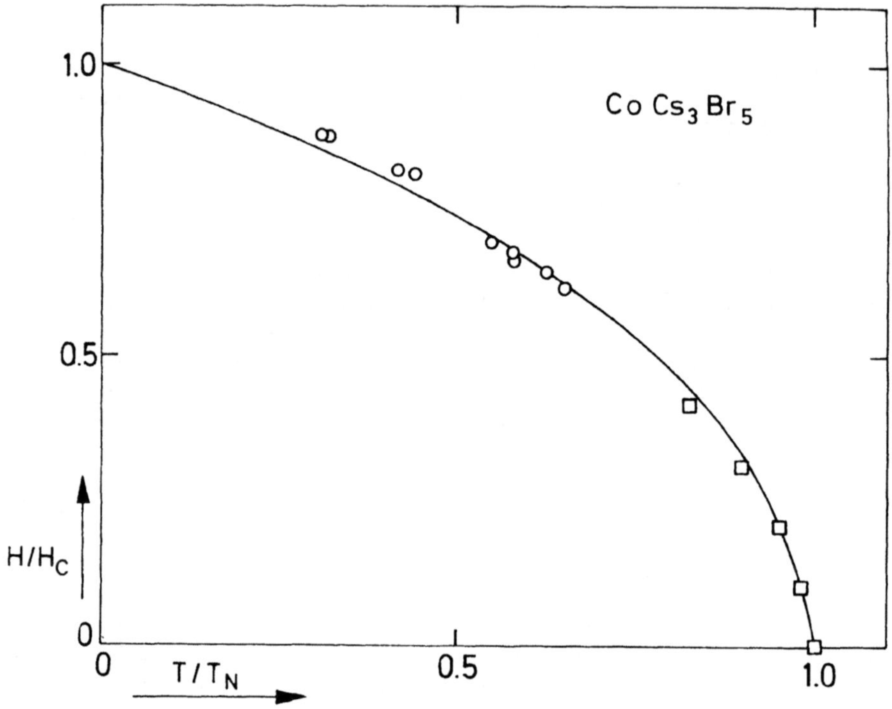


Fig. 16. Phase diagram of the $Cs_3CoBr_5$ on a reduced temperature ($T_N = 0.282$ K) and field ($H_c = 2100$ Oe) scale. The different symbols are explained in the text and the full line is the prediction of the $S = 1/2$ Ising antiferromagnet for a s.p. lattice derived by Bienenstock [86]. Reproduced by permission of Elsevier Science Publishers.

characterized by a strong uniaxial anisotropy. Indeed, the predictions for the $S = 1/2$ Ising model on the s.p. lattice fit the data nicely over a wide temperature range. In particular, both compounds appear to follow the predicted asymptotic behaviour for $0.90 < T/T_N < 1.10$. Furthermore, the sublattice magnetization, in the same temperature range, fits the asymptotic form (4) with $\beta = 1/8$ [145, 200].

Although there are many predictions available for arbitrary spin values, thin films, further-neighbour interaction and so on, up to day appropiate magnetic systems in which these effects might be tested experimentally have not been found to our knowledge.

## 5.2. HEISENBERG MODEL

Whereas the number of highly anisotropic magnets is limited, for the theoretically less well-known Heisenberg case there are many experimental realizations. For very symmetric crystal field, the ground multiplets of some of the transition metal

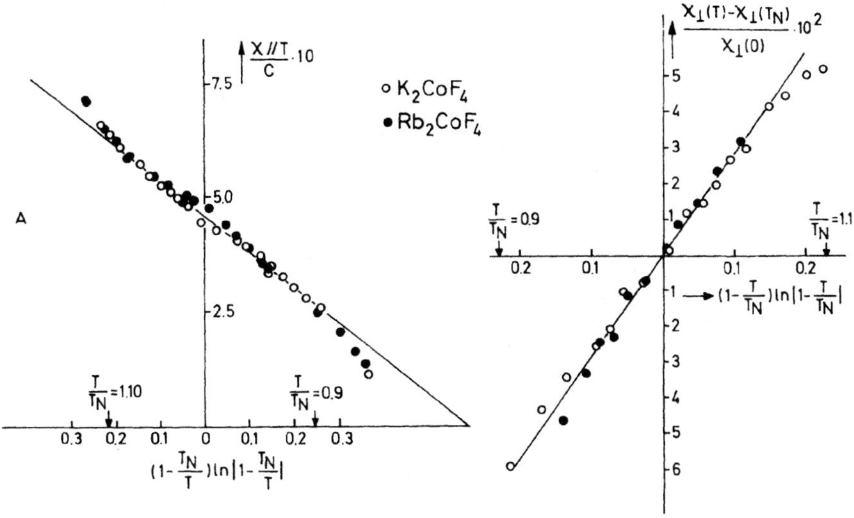


Fig. 17. Magnetic susceptibilities parallel, $\chi_{\|}$, and perpendicular, $\chi_{\perp}$, to the easy axis in the highly uniaxial anisotropic $Co^{2+}$ compounds $K_2CoF_4$ and $Rb_2CoF_4$ as given by Breed *et al.* [196]. The constants $C = Ng_{\|}^2\mu_B^2S^2/3k_B$, $S = 1/2$, $g_{\|} = 41$ for the $K_2CoF_4$ and $g_{\|} = 49.4$ for the $Rb_2CoF_4$ have been used. Reproduced by permission of Elsevier Science Publishers.

ions (e.g. $Mn^{2+}$, $Ni^{2+}$, . . .) have zero angular momentum, so that these materials will have nearly isotropic properties, with an effective spin value equal to their spin multiplicity. Then, by choosing compounds with the appropriate spatial distributions of the magnetic ions, several series of 2-d magnetic insulators have been found [145]. Furthermore, the same characteristics also may be found for the rare earth ions Eu, Sm, Nd, etc., and Heisenberg like effective-spin Hamiltonians have been used in the description of their magnetic properties [145].

### 5.2.1. *Spin 1/2*

The series of layer-type copper compounds $(C_nH_{2n+1}NH_3)_2CuCl_4$ with $n = 0, 1, 2, 3, \ldots$ [145, 201], $K_2CuF_4$ [202], $Rb_2CuCl_4$ [203], etc., provide good approximations of the $S = 1/2$, 2-d Heisenberg magnet, with predominant ferromagnetic interlayer interactions, $J_{xy}$ on an s.p. magnetic lattice. At high temperatures, $2 < k_BT/J_{xy} < 10$, the $\chi$-data of most of these compounds fit the predictions for 2-d Heisenberg ferromagnets, whereas at lower temperatures some differences appear. This behaviour is due to the interlayer interactions, $J_z$, because for $T \to T_c^+$, the sign of $J_z$ should determine whether $\chi$ deviates upward or downwards from the pure 2-d behaviour, whereas the absolute value of $R \equiv J_z/J_{xy}$ fixes the crossover temperature, i.e. the temperature at which the difference with the 2-d predictions are experimentally appreciable (see Section 2.3.3).

Close to $T_c$, the presence of small XY- or Ising-like anisotropy fields of dipolar origin give rise to spin dimensionality crossover, which should be added to the above effects. Figure 18 shows the $\chi$ behaviour of some selected compounds which display Ising like behaviour in the critical region [204]. For $(C_nH_{2n+1}NH_3)_2CuCl_4$, $n = 1$, 3 and 10, the easy axis $\chi$-data of two different samples show clearly two regions in which they follow characteristic straight lines (Equation 3) with slopes $\gamma = 1.75$ (2-d-Ising) and $\gamma = 1.25$ (3-d-Ising). The pure 2-d isotropic magnet cannot order at any finite temperature and its behaviour therefore is very sensitive to the 3-d coupling or to anisotropy effects. For

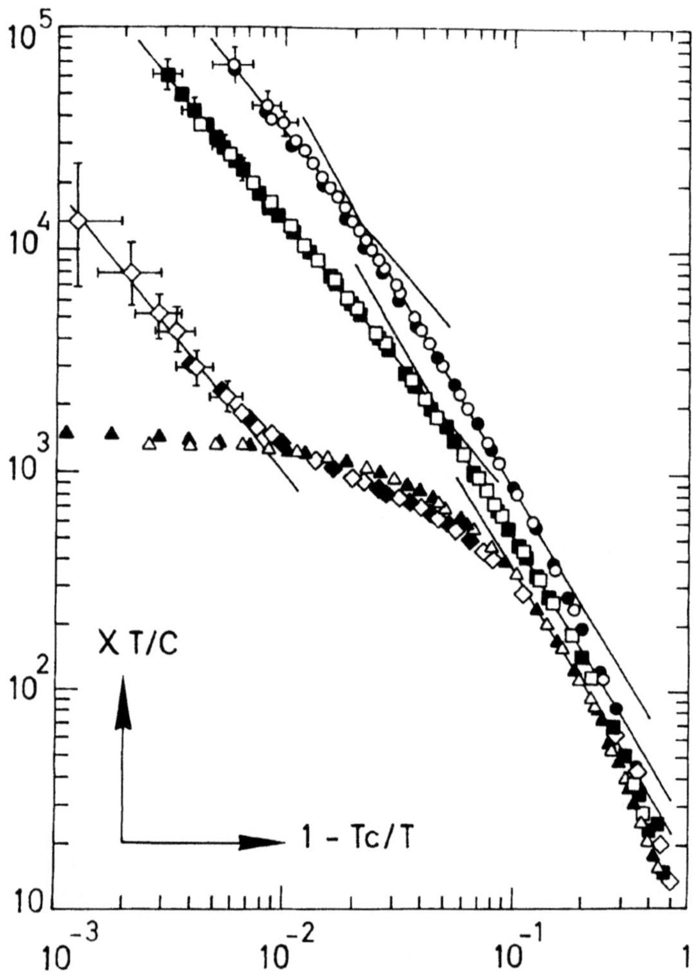


Fig. 18. Magnetic easy axis susceptibility of some 2-d ferromagnets in the s.p. lattice with uniaxial anisotropy. Measurements of two different samples of $(C_nH_{2n}H_{2n+}NH_3)_3CuCl_4$, $n = 1$ (□, ■), $n = 3$ (△, ▲) and $n = 10$ (○, ●). For the $n = 3$ compound also the next preferred direction are shown (◇, ◆). Taken from de Jongh [204]. Reproduced by permission of Elsevier Science Publishers.

anisotropy energies higher than the energy associated with the interlayer coupling the $\chi$ data closed enough to $T_c$ may show first a 2-d Ising or XY behaviour and second a crossover to the corresponding 3-d behaviour. This holds for the compounds in Figure 18. Contrariwise, if the 3-d coupling energy is higher than the anisotropy energies, a 3-d Heisenberg ($\gamma = 1.405$) [27] behaviour should be displayed first, whereafter the anisotropy should give a crossover to 3-d XY ($\gamma = 1.35$) [28] or Ising-type behaviour.

The study of 2-d Heisenberg $S = 1/2$ antiferromagnetis is of special interest in view of the possibility to detect at the lowest temperatures kinematic interaction effects [162]. Untill recently, the best examples found were the compounds $CuF_2 \cdot 2\ H_2O$ [205, 206] and $Cu(C_5H_5NO)_6(BF_4)_2$ [162, 163]. The $\chi$ measurements on $CuF_2 \cdot 2\ H_2O$ ($T_{max} = (26.\ \pm\ 1)$ K, $T_N = 10.9$ K) on powdered and single-crystal samples [205] show the main characteristics of the 2-d isotropic antiferromagnet (broad maxima well above $T_N$) but there are differences in the height of $\chi$ and another independent determination [207] has been used here. The data nicely follow the h.t.s. prediction with $J/k_B = -(14.\ \pm\ 1.)$K and $g = 2.20 \pm 0.02$ (see Figure 19). Moreover, one should note the difference at the lowest temperatures between the fitted $\chi$-powder ($\chi_p$) data and the usual 2-d spin-wave theory prediction (arrow in the figure, $\chi_p = 2/3\chi_\perp$) for the observed anisotropy $\alpha = H_A/H_E = 0.0037$ [145]. The 3-d coupling, $|J_z/J_{xy}| \simeq 0.015$, roughly estimated for this compound cannot be the only deviation responsible for such a big

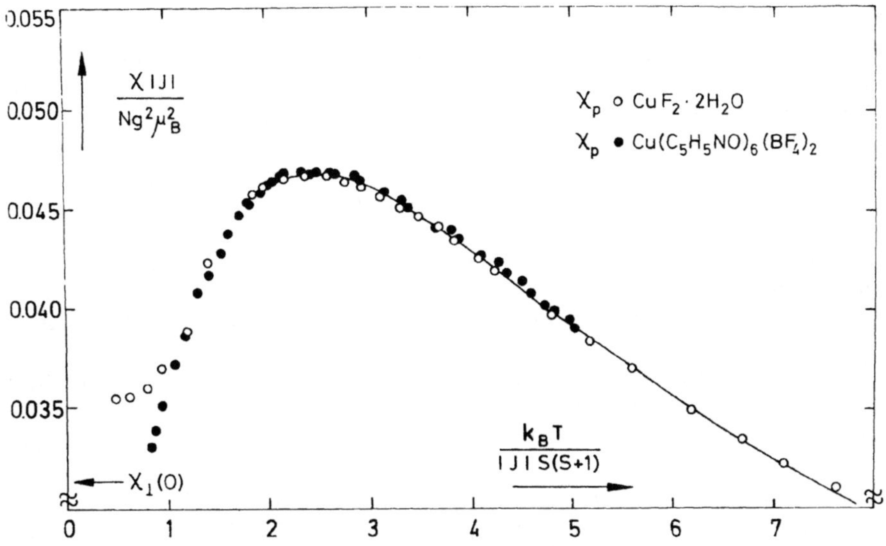


Fig. 19. Experimental powder susceptibility of $CuF_2 \cdot 2\ H_2O$ ($J/k_B = -14$ K, $g = 2.20$) and of $Cu(C_5H_5NO)_6(BF_4)_2$ ($J/k_B = -1.10$ K, $g = 2.20$) scaled to the h.t.s. prediction for $S = 1/2$, s.p. Heisenberg antiferromagnets. The arrow corresponds with the $\chi_\perp$ spin wave prediction for $\alpha = 0.0037$.

difference, which should be ascribed to the neglect of the kinematical interaction in the spin wave formalism used.

The $\chi$-data [208] of $Cu(C_5H_5NO)_6(BF_4)_2$ ($T_{max} = (1.9 \pm 0.1)$ K, $T_N = (0.62 \pm 0.01)$ K [162] are also displayed in Figure 19, with fitting parameters $J/k_B = -1.10$ K, $g = 2.20$. At the lowest temperatures the values of $\chi'_p = \chi_p|J|/Ng^2\mu_B^2 = 0.021$ is close to the theoretical one for small anisotropy ($\alpha = 0.001$) [209]. The $\chi$-measurements on single crystals [208] do not show a macroscopic easy axis, probably due to different local orientation of the Jahn—Teller elongated fluorine octahedra, which surround each copper ion, and to the presence of different crystallographic domains [209]. The comparisons then cannot go further. However the low value of the exchange interaction allows the direct observation of the broad maximum in the 2-d specific heat curve ($C/R = 0.44 \pm 0.01$ and $T(C_{max}) \simeq 1.4$ K), at temperatures far above $T_N$ at which a small lambda peak marks the onset of 3-d magnetic ordering. After subtraction of the phonon contribution the magnetic part of the specific heat has been displayed in Figure 20, together with the h.t.s. prediction, scaled with the previous $J$-value.

It has recently become apparent that the new high-$T_c$, layer-type oxide superconductors, notably $La_2CuO_4$, are most likely among the very best $S = 1/2$, 2-d Heisenberg antiferromagnets. A discussion has recently been given by de Jongh [208]. Since in these materials the exchange constants are so large (500—1000 K), it is only possible to measure the thermodynamic quantities considered here over very limited ranges of temperature.

### 5.2.2. *Spin 1*

The series of isomorphous compounds $X_2NiF_4$, X = Rb, Tl and K, together with the $BaNiF_4$, $Ba_2NiF_6$, etc. forms a representative set of isotropic, $S = 1$, s.p. lattice 2-d antiferromagnets, and may illustrate the behaviour for such a spin value.

For $Rb_2NiF_4$ there are data on both $\chi_{\parallel}$ and $\chi_{\perp}$ available [210], which allows the experimental determination of the temperature independent susceptibility, $\chi_{tip}$. Taking $\chi_{tip} = \chi_{\parallel}(T \to 0) = 1.1 \times 10^{-6}$ emu/g, the measurements fit the theoretical predictions quite well with the parameters $J/k_B = -48$ K, $g_{\parallel} = g_{\perp} = 2.43$ [148], as shown in Figure 21. However, the somewhat high value of $g$ should be noted. The limiting value of $\chi'_{\perp}(T \to 0) = 0.0450$ found from the fit is lower than predicted (0.0476) for the experimental value of the anisotropy $\alpha = 0.01$ [210], but small changes in $g_{\perp}$ may account for this disagreement.

The compound $K_2NiF_4$ is one of the best investigated magnetic systems [145], because of its small coupling between layers, $|J_z/J_{xy}| < 0.0037$ [211], small anisotropy, $\alpha = 0.002$ [210], and weak interaction between second neighbours within the layers [212], so that it approximates quite well the isotropic nearest-neighbour 2-d model. Following the same procedure as for the Rb isomorph, the $\chi$-data [210] have been scaled with $J/k_B = -51.5$ K, $g_{\parallel} = 2.30$, $g_{\perp} = 2.34$ and the results are displayed in Figure 21. The $J$ value agrees with that deduced from the sublattice magnetization with spin-wave theory [213]; moreover, the fit of $\chi'_{\perp}(T \to 0)$ leads to $g_{\perp} = 2.25$ for the same $J$. If the scatter of the experimental data is considered, the low-temperature fit is certainly correct.

The compound $BaNiF_4$ has the same magnetic characteristics although the

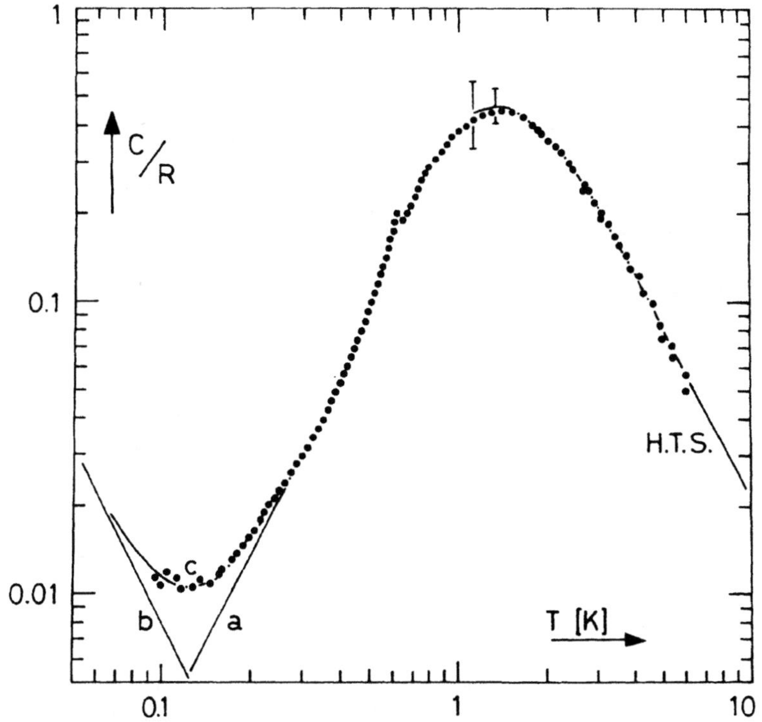


Fig. 20 . Magnetic specific heat of $Cu(C_5H_5NO)_6(BF_4)_2$ scaled with $J/k = -1.10$ K to the h.t.s. and spin wave (curve a) predictions. The stimated hyperfine contribution $bT^{-2}$ with $b = (8 \pm 0.5) \times 10^{-5}$ K$^2$ is depicted by curve b and the curve c is the addition of curves a and b. Taken from Algra *et al.* [163]. Reproduced by permission of Elsevier Science Publishers.

crystal structure is a bit different. Proceeding in the same way, $\chi_{\parallel}$ and $\chi_{\perp}$ [214] ($\chi_{\text{tip}} = 0.53 \times 10^{-4}$ emu/mol) have been fitted to the theory with $J/k_B = -33$ K, $g_{\perp} = g_{\parallel} = 2.43$, and are displayed in Figure 21. Once more, the value of $\chi'_{\perp}(T \to 0) = 0.0455$ is lower than the theoretical spin wave prediction of 0.0483 for the deduced anisotropy $\alpha = 0.03$ [145]. Small disagreements for the $g$-values are found for all these compounds, the $g$-factors being somewhat higher than usual [145], which may be due to the use of a constant $\chi_{\text{tip}}$ and/or to covalency effects present in the $Ni^{2+}$ ions.

Finally $Ba_2NiF_6$ offers another magnetostructural approximation of the 2-d, $S = 1$, isotropic antiferromagnet with a small 3-d coupling. The $\chi_{\parallel}$ and $\chi_{\perp}$ data [215], after a correction similar as above with $\chi_{\text{tip}} = 1.2 \times 10^{-6}$ emu/g, also fit the h.t.s. predictions quite well for $J/k_B = -38$ K, $g_{\parallel} = 2.23$, $g_{\perp} = 2.10$. The anisotropy is the highest of all the Ni salts considered here ($\alpha = 0.07$).

The presence of varying axial anisotropies gives rise to spin-dimensionality crossover, clearly displayed in the paramagnetic region by optical birefringence [216] and specific heat measurements [217]. The temperature dependence of the

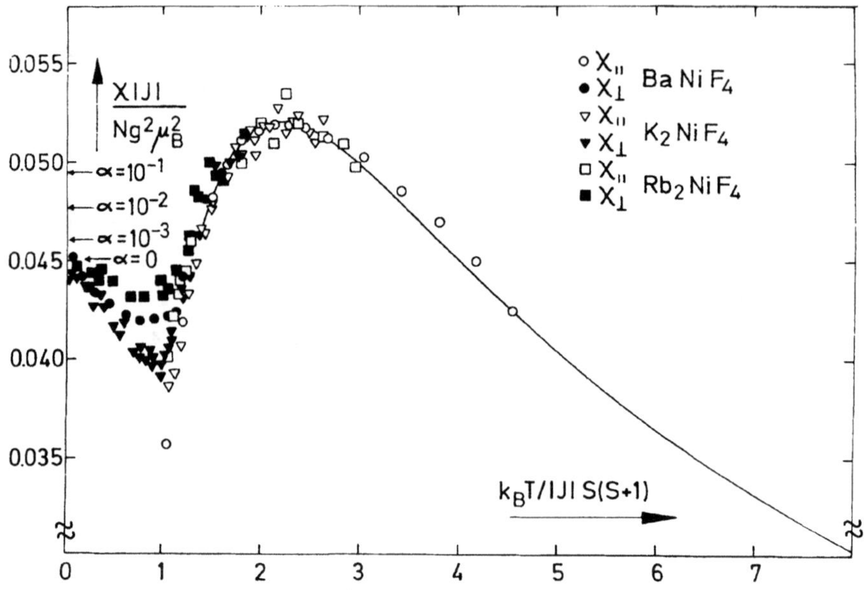


Fig. 21. Corrected magnetic susceptibilities of $Rb_2NiF_4$ ($J/k_B = -48$ K, $g_{\|} = g_{\perp} = 2.43$), $K_2NiF_4$ ($J/k_B = -51.5$ K, $g_{\|} = 2.30$, $g_{\perp} = 2.34$) and $BaNiF_4$($J/k_B = -33$ K, $g_{\|} = g_{\perp} = 2.43$), scaled to the 2-d, $S = 1$, Heisenberg antiferromagnet on the s.p. lattice. The arrows indicated the values of $\chi_{\perp}$ deduced from spin wave theory for different anisotropies. Taken from Navarro [148].

birefringence is the same as for the magnetic energy, which is proportional to the nearest-neighbour correlation function $\langle \mathbf{S}_i \mathbf{S}_j \rangle$, so that its temperature derivative shows similar characteristics as the specific heat [216]. For $K_2NiF_4$, which has the smaller anisotropy, the round maximum of the short-range ordering is clearly displayed and furthermore a small peak appears at $T_N$. In $Rb_2NiF_4$, the anisotropy is almost 2.5 times higher and in the derivative of the birefringence the peak at $T_N$ is higher and closer to the maximum. Finally, in $BaNiF_4$, which has an anisotropy an order of magnitude higher than for $K_2NiF_4$, the maximum has disappeared and a peak around $T_N$ similar in shape to the one of $K_2CoF_4$ is detected [216]. Due to the high value of the phonon contribution, the direct measurements of the specific heat [217] allow only the observation of the peaks, the size and shape of which show the same tendency as in the optical study. All these facts prove clearly that the increasing axial anisotropy on such compounds should give Ising-like behaviour in a wider temperature region. Furthermore, the sublattice magnetization measured in the $K_2NiF_4$, and for $T \to T_c^-$ clearly follows the 2-d Ising critical behaviour [145].

### 5.2.3. *Spin 3/2 and Spin 2*

In 2-d isotropic magnets, a spin value of 3/2 is found e.g. in compounds containing the transition metal ions $Cr^{3+}(3d^3)$ and $Mn^{4+}(3d^3)$ whereas $Cr^{2+}(3d^4)$ and $Fe^{2+}(3d^6)$ salts have spin 2. The $Cr^{3+}$ compounds tend to be ferromagnetic and the series of compounds with hexagonal structure $CrX_3$, X = I, Cl, Br [218—222] and $AgCrSe_2$, $NaCrS_2$ and $NaCrSe_2$ [223, 224] provide isotropic ferromagnets with $S = 3/2$ in two different 2-d lattices.

The chromiun trihalides have layers of $Cr^{3+}$ ions with a main ferromagnetic interaction between three nearest neighbours, $J_{xy}$, (h.c. lattice), separated by two closely-packed layers of halogen ions, which gives antiferro- ($T_N(\mathrm{Cl}) = 16.8$ K [218]) and ferromagnetic ($T_c(\mathrm{Br}) = 32.5$ K [219] and $T_c(\mathrm{I}) = 68$ K [220]) interlayer interactions, $J_z$. Most of the analyses for these compounds have been performed using data of the ordered phase and spin wave predictions with the following results $CrCl_3$: $J_{xy}/k_B = +5.25$ K, $J_z/J_{xy} \simeq -6 \times 10^{-4}$ [221]; $CrBr_3$: $J_{xy}/k_B = (+8.1 \pm 0.2)$ K, $J_z/J_{xy} \simeq 6 \times 10^{-3}$ [222]; and $CrI_3$: $J_{xy}/k_B = 13.5$ K, $J_z/J_{xy} = +0.1$ [221]. Comparisons with h.t.s. predictions are difficult, due to the 3-d coupling and the axial anisotropy present in the salts, but qualitatively a similar behaviour as in the Cu ferromagnetic compounds is observed.

In $AgCrSe_2$, $NaCrS_2$ and $NaCrSe_2$, [223, 224] the $Cr^{3+}$ ions are ferromagnetically coupled with six nearest-neighbours into p.t. layers, separated by two layers of non-magnetic metallic ions and another of chalcogen ions, which give an antiferromagnetic coupling [223]. The experimental data are very scarce, but preliminary analyses using molecular field theory give a ratio $J_z/J_{xy} \simeq -0.2$ for $AgCrSe_2$ and $NaCrS_2$, and an order of magnitude lower for $NaCrSe_2$, which values may be considered as upper bounds. However, the relatively strong interlayer coupling and the absence, up to date, of dimensionality crossover h.t.s. predictions for this lattice preclude further quantitative analysis.

$Ca_2MnO_4$ has $Mn^{4+}$ ions and the crystal structure of $K_2NiF_4$. It shows $S = 3/2$, 2-d (s.p.) isotropic behaviour with antiferromagnetic interactions [225, 226]. The $\chi_p$-data [225] show a broad maximum at $T_{max} \simeq 220$ K and the appearance of weak-ferromagnetism at the ordering temperature $T_N = 114$ K. The experimental data, scaled with $J/k_B = -27$ K and $g = 2$, fit the h.t.s. predictions when a very high $\chi_{tip} = 2.30 \times 10^{-4}$ emu/mol is considered. For this compound the presence of a strongly temperature dependent nonnegligible electrical conductivity may explain the inadequacy of the localized electron model here used. Also the measured moment $(2.0 \pm 0.3)\mu_B$ [226], which is lower than the theoretical value, $gS\mu_B = 3$ $\mu_B$($g = 2$ is expected for $Mn^{4+}$) or than the one corrected for the zero-point spin reduction, $2.6\mu_B$, which points to the existence of a nonnegligible electronic delocalization (covalency effects).

The $Cr^{2+}$ members of the family of isostructural compounds $X_2CrCl_4$, X = K, Rb and Cs [227] with the $K_2NiF_4$ structure, have effective spin 2 and show 2-d ferromagnetic characteristics. The fit of the polycrystalline $\chi$-data in the range $100\ \mathrm{K} < T < 320$ K to the h.t.s. predictions, with fixed $g = 1.98$, yields $J_{xy}/k =$ 10.1, 12.2 and 9.4 K for the K, Rb and Cs salts, respectively [227]. However, due to the proximity of the critical temperatures, $T_c(\mathrm{Rb}) = (63 \pm 2)$ K, $T_c(\mathrm{Cs}) \simeq 58$ K, to the fitting interval, the values of $J_{xy}$ are probably overestimated. The 3-d coupling and the XY anisotropy found in the series may explain the difference between the above value for $J_{xy}$ for the Rb salt and that deduced from the spin-wave dispersion relation $J/k_B = (7.7 \pm 0.1)$ K [228].

The $Fe^{2+}$ compounds for highly symmetric crystal fields and strong magnetic interactions may show isotropic magnetic properties. However, the spin-orbit coupling and any small crystal distortion split the ground manifold, giving a substantial anisotropy, as is nearly always found in the $Fe^{2+}$ salts. All the compounds available in the literature, namely $BaFeF_4$ [229], $X_2FeF_4$; X = Rb, K [230, 231] $Ba_2FeF_6$ [232] $(C_nH_{2n+1}NH_3)_2FeCl_4$ [233, 234] $((CH_3)_2NH_2)_2FeCl_4$ [235], etc. show the same type of competition of effects and, among them, the $BaFeF_4$ and $X_2FeF_4$ salts have been selected to illustrate the ensuing difficulties of the analysis.

For $BaFeF_4$ the $\chi$-data on a single crystal [229], and for temperatures above the antiferromagnetic region ($T_N$ = 54.2 K), show maxima with different height and position for each orthorhombic direction. Thus, in the easy axis ($b$-direction) the maximum is found at (80 ± 5) K with $\chi_{max} = 17.6 \times 10^{-3}$ emu/mol whereas in the perpendicular directions the values are $T_{max} = (65 \pm 5)$ K, and $\chi_{max} = 19.3 \times 10^{-3}$ and $16.7 \times 10^{-3}$ emu/mol for the $a$ and $c$ directions, respectively. In addition $\chi_{\parallel}$ does not go to zero when $T \to 0$, giving a $\chi_{tip} = (1.0 \pm 0.2) \times 10^{-3}$ emu/mol which has been accounted for in all the following fits. Due to the large axial anisotropy, estimated as $\alpha = 0.2$ [145] comparisons with both Heisenberg and Ising $S = 2$ models may be meaningful. From the Heisenberg model predictions (Table IX) the values $J/k_B = (-6.4 \pm 0.4)$ K and $g_{\parallel} = 2.28 \pm 0.06$ are deduced. With the Ising values (see Tables II and V) $J/k_B = (-5.8 \pm 0.4)$ K and $g_{\parallel} = 2.13$ are obtained from $T_{max}$, $\chi_{max}$ and $J/k_B = -5.1$ K and $g_{\parallel} = 1.99$ from $T_N$, $\chi_{max}$. These results may be compared with the measured moment [229] $(4.2 \pm 0.2)\mu_B$, which should be equal to $g_{\parallel}\mu_B S$ yielding $g_{\parallel} = 2.1 \pm 0.1$. If the zero-point spin reduction is taken into account the $g_{\parallel}$ value is slightly modified to 2.2 ± 0.1. This value agrees with that deduced from $T_{max}$, $\chi_{max}$ in both models. Moreover, from the experimental $\chi_{\perp}(T \to 0) = 19.1 \times 10^{-3}$ emu/mol, using $g = 2.1$ and the molecular field predictions $\chi_{\perp} z|J|/Ng^2\mu_B^2 = 0.25$, the value $J/k_B = -5.7$ K is derived. In fact, small changes in position and height of the antiferromagnetic $\chi$-maximum gives results for both models, and presumably for the more appropriate Ising—Heisenberg one, which, given the uncertainties involved, are not conclusive although the values deduced from the Ising $T_{max}$, $\chi_{max}$ seem to be the most reliable. However, the values deduced from the Ising critical temperature are not adequate, because its position is strongly dependent of the anisotropy and, by similarity with the results for $S = 1/2$ (Table XII), rapid $T_c$ changes are expected for $\alpha < 0.4$ ($b/a > 0.6$).

The compounds $Rb_2FeF_4$ and $K_2FeF_4$ [230, 231] also show 2-d antiferromagnetic properties but with planar anisotropies $\alpha = 0.2$ and $\alpha = 0.3$, respectively. The $Rb_2FeF_4$ $\chi$-data [230] show a maximum at $T_{max} = (90 \pm 5)$ K for all different directions, but with different heights ($\chi_{max} = 14.0 \times 10^{-3}$ emu/mol in the [100] and [110] directions and $\chi_{max} = 12.8 \times 10^{-3}$ emu/mol in the [001] one), the ordering temperature being $T_N$ = 56.8 K. Due to the measured planar anisotropy $\alpha = 0.2$ [231] the $S = 2$ Heisenberg and XY models are the expected limits, but up till now, no predictions are available for such spin value in the XY case and the comparison is restricted to the Heisenberg limit. Using the $T_{max}$, $\chi_{max}$ predictions, which, as seen before, do not change drastically from the Ising to Heisenberg limit, the value $J/k_B = (-7.3 \pm 0.4)$ K, with $g = 2.22 \pm 0.06$, is derived, in agreement with the values deduced from the spin wave dispersion relation [231] $J/k_B = -7.27$ K. For isomorphous $K_2FeF_4$ the planar anisotropy is higher, $\alpha \simeq 0.3$, and so is the ordering temperature, $T_N = 67.2$ K. Its behaviour is similar to the Rb salt.

The other $Fe^{2+}$ salts have either axial anisotropy, as in $Ba_2FeF_6$ [232], or planar anisotropy, as in $(C_nH_{2n+1}NH_3)_2FeCl_4$, $n = 1, 2, 3, 4$ and $((CH_3)_2NH_2)_2FeCl_4$ [234, 235], and may be analysed in the same way as above if the splitting of the ground manifold is smaller than the exchange constant. If this is not the case, an effective spin 1/2 XY or Ising model should be more adequate, as happens in the abovementioned Fe(triazole)$_2$(NCS)$_2$ [197].

### 5.2.4. *Spin 5/2*

Many $Mn^{2+}$ or $Fe^{3+}$ compounds closely approximate the 2-d Heisenberg antiferromagnet with $S = 5/2$. As regards the s.p. lattice one has $X_2MnF_4$, X = K, Rb [236, 237], $BaMnF_4$ [238], $X_2MnCl_4$; X = Cs, Rb [239], $(C_nH_{2n+1}NH_3)_2MnCl_4$ $n = 1, 2, 3$ [240], Mn(triazole)$_2$(NCS)$_2$ [241], etc. and the iron salts $XFeF_4$, X = Rb, Cs, K [242–244], Na and $NH_4$ [245]. In particular the compounds $K_2MnF_4$ and $Rb_2MnF_4$ belong to the first 2-d systems recognised as such [236], and since that date have been extensively studied. To illustrate the behaviour of such antiferromagnets, we show in Figure 22 the $\chi$ data [236] for $K_2MnF_4$, scaled upon the h.t.s. predictions with $J/k_B = -4.20$ K and $g = 2$. The minute value for the interlayer interaction $|J_z/J_{xy}| < 10^{-6}$ and the small anisotropy $\alpha = 3.9 \times 10^{-3}$ form the basis of the excellent agreement that is found. The fitting parameters derived in the paramagnetic phase may be checked using spin wave theory in the ordered phase and, with this purpose, the predictions for $\chi_\perp$ have been included for the experimental anisotropy value. To show the effect of the anisotropy, the data for $BaMnF_4$ [238], which has an anisotropy $\alpha = 3.1 \times 10^{-4}$, have been included. The data, scaled with $J/k_B = -2.80$ K and $g = 2$, also fit the h.t.s. prediction at temperatures around the maximum and above. However, below $T_N$ $\chi_\perp$ shows a minimum, deeper than the one for the $K_2MnF_4$, which is clearly related to the anisotropy. For comparable 3-d coupling, a smaller anisotropy should give a deeper minimum.

In the paramagnetic phase the comparison with theory is also valid for powdered samples, as has been done for $Rb_2MnCl_4$ and $Cs_2MnCl_4$ [239] in Figure 22. The original data pointed to the presence of a small amount of paramagnetic impurity. After correction for this the data were scaled, with $g = 2$, $J/k_B = -6.10$ K and $J/k_B = -4.95$ K for the Rb and Cs salts, upon the theoretical curve. Both compounds have about the same anisotropy [145] $\alpha = 1.5 \times 10^{-3}$ and follow the spin wave predictions for $\chi_p$ at the lowest temperatures. It is noted that the temperature range in which a good agreement with spin wave theory holds is wider for $\chi_p$ than for $\chi_\parallel$ or $\chi_\perp$, due to cancellation effects in the derivation of $\chi_p$.

Analogous comparisons with varying anisotropy and 3-d coupling have been

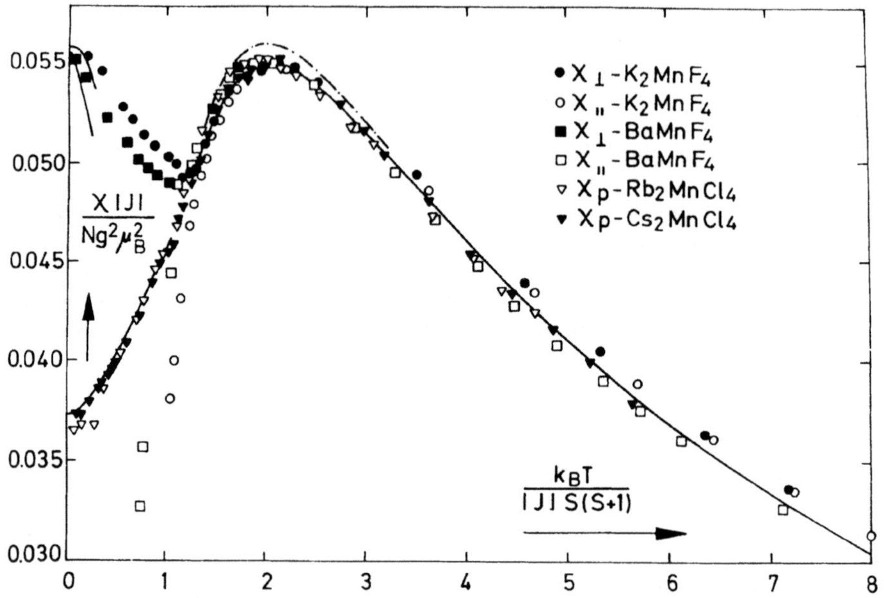


Fig. 22. Experimental susceptibilities of some manganese layer-type compounds compared with the 2-d, $S = 5/2$ Heisenberg antiferromagnet on the s.p. lattice. The dot-dashed line gives the analogous prediction for $S = \infty$, and the full lines at the lowest temperatures are the spin-wave predictions for $\chi_p$ and $\chi_\perp$ with the appropriate anisotropy values. Data are for: $K_2MnF_4$ (○, $\chi_\parallel$; ●, $\chi_\perp$ with $J/k_B = -4.20$ K), $BaMnF_4$ (□, $\chi_\parallel$; ■, $\chi_\perp$ with $J/k_B = -2.80$ K), $Rb_2MnCl_4$ (▽, $\chi_p$ with $J/k_B = -6.10$ K), and $Cs_2MnCl_4$ (▼, $\chi_p$ with $J/k_B = -4.95$ K). The value $g = 2$ has been taken for all the compounds.

performed for other $Mn^{2+}$ or $Fe^{3+}$ salts [240—245]. Most of these have critical temperatures higher than 30 K and in this respect the Mn(triazole)$_2$(NCS)$_2$ [241] should be mentioned because the low values for $T_N = 3.29$ K and for the exchange constant $J/k_B = -0.25$ K enable an accurate experimental determination of the magnetic specific heat, as well as the achievement of magnetic saturation in the relatively low field $H_s = 74$ kOe.

The ilmenite compound $MnTiO_3$ [246—248] forms layers of $Mn^{2+}$ ions, separated by two layers of oxygen and one of $Ti^{4+}$, giving a magnetic h.c. lattice. The $\chi$ measurements on powdered samples [246] and on a single crystal [247—248] show a broad maximum at $T_{max} \simeq 100$ K and antiferromagnetic 3-d order at $T_N = 64$ K [247]. The absolute value of the $\chi$-maximum found in the literature varies broadly, probably due to the presence of different amounts of $Mn^{3+}$ ions [248], but from the position of the maximum the value $J/k_B = -7.29$ K is derived which is different from the previous estimate of $J/k_B = -8.2$ to $-9$ K [145] or $J/k_B = -11$ K [248]. In Figure 23 the data [248] have been scaled with $J/k_B = -7.29$ K to the theoretical maximum. For comparative purposes the spin wave predictions for $\chi_\perp$ for the experimental anisotropy $\alpha = 1.5 \times 10^{-3}$ have been included.

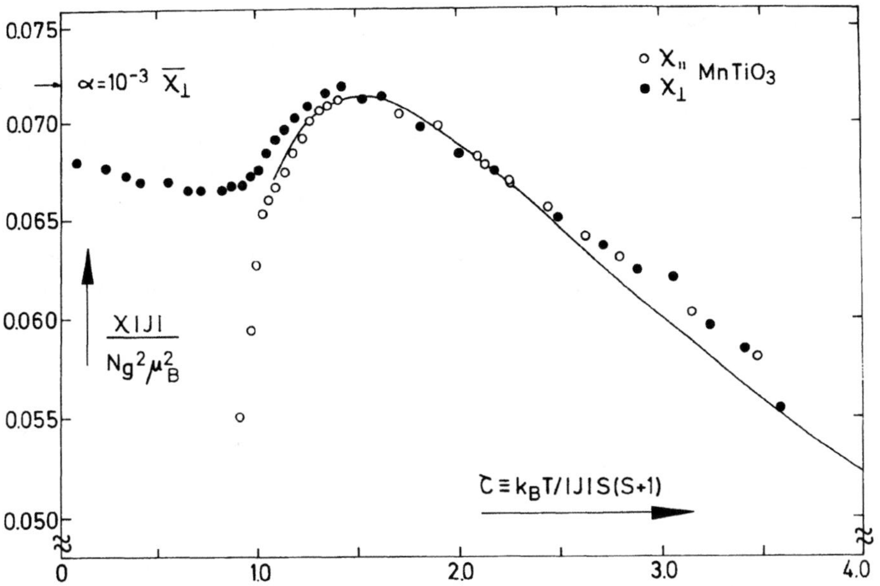


Fig. 23. Magnetic susceptibility of the ilmenite $MnTiO_3$ ($J/k_B$ = −7.29 K) compared with the prediction for $S = 5/2$ Heisenberg antiferromagnet in the h.c. lattice. The $\chi_\perp$ value at $T = 0$ deduced from spin wave theory with $\alpha = 0.001$ has been marked with an arrow.

### 5.2.5. *Restricted dimensionality*

The layered arrangement of the 3-d metal ions in the tetragonal structure of the $A_2MX_4$ compounds has yielded excellent examples of 2-d magnet with different $S$ values. Moreover, by changing the proportion of the M-metal with respect to the A and X ions several structures may be derived which may have one, two, three, . . . s.p. layers of M ions stacked one upon the other, yielding single, double, triple, . . . magnetic layers with the same distance between the sheets. Thus these materials provide experimental approximations of ultrathin magnetic films.

The first magnetic study on such a type of structure was performed in the family $Ca_2MnO_4$, $Ca_3Mn_2O_7$, $Ca_4Mn_3O_{10}$, $CaMnO_3$ [225] which, except for the last compound, show qualitatively the expected behaviour for 2-d systems of restricted dimensionality. However, as pointed out before, the non-negligible conductivity renders impossible further comparisons. In the fluorides and chlorides, on the other hand, the electrons are localized and therefore may be analysed with the above models. The magnetic double-layer systems, $Rb_3Mn_2F_7$, $K_3Mn_2F_7$ [168, 249–251], $Rb_3Mn_2Cl_7$ [252], $K_3Ni_2F_7$ [253] and $K_3Fe_2F_7$ [254] have been synthesized and studied magnetically.

The comparison of the $\chi_p$ data for $Rb_3Mn_2F_7$ and $K_3Mn_2F_7$ [168] with the h.t.s. predictions for $S = 5/2$ double-layer Heisenberg antiferromagnets gives results analogous to those obtained with the monolayer compounds, showing a broad maximum well above $T_N$, in contrast to the 3-d antiferromagnets. The fit obtained for the single crystal susceptibility of $K_3Mn_2F_7$ [249] is shown in Figure 24, where the theoretical prediction has been scaled upon the data with $J/k_B = (-4.02 \pm 0.12)$ K (h.t.s.). $J/k_B = (-3.81 \pm 0.08)$ K ($\chi_\perp$-spin wave) and $J/k_B = (-3.85 \pm 0.1)$ K ($\chi_\parallel$-spin wave). The slow convergence of the 2-d restricted dimensionality system to the bulk s.c. behaviour observed in the antiferromagnetic susceptibility (cf. Figure 14) [168] has been also verified in the sublattice magnetization measurements [250] and in other characteristic properties [251] found for $K_3Mn_2F_7$ and related compounds [252—254]. In fact, the magnetic behavior of these compounds is closer to that of $K_2MnF_4$, $Rb_2MnF_4$, etc. than to that of $KMnF_3$, $RbMnF_3$, etc. Although triple layer compounds exist [225] to date there are no conclusive magnetic studies of their behaviour.

## 5.3. XY AND ISING—HEISENBERG MODELS

Magnetic 2-d systems which should be described with the XY model are probably more difficult to find, and few compounds have been reported. The physical mechanism responsible for the planar anisotropy is analogous to that for the

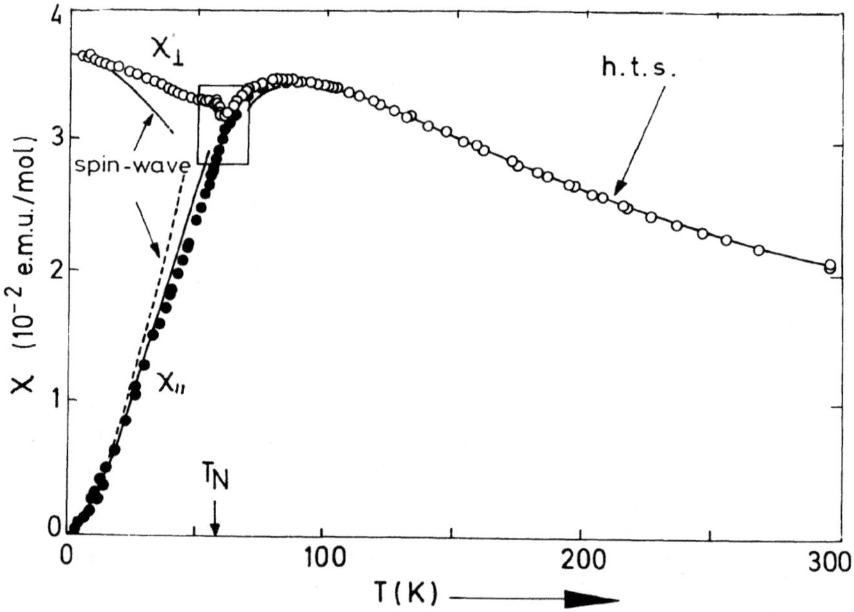


Fig. 24. Molar susceptibility, corrected for diamagnetism, of $K_3Mn_2F_7$ for the parallel (○) and perpendicular (●) directions. The full curves are theoretical predictions scaled with $J/k_B = -4.02$ K (h.t.s.), $-3.81$ K ($\chi_\perp$) and $-3.85$ K ($\chi_\parallel$ and $-3.85$ K ($\chi_\parallel$) for the $S = 5/2$ Heisenberg antiferromagnet on the double-layer lattice, as explained in the text. Taken from Arts *et al.* [249]. Reproduced by permission of Pergamon Press.

Ising anisotropy. For some crystal field symmetries, the lowest doublet may be described by an effective $S = 1/2$ Hamiltonian with strong planar anisotropy. The $CoX_2 \cdot 6\ H_2O$, X = Cl, Br [255—258], $Co(triazole)_2(NCS)_2$ [259] and $BaCo_2(AsO_4)_2$ [260] resemble the 2-d, XY model with different degree of approximation.

The compounds $CoCl_2 \cdot 6\ H_2O$, $T_N = 2.29$ K, and $CoBr_2 \cdot 6\ H_2O$, $T_N = 3.15$ K, [255] have monoclinic structure formed by layers of cobalt ions octahedrally coordinated by two clorine ions and four water molecules, in which each cobalt has antiferromagnetic interactions with four neighbours of the layer (s.p. lattice). The reported $\chi$ data [257, 258] clearly indicate a planar anisotropy, mainly due to crystal field effects [258]. In such a case, the anisotropy of the effective exchange parameters is related to the effective $g$-values [256], and $a/b \simeq (g_z/g_x)^2$ is about 0.29 and 0.19 [258] for the clorine and bromide compound, respectively. If $a/b$ is nonnegligible, deviations from the predictions for the pure XY model ($a/b = 0$) should result. However, the antiferromagnetic $\chi$ is not very sensitive to $a/b$ because for the $S = 1/2$ s.p. lattice and $a = 0$, $\tau_{max} = k_B T_{max}/|J| S(S+1) = 2.30$, $\chi'_{max} = 0.052$ (XY model); for $b = 0$ $\tau_{max} = 2.3249$, $\chi'_{max} = 0.0536$ (Ising model) and for $a = b = 1$, $\tau_{max} = 2.495$ $\chi'_{max} = 0.0469$. Therefore, in the intermediate XY-Heisenberg case with $a/b \approx 0.2$ only small differences ($\approx \pm 2\%$) with the $\tau_{max}$, $\chi'_{max}$ predictions for the XY model are expect. Scaling with $J/k_B = (-2.05 \pm 0.05)$ K, $g_x = 5.2$, $g_y = 5.1$ and $J/k_B = (-2.45 \pm 0.05)$ K $g_x = g_y = 5.0$ [258] for the Cl and Br compounds, respectively, the data indeed nicely fit the prediction for the pure XY model. Moreover, the agreement of the derived $g$-values with those deduced by other methods gives an independent check of the consistency of the above comparisons.

Whereas the antiferromagnetic $\chi$ data are not sensitive to the $a/b$ ratio, the specific heat changes drastically. The absence of singular behaviour in $C(T)$ for the XY and Heisenberg models implies an extreme sensitivity to an Ising-like anisotropy in the easy plane, as this leads to a lambda type specific heat peak at $T_c$. In Figure 25, the magnetic specific heat of $CoCl_2 \cdot 6\ H_2O$ scaled with $J/k_B = -2.05$ K is displayed together with the predictions for $S = 1/2$ for the s.p. Ising, Heisenberg and XY models. For $k_B T/|J| > 1.5$ the data nicely follow the predictions for the XY model, whereas at low temperatures deviations occur and a lambda peak appears. The Ising-like characteristics in the critical region is clearly seen by the fact that the data around $T_c$ are comparatively close to the 2-d Ising predictions.

In $Co(triazole)_2(NCS)_2$ [259] the cobalt ions are also octahedrally coordinated and a 2-d layered structure with four interacting neighbours is formed. The compound orders antiferromagnetically at $T_N = 5.71$ K with a small canting and the $\chi$ and $C(T)$ data scaled with $J/k_B = -5.0$ K, $g_x = 5.05 \pm 0.10$, $g_y = 4.5$ and $g_z = 2.5$ fit the theoretical predictions for the $S = 1/2$, XY model. The magnetic specific heat below $T_N$ has qualitatively the same behaviour as for $CoCl_2 \cdot 6\ H_2O$, but the larger differences between $g_x$ and $g_y$ produce a higher Ising-type anisotropy so that deviations from the Ising predictions are smaller.

Lastly, $BaCo_2(AsO_4)_2$ [260] shows 2-d behaviour with strong planar anisotropy,

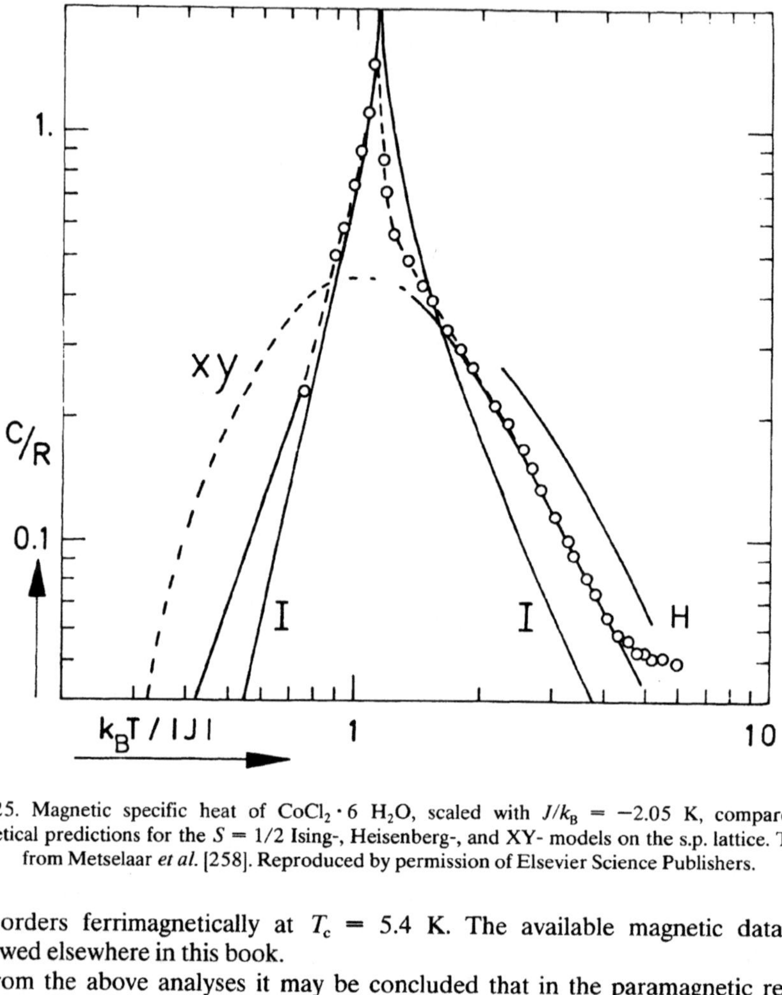


Fig. 25. Magnetic specific heat of $CoCl_2 \cdot 6\ H_2O$, scaled with $J/k_B = -2.05$ K, compared to theoretical predictions for the $S = 1/2$ Ising-, Heisenberg-, and XY- models on the s.p. lattice. Taken from Metselaar *et al.* [258]. Reproduced by permission of Elsevier Science Publishers.

and orders ferrimagnetically at $T_c = 5.4$ K. The available magnetic data are reviewed elsewhere in this book.

From the above analyses it may be concluded that in the paramagnetic region for the 2-d systems with XY anisotropy and $a/b < 0.2$ the comparison of $\chi$ with the XY predictions gives physically meaningful results, whereas the specific heat appears to behave much more like the 2-d Ising model. Whether this is due solely to a spin dimensionality crossover from XY to Ising-like behaviour as a consequence of small Ising-type anisotropy within the easy plane is an interesting problem that should be investigated further. The point is that the possibility of an intrinsic specific heat anomaly in the 2-d XY model cannot be discarded in view of recent theoretical results (see the chapter by Regnault and Rossat-Mignod).

**Acknowledgements**

The author wishes to thank Prof. Dr. D. González for his interest, encouragement and a careful reading of the manuscript, and Prof. Dr. L. J. de Jongh for many valuable discussions and criticisms. The financial support of the C.A.I.C.Y.T. (project 1423/82) and C.I.C.Y.T. (project PB86-0187) is gratefully acknowledged.

## Appendix A: Series Expansions for the Ising Model

TABLE A-I

High-temperature susceptibility coefficients for the 2-d $S = 1/2$ Ising model

$$k_B T\chi/Nm^2 = 1 + \sum_{n=1}^{\infty} a_n w^n \quad w = \tanh(J/2k_B T)$$

(Coefficients taken from Skyes *et al.* [55] and Nickel [56]. Reprinted by permission of IOP Publishing Ltd.)

| $n$/lat. | p.t. | s.p. | h.c. | $n$/lat. | s.p. | h.c. |
|---|---|---|---|---|---|---|
| 1 | 6 | 4 | 3 | 22 | 2350157268 | 1348998 |
| 2 | 30 | 12 | 6 | 23 | 5855734740 | 2403840 |
| 3 | 138 | 36 | 12 | 24 | 14569318492 | 4299018 |
| 4 | 606 | 100 | 24 | 25 | 36212402548 | 7677840 |
| 5 | 2586 | 276 | 48 | 26 | 89896870204 | 13635630 |
| 6 | 10818 | 740 | 90 | 27 | 222972071236 | 24206220 |
| 7 | 44574 | 1972 | 168 | 28 | 552460084428 | 43092888 |
| 8 | 181542 | 5172 | 318 | 29 | 1367784095156 | 76635984 |
| 9 | 732678 | 13492 | 600 | 30 | 3383289570292 | 135698970 |
| 10 | 2935219 | 34876 | 1098 | 31 | 8363078796612 | 240199320 |
| 11 | 11687202 | 89764 | 2004 | 32 | 20656054608404 | 426144654 |
| 12 | 46296210 | 229628 | 3696 | 33 | 50987841944612 | |
| 13 | 182588850 | 585508 | 6792 | 34 | 125771030685740 | |
| 14 | 717395262 | 1486308 | 12270 | | | |
| 15 | 2809372302 | 3763460 | 22140 | | | |
| 16 | 10969820358 | 9497380 | 40224 | | | |
| 17 | — | 23918708 | 72888 | | | |
| 18 | — | 60080156 | 130650 | | | |
| 19 | — | 150660388 | 234012 | | | |
| 20 | — | 377009364 | 421176 | | | |
| 21 | — | 942106116 | 756624 | | | |

TABLE A-II

High-temperature internal energy coefficients for the $S = 1/2$ Ising model

$$-2U/NJ = 1/2zw + \sum_{n=2}^{\infty} u_n w^n$$

(Coefficients taken from Domb [22]. Note that for loosely-packed lattices only even powers of $u$ appear).

| $n$/lat. | h.c.($u_{2n}$) | s.p.($u_{2n}$) | p.t.($u_n$) |
|---|---|---|---|
| 1 | 1.5 | 2 | 3 |
| 2 | 0 | 4 | 6 |
| 3 | 3 | 8 | 12 |
| 4 | −3 | 24 | 24 |
| 5 | 15 | 84 | 54 |
| 6 | −24 | 328 | 138 |
| 7 | 93 | 1372 | 378 |
| 8 | −180 | 6024 | 1080 |
| 9 | 639 | 27412 | 3186 |
| 10 | −1368 | 128228 | 9642 |
| 11 | 4653 | 613160 | 29784 |
| 12 | −10605 | 2985116 | 93552 |
| 13 | 35169 | 14751592 | 297966 |
| 14 | −83664 | 73825416 | 960294 |
| 15 | 272835 | 373488764 | 3126408 |
| 16 | −669627 | 1907334616 | 10268688 |
| 17 | 2157759 | 9820757380 | 33989388 |
| 18 | −5423280 | 50934592820 | 113277582 |
| 19 | 17319837 | 265877371160 | 379833906 |
| 20 | −44354277 | 1395907472968 | 1280618784 |

TABLE A-III

Low-temperature susceptibility coefficients for ferro- and antiferromagnetic interaction in the 2-d $S = 1/2$ Ising model

$$k_B T \chi^{F,AF}/4Nm^2 = u^{z/2} \sum_{n=0}^{\infty} b_n^{F,AF} u^n, \quad u = \exp(-2J/k_B T)$$

(Coefficients taken from Sykes *et al.* [67, 69] and from Baxter and Enting [71].)

| | FERRO | | | ANTIFERRO | |
|---|---|---|---|---|---|
| *n*/lattice | h.c.(*) | s.p. | p.t. | h.c.(*) | s.p. |
| 0 | 1 | 1 | 1 | 1 | 1 |
| 1 | 6 | 8 | 0 | 0 | 0 |
| 2 | 27 | 60 | 12 | 3 | 4 |
| 3 | 122 | 416 | 4 | 2 | 8 |
| 4 | 516 | 2791 | 129 | 12 | 39 |
| 5 | 2148 | 18296 | 122 | 24 | 152 |
| 6 | 8792 | 118016 | 1332 | 80 | 672 |
| 7 | 35622 | 752008 | 960 | 222 | 3016 |
| 8 | 143079 | 4746341 | 10919 | 687 | 13989 |
| 9 | 570830 | 29727472 | 11372 | 2096 | 66664 |
| 10 | 2264649 | 185016612 | 132900 | 6585 | |
| 11 | 8942436 | 1145415208 | 126396 | 20892 | |
| 12 | 35169616 | 7059265827 | 1299851 | 67216 | |
| 13 | 137839308 | 43338407712 | 1349784 | 218412 | |
| 14 | 2154378816 | 265168691392 | 50369760 | 2865216 | |
| 15 | 8395571712 | 1617656173824 | 56095776 | 9470784 | |
| 16 | 32639750676 | 9842665771649 | 484296732 | 31525524 | |
| 17 | 126625222200 | 59748291677832 | 571273344 | 105594912 | |
| 18 | 490300659692 | 361933688520940 | 4628107216 | 355673804 | |
| 19 | | 2188328005246304 | | | |
| 20 | | 13208464812265546 | | | |
| 21 | | 79600379336505342 | | | |

* The expansion variable is $z = u^{1/2}$.

TABLE A-IV

Coefficients of the spontaneous magnetization in the 2-d $S = 1/2$ Ising Model

$$M/Nm = 1 - 2u^{z/2} - u^{z-1} \sum_{n=1}^{\infty} b_n u^n$$

(Data taken from Sykes *et al.* [67]. Tables A-II, A-III and A-IV reprinted by permission of Academic Press.)

| *n*/lat. | h.c.(*) | s.p. | p.t. |
|---|---|---|---|
| 0 | 6 | 8 | 12 |
| 1 | 18 | 34 | −2 |
| 2 | 54 | 152 | 78 |
| 3 | 168 | 714 | −24 |
| 4 | 534 | 3472 | 548 |
| 5 | 1732 | 17318 | −228 |
| 6 | 5706 | 88048 | 4050 |
| 7 | 19038 | 454378 | −2030 |
| 8 | 64176 | 2373048 | 30960 |
| 9 | 218190 | 12515634 | −17670 |
| 10 | 747180 | 66551016 | 242402 |
| 11 | 2574488 | 356345666 | −152520 |
| 12 | 8918070 | 1919453984 | 1932000 |
| 13 | 31036560 | 10392792766 | −1312844 |
| 14 | 108457488 | 56527200992 | 15612150 |
| 15 | 380390574 | | −11297052 |
| 16 | 1338495492 | | |

* The expansion variable is $z = u^{1/2}$.

TABLE A-V

High-temperature susceptibility series at zero-field, for the general Ising model on the s.p. and p.t. lattices

$$3\chi k_B T/Ng^2\mu_B^2 X = \sum_{n=0}^{\infty} h_n(X)K^n, \quad h_n(X) = 1/D_n \sum_{i=1}^{n} b_{n,i}X^i$$

where $X = S(S+1)$ and $K = 2J/k_B T$. For each lattice and given order, the first number listed in $D_n$ (Data taken from Camp and Van Dyke [99] Reprinted by permission of the American Physical Society.)

| $n$, $i$/lat. | s.p. | p.t. | $n$, $i$/lat. | s.p. | p.t. |
|---|---|---|---|---|---|
| 1 | 3 | 1 | 8 | 70727580000 | 110120680000 |
| 1 | 4 | 2 | 1 | −212625 | −496125 |
| 2 | 45 | 5 | 2 | 13908510 | 95829480 |
| 1 | −6 | −1 | 3 | −286090380 | −3418716780 |
| 2 | 68 | 18 | 4 | 2834637552 | 51936829488 |
| 3 | 675 | 75 | 5 | −16115073984 | −420350361696 |
| 1 | 6 | 1 | 6 | 57157269408 | 1974825335232 |
| 2 | −216 | −66 | 7 | −125021664576 | −5447227764544 |
| 3 | 1144 | 464 | 8 | 136519395712 | 7291822764928 |
| 4 | 28350 | 6300 | 9 | 2334010140000 | 18153412200000 |
| 1 | −45 | −15 | 1 | 1771875 | 20671875 |
| 2 | 1806 | 1116 | 2 | −127575000 | −5455107000 |
| 3 | −16236 | −15956 | 3 | 2915701380 | 219572014140 |
| 4 | 50744 | 64904 | 4 | −32655191040 | −3736467377712 |
| 5 | 1190700 | 661500 | 5 | 216201641904 | 34847273833200 |
| 1 | 270 | 225 | 6 | −923056564608 | −198561705025920 |
| 2 | −12960 | −23652 | 7 | 2629938997056 | 724887263981120 |
| 3 | 190152 | 549228 | 8 | −4805264923392 | −1645784670954240 |
| 4 | −1024032 | −4010864 | 9 | 4468880440064 | 1864771814420736 |
| 5 | 2235808 | 11092044 | 10 | 6371847682200000 | 283193230320000 |
| 6 | 35721000 | 3969000 | 1 | −1616162625 | −107744175 |
| 1 | −1890 | −315 | 2 | 125415722250 | 34064787330 |
| 2 | 100260 | 38070 | 3 | −3025742906700 | −1471018205196 |
| 3 | −1630728 | −1024404 | 4 | 35695584786600 | 27028694292768 |
| 4 | 11497632 | 10828976 | 5 | −253070383970592 | −277966027290816 |
| 5 | −41433696 | −51683088 | 6 | 1198433342376000 | 1810359274718304 |
| 6 | 68310016 | 106529088 | 7 | −3998236247726400 | −7925971293760704 |
| 7 | 13395375000 | 59535000 | 8 | 9444732079338240 | 23678823685360128 |
| 1 | 141750 | 945 | 9 | −14794167627828480 | −45666137074603008 |
| 2 | −8505000 | −155790 | 10 | 11974432007975936 | 44769163917919744 |
| 3 | 163501200 | 5059764 | | | |
| 4 | −1483268400 | −67444248 | | | |
| 5 | 7190638800 | 442284696 | | | |
| 6 | −19711689600 | −1550331552 | | | |
| 7 | 25954467200 | 2524174144 | | | |

## Appendix B.: Series Expansions for the Heisenberg Model

TABLE B-I

High-temperature susceptibility coefficients for the $S = 1/2$ Heisenberg model on 2-d lattices

$$\chi = Ng^2\mu_B^2 S(S+1)/3k_B T\left(1 + \sum_{n=1}^{\infty} a_n K^n\right)$$

(Coefficients taken from Rushbrooke *et al.* [27] and Yamaji and Kondo [137] and reprinted by permission of Academic Press and the Physical Society of Japan.)

| $n$/lat. | p.t. | s.p. | h.c. |
|---|---|---|---|
| 1 | 3 | 2 | 1.5 |
| 2 | 6 | 2 | 0.75 |
| 3 | 8.5 | 1.3333333333 | −0.125 |
| 4 | 9.375 | 1.0833333333 | 0.3125 |
| 5 | 11.025 | 1.1833333333 | 0.66875 |
| 6 | 16.964583333 | 0.50972222222 | −0.59270833 |
| 7 | 21.152678571 | −0.32182539683 | −0.7399553571 |
| 8 | 8.8058779762 | 0.40739087302 | 1.484784226 |
| 9 | −9.6784556878 | 1.0672839506 | |
| 10 | | −0.69281883818 | |

TABLE B-II

High-temperature susceptibility series for the general spin Heisenberg model on 2-d lattices.

$$\chi = Ng^2\mu_B^2X/3k_BT\left(1+\sum_{n=1}^{\infty}a_n(X)K^n\right) \qquad a_n(X)=\sum_{i=1}^{n}a'_nX^n$$

(Data taken from Rushbrooke and Wood [26], Stephenson *et al.* [136] and Yamaji and Kondo [137].)

| $n$/lat. | h.c. | s.p. | p.t. |
|---|---|---|---|
| 1 | 2.0 | 2.6666... | 4.0 |
| 2 | −1.0 | −1.3333... | −2.0 |
| | 2.6666... | 5.3333... | 13.3333... |
| 3 | 0.5333... | 0.7111... | 1.0666... |
| | −3.2 | −6.0444... | −16.1777... |
| | 3.0222... | 9.9555... | 39.8222... |
| 4 | −0.3333... | −0.4444... | −0.6666... |
| | 3.4222... | 6.4 | 16.7111... |
| | −5.7185185.. | −17.85679 | −81.8962962.. |
| | 3.3185185.. | 16.908642 | 110.6962962.. |
| 5 | 0.243809524.. | 0.325079365.. | 0.487619048.. |
| | −3.32952381.. | −6.21714286.. | −16.2946030.. |
| | 9.574179895.. | 27.6001411.. | 129.720212.. |
| | −8.070546738.. | 43.2084656.. | −329.535662.. |
| | 3.679717813.. | 27.2404468.. | 292.309559.. |
| 6 | −0.2044... | −0.272592632.. | −0.4088... |
| | 3.3465502.. | 6.25439152.. | 16.5875132.. |
| | −1.39715697.. | −39.1263492.. | −181.352240.. |
| | 17.5785538.. | 87.3315930.. | 714.219118.. |
| | −11.0654909.. | −90.3590201.. | −1152.81721.. |
| | 3.57587301.. | 42.2121576.. | 741.855195.. |
| 7 | 0.193015873.. | 0.257354497.. | 0.386031846.. |
| | −3.592126985.. | −6.723724868.. | 18.104211640.. |
| | 18.890271605.. | 5.235835861.. | 243.85721341.. |
| | 34.984183422.. | −159.70991182.. | −1322.7384926.. |
| | 27.065766804.. | 227.83217402.. | 3186.8343861.. |
| | −14.451985107.. | −172.89887828.. | 3667.7067325.. |
| | 2.884577699.. | 63.06700216.. | 182.2051380.. |
| 8 | −0.202355174.. | −0.269817359.. | −0.404482782.. |
| | 4.157464827.. | 7.795048314.. | 21.326326851.. |
| | −25.458338221.. | −70.517926962.. | −332.957300122.. |
| | 62.729007787.. | 274.077823598.. | 2259.611626251.. |
| | 64.363864952.. | −509.010466537.. | −7397.774723637.. |
| | 40.577429721.. | 517.835575876.. | 12335.065873521.. |
| | −15.177560911.. | −309.362558141.. | −10884.346727490.. |
| | 2.679563655.. | 91.663819191.. | 4351.677538847.. |

TABLE B-III

High-temperature susceptibility coefficients for the $S = \infty$ Heisenberg model. (Coefficients as defined in Table B-I and taken from Rushbrooke and Wood [26], Camp and Van Dycke [45] and Stanley [141] and reprinted by permission of IOP Published Ltd. and the American Physical Society.)

| $n$/lat. | p.t. | s.p. | h.c. |
|---|---|---|---|
| 1 | 4.0 | 2.6666667 | 2.0 |
| 2 | 13.3333333 | 5.3333333 | 2.6666667 |
| 3 | 39.8222222 | 9.9555556 | 3.0222222 |
| 4 | 110.696296 | 16.9086420 | 3.3185185 |
| 5 | 292.309559 | 27.2404468 | 3.6797178 |
| 6 | 741.855191 | 42.2121576 | 3.5758730 |
| 7 | 1822.0514 | 63.0670021 | 2.88457770 |
| 8 | 4351.6776 | 91.6638192 | 2.67956366 |
| 9 | 10144.7594 | 129.4967 | 3.2328 |
| 10 | 23151.5426 | 179.050402 | |

TABLE B-IV

High-temperature specific heat coefficients for the $S = 1/2$ Heisenberg model in 2-d lattices

$$C(T) = 2/3Nk_BzS^2(S+1)^2K^2\left(1 + \sum_{n=1}^{\infty} c_nK^n\right)$$

(Coefficients taken from Baker *et al.* [133] and Yamaji and Kondo [137].)

| $n$/lat. | h.c. | s.p. | p.t. |
|---|---|---|---|
| 1 | −1.0 | −1.0 | 1.0 |
| 2 | −1.35 | −1.75 | −4.25 |
| 3 | 4.1666.. | 4.1666.. | −7.50 |
| 4 | −13.7589286.. | 1.35416666.. | 14.229166.. |
| 5 | −13.4291666.. | −10.82083333.. | 42.4375 |
| 6 | −2.7142363.. | 3.33090288.. | −32.85590278.. |
| 7 | 30.8652777.. | 22.28908731.. | −201.9724207.. |
| 8 | | −19.89351439.. | 0.810106648.. |

TABLE B-V

High-temperature specific heat coefficients for the general spin Heisenberg model on 2-d lattices

$$C(T) = 2/3Nk_BzS^2(S+1)^2K^2\left(1 + \sum_{n=1}^{\infty} c_n(X)K^n\right) \qquad c_n(X) = \sum_{i=1}^{n} c'_iX^i$$

(As taken from Rushbrooke and Wood [26] and Yamaji and Kondo [137].)

| $n$/lat. | h.c. | s.p. | p.t. |
|---|---|---|---|
| 1 | −1.0 | −1.0 | −1.0 |
| | 0.0 | 0.0 | 2.6666... |
| 2 | 0.8 | 0.8 | 0.8 |
| | −3.4666... | −4.8 | −10.1333... |
| | 0.8 | 1.8666... | 4.5333... |
| 3 | −0.6666... | −0.6666... | −0.6666... |
| | 5.7777... | 8.0 | 15.1111... |
| | 0.8888... | −2.074074.. | −35.851851.. |
| | 0.0 | 0.0 | 4.740740.. |
| 4 | 0.609523810.. | 0.609523810.. | 0.609523810.. |
| | −7.434920635.. | −10.323809524.. | −19.034920634.. |
| | −19.154497354.. | 25.120634921.. | 110.557671958.. |
| | 4.317460317.. | −12.867724868.. | −81.845502646.. |
| | 0.507936508.. | −0.677248677.. | 0.744973545.. |
| 5 | −0.6133... | −0.6133... | −0.6133... |
| | 9.1911... | 12.8 | 23.6088... |
| | −26.816296396.. | −57.388148148.. | −212.442962963.. |
| | −5.629629630.. | 28.855308642.. | 442.864197531.. |
| | −7.111... | 0.9481481.... | −124.365432099.. |
| | 0.0 | 0.0 | −11.124938272.. |
| 6 | 0.6755... | 0.6755... | 0.6755... |
| | −11.690666610.. | −16.327111.. | −30.359703704.. |
| | 48.651061687.. | 101.86903704.. | 355.71358025.. |
| | −31.607835378.. | −159.06923457.. | −1344.1733004.. |
| | −12.772082306.. | 78.110200277.. | 1213.0217613.. |
| | −14.798134431.. | 1.778304529.. | −87.887012339. |
| | −6.184032922.. | −2.415319626.. | −37.808636467.. |
| 7 | −0.809433212.. | −0.809433212.. | −0.809433212.. |
| | 15.624852863.. | 21.877552849.. | 41.146017042.. |
| | −80.329263989.. | −167.271656761.. | −575.644016680.. |
| | 116.640653125.. | 431.175595349.. | 3064.523867900.. |
| | 22.769573518.. | −264.502308974.. | −5620.704614456.. |
| | 28.833335686.. | 7.718828930.. | 2347.960704228.. |
| | 10.601199295.. | 4.140547913.. | 183.897740929.. |
| | 0.0 | 0.0 | −94.054002970.. |

TABLE B-VI

High-temperature specific heat coefficients for the $S = \infty$, Heisenberg model on 2-d lattices. (Coefficients as defined in Table B-V and taken from Rushbrooke and Wood [26], Yamaji and Kondo [137] and Lambeth and Stanley [176].)

| $n$/lat. | p.t. | s.p. | h.c. |
|---|---|---|---|
| 1 | 2.6666666667 | 0.0 | 0.0 |
| 2 | 4.533333333 | 1.866666667 | 0.8 |
| 3 | 4.74074074074 | 0.0 | 0.0 |
| 4 | 0.74497354497 | −0.677248677 | 0.50793657936 |
| 5 | −11.1249382716 | 0.0 | 0.0 |
| 6 | −37.808636467 | −2.4153196263 | −6.18403292181 |
| 7 | −94.05400296958 | 0.0 | 0.0 |
| 8 | — | −3.989975248 | — |

## Appendix C: Series Expansions for the XY Model

TABLE C-I

High-temperature series coefficients of the zero-field mean square fluctuation of the magnetization, $Y_\perp$, and of the susceptibility, $\chi_\perp$, for the $S = 1/2$ XY model on the s.p. and p.t. lattices.

$$Y_\perp = \langle M_\perp^2 \rangle / Nm^2 = \sum_{n=0}^{\infty} a_n K^n \quad \text{and} \quad \chi'_\perp = \chi_\perp T/C = \sum_{n=0}^{\infty} b_n K^n$$

(Data taken from Rogiers *et al.* [184], Obokata *et al.* [140] and Jou and Chen [187].)

| $n$ | $Y_\perp$(p.t.) | $Y_\perp$(s.p.) | $\chi'_\perp$(s.p.) | $\chi'_\perp$(p.t.) | $\chi'_\parallel$(s.p.) |
|---|---|---|---|---|---|
| 0 | 1 | 1 | 1 | 1 | 1 |
| 1 | 3 | 2 | 2 | 0 | 0 |
| 2 | 7.5 | 3 | 2.666 | −1.5 | −1 |
| 3 | 14.5 | 3.3333 | 3.1666 | −1.0 | 0 |
| 4 | 24.385 | 3.25 | 3.3666 | 1.75 | 0.8333 |
| 5 | 39.8625 | 3.275 | 3.397222 | 5.125 | 0 |
| 6 | 66.6875 | 3.430555556 | 3.3908730 | −1.595833 | −0.63611 |
| 7 | 110.75907738 | 3.391964286 | — | −6.2125 | — |
| 8 | 175.09226190 | 3.099007937 | | | |
| 9 | 262.6967552 | 2.728331680 | | | |
| 10 | 386.2849611 | 2.450618386 | | | |
| 11 | 574.4431813 | 2.257487411 | | | |
| 12 | 861.4927498 | 2.012109617 | | | |

TABLE C-II

High-temperature series coefficients of $\chi_\perp$ on the 2-d classical XY($\infty$) and Planar Rotator models.

$$\chi_\perp T/C = \sum_{n=0}^{\infty} a_n K^n \qquad K = 2J/k_b T$$

(Data taken from Camp and Van Dycke [45], Ferer and Velgakis [157] and Butera *et al.* [185] and reprinted by permission of the publishers.)

| | XY($\infty$) | | PR | |
|---|---|---|---|---|
| *n*/lat. | s.p. | p.t. | s.p. | p.t. |
| 0 | 1.0 | 1.0 | 1.0 | 1.0 |
| 1 | 1.3333333 | 2.0 | 2.0 | 3.0 |
| 2 | 1.4222222 | 3.466666 | 3.0 | 7.5 |
| 3 | 1.4696296 | 5.582222 | 4.25 | 16.875 |
| 4 | 1.4052010 | 8.568888 | 5.50 | 35.625 |
| 5 | 1.3113899 | 12.7001220 | 6.8541667 | 72.0625 |
| 6 | 1.1818169 | 18.3262955 | 8.2656250 | 141.2734375 |
| 7 | 1.0451803 | 25.88777622 | 9.7220052 | 270.1728516 |
| 8 | 0.9065299 | 35.93996240 | 11.2050781 | 506.3834635 |
| 9 | — | 49.17983102 | 12.67555338542 | 933.5703775 |
| 10 | | 66.47874172 | 14.15201280382 | 1697.512101 |
| 11 | | 88.92278656 | 15.60190022786 | 3050.264278 |
| 12 | | 117.8618526 | 17.01930067274 | 5424.862134 |
| 13 | | | 18.39246629987 | |
| 14 | | | 19.71450651562 | |
| 15 | | | 20.97145583863 | |

TABLE C-III

High-temperature series coefficients of the partition function of the $S = 1/2$ XY model for 2-d lattices

$$(\ln \Lambda)/N = \sum_{n=2}^{\infty} \lambda_n K^n$$

(Data taken from Rogiers *et al.* [184] and Betts *et al.* [181].)

| $n$/lat. | p.t. | s.p. | h.c. |
|---|---|---|---|
| 2 | 0.75 | 0.5 | 0.375 |
| 3 | 0.5 | 0.0 | 0.0 |
| 4 | 0.03125 | −0.0208333 | −0.078125 |
| 5 | −0.375 | 0.0 | 0.0 |
| 6 | −0.33125 | −0.0152777 | 0.04791666 |
| 7 | 0.2020833 | 0.0 | 0.0 |
| 8 | 0.644652158 | 0.0184466181 | −0.040350632 |
| 9 | 0.266774140 | 0.0 | 0.0 |
| 10 | −0.804054646 | −0.015259590 | 0.072247437 |
| 11 | −1.244971892 | 0.0 | |
| 12 | 0.120835886 | 0.009194996 | |
| 13 | 2.253582554 | 0.0 | |

## References


1. P. Weiss, *J. De Physique* **6** (1907) 661.
2. P. Weiss, *Ann. Phys.* **17** (1932) 97.
3. L. Neel, *Ann. Phys.* **18** (1932) 1.
4. For a comprehensive general view of the effective field theory see, e.g. J. S. Smart (1966) in *Effective Field Theories of Magnetism*, W. B. Saunders Company.
5. E. Ising, *Z. Phys.* **31** (1925) 253.
6. W. Heisenberg, *Z Phys.* **49** (1928) 619.
7. T. Arai, *Phys. Rev.* **126** (1962) 471.
8. T. Arai, *Phys. Rev.* **134A** (1964) 824.
9. A review of the exact results available for Hamiltonian (1) can be found in *Phase Transitions and Critical Phenomean* (1972) Vol. I, C. Domb and M. S. Green Eds. (Academic Press).
10. C. Domb, *Adv. Phys.* **9** (1960) 149 and 245.
11. M. E. Fisher, *Rep. Prog. Phys.* **30** (1967) 615.
12. H. A. Kramers, *Commun. Kamerlingh Onnes Lab., Leiden Suppl.* **83** (1936) 1.
13. W. Opechowsky, *Physica* **4** (1937) 181.
14. H. A. Kramers and G. H. Wannier *Phys. Rev.* **60** (1941) 252 and 263.
15. L. Onsager *Phys. Rev.* **65** (1944) 117.
16. C. Domb *Proc. R. Soc.* **A199** (1949) 199.
17. A. J. Wakefield, *Proc. Cam. Phil. Soc.* **47** (1951) 419 and 799.
18. C. Domb and R. B. Potts, *Proc. Roy. Soc.* **A210** (1951) 125.
19. C. Domb in: 'Changement de phases' *Comptes Rendus de la 2 ème. Réunion de Chimie Physique*, Paris, 1952, p. 192.
20. C. Domb and M. F. Sykes, *Proc. Roy. Soc.* **A235** (1956) 247.
21. C. Domb and M. F. Sykes, *Proc. Roy. Soc.* **A240** (1957) 214.

22. C. Domb in: *Phase Transitions and Critical Phenomena* Vol. III, Chap. 6, C. Domb and M. S. Green, Eds., Academic Press, 1974.
23. V. Zehler, *Z. Naturf.* **5a** (1950) 344.
24. H. A. Brown and J. M. Luttinger, *Phys. Rev.* **100** (1955) 685.
25. H. A. Brown, *Phys. Rev.* **104** (1956) 624.
26. G. S. Rushbrooke and P. J. Wood, *Proc. Phys. Soc. London* **68** (1956) 1161, and *Mol. Phys.* **1** (1958) 257.
27. G. S. Rushbrooke, G. A. Baker and P. J. Wood in: *Phase Transitions and Critical Phenomena* Vol. III, Chap. 5; C. Domb and M. S. Green, Eds., Academic Press, 1974.
28. D. D. Betts in: *Phase Transitions and Critical Phenomena* Vol. III, Chap. 8, C. Domb and M. S. Green, Eds., Academic Press, 1974.
29. H. E. Stanley in: *Phase Transitions and Critical Phenomena* Vol. III, Chap. 7, C. Domb and M. S. Green, Eds., Academic Press, 1974.
30. K. G. Wilson, *Phys. Rev.* **B4** (1971) 3174 and 3184.
31. H. De Raedt and A. Lagendijk, *Phys. Rep.* **127** (1985) 233.
32. G. A. Baker Jr., B. G. Nickel and D. I. Merion, *Phys. Rev.* **B17** (1978) 1365: J. C. Le Guillou and J. Zinn-Justin, *Phys. Rev.* **B21** (1980) 3976.
33. D. S. Gaunt in: *Phase Transitions Carges 1980*; H. Levy, J. C. Le Guillou and Z. Zinn-Justin Eds., Plenum Press, 1982, p. 217.
34. J. Zinn-Justin, *J. Physique* **42** (1981) 783: B. G. Nickel and M. Dixon *Phys. Rev.* **B26** (1982) 3965: J. Adler, M. Moshe and V. Privman, *Phys. Rev.* **B26** (1982) 3958: J. H. Chen, M. E. Fisher and B. G. Nickel, *Phys. Rev. Lett.* **48** (1982) 630.
35. A. J. Guttmann, *Phys. Rev.* **B33** (1986) 5089; *J. Phys.* **A20** (1987) 1855: P. R. Graves-Morris, *J. Phys.* **A21** (1988) 1867.
36. J. O. Indekeu, A. L. Stella, J. Rogiers, *Phys. Rev.* **B32** (1985) 7333.
37. D. S. Gaunt and A. J. Guttmann In: *Phase Transitions and Critical Phenomena* Vol. III, Chap. 4, C. Domb and M. S. Green, Eds. Academic Press, 1974.
38. C. J. Pearce, *Adv. Phys.* **27** (1978) 89.
39. G. A. Baker Jr., *Phys. Rev.* **124** (1961) 768.
40. G. A. Baker Jr. in: *Advances in Theoretical Physics* Vol. I, pp. 1—58. K. A. Brueckner, Eds., Academic Press, 1965.
41. G. A. Baker Jr. in: *The Padé Approximant in Theoretical Physics*, Chap. 1, G. A. Baker and J. L. Gammel, Eds. Academic Press, 1971.
42. D. L. Hunter and G. A. Baker Jr., *Phys. Rev.* **B7** (1973) 3346.
43. G. A. Baker Jr., G. S. Rushbrooke and H. E. Gilbert, *Phys. Rev.* **135** (1964) A1272.
44. J. M. Kosterlitz and D. J. Thouless, *J. Phys.* **C6** (1973) 1181.
45. W. J. Camp and J. P. Van Dyke, *J. Phys.* **C8** (1975) 336.
46. G. A. Baker Jr. and D. L. Hunter, *Phys. Rev.* **B7** (1973) 3377.
47. D. L. Hunter and G. A. Baker Jr., *Phys. Rev.* **B19** (1979) 808.
48. M. E. Fisher and J. H. Chen in: *Phase Transitions, Carges 1980*; H. Levy J. C. Le Guillou and J. Zinn-Justin, Eds., Plenum Press, 1982, p. 169.
49. L. Onsager, *Nuovo Cim.* **6** (Suppl.) (1949) 261.
50. C. N. Yang, *Phys. Rev.* **85** (1952) 808.
51. M. E. Fisher, *J. Math. Phys.* **4** (1963) 124.
52. H. N. V. Temperley in: *Phase Transitions and Critical Phenomena* Vol. I, Chap. 6, C. Domb and M. S. Green Eds. Academic Press, 1972.
53. J. H. Barry and M. Khatun, *Phys. Rev.* **B35** (1987) 8601.
54. J. E. Brooks and C. Domb, *Proc. Roy. Soc.* **A207** (1951) 343.
55. M. F. Sykes, *J. Math. Phys.* **2** (1961) 52: M. F. Sykes and M. E. Fisher, *Physica* **28** (1962) 919: M. F. Sykes, D. S. Gaunt, P. D. Roberts and J. A. Wyles, *J. Phys.* **45** (1972) 624.
56. B. G. Nickel in: *Phase Transitions, Carges 1980*, H. Levy, J. C. Le Guillou and Z. Zinn-Justin Eds., Plenum Press, 1982, p. 291.
57. C. Domb and M. F. Sykes, *J. Math. Phys.* **2** (1961) 63.
58. S. Gartenhaus and W. Scott McCullough, *Phys. Lett.* **A127** (1988) 315; *Phys. Rev.* **B38** (1988) 11688.

59. E. Barouch, B. M. McCoy and T. T. Wu, *Phys. Rev. Lett.* **31** (1973) 1409: T. T. Wu, B. M. McCoy, G. A. Tracy and E. Barouch, *Phys. Rev.* **B13** (1976) 316.
60. A. J. Guttmann, *Phys. Rev.* **B9** (1974) 4991: *J. Phys.* **A8** (1975) 1236.
61. X. P. Kong, H. Au-Yang and J. H. H. Perk, *Phys. Lett.* **A116** (1986) 54.
62. M. F. Sykes and M. E. Fisher, *Phys. Rev. Lett.* **1** (1958) 321.
63. M. E. Fisher, *Physica* **25** (1959) 521.
64. J. Ashkin and W. E. Lamb, *Phys. Rev.* **64** (1943) 159.
65. M. F. Sykes, J. W. Essam and D. S. Gaunt, *J. Math. Phys.* **6** (1965) 283.
66. M. F. Sykes, D. S. Gaunt, J. W. Essam and D. L. Hunter, *J. Math. Phys.* **14** (1973) 1060.
67. M. F. Sykes, D. S. Gaunt, S. R. Mattingly, J. W. Essam, and C. J. Elliot, *J. Math. Phys.* **14** (1973) 1066.
68. M. F. Sykes, D. S. Gaunt, J. L. Martin S. R. Mattingly and J. W. Essam, *J. Math. Phys.* **14** (1973) 1071.
69. M. F. Sykes, M. G. Watts and D. S. Gaunt, *J. Phys.* **A8** (1975) 1441 and 1448.
70. M. F. Sykes, S. McKenzie, M. G. Watts and D. S. Gaunt, *J. Phys.* **A8** (1975) 1461.
71. R. J. Baxter and I. G. Enting, *J. Stat. Phys.* **21** (1979) 103.
72. M. F. Sykes and D. L. Hunter, *J. Phys.* **A13** (1980) 1121.
73. J. W. Essam and M. E. Fisher, *J. Chem. Phys.* **38** (1963) 802.
74. M. E. Fisher, *Phys. Rev.* **113** (1959) 969.
75. J. W. Essam and D. L. Hunter, *J. Phys.* **C1** (1968) 392.
76. D. S. Gaunt, M. E. Fisher, M. F. Sykes and J. W. Essam, *Phys. Rev. Lett.* **13** (1964) 713.
77. C. Domb and D. L. Hunter, *Proc. Phys. Soc.* **86** (1965) 1147.
78. D. S. Gaunt, *Proc. Phys. Soc.* **92** (1967) 150.
79. C. Domb, *J. Appl. Phys.* **39** (1968) 620.
80. D. S. Gaunt and C. Domb, *J. Phys.* **C3** (1970) 1442.
81. D. S. Gaunt and M. F. Sykes, *J. Phys.* **C5** (1972) 1429.
82. D. D. Betts and L. Filipow, *Can. J. Phys.* **50** (1972) 3117.
83. G. A. Baker Jr., and D. Kim, *J. Phys.* **A13** (1980) L103.
84. I. G. Enting and R. J. Baxter, *J. Phys.* **A13** (1980) 3723.
85. C. Domb, *J. Phys.* **A9** (1976) 283.
86. A. Bienenstock, *J. Appl. Phys.* **37** (1966) 1459.
87. A. Bienenstock and J. Lewis, *Phys. Rev.* **160** (1967) 393.
88. D. C. Rapaport and C. Domb, *J. Phys.* **C4** (1971) 2684.
89. R. B. Griffiths, *Phys. Rev. Lett.* **24** (1970) 1479.
90. M. Ferer, *Phys. Rev.* **B18** (1978) 6265.
91. R. J. Elliott and C. Wood, *J. Phys.* **C4** (1971) 2359.
92. A. Yanase, Y. Takeshige and M. Suzuki, *J. Phys. Soc. Japan* **41** (1976) 1108.
93. J. Oitmaa and M. Plischke, *J. Phys.* **C9** (1976) 2093.
94. C. Domb and M. F. Sykes, *Phys. Rev.* **128** (1962) 168.
95. P. F. Fox and D. S. Gaunt, *J. Phys.* **C3** (1970) L88: *J. Phys.* **C5** (1972) 3085.
96. A. J. Guttmann, C. Domb, P. F. Fox, *J. Physique* **32** (1971) C1-354.
97. K. De'Bell and D. D. Betts, *Physica* **A119** (1983) 78.
98. P. F. Fox and A. J. Guttmann, *J. Phys.* **C6** (1973) 913: J. Adler and I. G. Enting, *J. Phys.* **A17** (1984) 2233.
99. J. P. Van Dyke and W. J. Camp, *Phys. Rev.* **B9** (1974) 3121: W. J. Camp and J. P. Van Dyke, *Phys. Rev.* **B11** (1975) 2579.
100. J. Oitmaa and M. J. Velgakis, *J. Phys.* **A19** (1986) L553.
101. D. W. Wood and N. W. Dalton, *J. Phys.* **C5** (1972) 1675.
102. S. L. Schofield and R. G. Bowers, *J. Phys.* **A14** (1981) 2163: B. Y. Yousif and R. G. Bowers, *J. Phys.* **A16** (1983) 3361: R. G. Bowers and B. Y. Yousif, *J. Phys.* **A17** (1984) 895.
103. R. R. P. Singh and S. Chakravarty, *Phys. Rev. Lett.* **57** (1986) 245: *Phys. Rev.* **B36** (1987) 546 and 559.
104. P. W. Kasteleyn, *J. Math. Phys.* **4** (1963) 287.
105. L. E. Ballentine, *Physica* **30** (1964) 1231.

106. G. A. T. Allan, *Phys. Rev.* **B1** (1970) 352.
107. T. W. Capehart and M. E. Fisher, *Phys. Rev.* **B13** (1976) 5021.
108. J. Oitmaa and I. G. Enting, *J. Phys.* **A8** (1975) 1097.
109. K. Binder and P. C. Hohenberg, *Phys. Rev.* **B6** (1972) 3461: *Phys. Rev.* **B9** (1974) 2194.
110. I. G. Enting and A. J. Guttmann, *J. Phys.* **A13** (1980) 1043.
111. A. J. Guttmann, G. M. Torrie and S. G. Whittington, *J. Magn. Magn. Mat.* **15—18** (1980) 1091.
112. C. Domb and N. W. Dalton, *Proc. Phys. Soc.* **89** (1966) 859.
113. N. W. Dalton and D. W. Wood, *J. Math. Phys.* **10** (1969) 1271.
114. M. Plischke and J. Oitmaa, *Phys. Rev.* **B19** (1979) 487: J. Oitmaa *Can. J. Phys.* **59** (1981) 15: *J. Phys.* **A14** (1981) 1159: *J. Phys.* **A15** (1982) 573.
115. J. Oitmaa and M. J. Velgakis, *J. Phys.* **A20** (1987) 1269.
116. R. Abe, *Prog. Theor. Phys.* **44** (1970) 339.
117. J. Oitmaa and I. G. Enting, *Phys. Lett.* **36A** (1971) 91.
118. G. Paul and H. E. Stanley, *Phys. Rev.* **B5** (1972) 2578: F. Harbus and H. E. Stanley, *Phys. Rev.* **B7** (1973) 365.
119. R. Krasnow, F. Harbus, L. L. Liu and H. E. Stanley, *Phys. Rev.* **B7** (1973) 370.
120. L. L. Liu and H. E. Stanley, *Phys. Rev.* **B8** (1973) 2279: F. Harbus and H. E. Stanley, *Phys. Rev.* **B8** (1973) 2268.
121. L. J. de Jongh and H. E. Stanley, *Phys. Rev. Lett.* **36** (1976) 817.
122. R. Navarro and L. J. de Jongh, *Physica* **B94** (1978) 67.
123. C. A. W. Citteur and P. W. Kasteleyn, *Physica* **68** (1973) 43 and 491.
124. F. Harbus and H. E. Stanley, *Phys. Rev.* **B8** (1973) 1141: S. Redner and H. E. Stanley, *J. Phys.* **C10** (1977) 4765: *Phys. Rev.* **B16** (1977) 4901.
125. W. Selke, *Modulated Structure Materials*, T. Tsakalatos, Ed., Dordrecht, Martinus Nijhoff, pp. 23, 1984.
126. J. Oitmaa, *J. Phys.* **A18** (1985) 365.
127. M. E. Fisher and W. Selke, *Phys. Rev. Lett.* **44** (1980) 1502: *Phil. Trans. R. Soc.* **302** (1463) (1981) 1: V. L. Pokrovsky and G. V. Uimin, *J. Phys.* **C15** (1982) L353: *Sov. Phys. JETP* **55** (1982) 950: J. Smith and J. Yeomans, *J. Phys.* **C16** (1983) 5305.
128. J. Oitmaa, M. J. Velgakis, M. T. Bachelor and M. N. Barber, *J. Magn. Magn. Mat.* **54—57** (1986) 665: J. Oitmaa and M. J. Velgakis, *J. Phys.* **A20** (1987) 1495.
129. N. D. Mermin and H. Wagner *Phys. Rev. Lett.* **17** (1966) 1133.
130. F. Keffer in: *Encyclopedia of Physics* Vol. XVIII, 2; H. D. J. Wijn Ed. Springer Verlag, 1966.
131. A. L. Fetter and J. D. Walecka, in: *Quantum Theory of Many-Particle Systems*, McGraw-Hill, 1971.
132. C. Domb, *Phil. Mag. Suppl.* **9** (1960) 149.
133. G. A. Baker Jr., H. E. Gilbert, J. Eve and G. S. Rushbrooke, *Phys. Lett.* **25A** (1967) 207.
134. M. Wortis in: *Phase Transitions and Critical Phenomena* Vol. III, Chap. 3, C. Domb and M. S. Green, Eds. Academic Press, 1974.
135. G. S. Rushbrooke and P. J. Wood, *Mol. Phys.* **6** (1963) 409.
136. P. L. Stephenson, K. Pirnie, P. J. Wood and J. Eve, *Phys. Lett.* **27A** (1968) 2.
137. K. Yamaji and J. Kondo, *J. Phys. Soc. Japan* **35** (1973) 25.
138. J. Gammel, W. Marshall and L. Morgan, *Proc. Roy. Soc.* **275** (1963) 257.
139. H. E. Stanley and T. A. Kaplan, *Phys. Rev. Lett.* **17** (1966) 913.
140. T. Obokata, I. Ono and T. Oguchi, *J. Phys. Soc. Japan* **23** (1967) 516.
141. H. E. Stanley, *Phys. Rev.* **158** (1967) 546: *Phys. Rev. Lett.* **20** (1967) 589.
142. See for example A. Nymeyer, A. J. Guttmann, *J. Phys.* **A18** (1985) 495.
143. M. A. Moore, *Phys. Rev. Lett.* **23** (1969) 861.
144. N. W. Dalton and D. W. Wood, *Proc. Phys. Soc.* **90** (1967) 459.
145. An excellent review of experimental data up to 1974 on 2-d magnets may be found in: L. J. de Jongh and A. R. Miedema, *Adv. Phys.* **23** (1974) 1.
146. M. E. Lines, *J. Appl. Phys.* **40** (1969) 1352: *J. Phys. Chem. Solids* **31** (1970) 101: *Phys. Rev.* **B3** (1971) 1749.

147. A. Danielian and K. W. H. Stevens, *Proc. Phys. Soc.* **B77** (1961) 116.
148. R. Navarro, Ph.D. Thesis, University of Zaragoza, Spain, 1976.
149. N. W. Dalton and D. W. Wood, *Phys. Lett.* **28A** (1968) 417.
150. D. Jasnow and M. Wortis, *Phys. Rev.* **176** (1968) 739.
151. D. D. Betts, C. J. Elliott and R. V. Ditzian, *Can. J. Phys.* **49** (1971) 1327.
152. V. Mubayi and R. V. Lange, *Phys. Rev.* **178** (1969) 882.
153. A. Oguchi, *Progr. Theor. Phys. (Kyoto)* **46** (1971) 63.
154. V. L. Berezinskii *Sov. Phys. JETP* **34** (1972) 610.
155. R. Navarro and L. J. de Jongh, *Physica* **B84** (1976) 229.
156. A. J. Guttmann, *Statphys.* **13** (2) (1977) 610: *J. Phys.* **A11** (1978) 545.
157. M. Ferer and M. J. Velgakis, *Phys. Rev.* **B27** (1983) 314: M. Ferer, A. H. Aidnejad and M. J. Velgakis *J. Appl. Phys.* **55** (1984) 2438.
158. E. Praveczki, *J. Phys.* **C13** (1980) 2161.
159. J. Tobochnik and G. V. Chester, *Phys. Rev.* **B20** (1979) 3761.
160. S. M. Shenker and J. Tobochnik, *Phys. Rev.* **B22** (1980) 4462.
161. D. P. Landau and K. Binder, *Phys. Rev.* **B24** (1981) 1391.
162. R. Navarro, H. A. Algra, L. J. de Jongh, R. L. Carlin and C. J. O'Connor, *Physica* **B86–88** (1977) 693.
163. H. A. Algra, L. J. de Jongh and R. L. Carlin, *Physica* **B93** (1978) 24.
164. G. A. Baker Jr., J. Eve and J. G. Rushbrooke, *Phys. Rev.* **B2** (1970) 706.
165. R. L. Stephenson and P. J. Wood, *Phys. Rev.* **173** (1968) 475.
166. M. E. Fisher and D. Jasnow, *Phys. Rev.* **B3** (1971) 907.
167. D. S. Ritchie and M. E. Fisher, *Phys. Rev.* **B7** (1973) 480.
168. R. Navarro, J. J. Smit, L. J. de Jongh, W. J. Crama and D. J. W. IJdo, *Physica* **B83** (1976) 97.
169. R. Navarro and L. J. de Jongh, *Physica* **B98** (1979) 1.
170. M. E. Fisher and M. N. Barber, *Phys. Rev. Lett.* **28** (1972) 1516.
171. K. Ohno and Y. Okabe, *Phys. Lett.* **95A** (1983) 38.
172. A. J. Bray and M. A. Moore, *J. Phys.* **A10** (1977) 1927.
173. P. J. Wojtowicz and R. I. Joseph, *Phys. Rev.* **135A** (1964) 1314.
174. P. J. Wojtowicz, *J. Appl. Phys.* **35** (1964) 991.
175. N. W. Dalton, *Proc. Phys. Soc.* **88** (1966) 659.
176. D. N. Lambeth and H. E. Stanley, *Phys. Rev.* **B12** (1975) 5302.
177. J. A. Puertolas, R. Navarro, F. Palacio, J. Bartolome, D. Gonzalez and R. L. Carlin, *J. Mag. Magn. Mat.* **31–34** (1983) 1243: *Phys. Rev.* **B31** (1985) 516.
178. See for example F. Falo, R. Navarro, M. Floría, *J. Phys.* **C21** (1988) 445.
179. B. I. Halperin and P. C. Hohenberg, *Phys. Rev.* **188** (1969) 898.
180. H. Falk and L. W. Bruch, *Phys. Rev.* **180** (1969) 442.
181. D. D. Betts, C. J. Elliott and M. H. Lee, *Can. J. Phys.* **48** (1970) 1566: D. D. Betts, J. T. Tsai and C. J. Elliott, *Proc. Int. Conf. Magn.*, Moscow, 1973.
182. I. H. Lee and R. E. Shrock, *J. Phys.* **A21** (1988) 2111.
183. L. J. de Jongh, D. D. Betts and D. J. Austen, *Solid St. Commun.* **15** (1974) 1711.
184. J. Rogiers, T. Lookmann, D. D. Betts and C. J. Elliott, *Can. J. Phys.* **56** (1978) 409: J. Rogiers, E. W. Grundke and D. D. Betts, *Can. J. Phys.* **57** (1979) 1719.
185. P. Butera, M. Comi and G. Marchesi, *Phys. Rev.* **B33** (1986) 4725.
186. C. Twining, *J. Phys.* **A21** (1988) 1735.
187. D. C. Jou and H. H. Chen, *J. Phys.* **C6** (1973) 2713.
188. M. H. Lee, *J. Math. Phys.* **12** (1971) 61: *Phys. Rev.* **B12** (1975) 276: *Physica* **119A** (1983) 504.
189. D. C. Jou and H. H. Chen: *Phys. Lett.* **43A** (1973) 79.
190. T. Ishikawa and T. Oguchi: *J. Phys. Soc. Japan* **31** (1981) 1021.
191. A. Abraham and B. Bleaney in: *Electron Paramagnetic Resonance of Transition Ions*, Clarendon Press, Oxford, 1970.
192. L. J. de Jongh, *J. Appl. Phys.* **49** (1978) 1305.

193. H. W. J. Blote and W. J. Huiskamp, *Phys. Lett.* **29A** (1969) 304: H. W. J. Blote, Ph.D. Thesis, University of Leiden, Netherlands, 1972.
194. R. F. Wielinga, H. W. J. Blote, J. A. Roest and W. J. Huiskamp (1967) *Physica* **34**, 223.
195. K. W. Mess, E. Langendijk, D. A. Curits and W. J. Huiskamp, *Physica* **34** (1967) 126.
196. D. J. Breed, K. Gilijamse and A. R. Miedema, *Physica* **45** (1969) 205.
197. D. W. Engelfriet, W. L. Groeneveld and G. M. Nap, *Z. Naturforsh.* **35a** (1980) 852.
198. H. Hulsing, H. G. Kahle, M. Schwab and H. J. Schwarzbauer, *J. Magn. Magn. Mat.* **9** (1978) 18.
199. R. P. Van Stapele, J. C. M. Henning, G. E. G. Hardeman and P. F. Bongers, *Phys. Rev.* **150** (1966) 310.
200. E. J. Samuelsen, *Proc. Int. Conf. Magn.* Moscow, 1973.
201. A. R. Miedema, P. Bloembergen, J. H. P. Colpa, F. W. Gorter, L. J. de Jongh and L. Noordermeer, *A.I.P. Conf. Proc.* **18** (1974) 806.
202. I. Yamada, *J. Phys. Soc. Japan* **33** (1972) 979: K. Hirakawa and H. Ikeda, *J. Phys. Soc. Japan* **35** (1973) 1328: A. Dupas and J. P. Renard, *J. Phys.* **31** (1976) C1—215.
203. H. T. Witteveen, *Physica* **71** (1974) 204.
204. L. J. de Jongh *Physica* **82B** (1976) 247.
205. R. M. Bozorth, J. W. Nielsen, *Phys. Rev.* **110** (1958) 879: S. Tazawa, K. Nagata, M. Date, *J. Phys. Soc. Japan* **20** (1965) 181.
206. R. M. Clay and L. A. K. Staveley, *Proc. Low. Temp. Calorimetry Conf.* Helsinki, 1966 (*Ann. Acad. Sci. Fennicae* **A-VI**, 210 p. 194).
207. L. J. de Jongh, unpublished results.
208. R. Burriel, A. Lambrecht, R. L. Carlin, *J. Appl. Phys.* to be published; L. J. de Jongh, *Solid St. Commun.* **65** (1988) 963.
209. J. S. Wood, C. P. Keijzers, E. de Boer and A. Buttafava, *Inorg. Chem.* **19** (1980) 2213.
210. M. Matsuura, K. Gillijamse, J. E. W. Sterkenburg and D. J. Breed, *Phys. Lett.* **33A** (1970) 363.
211. J. Skalyo Jr., G. Shirane, R. J. Birgeneau and H. J. Guggenheim, *Phys. Rev. Lett.* **23** (1969) 1394.
212. Y. Yamaguchi and N. Sakamoto, *J. Phys. Soc. Japan* **27** (1969) 1444.
213. H. W. de Wijn, L. R. Walker and R. E. Walstedt, *Phys. Rev.* **B8** (1973) 285.
214. D. E. Cox, M. Eibschutz, H. J. Guggenheim and L. Holmes, *J. Appl. Phys.* **41** (1970) 943.
215. Y. Yamaguchi and T. Sakuraba, *J. Phys. Soc. Japan* **34** (1973) 834.
216. K. Iio, M. Sakatani, K. Nayata, *J. Phys. Soc. Japan* **45** (1978) 1567.
217. I. Hatta and H. Ikeda, *J. Phys. Soc. Japan* **48** (1980) 77.
218. J. W. Cable, M. K. Wilkinson and E. O. Wollan, *J. Phys. Chem. Solids* **19** (1961) 29: W. N. Hansen and M. Griffel, *J. Chem. Phys.* **28** (1958) 902.
219. J. F. Dillon Jr., *J. Phys. Soc. Japan* **19** (1964) 1662: L. D. Jennings and W. N. Hansen, *Phys. Rev.* **139** (1965) A1694.
220. W. N. Hansen and M. Griffel, *J. Chem. Phys.* **30** (1959) 913.
221. A. Narath and H. L. Davis, *Phys. Rev.* **137** A163: (1965) A163: A. Narath, *Phys. Rev.* **140** (1965) A854.
222. E. J. Samuelsen, R. Silberglitt, G. Shirane and J. P. Remeika, *Phys. Rev.* **B3** (1971) 157.
223. P. F. Bongers, C. F. Van Bruygen, J. Koopstra, W. P. F. A. M. Omloo, G. A. Wiegers and F. Jellinek, *J. Phys. Chem. Solids* **29** (1968) 977.
224. K. W. Blazey and H. Rohrer, *Phys. Rev.* **185** (1969) 712.
225. J. B. Mac-Chesney, H. J. Williams, J. F. Potter and R. C. Sherwood, *Phys. Rev.* **164** (1967) 779.
226. D. E. Cox, G. Shirane, R. J. Birgeneau and J. B. MacChesney, *Phys. Rev.* **188** (1969) 930.
227. A. K. Gregson, P. Day, D. H. Leech, M. J. Fair and W. E. Gardner, *J. Chem. Soc.* (*Dalton*) (1975) 1306.
228. M. T. Hutchings, M. J. Fair, P. Day, P. J. Walker, *J. Phys.* **C9** (1976) L55.
229. M. Eibschutz, L. Holmes, H. J. Guggenheim and D. E. Cox, *J. Phys.* **32** (1971) C1—759.

230. G. K. Wertheim, H. J. Guggenheim, H. J. Levinstein, D. N. E. Buchanan and R. C. Sherwood, *Phys. Rev.* **173** (1968) 614.
231. M. P. H. Thurlings, A. Van der Pol and H. W. de Wijn, *Solid St. Commun.* **24** (1977) 829: F. Macco, W. Lehmann, W. Breitling, A. E. Slawska-Waniewska, R. Weber and V. Durr, *Solid St. Commun.* **26** (1978) 429.
232. H. G. von Schnering, *Z. Anorg. Allg. Chem.* **353** (1968) 1 and 13.
233. M. F. Mostafa and R. D. Willett, *Phys. Rev.* **B4** (1971) 2213.
234. R. D. Willet, and R. G. Gerstein, *Phys. Lett.* **44A** (1973) 153.
235. M. F. Mostafa, M. A. Semary and M. A. Ahmed, *J. Magn. Magn. Mat.* **15–18** (1980) 448.
236. D. J. Breed, *Phys. Lett.* **23** (1966) 181: *Physica* **37** (1967) 35.
237. R. J. Birgeneau, H. J. Guggenheim and G. Shirane, *Phys. Rev.* **B1** (1970) 2211.
238. L. Holmes, M. Eibschutz, H. J. Guggenheim, *Solid St. Commun.* **7** (1969) 973.
239. H. T. Witteveen, *J. Solid. St. Chem.* **11** (1974) 245.
240. W. D. Van Amstel and L. J. de Jongh, *Solid St. Commun.* **11** (1972) 1423.
241. D. W. Engelfriet and W. L. Groeneveld, *Z. Naturforsh.* **33a** (1978) 848.
242. M. Eibschutz, H. J. Guggenheim, L. Holmes, *J. Appl. Phys.* **42** (1981) 1485.
243. M. Eibschutz, H. J. Guggenheim, L. Holmes and J. L. Bernstein, *Solid St. Commun.* **11** (1972) 457.
244. G. Heger, R. Geller and D. Babel, *Solid St. Commun.* **9** (1971) 335.
245. J. M. Dance, F. Menil, D. Hanzel, R. Sabatier, A. Tressaud, G. Le Flem and P. Hagenmuller, *Physica* **86–88B** (1977) 699: J. M. Dance, R. Sabatier, F. Menil, M. Wintenberger, J. C. Cousseins, G. Le Flem and A. Tressaud, *Solid St. Commun.* **19** (1976) 1059.
246. Y. Ishikawa and S. Akimoto, *J. Phys. Soc. Japan* **13** (1958) 1298: G. S. Heller, *Proc. Int. Conf. Mag* Nottingham p. 443, 1964: A. Sawaoka, S. Miyahara, S. Akimoto and H. Fujisawa, *J. Phys. Soc. Japan* **21** (1966) 185.
247. J. Akimitsu, Y. Ishikawa and Y. Endoh, *Solid St. Commun.* **8** (1971) 87.
248. H. Yamauchi, H. Hiroyoshi, M. Yamada, H. Watanabe and H. Takei, *J. Magn. Magn. Mat.* **31–34** (1983) 1071.
249. A. F. M. Arts, C. M. J. van Uijen, J. A. van Luijk, H. W. de Wijn and C. J. Beers, *Solid St. Commun.* **21** (1977) 13.
250. A. F. M. Arts and H. W. de Wijn, *Physica* **86–88B** (1977) 647: *Phys. Rev.* **B15** (1977) 4348.
251. C. M. J. van Uijen, E. Frikkee and H. W. de Wijn, *Phys. Rev.* **B19** (1979) 509.
252. E. Gurewitz, J. Makovsky and H. Shaked, *Phys. Rev.* **B14** (1976) 2071.
253. J. Ferguson, E. R. Kransz, G. B. Robertson and H. J. Guggenheim, *Chem. Phys. Lett.* **18** (1972) 551.
254. M. P. H. Thurlings, A. M. van Diepen and H. W. de Wijn, *J. Magn. Magn. Mat.* **15** (1980) 637.
255. H. Forstat, G. Taylor and R. D. Spence, *Phys. Rev.* **116** (1959) 897: W. K. Robinson and S. A. Friedberg, *Phys. Rev.* **117** (1960) 402.
256. N. Uryu, J. Skalyo Jr., S. A. Friedberg, *Phys. Rev.* **144** (1960) 689: J. P. A. M. Hijmans, Q. A. G. Vlimmeren and W. J. M. de Jonge, *Phys. Rev.* **B12** (1975) 3859.
257. J. W. Metselaar, D. de Klerk, *Physica* **63** (1973) 191: **65**, 208.
258. J. W. Metselaar, L. J. de Jongh and D. de Klerk, *Physica* **79B** (1975) 53.
259. D. W. Engelfriet, W. L. Groeneveld, H. A. Groenendijk, J. J. Smit and G. M. Nap, *Z. Naturforsh.* **35a** (1980) 115.
260. L. P. Regnault, P. Burlet, J. Rossat-Mignod, *Physica* **86–88B** (1977) 660: L. P. Regnault and J. Rossat-Mignod, *J. Magn. Magn. Mat.* **14** (1979) 194.

# SPIN WAVES IN TWO-DIMENSIONAL MAGNETIC SYSTEMS: THEORY AND APPLICATIONS

A. F. M. ARTS AND H. W. DE WIJN

*Faculty of Physics and Astronomy, University of Utrecht, P.O. Box 80.000, 3508 TA Utrecht, The Netherlands*

## 1. Introduction

This chapter is concerned with the elementary excitations, and the thermodynamic quantities derived from them, of two-dimensional (2-d) ordered magnetic systems in which the primary interaction is nearest-neighbor Heisenberg exchange. It has since long been recognized that the excitation spectrum of such systems comprises energies that, irrespective of the dimensionality, are substantially below, say, $2z|J|S$, in which $J$ represents the exchange and $z$ the magnetic coordination number. In the classical picture, the transverse components of the spins take part in collective precessions, and the long-wavelength modes of these precessions have energies near zero. These concepts were first introduced by Bloch [1]. Presently, assuming the presence of translational invariance, we commonly refer to these motions as spin waves, or the quantum-mechanical equivalent, magnons.

It is unfortunate that to this day no satisfactory way has been found to deal with even the simplest model involving Heisenberg exchange in complete generality. A formidable obstacle in the development of the theory is that the true excitations of the system cannot be directly associated with, for instance, a set of Bose ladders. This is already the case for ferromagnets, in which at zero temperature all spins are aligned along some preferential axis. To make matters worse, in antiferromagnets even the ground state itself is not precisely known. It differs from the Néel state, i.e., the state characterized by full staggered (antiparallel) arrangement of the spins. To overcome the associated mathematical problems, one has to resort to approximative approaches, inevitably at the sacrifice of important aspects. The most successful of these is the low-temperature expansion in local spin deviation operators, followed by a transformation to the reciprocal lattice. This so-called spin wave expansion however fails to treat the fluctuations in the critical regime. Perhaps the most transparant of the various expansions is the one to successive orders of the parameter $1/2S$ introduced by Holstein and Primakoff (HP) in 1940 [2]. The HP expansion will be used throughout this chapter; for a review of other expansions, in particular the Dyson—Maleev expansion [3, 4], as well as the application of spin wave theory to many other magnetic substances not treated here, the reader is referred to the monumental work of Keffer [5].

The first significant term of any expansion results in the *linear* spin wave theory [6—8]. Here, the spin Hamiltonian is reduced to a form that is bilinear in Bose creation and destruction operators of spin deviations, which, under conditions



*L. J. De Jongh (Ed.), Magnetic Properties of Layered Transition Metal Compounds* 191—229.

usually encountered in real systems, allows canonical transformation to normal modes. The linear theory is lend considerable legitimization by its success to describe the dependence of the experimental excitations on the wave vector at low temperatures, i.e., the form of the dispersion relation. As it turns out, the effects of higher orders may to a great extent be stored in modified spin Hamiltonian parameters, notably the exchange and anisotropy constants. In the cases at hand below, these effects are of order 10%, so that the parameters derived from experiment with the linear theory may not directly be identified with the true values.

The higher-order terms of the expansion successively take into account various interactions among the linear spin waves. These interactions may be subdivided in kinematical and dynamical ones [3]. The former take into account that the number of spin deviations at any site cannot exceed twice the spin quantum number $S$, and tends to increase the energies. Note here that each excited spin wave reduces the total $\langle S^z \rangle$ of the specimen by one unit. The dynamical interactions cause modifications of the dispersion relation because of the presence of other spin waves, such as due to thermal excitation. The dynamical interactions, which in most cases dominate the kinematical ones, are attractive, i.e., lower the energies. Both interactions are, of course, contained in the spin wave expansion. In the absence of transverse external fields, then, the lowest-order terms beyond the ones leading to the linear theory are products of four spin deviation operators. In a scheme first devised by Oguchi [9], they may straightforwardly be decoupled in the random-phase approximation (RPA) to products bilinear in the operators, and subsequently be treated as a first-order perturbation, while inserting the numbers of spin waves of the various wave vectors **k** already present at the given temperature. This procedure is often referred to as renormalization, despite the improper usage of the term.

An important quantity in the derivation of thermodynamic quantities, such as the sublattice magnetization, from the spin wave dispersion law is the density of states. In isotropic systems there is no gap in the excitation spectrum at zero wave vector, while as the dimensionality decreases, the low-**k** density of states increases. In 2-d this leads to a cataclysmic excitation of spin waves at any finite temperature, i.e., the system cannot sustain order. This, in effect, is the spin wave formulation of the famous theorem by Mermin and Wagner [10], stating that 2-d spin arrays with isotropic interactions cannot sustain long-range order except at zero temperature. In the presence of anisotropy such as to leave the system with an easy axis, a gap appears, inhibiting the disordering at sufficiently low temperatures.

As we will see below, spin wave theory has been extremely successful in accounting for the low-temperature properties of a number of 2-d magnetic systems with anisotropy weak compared to the exchange. The reason for this is twofold. First, the systems themselves are of high symmetry, such as to have square-lattice magnetic layers, and exhibit well-defined interactions. The interactions among the layers are weak. Second, it is feasible to work out the theory with inclusion of the first-order corrections in $1/2S$ without further approximation by exact evaluation of the relevant summations over the 2-d Brillouin zone. The simplicity of the magnetic lattice also offers substantial advantages from the experimental point of view. On the other hand the theory is not as successful in 2-d as it has been in 3-d at higher temperatures. Generally speaking, the spin wave approach fails in 2-d above $\frac{1}{2}T_N$, as compared to, say, 0.9 $T_N$ in 3-d [11]. This is clearly due to the wider temperature range of the critical fluctuations in 2-d systems, and there is no hope that the situation near $T_N$ can be mended by including the next few orders in $1/2S$.

## 2. Magnetic Structures and Spin Hamiltonians

Since the pioneering work by Breed [12], almost all of the experimental work on spin waves and their associated properties in 2-d magnetic insulators has been focused on compounds with the $K_2NiF_4$ layered perovskite crystal structure. The structure, shown in Figure 1, is built up of magnetic $NiF_2$ layers separated by sheets of nonmagnetic KF. The magnetic ions in a $NiF_2$ layer are situated on a simple-quadratic lattice, and interact with strong nearest-neighbor exchange. The in-layer exchange is generally antiferromagnetic in Mn, Ni, and Fe compounds,

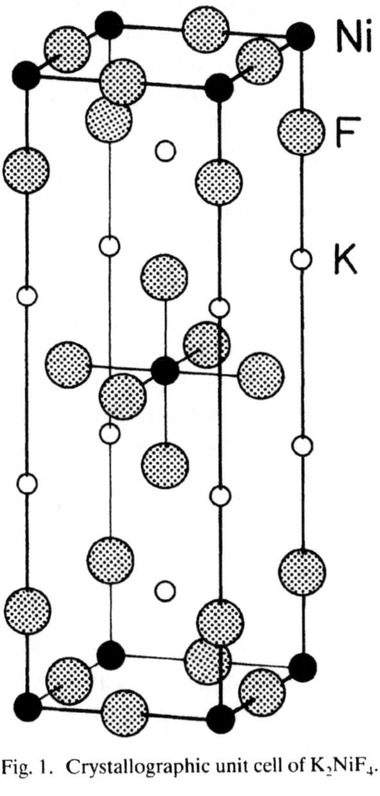


Fig. 1. Crystallographic unit cell of $K_2NiF_4$.

such as $K_2FeF_4$. As concerns the Cu compounds, the in-layer exchange is generally ferromagnetic in the fluorides, such as $K_2CuF_4$, and antiferromagnetic in the oxides, such as $La_2CuO_4$. The interactions between nearest-neighbor magnetic layers are substantially weaker than the primary exchange, in the case of antiferromagnetic intralayer ordering not only because of the larger distances involved, but also because of the crystalline symmetry. Adjacent layers are shifted over half a 2-d unit cell, and accordingly every exchange interaction between spins in these layers is, at least in the Néel state, balanced by the interaction along a similar path with one of the opposite spins. The exchange between next-nearest-neighbor layers does not vanish for reasons of symmetry, but is reduced by many orders of magnitude because at least five intervening ligands are involved. As concerns the magnetic ions, the same structure is found in the series of compounds $(C_nH_{2n+1}NH_3)_2XCl_4$, with X a transition metal ion. Here, quadratic $XCl_2$ layers are separated by the nonmagnetic alkyl ammonium groups, with the distance between successive layers increasing with $n$. A review of the magnetic properties of low-dimensional systems, including those of the $K_2NiF_4$ structure and the alkyl ammonium compounds, has been given by de Jongh and Miedema [13].

Weak anisotropic interactions of dipolar or single-ion nature, additional to the Heisenberg exchange, force the spins to order along the c axis or some other preferential direction. They have a profound effect on the dispersion near the center of the Brillouin zone. To find the single-ion anisotropies, as well as any anisotropy of the exchange by mediation of the spin-orbit coupling, we consider the spin Hamiltonians of relevant transition metal ions in the predominantly octahedral environment of ligands. In the case of $Mn^{2+}$, the ground state has zero orbital angular momentum ($L = 0$), combined with a large spin angular momentum ($S = 5/2$). The crystal-field anisotropies involve higher multiplets, and are weak. The anisotropy in effect arises almost exclusively from the magnetic dipolar interaction, which is several orders of magnitude weaker than the exchange between $Mn^{2+}$ neighbors, and favors alignment of the spins parallel to the tetragonal axis. In a spin wave description, however, dipolar interactions, which involve summations over many pairs of spins, are unfortunately cumbersome to handle. For mathematical tractability, therefore, the anisotropy is represented by a staggered magnetic field $H_A$ acting on the individual spins, despite its failure to account for the temperature dependence of the anisotropy as a result of spin wave interactions. In the presence of an external magnetic field $\mathbf{H}$ along the tetragonal axis the spin Hamiltonian is then given by

$$\mathscr{H} = 2|J| \sum_{\langle l,m \rangle} \mathbf{S}_l \cdot \mathbf{S}_m - g\mu_B(H + H_A) \sum_l S_l^z - g\mu_B(H - H_A) \sum_m S_m^z, \quad (1)$$

where $g$ is the $g$-factor, and $l$ and $m$ refer to sites on the spin-up and spin-down sublattices, respectively. The exchange interaction $J$ is restricted to nearest-neighbor pairs, denoted by $\langle l, m \rangle$. Values for $J$ and $H_A$ for various $Mn^{2+}$ compounds are given in Table I.

Once layer-type magnetic systems order, they also do so in 3-d by virtue of the

TABLE I
Parameters of selected 2-d magnetic systems.

| Compound | $T_N$ (K) | $J/k_B$ (K) | $E_{k=0}(T=0)/k_B$ (K) | $\alpha$ | $H_A$ (kG) | $D/hc$ ($cm^{-1}$) | $E/hc$ ($cm^{-1}$) |
|---|---|---|---|---|---|---|---|
| Antiferromagnetic: | | | | | | | |
| $K_2MnF_4$ | 42.14[a] | $-4.20 \pm 0.03$[b] | $7.40 \pm 0.05$[c] | 0.0038 | 2.35 | | |
| $Rb_2MnF_4$ | 38.4[d] | $-3.69 \pm 0.05$[b] | $7.28 \pm 0.05$[c] | 0.0047 | 2.59 | | |
| $K_2NiF_4$ | 97.1[d] | $-51.1 \pm 0.4$[b] | $27.5 \pm 0.3$[e] | 0.0021 | 5.73 | −0.46 | |
| $Rb_2NiF_4$ | 94.5[f] | $-41 \pm 2$[g] | $49.6 \pm 0.3$[h] | 0.011 | 24 | −2.4 | |
| $K_2FeF_4$ | 63.0[i] | $-7.85 \pm 0.15$[i] | $26.9 \pm 1.1$[i] | | | +4.0 | −0.34 |
| $Rb_2FeF_4$ | 56.3[d] | | | | | | |
| $Rb_2MnCl_4$ | 56.0[j] | −5.6[j] | 7.51[j] | 0.0022 | 1.83 | | |
| Ferromagnetic: | | | | | | | |
| $K_2CuF_4$ | 6.25[k] | 11.4[l] | $2J'/k_B = 0.017$ K[m] | | $2J_A/k_B = -0.178$ K[n] | | |
| $Rb_2CuF_4$ | 6.05[o] | | | | | | |

[a] Neutron diffraction R. J. Birgeneau *et al.*, *Phys. Rev. B* **8** (1973) 304.
[b] NMR H. W. de Wijn *et al.*, *Phys. Rev. B* **8** (1973) 285.
[c] AFMR H. W. de Wijn *et al.*, *Phys. Rev. B* **8** (1973) 299.
[d] Neutron diffraction R. J. Birgeneau *et al.*, *Phys. Rev. B* **1** (1970) 2211.
[e] AFMR R. J. Birgeneau *et al.*, *Solid State Commun.* **8** (1970) 13.
[f] Birefringence K. Iio *et al.*, *J. Phys. Soc. Jpn.* **45** (1978) 1567.
[g] Susceptibility J. Ferré *et al.*, *J. Phys. C* **6** (1973) 1623.
[h] AFMR K. Nagata and Y. Tomono, *J. Phys. Soc. Jpn.* **36** (1974) 78.
[i] Neutron diffraction M. P. H. Thurlings *et al.*, *Phys. Rev. B* **25** (1982) 4750.
[j] Neutron diffraction B. Schröder *et al.*, *Phys. Stat. Sol. (b)* **97** (1980) 501.
[k] Specific heat I. Yamada, *J. Phys. Soc. Jpn.* **33** (1972) 979.
[l] Neutron diffraction K. Hirahawa *et al.*, *J. Phys. Soc. Jpn.* **52** (1983) 4220.
[m] Parallel pumping H. Yamazaki, *J. Phys. Soc. Jpn.* **37** (1974) 667.
[n] FMR A. S. Borovik-Romanov *et al.*, *J. Phys. C* **13** (1980) 879.
[o] Susceptibility C. Dekker *et al.*, *Phys. Rev. B* **38** (1988) 11512.

small interlayer interactions. In $K_2MnF_4$ the ordering between spins in next-nearest-neighbor layers is ferromagnetic [14]. In the isomorph $Rb_2MnF_4$, ferromagnetic and antiferromagnetic interlayer orderings occur next to each other [15], indicating that local lattice imperfections and impurities can tip the subtle balance between ferromagnetic and antiferromagnetic interlayer interactions. It is of interest to note that the transition temperatures of the two phases are the same, $T_N = 38.3 \pm 0.1$ K. Similar effects are observed in $Rb_2MnCl_4$ and $Cs_2MnCl_4$ [16], where the order between next-nearest-neighbor layers appears to depend on the sample preparation, with recovery of the ferromagnetic arrangement after annealing.

As regards $K_2NiF_4$ and $Rb_2NiF_4$, the orbital momentum ($L = 3$) of $Ni^{2+}$ is quenched by the octahedral crystalline field. The ground level is an orbital singlet with an effective spin equal to the real spin $S = 1$. The small tetragonal distortion of the fluorine octahedron along the $c$ axis splits this spin triplet in a singlet and a doublet, i.e., the magnetic anisotropy is of the form $D(S^z)^2$, which may be converted to a staggered anisotropy field $H_A = -2DS(1 - 1/2S)$. In $K_2NiF_4$ the single-ion contribution is nearly an order of magnitude larger than the dipolar anisotropy, but still much weaker than the antiferromagnetic exchange (Table I). In fact, in $K_2NiF_4$ the anisotropy constant $\alpha$, i.e., the ratio between anisotropy and exchange fields, is among the lowest of all layer-type $Ni^{2+}$ compounds studied so far ($\alpha \approx 2 \times 10^{-3}$). The spins order along the $c$ axis ($D < 0$). The coupling between spins in next-nearest layers is ferromagnetic [15].

An interesting contrast to the Mn- and Ni-based compounds form $K_2FeF_4$ and $Rb_2FeF_4$. The ground state of free $Fe^{2+}$ has $L = 2$ and $S = 2$, and is split by the octahedral field in a doublet and a triplet, the latter lying lowest. A small tetragonal distortion further separates the orbital triplet into a ground singlet and a doublet at a distance $\delta \approx 700$ K. In contrast to $K_2NiF_4$ the primary anisotropy leaves the spins in the plane. Since there is no dipolar anisotropy within the plane, $K_2FeF_4$ and $Rb_2FeF_4$ would be ideal candidatès for a 2-d XY systems, were it not that the symmetry of the crystal and the spin value $S = 2$ allow crystal-field anisotropy terms quartic in the spin components. The various contributions to the single-ion anisotropy, as well as the nearest-neigbor exchange in the orbital ground state, can be found by deriving the spin Hamiltonian for a pair of neighboring $Fe^{2+}$ in the tetragonally distorted octahedral field [17]. The spin-orbit coupling and the Heisenberg exchange between pairs is taken as a perturbation. Perturbation theory up to fourth order and integration over orbital variables leads to the spin Hamiltonian

$$\mathscr{H} = 2|J| \sum_{\langle l, m\rangle} \mathbf{S}_l \cdot S_m + \sum_i \{D(S_i^z)^2 + a(S_i^z)^4 + \tfrac{1}{2}e[(S_i^+)^4 + (S_i^-)^4]\}. \quad (2)$$

Note that the exchange interaction has retained the scalar form. Using reasonable estimates for the crystal-field parameters and the spin-orbit coupling, one finds $D/k_B \sim 6$ K, $a/k_B \sim 0.6$ K, and $e/k_B \sim -0.03$ K [17]. The in-plane anisotropy is entirely provided by the fourth-order perturbation in the spin-orbit coupling. The $e$ term orients the spins along the magnetic $a$ or $b$ axis, without preference. Accordingly, there are two domains. It is of interest to note that the stacking of nearest-neighbor layers is unique, in contrast to the systems with order along the $c$ axis. The coupling presumably is by dipolar fields, which indeed favor ferromagnetic alignment in planes spanned by the spin direction and the $c$ axis. These effects may be enhanced by minute distortions from the tetragonal symmetry, which have been suggested to occur in $Rb_2FeF_4$ [15, 18]. These distortions would destroy the cancellation of nearest-neighbor inter-planar exchange interactions.

We next turn our attention to the Cu-based fluorides, which order ferromagnetically within the layers. The ground state of free $Cu^{2+}$ has $L = 2$ and $S = 1/2$. Like the case of $Fe^{2+}$, the octahedral field gives rise to a doublet and a triplet, but the former is lower. The twofold orbital degeneracy of the ground state makes $Cu^{2+}$ a

Jahn—Teller ion. In both the fluorides and organic chlorides a cooperative Jahn—Teller ordering of *d* orbitals occurs along the crystallographic [100] and [010] axes [19, 20]. The ordering is accompanied by a displacement of the in-layer $F^-$ or $Cl^-$ ligands. This special arrangement of the *d* orbitals is responsible for the ferromagnetic exchange observed in all these compounds [21]. An additional effect of the orbital ground doublet is the occurrence of significant anisotropy in the exchange interaction. In $K_2CuF_4$ this is the dominant term in the anisotropy, which forces the spins to lie within the basal plane [22]. Because the spin value is $S = 1/2$, no crystalline (single-ion) anisotropy is expected. The planar anisotropy appears to be very small, of the order of a few Gauss [23]. Quite similar results have recently been obtained for $Rb_2CuF_4$ [24]. The 3-d ordering is accordingly not induced by anisotropy, but by nearest-neighbor inter-planar interactions which no longer cancel by virtue of symmetry.

Of particular interest, recently, has become the magnetism of the $Cu^{2+}$-based oxides, notably $La_2CuO_4$ ($T_N \approx 300$ K), because of the possible connection with the mechanism of high-$T_c$ superconductivity occurring in their oxygen-deficient counterparts. The magnetic order within the $Cu_2O$ layers appears to be antiferromagnetic [25], with a magnetic structure identical to that of $K_2FeF_4$, except for small rhombic distortions and canting [26]. In combination with the weak anisotropy parameter ($\alpha \sim 10^{-5}$), this nevertheless results in a near ideal realization of the 2-d, $S = \frac{1}{2}$ Heisenberg antiferomagnet. The parent compound of another family of high-$T_c$ superconductors, $YBa_2Cu_3O_6$ ($T_N \approx 500$ K), has quite a similar magnetic structure, except that each layer is made up of a pair of coupled layers [27]. As such it is related to the antiferromagnetic double layer $K_3Mn_2F_7$, the spin wave theory of which has been developed in detail [28], but will not be treated here.

## 3. Spin Wave Theory of Model Systems

In this section we develop the 2-d spin wave formalism for application to simple model systems. Treatments applicable to low-symmetry systems [29—31] and strongly anisotropic magnets [32] have been given in the literature, yet the various aspects of 2-d spin wave theory are best illustrated with model Hamiltonians such as the ones presented in Section 2. Moreover, the few systems studied in detail are to a great extent described by these models.

We first consider a quadratic-layer antiferromagnet with isotropic exchange between nearest neighbors and a weak anisotropy leaving the magnetization along the *c* axis. The anisotropy is represented by a staggered magnetic field $H_A$ along this axis. Our starting point is Equation (1) of Section 2. The standard way to proceed is to apply a suitable transformation that expresses the deviations from the completely aligned Néel state in boson annihilation and creation operators. We use the HP transformation, primarily because it is unitary. Upon defining the boson operators $a_l^\dagger$ and $b_m^\dagger$, creating spin deviations on the spin-up and spin-down sublattices, respectively, we have

$$S_l^z = S - a_l^\dagger a_l, \qquad S_m^z = -S + b_m^\dagger b_m,$$
$$S_l^+ = (2S)^{\frac{1}{2}} f_l a_l, \qquad S_m^+ = (2S)^{\frac{1}{2}} b_m^\dagger f_m,$$
$$S_l^- = (2S)^{\frac{1}{2}} a_l^\dagger f_l, \qquad S_m^- = (2S)^{\frac{1}{2}} f_m b_m, \tag{3}$$

with

$$f_l = (1 - a_l^\dagger a_l/2S)^{\frac{1}{2}}, \qquad f_m = (1 - b_m^\dagger b_m/2S)^{\frac{1}{2}}. \tag{4}$$

In the HP approach the unphysical states, i.e. states for which the number of local spin deviations exceeds $2S$, are projected out by the factors $f$, which, however, need to be expanded. In the lowest-order approximation $f = 1$. This leads to the so-called linear spin wave theories, in which interactions among spin waves are ignored entirely. Recently, Goldhirsh [33] has devised a method that avoids the problem of the unphysical states by a multivalued transformation between spin space and Bose space. As it turns out, however, the HP and Goldhirsh transformations do not differ from one another until the $(2S + 1)$th order. This implies that for $S \geqslant 1$ the differences occur beyond the first-order corrections to the linear theory discussed below.

We now proceed with solving Equation (1). Upon substituting Equation (3) expanded up to order $1/2S$, the spin Hamiltonian can be separated into a bilinear part and a part containing four-operator terms. The bilinear part reads

$$\mathscr{H}_0 = 2|J|S \sum_{\langle l,m \rangle} [(1+\alpha)(a_l^\dagger a_l + b_m^\dagger b_m) + a_l b_m + a_l^\dagger b_m^\dagger + h(a_l^\dagger a_l - b_m^\dagger b_m)], \tag{5}$$

where $\alpha = g\mu_B H_A/2z|J|S$ is the anisotropy constant, with $z$ the number of magnetic nearest neighbors, while $h = g\mu_B H/2z|J|S$. The four-operator part $\mathscr{H}_1$ is given by

$$\mathscr{H}_1 = -\frac{|J|}{2} \sum_{\langle l,m \rangle} (a_l^\dagger a_l a_l b_m + a_l b_m^\dagger b_m b_m + b_m^\dagger a_l^\dagger a_l^\dagger a_l + b_m^\dagger b_m^\dagger b_m a_l^\dagger + 4 a_l^\dagger a_l b_m^\dagger b_m). \tag{6}$$

To deal with $\mathscr{H}_0$, we benefit from the translational invariance of the system by applying a Fourier transformation that expresses the local operators in two sets of spin waves, each set restricted to one of the sublattices. That is,

$$a_{\mathbf{k}} = N_0^{-\frac{1}{2}} \sum_l a_l e^{i\mathbf{k}\cdot\mathbf{r}_l}, \qquad b_{\mathbf{k}}^\dagger = N_0^{-\frac{1}{2}} \sum_m b_m^\dagger e^{i\mathbf{k}\cdot\mathbf{r}_m}, \tag{7}$$

with $N_0$ the number of magnetic unit cells. $\mathscr{H}_0$ is subsequently diagonalized by the $2 \times 2$ Bogoliubov transformation

$$\alpha_{\mathbf{k}} = u_{\mathbf{k}} a_{\mathbf{k}} + v_{\mathbf{k}} b_{\mathbf{k}}^\dagger, \qquad \beta_{\mathbf{k}}^\dagger = v_{\mathbf{k}} a_{\mathbf{k}} + u_{\mathbf{k}} b_{\mathbf{k}}^\dagger, \tag{8}$$

chosen such that $\alpha_{\mathbf{k}}$ and $\beta_{\mathbf{k}}$ are normal modes, to obtain the spin wave energies to lowest order in $1/2S$.

To evaluate the effects of $\mathscr{H}_1$ on the dispersion law, i.e., corrections of order

$1/2S$, one may follow two ways: (i) After Oguchi [9] one inserts the inverses of Equations (7) and (8) into $\mathcal{H}_1$, and subsequently applies random phase decoupling to products such as $\alpha_{\mathbf{k}}^{\dagger}\alpha_{\mathbf{k}'}\beta_{\mathbf{k}''}^{\dagger}\beta_{\mathbf{k}'''}\delta_{\mathbf{k}-\mathbf{k}'-\mathbf{k}''+\mathbf{k}''',0}$. The contribution of $\mathcal{H}_1$ to the spin wave energies is then derived upon retaining the diagonal terms, and substituting the numbers of thermally excited spin waves $n_{\mathbf{k}}^{(1)}$ and $n_{\mathbf{k}}^{(2)}$ for $\langle \alpha_{\mathbf{k}}^{\dagger}\alpha_{\mathbf{k}}\rangle$ and $\langle \beta_{\mathbf{k}}^{\dagger}\beta_{\mathbf{k}}\rangle$, respectively. (ii) The second way is a shortcut based on the notion that the transformations are linear. The decoupling is done directly in $\mathcal{H}_1$, i.e., the decoupling is applied to the local spin deviation operators prior to passage to **k** space [34]. For example, for the first term in Equation (6) we set

$$a_l^{\dagger}a_l a_l b_m = 2a_l^{\dagger}a_l\langle a_l b_m\rangle + 2a_l b_m\langle a_l^{\dagger}a_l\rangle. \tag{9}$$

Note here that the operator combinations $\langle a_l b_m^{\dagger}\rangle$, $\langle a_l a_l\rangle$, and $\langle b_m^{\dagger} b_m^{\dagger}\rangle$, etc., do not contribute because they are off-diagonal after transformation to normal modes. By addition of the decoupled Equation (6) to $\mathcal{H}_0$, one arrives at an effective two-operator Hamiltonian, which can be diagonalized with Equations (7) and (8), however with coefficients $u_{\mathbf{k}}$ and $v_{\mathbf{k}}$ modified with respect to diagonalization of Equation (6) alone. These modifications are, of course, of order $1/2S$, and must be determined recurrently. In method (ii) the renormalization of the spin wave energies is **k** independent, while the first approach leads to a weak **k** dependence, but up to order $1/2S$ the results are identical. In the following discussion we follow method (i). In Section 6 we return to method (ii), which is more suited to dealing with nontranslation-invariant problems.

The dispersion relation of the two spin wave branches is, up to order $1/2S$, then given by [35]

$$E_{\mathbf{k}}^{(1,2)}/2z|J|S = [(1+\alpha)^2 - \gamma_{\mathbf{k}}^2]^{\frac{1}{2}} \mp h \pm (p^{(1)} - p^{(2)}) - \\ - g_{\mathbf{k}}(R_0 + R_1), \tag{10}$$

where

$$g_{\mathbf{k}} = (1 + \alpha - \gamma_{\mathbf{k}}^2)/[(1+\alpha)^2 - \gamma_{\mathbf{k}}^2]^{\frac{1}{2}}, \tag{11}$$

and where

$$\gamma_{\mathbf{k}} = \frac{1}{z}\sum_{\boldsymbol{\delta}} e^{i\mathbf{k}\cdot\boldsymbol{\delta}} \tag{12}$$

is a geometrical factor, with $\boldsymbol{\delta}$ denoting a nearest-neighbor displacement in the quadratic layer. In the case of the quadratic-layer antiferromagnets $\gamma_{\mathbf{k}} = \cos\frac{1}{2}k_x a \cos\frac{1}{2}k_y a$, with $a$ the magnetic lattice constant. The quantities $R_0$ and $R_1$ represent the temperature-independent and temperature-dependent renormalizations, respectively. They read

$$R_0 = \frac{1}{2SN_0}\sum_{\mathbf{k}} (g_{\mathbf{k}} - 1), \tag{13}$$

$$R_1 = \frac{1}{2SN_0}\sum_{\mathbf{k}} g_{\mathbf{k}}(n_{\mathbf{k}}^{(1)} + n_{\mathbf{k}}^{(2)}), \tag{14}$$

in which the summations run over the first 2-d Brillouin zone, and

$$n_{\mathbf{k}}^{(1,2)} = [\exp(E_{\mathbf{k}}^{(1,2)}/k_B T) - 1]^{-1} \tag{15}$$

are the Bose occupation numbers of the two branches. The term $p^{(1)} - p^{(2)}$ in Equation (10), with

$$p^{(1,2)} = \frac{1}{2SN_0} \sum_{\mathbf{k}} n_{\mathbf{k}}^{(1,2)} \tag{16}$$

is proportional to the difference of the up and down sublattice magnetizations. It is a generalization of the more common, but in the case of 2-d systems precarious, renormalization of the spin wave energies according to the parallel susceptibility, and vanishes in zero field.

In the case of $K_2MnF_4$ in a field of 4.5 T, which is not far from the spin-flop field $H_{sf} = 5.4$ T, the temperature dependences of $R_1$, $p^{(1)}$, and $p^{(2)}$ are shown in Figure 2. As it turns out, even close to $T_N$ the effects of renormalization do not exceed, say, 8%. At temperatures up to about $\frac{1}{2}T_N$, the dominant contribution to the renormalization arises from $R_0$, which is virtually independent of the temperature and the field. From Equations (10) and (11) we see that the first-order corrections of the magnon energies vary by a factor of about 2 only, viz., from $-\frac{1}{2}R_0$ at the bottom of the band ($\alpha \ll 1$) to $-R_0/(1+\alpha)$ at the zone boundary. In the regime below $\frac{1}{2}T_N$, therefore, the effects of $R_0$ can to a very good approximation be simulated by effective values of $J$ and $\alpha$, while using the zero-order dispersion relation [36]. That is, in zero external field,

$$E_{\mathbf{k}} = 2z|J_s|S[(1+\alpha_s)^2 - \gamma_{\mathbf{k}}^2]^{\frac{1}{2}}, \tag{17}$$

where $J_s$ and $\alpha_s$ are obtained by equating Equations (10) and (17) at the zone boundary and at $k = 0$, yielding in case $\alpha \ll 1$ and $R_0 \ll 1$

$$J_s = J(1 - R_0), \qquad \alpha_s = \alpha/(1 - R_0). \tag{18}$$

Once the dispersion relation is calculated, it is straightforward to derive the various thermodynamic quantities. Carrying out the spin wave transformations in reverse order, one finds the magnetizations $M_{\uparrow,\downarrow} = -g\mu_B N_0 \langle S_{\uparrow,\downarrow}^z \rangle$ of the spin-up (↑) and spin-down (↓) sublattices from

$$\begin{aligned} \langle S_{\uparrow}^z \rangle &= +S - \Delta_0 - \Delta S(T) + \Delta S(H, T), \\ \langle S_{\downarrow}^z \rangle &= -S + \Delta_0 + \Delta S(T) + \Delta S(H, T). \end{aligned} \tag{19}$$

Here $\Delta_0$ is the zero-point spin reduction

$$\Delta_0 = \frac{1}{2N_0} \sum_{\mathbf{k}} (f_{\mathbf{k}} - 1), \tag{20}$$

with

$$f_{\mathbf{k}} = \frac{1+\alpha}{[(1+\alpha) - \gamma_{\mathbf{k}}^2]^{\frac{1}{2}}}, \tag{21}$$

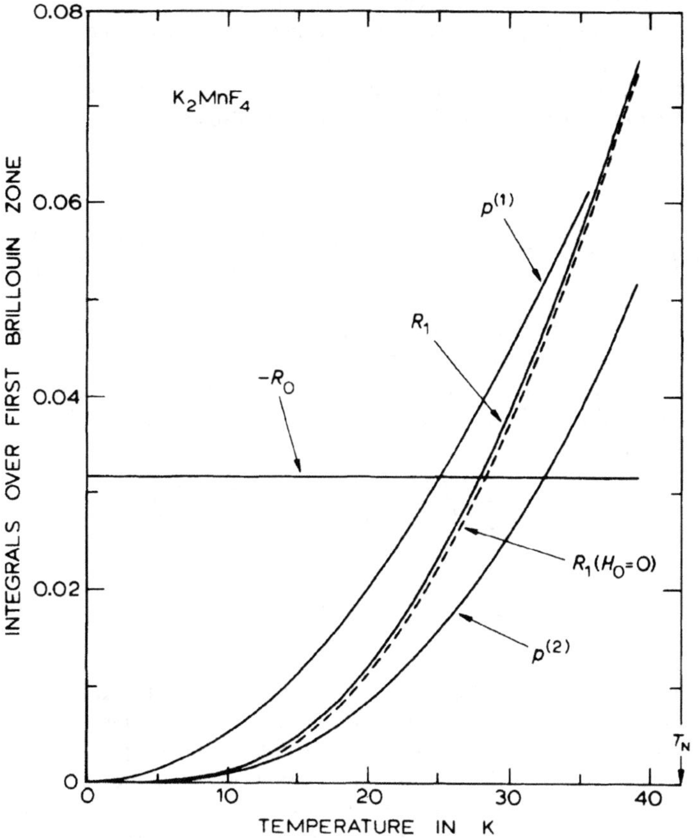


Fig. 2. Integrals over the first Brillouin zone, occurring in spin wave theory, vs the temperature for $K_2MnF_4$. The full curves have been computed at 45 kG; the dashed curve applies to zero external field. After H. W. de Wijn *et al.*, *Phys. Rev. B* **8** (1973) 199.

which for a simple quadratic lattice in the limit $\alpha = 0$ equals $\Delta_0 = 0.197$. This result is somewhat reduced in the presence of anisotropy, becoming $\Delta_0 = 0.170$ for $\alpha = 0.0038$ appropriate to $K_2MnF_4$. Note that all parts of the Brillouin zone contribute. The quantity $\Delta S(T)$ reads

$$\Delta S(T) = \frac{1}{2N_0} \sum_{\mathbf{k}} f_{\mathbf{k}}(n_{\mathbf{k}}^{(1)} + n_{\mathbf{k}}^{(2)}), \tag{22}$$

which is only weakly dependent on the external field through the sum $n_{\mathbf{k}}^{(1)} + n_{\mathbf{k}}^{(2)}$. The field dependence of the magnetization predominantly resides in the term

$$\Delta S(H, T) = \frac{1}{2N_0} \sum_{\mathbf{k}} (n_{\mathbf{k}}^{(1)} - n_{\mathbf{k}}^{(2)}), \tag{23}$$

which reflects the unbalance between the two sublattices.

Differentiation of Equation (23) with respect to $H$ yields the parallel susceptibility per spin, $\chi_{\parallel}$. The expression for general $H$ is quite involved. In zero field we have, with inclusion of the field-dependent renormalization contained in the term $p^{(1)} - p^{(2)}$ in the dispersion law Equation (10) (see also [37])

$$\chi_{\parallel}(H=0)$$

$$= \frac{g^2\mu_B^2}{k_B T}\frac{1}{N_0}\sum_{\mathbf{k}} n_{\mathbf{k}}^2 \exp(E_{\mathbf{k}}/k_B T) \Bigg/ \left(1 + \frac{2z|J|}{k_B T}\sum_{\mathbf{k}} n_{\mathbf{k}}^2 \exp(E_{\mathbf{k}}/k_B T)\right). \tag{24}$$

The above analysis but with a transverse magnetic field added to the Hamiltonian Equation (5) yields the perpendicular susceptibility $\chi_{\perp}$. Here, one-boson terms are eliminated by applying a shift of the origin of the $k = 0$ spin waves [7] or, alternatively, of the local spin deviation operators. The result per spin is

$$\chi_{\perp}(H=0) = \frac{g^2\mu_B^2}{2z|J|(1+\frac{1}{2}\alpha)}\left(1 - \frac{1}{2N_0 S}\sum_{\mathbf{k}} (f_{\mathbf{k}} - g_{\mathbf{k}})(2n_{\mathbf{k}} + 1)\right)$$

$$= \frac{g^2\mu_B^2}{2z|J|(1+\frac{1}{2}\alpha)}\left(1 - \frac{\Delta_0 + \Delta S(T)}{S} + R_0 + R_1\right). \tag{25}$$

At zero temperature this reduces to the expression given by Keffer [5] and Breed [12]. A general treatment of the odd-boson terms can be found in [38]. A perturbative approach leading to numerical results for $\chi_{\perp}$ is reported in [37].

With regard to the evaluation of the summations over the first Brillouin zone occurring in the above spin wave expressions, we note that all functions to be summed depend ultimately on $\mathbf{k}$ through $\gamma_{\mathbf{k}}$ only. In the case of the 2-d quadratic lattice, these summations can be evaluated rigorously by making use of the exact relation [36]

$$\frac{1}{N_0}\sum_{\mathbf{k}} F(\gamma_{\mathbf{k}}) = \frac{4}{\pi^2}\int_0^1 dz\, K((1-z^2)^{\frac{1}{2}})F(z). \tag{26}$$

Here, $K(m) = \int_0^{\pi/2} dx(1 - m^2 \sin^2 x)^{-\frac{1}{2}}$ is a complete elliptic integral of the first kind, for which fast computer routines based on repeatedly taking arithmetic and geometric means are available. For the sake of the computations, therefore, Equation (26) in effect reduces the 2-d summations to integrals over one variable. Some caution should, however, be excercised near $z = 0$, i.e., the zone boundary, where $K((1-z^2)^{\frac{1}{2}})$ diverges in a logarithmic way. Near the divergence one should utilize suitable approximations, such as $\int_0^{\varepsilon} dz\, K((1-z^2)^{\frac{1}{2}})F(z) = [(1 + \ln 4)\varepsilon - \varepsilon \ln \varepsilon]F(0)$.

For the quadratic-layer antiferromagnets, useful, analytic approximations of

$\Delta S(T)$ and $\chi_{\|}$, valid at low temperatures, may be derived by expanding the trigonometric functions in $\gamma_{\mathbf{k}}$ to obtain $\gamma_{\mathbf{k}} = 1 - \frac{1}{8} a^2k^2$, replacing the summations over the zone by integrals, and extending the upper limits of the integrations to infinity. This long-wavelength approximation, also called the $k^2$ approximation, yields for the decrement of the magnetization in zero field [39]

$$\Delta S(T) = \frac{-(1 + \alpha_s)k_B T}{4\pi |J_s| S} \ln [1 - \exp(-E_{k=0}/k_B T)], \tag{27}$$

and for the parallel susceptibility per spin

$$\chi_{\|}(T) = \frac{2g^2\mu_B^2 k_B T}{(2z|J_s|S^2)} \left( \frac{E_{k=0}}{k_B T} [\exp(E_{k=0}/k_B T) - 1]^{-1} - \right.$$

$$\left. - \ln [1 - \exp(-E_{k=0}/k_B T)] \right). \tag{28}$$

From Equations (27) and (28) it is apparent that the temperature variation of the thermodynamic quantities is primarily controlled by the $k = 0$ gap of the spin wave spectrum, which in the same approximation amounts to $E_{k=0} = 2z|J_s|S(2\alpha_s + \alpha_s^2)^{\frac{1}{2}}$. Although the anisotropy is quite small in most of the 2-d compounds studied (see Table I), the interplay with the exchange leads to so large a gap that the number of excitations remains finite even at relatively high temperatures.

It is, therefore, expedient to make a comparison with the coresponding results for quadratic-layer ferromagnets, in which the $k = 0$ gap is of order $2zJS\alpha$ only. Starting with a spin Hamiltonian analogous to Equation (1), i.e., an isotropic exchange between neighboring spins and an anisotropy field $H_A$, the analysis follows the lines of the antiferromagnetic case, except that the quadratic part is already diagonal following spatial Fourier transformation to spin waves. The dispersion relation becomes [40]

$$E_{\mathbf{k}}/2zJS = (1 - R')(1 - \gamma_{\mathbf{k}}) + \alpha + h, \tag{29}$$

in which the renormalization is expressed by

$$R' = \frac{1}{N_0 S} \sum_{\mathbf{k}} (1 - \gamma_{\mathbf{k}}) n_{\mathbf{k}}. \tag{30}$$

Contrary to the antiferromagnetic case, there is no temperature-independent renormalization of the spin wave energies. The gap in the spin wave spectrum simply amounts to $g\mu_B(H + H_A)$. In the $k^2$ approximation we find for the decrement of the magnetization

$$\Delta S = \frac{-k_B T}{\pi z J S(1 - R)} \ln \{1 - \exp[-g\mu_B(H + H_A)/k_B T]\}, \tag{31}$$

which closely resembles Equation (27), but because of the different dispersion grows nearly linearly with temperature, at least for $k_B T \gg g\mu_B(H + H_A)$. This is the 2-d analogue of the Bloch $T^{3/2}$ law in 3-d. Particularly simple is the result for the parallel susceptibility per spin, for which we have

$$\chi_{\parallel} = \frac{g^2\mu_B^2}{\pi zJS(1-R)}\, n_{k=0}\,, \tag{32}$$

or, in case $k_B T \gg g_B(H + H_A)$,

$$\chi_{\parallel} = \frac{g\mu_B k_B T}{\pi zJS(1-R)(H+H_A)}\,. \tag{33}$$

## 4. Dispersion Relation

Probably the most direct way to observe spin waves and their dispersion is inelastic neutron scattering. The earliest 2-d magnetic compounds studied are the easy-axis antiferromagnets with small anisotropy $K_2NiF_4$ by Skalyo *et al.* [41] and $K_2NiF_4$ by Birgeneau *et al.* [14]. As an example we reproduce in Figure 3 some results for inelastic neutron scattering in $K_2MnF_4$, viz., the dispersion for spin waves propagating along the $a^*$ and $c^*$ axes in reciprocal space at 4.5 K and $T_N$. Scans taken at constant momentum transfer **Q** reveal sharp excitations, which shift towards lower energies with increasing temperature. Above the transition temperature $T_N = 42.14$ K the excitations broaden, yet remain well defined up to 1.1 $T_N$ for nonzero $k$. The genuine 2-d character of the magnetism manifests itself in the absence of any dispersion along the $c^*$ axis. In fact, the data imply that the interlayer exchange is smaller than the intralayer exchange by a factor of at least 250. The solid line in the upper frame of Figure 3 is a least-squares adjustment of the dispersion relation Equation (17) to the low temperature data with $J_s$ and $\alpha_s$ as adjustable parameters. An excellent fit was obtained for $J_s/k_B = -4.36 \pm 0.05$ K and $\alpha_s = (3.64 \pm 0.06) \times 10^{-3}$. In connection with Equation (17) we have argued that $J_s$ and $\alpha_s$ are not the true exchange and anisotropy constants, but are modified parameters that include the renormalization, i.e., at low temperatures the quantity $R_0$. For $K_2MnF_4$, $R_0 = -0.0136$ from calculations (see Figure 2), so that one obtains with Equations (18) for the intralayer exchange $J/k_B = -4.23 \pm 0.05$ K and for the anisotropy parameter $\alpha = (3.76 \pm 0.06) \times 10^{-3}$. In the case when second- and further-neighbor intralayer exchange interactions are relevant, the constant $J_s$ in Equation (17) is in the long-wavelength approximation to be identified with $J_{1s} - 2J_{2s} - 4J_{3s}$ in an obvious notation. In $Rb_2MnF_4$, $J_{1s}$ and $J_{2s}$ have been determined separately from an analysis of the dispersion measured with inelastic neutron diffraction over the entire zone [42] confirming the identification and yielding $J_{2s}/2J_{1s} = 0.0088 \pm 0.0013$. The effect of the temperature-dependent renormalization on the spin wave energies is seen from the data at $T_N$ in Figure 3 in relation to those at low temperature. At $T_N$ the energy of the $k = 0$ mode falls to zero, and the dispersion as a function of $k_x$ follows a simple sine curve. The maximum energy at the zone boundary indeed corresponds to the spin wave value $2z|J|S(1 + \alpha - R_0 - R_1)$, where $R_1 = +0.089$ at $T_N$. Apparently, spin wave theory with inclusion of the four-operator part of the Hamiltonian adequately

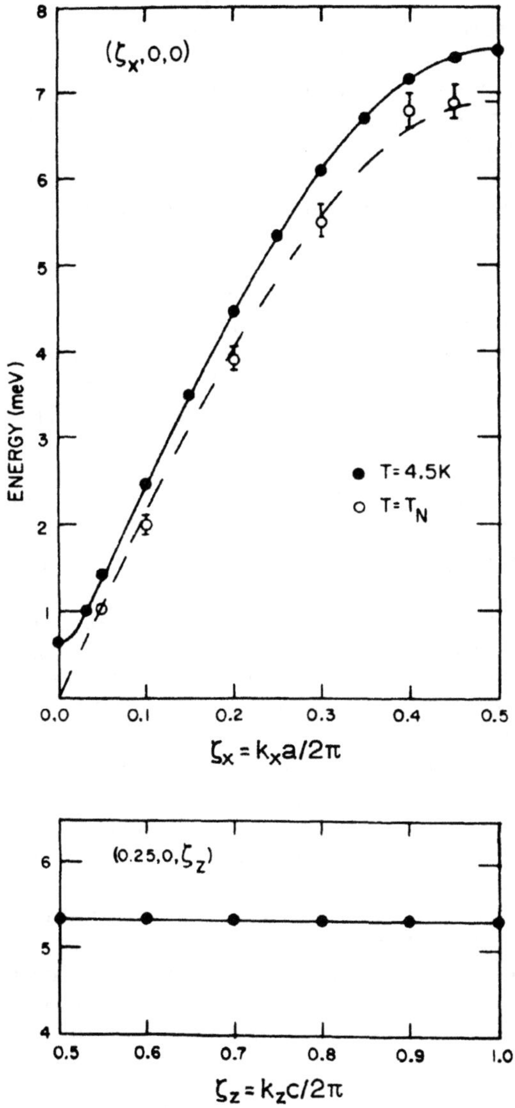


Fig. 3. Spin wave dispersion in $K_2MnF_4$ at 4.5 K and at $T_N$ along the $a^*$ axis (upper frame) and $c^*$ axis (lower frame) measured with inelastic neutron scattering. The solid curve in the upper frame represents a fit of the dispersion relation Equation (17), with $J_s$ and $\alpha_s$ as adjustable parameters. The dashed curve, the simple sine dependence to which Equation (10) reduces for $H = 0$ and $\alpha = 0$, accounts for the temperature-dependent renormalization. The absence of dispersion along $c^*$ demonstrates the 2-d character. After R. J. Birgeneau *et al.*, *Phys. Rev. B* **8** (1973) 304.

describes the spin wave excitations over the entire ordered temperature regime. Similar results have been obtained for $K_2NiF_4$ [36], $Rb_2MnCl_4$ [43], and $(CH_3NH_3)_2MnCl_4$ [43]. In $K_2NiF_4$, $R_0 = -0.0790$, while the temperature-dependent renormalization is small. It amounts to $R_1 = +0.0023$ at $\frac{1}{2}T_N$, increasing to $R_1 = +0.026$ at $T_N = 97.1$ K. The smallness of $R_1$ here is obviously related to the small number of excitations even near $T_N$.

More accurate determinations of the temperature variation of the uniform ($k = 0$) spin wave mode can be accomplished by use of antiferromagnetic resonance (AFMR). In the Ni-based fluorides the AFMR falls in the far infrared. In $K_2NiF_4$ at liquid helium temperatures, the $k = 0$ gap has been observed at $19.1 \pm 0.2$ cm$^{-1}$ [44] and $19.6 \pm 0.2$ cm$^{-1}$ [45], and in $Rb_2NiF_4$ at $34.5 \pm 0.2$ cm$^{-1}$ [45]. With increasing temperature the gap was found to decrease in nearly the same manner as the sublattice magnetization. For a calculation of the development of the gap with temperature on the basis of spin wave theory, one has to replace the temperature-independent staggered field $H_A$ in Equation (1) by a more realistic anisotropy energy, i.e., in Ni-based systems the single-ion anisotropy of the form $D(S^z)^2$. The HP formalism still allows the energy of the $k = 0$ mode to be calculated up to order $1/2S$. The results, reproduced in Figure 4 [45], show excellent agreement with experiment up to, say, 0.6 $T_N$. In the long-wavelength approximation it can be proved analytically that the $k = 0$ energy indeed scales with the sublattice magnetization.

In the Mn-based compounds the $k = 0$ gap is small enough to achieve a resonance in the microwave region by bringing down one of the spin wave branches with an external field [35, 46]. Experimental resonance fields for AFMR at 24.01 GHz in $K_2MnF_4$ and $Rb_2MnF_4$, taken from [35], are shown in Figure 5. The observed initial increase of the resonance field with increasing temperature is essentially due to the rising field-induced unbalance of the sublattice magnetizations when approaching spin flop, i.e., the term $p^{(1)} - p^{(2)}$ in the dispersion relation (Equation (10)). In first order this term is proportional to $H$, and it can therefore be visualized as an additional field screening the external field, in the way of the classical analogue of precessing sublattice magnetizations. In the latter case, the field for AFMR is given by $(1 - \chi_{\|}/2\chi_{\perp})H$. The classical result is, however, not suited for the evaluation of the field-dependent renormalization in 2-d systems, for one thing because it is not clear how the zero-point spin reduction, which for $S = 5/2$ already amounts to 7%, must be incorporated in the equations of motion of the magnetizations. Instead, the fully renormalized spin wave formalism, Equations (10—16), must be utilized to rework the measured AFMR frequencies to zero field. Given $J$, one first deduces $H_A$ from the data as a function of the temperature. Subsequently, setting $H = 0$ (and thus $p^{(1)} = p^{(2)}$), one recalculates the gap from Equation (10) with the $H_A(T)$ just derived. As in the Ni-based systems, the temperature variation of the $k = 0$ gap was found to closely follow the sublattice magnetization. Values of $E_{k=0}$ extrapolated to zero temperature are entered in Table I. The latter results are of course to a high degree independent of the method of data reduction. It may further be noticed that the thermal increase of the AFMR linewidth in $K_2MnF_4$ and $Rb_2MnF_4$ was observed to follow a $T^4$

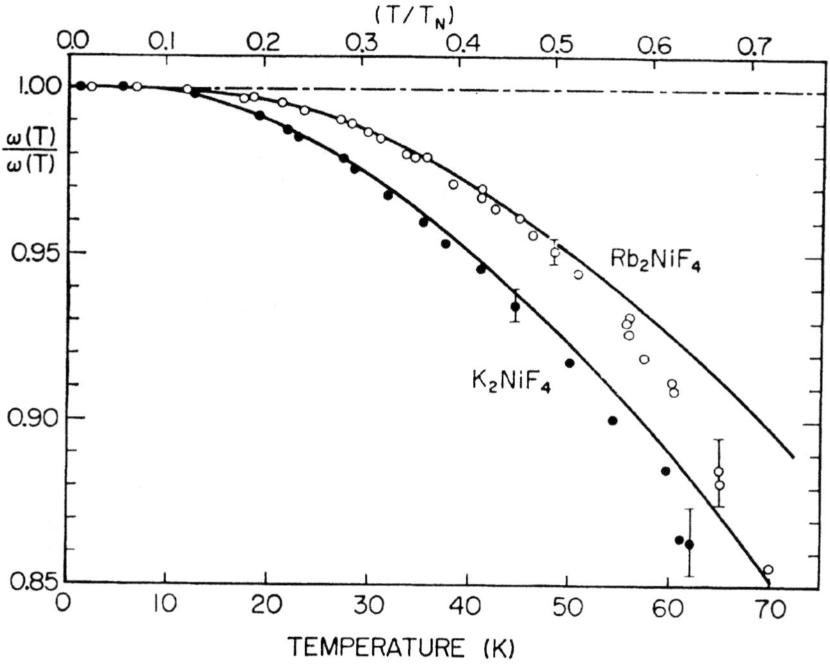


Fig. 4. Reduced AFMR frequencies in $K_2NiF_4$ and $Rb_2NiF_4$ vs the temperature measured with infrared absorption. The solid curves are calculated from renormalized spin wave theory with inclusion of $D(S^z)^2$ anisotropy. After K. Nagata and Y. Tomono, *J. Phys. Soc. Jpn.* **36** (1974) 78.

dependence [35]. A multimagnon theory, which comprises processes involving up to six magnons, was found to account for this broadening [47].

The finding that the gap varies with temperature as the sublattice magnetization has recently been put on a more firm basis by the development, up to first-order corrections in $1/2S$, of a spin wave theory in which dipolar interactions rather than a staggered field constitute the anisotropy [48]. With inclusion of a minute amount of $D(S^z)^2$ anisotropy such as is appropriate to $Mn^{2+}$, it was found that in $K_2MnF_4$ the calculated zero-temperature value of the gap matches the experimental result $E_k = 0(T = 0) = 7.40 \pm 0.05$ K. More importantly, however, a faithful renormalization with temperature of the spin waves in the zone center was achieved, despite the absence of temperature-dependent parameters. The spin wave gap was found to track the thermal reduction of the sublattice magnetization within a few tenths of a percent up to at least 35 K.

Another well studied compound is $K_2FeF_4$. Here the spins lie within the basal plane, with the crystallographic [110] direction as preferred axis. Adopting the direction of magnetization as the $x$ axis and using Equation (2), the spin

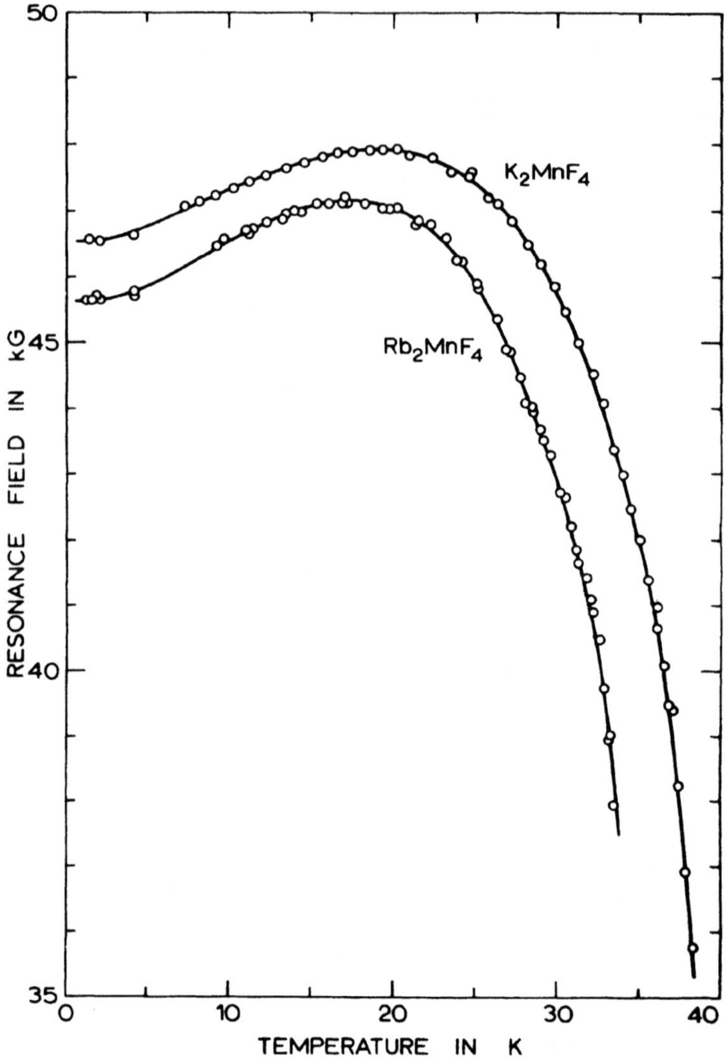


Fig. 5. Experimental resonance fields of AFMR in $K_2MnF_4$ and $Rb_2MnF_4$ vs the temperature. The microwave frequency is 24.01 GHz and the external field is parallel to the $c$ axis. After H. W. de Wijn *et al., Phys. Rev. B* **8** (1973) 299.

Hamiltonian may be rewritten as [17]

$$\mathscr{H} = 2|J| \sum_{\langle l,m \rangle} \mathbf{S}_l \cdot \mathbf{S}_m + \sum_i \{D(S_i^z)^2 + E[(S_i^x)^2 - (S_i^y)^2]\}, \tag{34}$$

where we have omitted the negligible quartic axial anisotropy and have decoupled the quartic in-layer anisotropy in spin space to quadratic terms according to

$$\tfrac{1}{2}e[(S_i^+)^4 + (S_i^-)^4] = 6e\langle (S^x)^2 - (S^y)^2\rangle\, [(S_i^x)^2 - (S_i^y)^2]$$
$$= E[(S_i^x)^2 - (S_i^y)^2]. \quad (35)$$

The factor $\langle (S^x)^2 - (S^y)^2\rangle$ in $E$ is in turn determined by self-consistency. The partial decoupling in spin space according to Equation (35), applied prior to further decoupling following the HP expansion, is a sort of compromise. It offers the great advantage of substantially improving the convergence, yet retains essential parts of the correlations. Further, Equation (34) in conjunction with Equation (35) still reflects the tetragonal symmetry (Both $E$ and $(S_i^x)^2 - (S_i^y)^2$ change sign upon interchanging $x$ and $y$). In an analysis which follows lines similar to the case of uniaxial anisotropy, the two-operator part of the Hamiltonian becomes

$$\mathscr{H}/2z|J|S = \sum_{\mathbf{k}} [(1 + A)(a_{\mathbf{k}}^\dagger a_k + b_{\mathbf{k}}^\dagger b_{\mathbf{k}}) + \gamma_{\mathbf{k}}(a_{\mathbf{k}}^\dagger b_{\mathbf{k}}^\dagger + a_{\mathbf{k}} b_{\mathbf{k}}) +$$
$$+ \tfrac{1}{2}B(a_{\mathbf{k}}^\dagger a_{-\mathbf{k}}^\dagger + a_{\mathbf{k}} a_{-\mathbf{k}} + b_{\mathbf{k}}^\dagger b_{-\mathbf{k}}^\dagger + b_{\mathbf{k}} b_{-\mathbf{k}})], \quad (36)$$

with $A = (D - 3E)/2z|J|$ and $B = (D + E)/2z|J|$. Diagonalization of Equation (36) by a 4 × 4 Bogoliubov transformation mixing $a_{\mathbf{k}}$, $a_{-\mathbf{k}}^\dagger$, $b_{\mathbf{k}}^\dagger$, $b_{-\mathbf{k}}$ leads to the unrenormalized expression for the spin wave dispersion

$$E_{\mathbf{k}}^{(1,2)}/2z|J|S = [(1 + A)^2 - (\gamma_{\mathbf{k}} \pm B)^2]^{\frac{1}{2}}. \quad (37)$$

Equation (37) represents two nondegenerate spin wave branches. The lower $k = 0$ gap, which has energy $2z|J|S[(A - B)(2 + A + B)]^{\frac{1}{2}}$, is, for not too large $A$ and $B$, primarily governed by the in-layer anisotropy $E$. The upper gap at $2z|J|S(A + B)(2 + A - B)^{\frac{1}{2}}$ is mainly determined by the axial anisotropy. The two branches meet at the zone boundary. In the semiclassical picture of precessing magnetic moments, the low-energy modes correspond to elliptical precessions of the spins with the long axis of the ellipse in the basal plane. The high-energy modes have the long axis of the elliptical precession perpendicular to the plane. The smaller the in-layer anisotropy, the larger the ellipticity becomes. On the other hand, when $B = 0$, we return to the uniaxial problem. The precession then is strictly circular about the $x$ axis and the two spin wave branches are degenerate for all $\mathbf{k}$. The first-order corrections to the spin wave energies can be calculated in a way similar to the case of the uniaxial antiferromagnet [17]. The resultant expressions turn out to be quite involved, and we will not present them here. The main result is that in $K_2FeF_4$ at low temperatures the excitations are substantially renormalized in the center of the Brillouin zone. The energies of the lower and upper $k = 0$ modes are reduced by 20 and 13%, respectively. The corrections decrease to 0.3% at the zone boundary.

Experimental results for the spin wave dispersion in $K_2FeF_4$, obtained by neutron diffraction [17] and Raman scattering [49], have been collected by Thurlings *et al.* Indeed, two well defined excitations are observed which, with increasing temperature, shift towards lower energies. The dispersion along the $a^*$ and $c^*$ directions at 4.2 K is shown in Figure 6. As in $K_2MnF_4$ and $K_2NiF_4$, the 2-d character of the magnetic lattice is reflected in the absence of dispersion along

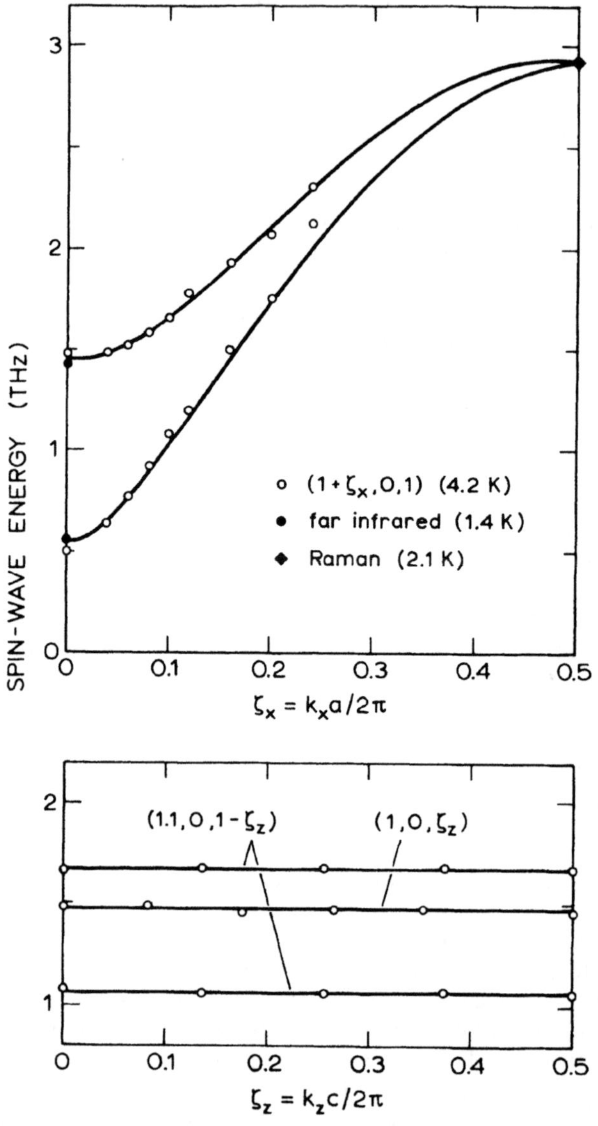


Fig. 6. Spin wave dispersion in $K_2FeF_4$ at 4.2 K along the $a^*$ axis (upper frame) and the $c^*$ axis (lower frame) measured with inelastic neutron scattering (open circles) and far-infrared absorption (closed circles), and derived from two-magnon Raman scattering (closed diamond). The solid curves in the upper frame represent a least-squares fit of the renormalized spin wave dispersion relation, with *J*, *D*, and *E* as adjustable parameters. After M. P. H. Thurlings *et al.*, *Phys. Rev. B* **25** (1982) 4750.

the $c^*$ axis. The lower gap is found at 17 ± 4 $cm^{-1}$, and the upper gap at 49 ± 2 $cm^{-1}$. Both values, within the limits of error, equal the more accurate results from one-magnon far-infrared absorption, 18.7 ± 0.8 $cm^{-1}$ [17] for the lower $k = 0$ mode, and 48.5 ± 0.3 $cm^{-1}$ [50] and 47.7 ± 0.08 $cm^{-1}$ [17] for the upper one. The data point at the zone boundary is derived from two-magnon Raman scattering, with account of magnon—magnon interactions [49]. The solid line in Figure 6 represents a least-squares adjustment of the first-order renormalized spin wave theory with $J$, $D$, and $E$ as adjustable parameters. An excellent fit is achieved simultaneously for the two branches for $J/k_B = -7.85 \pm 0.15$ K, $D/k_B = 5.7 \pm 0.1$ K, and $E/k_B = -0.49 \pm 0.03$ K. To illustrate what bearing first-order renormalization has on the parameters, the data have also been fitted with the unrenormalized expression, Equation (37). Despite the **k** dependence of the first-order correction there appears to be no marked deterioration of the fit. The results for the unrenormalized parameters are $J/k_B = -8.2 \pm 0.15$ K, $D/k_B = 4.3 \pm 0.1$ K, and $E/k_B = -0.29 \pm 0.03$ K. A comparison with the first-order renormalized parameters is interesting in that it gives some indication of the variations that could be introduced by corrections of still higher order in $1/2S$. These corrections may amount to as much as 1 K for $D/k_B$ and 0.2 K for $E/k_B$. In any case, however, the values for both anisotropy parameters are close to the estimates from a crystal-field calculation, which are $D/k_B = 6$ K and $E/k_B = 6e\langle(S^x)^2 - (S^y)^2\rangle = -0.3$ K.

A close analogy is supposed to exist between the spin wave dispersions of $La_2CuO_4$ and $K_2FeF_4$, which are nearly identical from the point of view of their magnetic structures. Indeed, with inelastic neutron scattering two $k = 0$ gaps have been observed, at 8 ± 2 and 20 ± 5 $cm^{-1}$ [51]. The zone boundary excitations have been observed with Raman scattering to have energies of order 2000 $cm^{-1}$ [52]. No detailed spin wave analysis has as yet been performed, i.e., with inclusion of the first-order corrections anticipated to be substantial for $S = \frac{1}{2}$. In any event, however, again because $S = \frac{1}{2}$, crystalline anisotropy is absent. The gaps are therefore entirely due to anisotropic and off-diagonal components of the exchange, with a minor contribution from dipolar anisotropy.

We finally turn our attention to a ferromagnetic case. Several neutron diffraction studies have been performed to study the excitations in $K_2CuF_4$($T_c = 6.25$ K) [53—55], the archetypal example of 2-d ferromagnetism [22, 23]. The spin wave dispersion along the $a^*$ direction at 1.1 K as measured by Hirakawa *et al.* [55] is presented in Figure 7. The sinusoidal curve in Figure 7(a), predicted for the nearest-neighbor Heisenberg 2-d ferromagnet, Equation (29), presents an excellent tracking of the data points for $J/k_B = 12.0$ K upon setting $R$, $h$, and $\alpha$ equal to zero and taking **k** along the $a^*$ axis. Closer inspection of the data, however, reveals that the spin wave energies in the regime $k_x < 0.1(2\pi/a)$ deviate from a simple-quadratic dependence on $k_x$. This is shown by the dashed line in Figure 7(b). To account for the departure the authors have included in the analysis, in addition to the primary intralayer exchange $J$, a small intralayer Ising exchange $J_A$ forcing the spins to lie within the basal plane, and a weak interlayer Heisenberg exchange $J'$ invoking the 3-d long-range order. The Hamiltonian thus

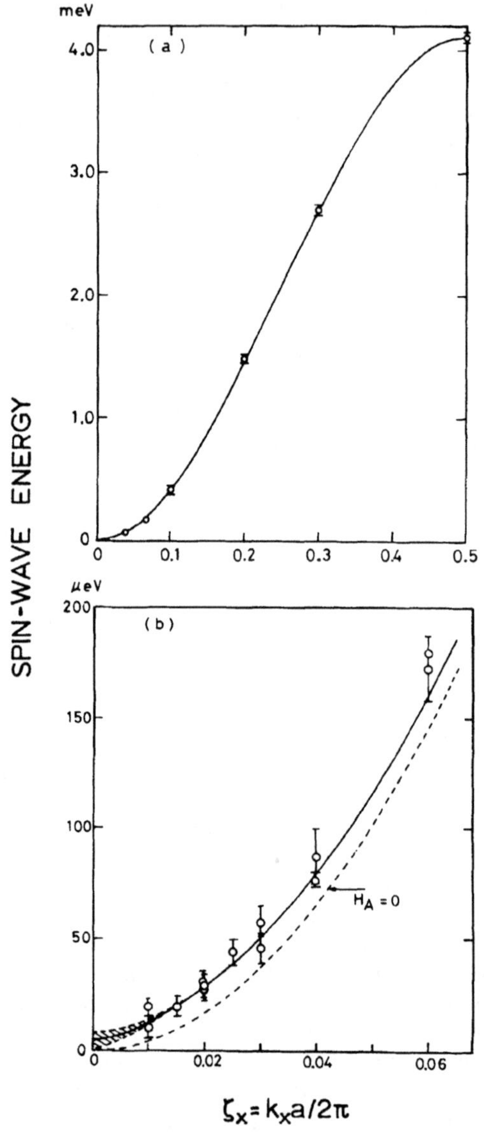


Fig. 7. Spin wave dispersion in $K_2CuF_4$ at 1.1 K along the $a^*$ axis measured with inelastic neutron scattering. Solid curve in upper frame represents a best fit of Equation (29), which however deviates from the data points at low energies (dashed curve in lower frame). Solid curve in lower frame represents a best fit of Equation (39), which includes anisotropic exchange and interlayer interactions. After K. Hirakawa *et al.*, *J. Phys. Soc. Jpn.* **52** (1983) 4220.

reads

$$\mathcal{H} = -2J \sum_{\langle i,j \rangle} \mathbf{S}_i \cdot \mathbf{S}_j - 2J_A \sum_{\langle i,j \rangle} S_i^z S_j^z - 2J' \sum_{\langle i,k \rangle} \mathbf{S}_i \cdot \mathbf{S}_k, \tag{38}$$

The unrenormalized spin wave dispersion at zero temperature and zero external field may then be written as (cf. Equation (29); see also [56])

$$E_{\mathbf{k}} = (A_{\mathbf{k}}^2 - B_{\mathbf{k}}^2)^{\frac{1}{2}}, \tag{39}$$

with

$$A_{\mathbf{k}} = 2zJS(1-\gamma_{\mathbf{k}}) + 2z'J'S(1-\gamma'_{\mathbf{k}}) - 2zJ_A S\gamma_{\mathbf{k}},$$
$$B_{\mathbf{k}} = 2zJ_A S\gamma_{\mathbf{k}}, \tag{40}$$

in which the $\gamma_{\mathbf{k}}'$s are summations over the relevant nearest neighbors. The solid line in Figure 7(b) then represents a fit to Equation (39), with the result $J/k_B = 11.4$ K. The parameters $J_A/k_B = -0.089$ K and $J'/k_B = 0.0088$ K were taken from ferromagnetic resonance [57] and parallel pumping experiments [58], respectively. No evidence has been found for any anisotropy within the layers, such as may be provided by dipolar interactions. The finite gap in Figure 7(b) at $k = 0$ is caused by the experimental condition that the scans were taken at $k_z = 0.4(2\pi/c)$. With increasing temperature the energies shift to lower values. Although no accurate analysis is available, the existing data confirm a **k**-independent renormalization below $T_c$, in conformity with Equation (29). The absence of anisotropy within the planes and the small XY-like anisotropy have interesting consequences for the magnetic behavior near the transition temperature. We here mention only a spin wave calculation for the lineshape of the excitations by Moussa and Villain [59].

## 5. Thermodynamic Properties

As we have learned in Section 4 from the results of neutron diffraction and AFMR, renormalized theories appear to adequately describe the energies of spin waves through the entire zone and all the way up to the transition temperature, with the important exception of the zone center. In the case of antiferromagnets, it was found that theories based on staggered anisotropy fields entirely fail to account for the drop of the $k = 0$ energy gap with temperature. Substantial improvements could be achieved by starting from more realistic Hamiltonians, indicating an upper temperature limit of, say, 0.6 $T_N$. Highly accurate results for the zone center are further available from thermodynamic properties, which emphasize the zone center through their dependence on Bose occupation numbers. As we will see below, in particular the sublattice magnetization provides a test on the limits of validity of 2-d spin wave theory at the least as precise as does AFMR.

For an accurate experimental verification of the spin wave predictions for thermodynamic quantities, the 2-d transition-metal fluorides are again ideal candidates. The out-of-layer $^{19}F^I$ nuclei (Figure 1), with nuclear moment $I = \frac{1}{2}$, neighbor the magnetic ions and resonate in strong hyperfine fields scaling with the sublattice magnetizations. That is, in zero magnetic field, $f(T) = A_{19}\langle S^z \rangle$, where $A_{19}$ is the transferred hyperfine coupling constant with additional contributions of order 1% of dipolar origin. The in-layer $^{19}F^{II}$ nuclei similarly probe the difference of the sublattice magnetizations. The resonance frequencies $f(T)$ can be measured with such a high accuracy that the experimental uncertainty resides primarily in the determination of the temperature. When propagating this error to the frequency, it is feasible to obtain overall inaccuracies of a few parts in $10^5$ at the lowest temperatures, increasing to a few parts in $10^4$ at $\frac{1}{2} T_N$.

Results for $K_2MnF_4$ in zero external field are shown in Figure 8 [35]. The solid line here is a least-squares fit of Equation (19) in conjunction with Equation (10) to the data in the absence of an external field ($n_{\mathbf{k}}^{(1)} = n_{\mathbf{k}}^{(2)}$). The nearest-neighbor exchange constant $J$, the staggered-anisotropy-field parameter at zero temperature $\alpha(T = 0)$, and, of course, $f(T = 0)$ are the adjustable parameters. The gap $E_{k=0}$ is, according to experiment, assumed to scale with the sublattice magnetization. The fit is one of a series carried out over temperature ranges extending from 1.5 K to a selection of upper limits. The noteworthy result is that excellent fits are achieved up to 18 K, but beyond this point the spin wave description of the sublattice magnetization breaks down even with the magnon energies renormalized throughout the entire Brillouin zone. To see what improvement has resulted from the first-order renormalizations $R_0$ and $R_1(T)$, and at the same time to estimate what improvements might be expected from higher-order corrections, fits have also been done based on the dispersion relation, Equation (17), of the linear theory with effective parameters $J_s$ and $\alpha_s$. The fitting range of the first-order corrected theory appeared to exceed that of the 'simple' theory by only 3 K. Similar conclusions have been arrived at from the sublattice magnetization in two other axial antiferromagnets, $Rb_2MnF_4$ and $K_2NiF_4$ [35], from the susceptibility in $K_2MnF_4$ (Figure 9) [60], from the parallel susceptibility in the planar antiferromagnet $K_2FeF_4$ [61], and, for that matter, from the sublattice magnetization [28] and the susceptibility [62] in the double-layer antiferromagnet $K_3Mn_2F_7$. It is not likely that further improvements in the theory, such as the inclusion of a more realistic anisotropy instead of a temperature-dependent staggered anisotropy field, would improve matters. In a recent reanalysis of the $K_2MnF_4$ data with inclusion of dipolar interactions [48], in which case the exchange constant is the only adjustable parameter left, the breakdown of the theory again occurs around $\frac{1}{2} T_N$.

The currently accepted explanation of the breakdown is that at about $\frac{1}{2} T_N$ weakly anisotropic 2-d systems are already entering a regime where the drop of the sublattice magnetization is in part determined by critical fluctuations, i.e., fluctuations in the $z$ components of the spins. The latter are not adequately treated by spin wave expansions. Support for the occurrence of such fluctuations at an early stage is found in the smallness of $T_N$ relative to, for instance, the molecular-field result for the critical temperature $\theta = 2zJS(S + 1)/3k_B$. In $K_2MnF_4$ and $K_2NiF_4$ we have $T_N/\theta = 0.43$ and 0.36, respectively. Conversely, the region of agreement might extend to a point closer to $T_N$ the closer $T_N$ is to $\theta$. An example of such a system is the organic compound $Mn(trz)_2(NCS)_2$ [63], which has the

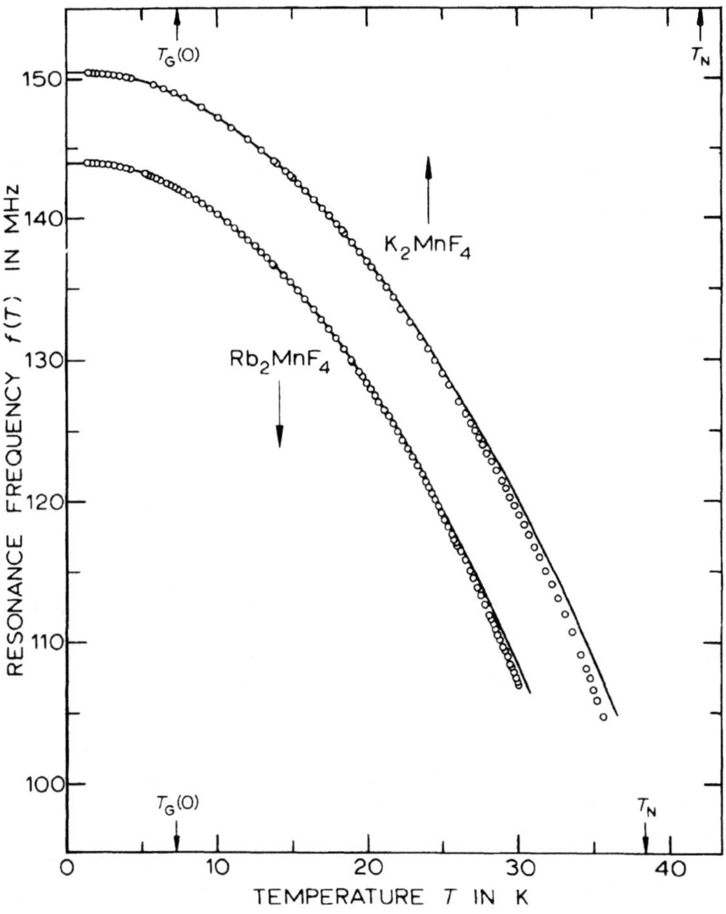


Fig. 8. Sublattice magnetizations in $K_2MnF_4$ and $Rb_2MnF_4$, as reflected in the NMR frequency $f(T)$ of the $^{19}F$ out-of-layer nuclei, vs the temperature in zero external field. The zero-temperature spin wave energy gaps and the Néel temperatures are indicated by $T_G(0)$ and $T_N$, respectively. The solid curves are fits of renormalized spin wave theory up to 18 and 17 K, respectively. After H. W. de Wijn *et al.*, *Phys. Rev. B* **8** (1973) 285.

magnetic structure of the tetrafluorides, but orders at $T_N = 3.29$ K, compared to $\theta = 5.8$ K. Indeed, from a proton NMR study excellent agreement was claimed between the sublattice magnetization and simple spin wave theory up to $T \sim 0.8\ T_N$. It should not be left unremarked that in general at the point of breakdown of the theory the thermal deviations from the ground state are quite small. In $K_2MnF_4$ and $Rb_2MnF_4$ the failure occurs at 7% decrement of the magnetization due to thermal motion added to a zero-point spin reduction of 7%. For $K_2NiF_4$ the corresponding values are 4% and 18%, respectively. It should also be noted that

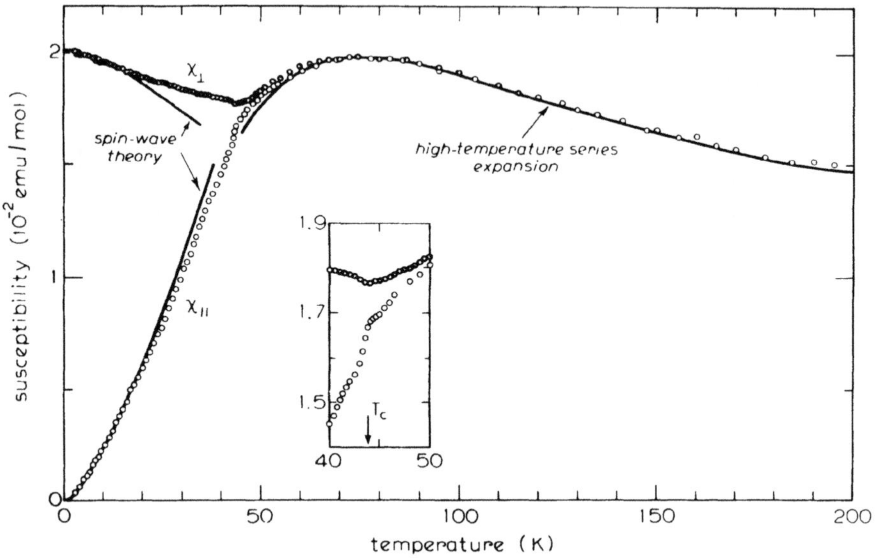


Fig. 9. The differential susceptibility of $K_2MnF_4$ vs the temperature in zero external magnetic field. The oscillating field (~2.5 G, 117 Hz) is applied parallel (open circles) or perpendicular (closed circles) to the $c$ axis. Solid curves below 40 K are calculated from renormalized spin wave expressions. Solid curve above $T_N$ is a fit of a high-temperature series expansion. After C.A.M. Mulder *et al.*, *Physica* **113B** (1982) 380.

half-way towards the transition the effects of the critical fluctuations are still minor, and could only be detected because of the high accuracy of the NMR technique. In $K_2MnF_4$ at 0.86 $T_N$, the highest temperature measured, the discrepancy between theory and experiment does not exceed 2%.

With an external magnetic field along the $c$ axis the degeneracy of the spin wave branches is lifted, and the lower branch will become more heavily populated than the upper one. The spin wave energies will, in addition to the Zeeman energy, $g\mu_B H$, be shifted by the first-order correction $p^{(1)} - p^{(2)}$ reflecting the unbalance of the sublattices (cf. Equation (23)). The temperature-dependent renormalization $R_1$ is only minutely affected by the field (see Figure 2). The reducing effect of a field on the sublattice magnetization when the field is pointing in an opposite sense to the magnetization is shown for $K_2MnF_4$ in Figure 10 [64]. The effects are relatively small, in units of the $^{19}F$ NMR frequency of order 1 MHz. For this reason the results have been plotted relative to the experimental zero-field resonance frequencies $f(T, H = 0)$, which are of order 150 MHz. The solid curves have been calculated using the renormalized spin wave theory with the values for $J$ and $\alpha(T = 0)$ derived from the zero-field data. The agreement is excellent for all fields up to temperatures of, say, 22 K. In contrast to the zero-field case, the first-order corrections now substantially improve the agreement. The improvement resides predominantly in the field dependent term $p^{(1)} - p^{(2)}$. This is clearly shown

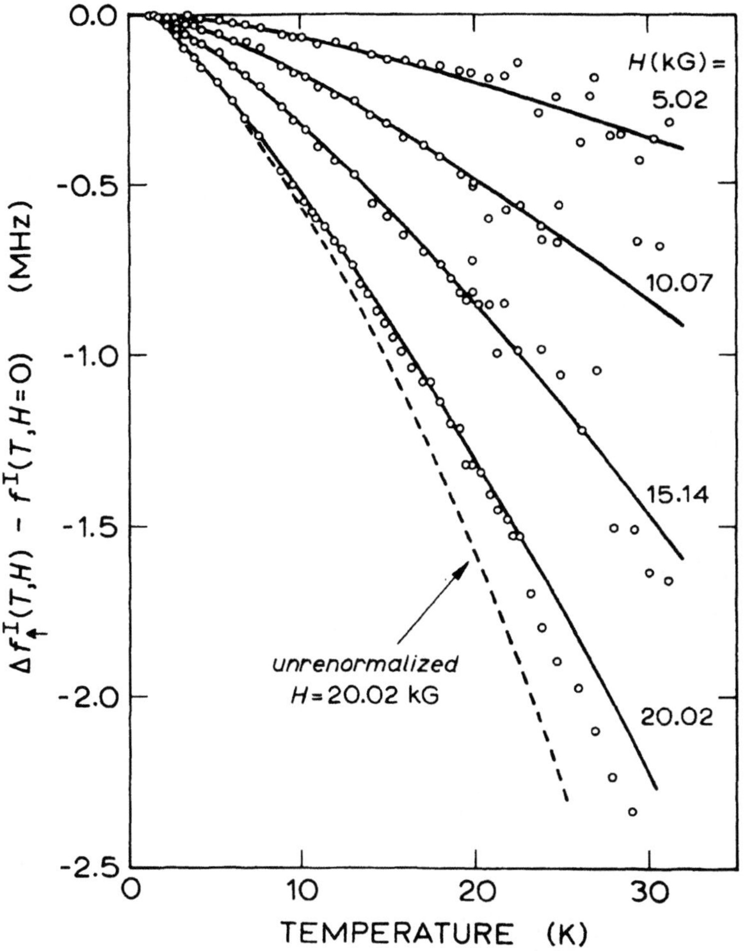


Fig. 10. Field-induced change of the magnetization of the spin-up sublattice in $K_2MnF_4$, as reflected in the NMR frequency of the out-of-layer $^{19}F^{I}$ nuclei, vs the temperature. The field is parallel to the $c$ axis. The data points have been corrected for the nuclear Zeeman shift. The solid lines have been calculated from renormalized spin wave theory. The dashed line has been calculated with the unrenormalized theory with effective parameters $J_s$ and $\alpha_s$. After A.F.M. Arts *et al.*, *Phys. Rev. B* **15** (1977) 4360.

by the dashed curve in Figure 10, which has been calculated on basis of the unrenormalized theory with the proper $J_s$ and $\alpha_s$. Data for the sublattice magnetization for fields close to spin flop ($H$ = 54.5 kG) are not available. However, the net magnetization, as reflected in the NMR frequency of $^{19}F$ nuclei situated between up and down spins (Figure 11), shows the increasing weight of the

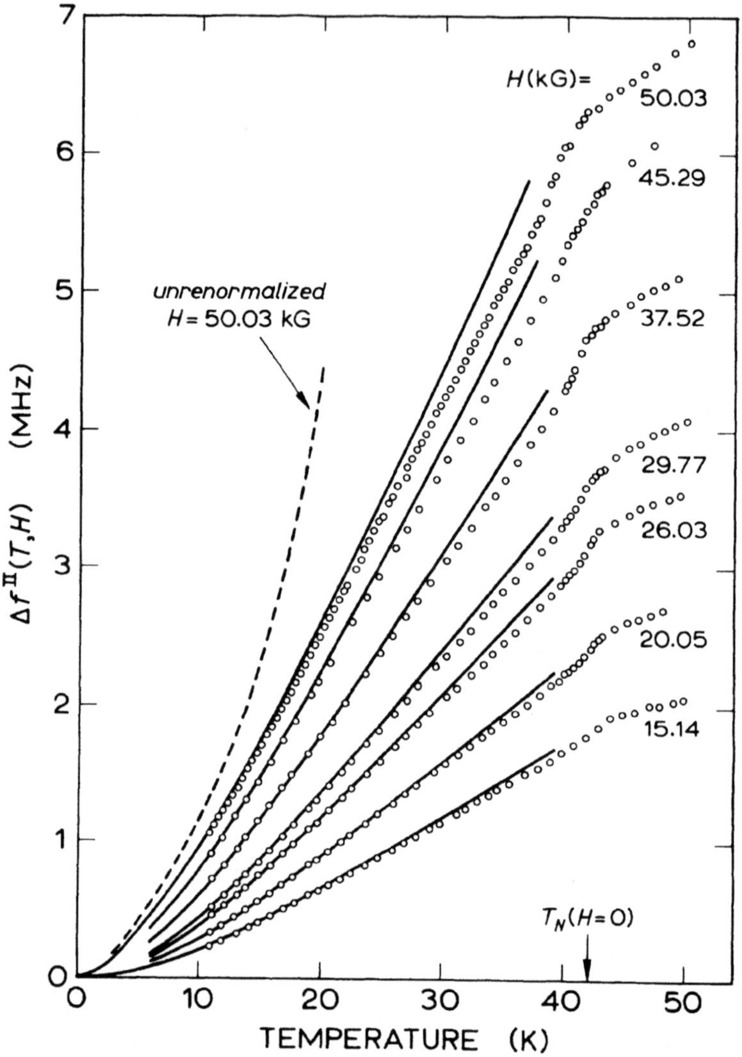


Fig. 11. Net magnetization, as reflected in the NMR frequency of the in-layer $^{19}F^{II}$ nuclei corrected for the Zeeman shift, vs the temperature for various fields parallel to the *c* axis. Solid lines represent renormalized spin wave theory. The dotted line results from the unrenormalized theory at 50.03 kG, with effective parameters $J_s$ and $\alpha_s$. After A.F.M. Arts *et al.*, *Phys. Rev. B* **15** (1977) 4360.

renormalizing corrections. The analysis including these corrections, represented by the solid lines, again agrees excellently with the data points up to $\frac{1}{2}T_N$. The results of the unrenormalized theory are clearly less impressive, with the range of agreement extending to steadily decreasing temperatures when the field is increased. At

15 kG this range still extends to 11 K, decreasing to only 4 K at 50 kG. In consideration of the substantial effects of the field-dependent renormalization, it is conceivable that further agreement in the region near spin-flop may be accomplished by extension to even higher-orders in 1/2S.

In the context of the sublattice magnetization of antiferromagnets we further discuss results on the zero-temperature reduction $\Delta_0$ of this quantity from the fully aligned value. $\Delta_0$, which, strictly speaking, is not a thermodynamic quantity, has long been regarded as a critical test of spin wave descriptions of antiferromagnets in general, because it explores the nature of the antiferromagnetic ground state. The latter, in turn, has been under considerable debate since rigorous proofs seem to ascertain that it would be a singlet, implying $\langle S_i^z \rangle = 0$ at any site. Real antiferromagnets, however, stabilize in Néel-like states, even when only minute amounts of anisotropy are present. It is just this sort of states on which spin wave theories rest by approximation with a limited number of excitations out of the Néel state. In the case of 3-d systems $\Delta_0$ is quite small, and no conclusive evidence is available. In the 2-d systems $K_2MnF_4$ and $Rb_2MnF_4$, on the other hand, $\Delta_0$ could, to sufficient accuracy, be extracted from the NMR of the $^{55}$Mn nuclei. Their hyperfine interaction is almost exclusively of intraatomic origin and, in contrast to the $^{19}$F hyperfine interaction, permits calibration in the nonmagnetic isomorphs. Although the NMR has escaped direct detection due to severe broadening by Suhl—Nakamura interaction, the resonance has been observed through its effect on the AFMR in a double-resonance experiment [65]. Both in $K_2MnF_4$ and $Rb_2MnF_4$ the reduction was found to be $\Delta_0 = 0.17 \pm 0.03$, in excellent agreement with the prediction from spin wave theory for the appropriate $\alpha$, $\Delta_0 = 0.170$. (For $\alpha = 0$, spin wave theory yields $\Delta_0 = 0.197$ for the quadratic layer, independent of $S$; cf. Equation (20).) In $K_2NiF_4$ an alternative route has been followed by noting that the normalized decrement of the frequency can be written as $[f(0) - f(T)]/f(0) = \Delta S(T)/(S - \Delta_0)$ [39]. A fit to the data may accordingly yield a value for $\Delta_0$, or rather the product $J(S - \Delta_0)$ (cf. Equation (27)). The method is particularly sensitive in the event that $\Delta_0$ is an appreciable fraction of the total spin, as is the case in $K_2NiF_4$ ($S = 1$). Inserting $J$ from other sources, one then finds $\Delta_0 = 0.20 \pm 0.03$, in close agreement with the spin wave value $\Delta_0 = 0.177$, calculated for $\alpha = 0.0021$ appropriate to $K_2NiF_4$. The question, of course, remains to what extent the spin wave approach is valid for $S = \frac{1}{2}$ systems, the case for which the kinematical, and for that matter the dynamical, corrections are largest. Kinematical interactions seem to reduce $\Delta_0$ by a few multiples of 0.01 [66]. A recent neutron scattering determination in $YBa_2Cu_3O_6$ resulted in $0.60 \pm 0.05\ \mu_B$ for the $Cu^{2+}$ moment at zero temperature, or $\Delta_0 = 0.20 \pm 0.03$ [67].

A spin wave analysis of the temperature and field dependences of the spontaneous magnetization in the ferromagnet $K_2CuF_4$ has been worked out in detail by Tsuru and Uryû [56]. The authors start from the Hamiltonian Equation (38), i.e., with inclusion of the additional exchange interactions necessary to achieve agreement with the dispersion, but augmented by the Zeeman energy $-g\mu_B H\ \Sigma_i S_i^z$. The Dyson—Maleev transformation is applied to produce a Hamiltonian expressed in boson operators. (HP would not have yielded different results up to the order considered.) The analysis, which further essentially follows the lines of Section 3 with inclusion of renormalization according to method (i), ultimately arrives at the first-order renormalized spin wave energies

$$E_{\mathbf{k}} = [(A_{\mathbf{k}} - \Delta_{\mathbf{k}})^2 - (B_{\mathbf{k}} - \Delta_{\mathbf{k}}^a)^2]^{\frac{1}{2}}, \tag{41}$$

where $A_{\mathbf{k}}$ and $B_{\mathbf{k}}$ are already specified in Equation (40), except for the addition of $g\mu_B H$ to $A_{\mathbf{k}}$. The renormalization is contained in $\Delta_{\mathbf{k}}$ and $\Delta_{\mathbf{k}}^a$, which read

$$\begin{aligned} \Delta_{\mathbf{k}} &= 2zJS(1-\gamma_{\mathbf{k}})f_1 + 2z'J'S(1-\gamma'_{\mathbf{k}})f_2 + 4zJ_A S\gamma_{\mathbf{k}} f_3, \\ \Delta_{\mathbf{k}}^a &= 4zJ_A\gamma_{\mathbf{k}} f_3. \end{aligned} \tag{42}$$

Here,

$$\begin{aligned} f_1 &= \frac{1}{N_0 S}\sum_{\mathbf{k}} (1-\gamma_{\mathbf{k}})\lambda_{\mathbf{k}} n_{\mathbf{k}}, \\ f_2 &= \frac{1}{N_0 S}\sum_{\mathbf{k}} (1-\gamma'_{\mathbf{k}})\lambda_{\mathbf{k}} n_{\mathbf{k}}, \\ f_3 &= \frac{1}{N_0 S}\sum_{\mathbf{k}} \lambda_{\mathbf{k}} n_{\mathbf{k}}, \end{aligned} \tag{43}$$

with

$$\lambda_{\mathbf{k}} = A_{\mathbf{k}}/(A_{\mathbf{k}}^2 - B_{\mathbf{k}}^2)^{\frac{1}{2}}. \tag{44}$$

The functions $f_1$, $f_2$, and $f_3$ depend on temperature and field. Note that Equation (41) reduces to the much more simple Equation (29) derived in Section 3 upon removal of the corrections to the primary exchange $J$. Also note that in $K_2CuF_4$, $\lambda_{\mathbf{k}}$ approximates to unity except at the zone center. Inspection of Equations (41—44) shows that, as in antiferromagnets, the renormalization can be simulated by using the unrenormalized expression, Equation (39) or Equation (41) with $\Delta_{\mathbf{k}} = \Delta_{\mathbf{k}}^a = 0$, upon inserting the effective exchange constants

$$J_s = J(1-f_1), \quad J'_s = J'(1-f_2), \quad J_{A,s} = J_A(1-f_3). \tag{45}$$

The spontaneous magnetization is given by

$$\langle S^z \rangle = S - \Delta_0 - \frac{1}{N_0}\sum_{\mathbf{k}} \lambda_{\mathbf{k}} n_{\mathbf{k}}, \tag{46}$$

where $\Delta_0 = (2N_0)^{-1}\ \Sigma_{\mathbf{k}}(\lambda_{\mathbf{k}} - 1)$ is the zero-point spin reduction, invoked by the antiferromagnetic exchange $J_A$. $\Delta_0$ is smaller by a factor of order $(J_A/J)^2$ than $\Delta_0$ in antiferromagnets, Equation (20), and accordingly entirely negligible.

The temperature and field dependencies of the spontaneous magnetization according in Equation (46) are compared with experimental data in Figures 12 and 13, both taken from [56]. The data have been collected by use of $^{19}F$ NMR at zero field over a range extending to 0.7 $T_N$ [68] and at 1.77 K up to 10 kG [69], respectively. Even more than in the case of antiferromagnets discussed above, first-order renormalization appears to have profound effects. Obviously, the near

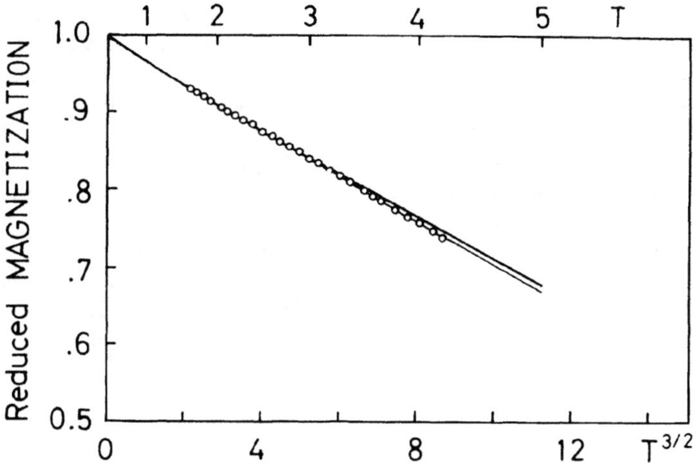


Fig. 12. Reduced spontaneous magnetization in $K_2CuF_4$ vs $T^{3/2}$, measured with $^{19}F$ NMR by H. Kubo, *J. Phys. Soc. Jpn.* **36** (1974) 675. The solid lines are calculated with renormalized spin wave theory, with $J/k_B$ = 11.2 K (thick line) and $J/k_B$ = 11.0 K (thin line). After K. Tsuru and N. Uryû, *J. Phys. Soc. Jpn.* **41** (1976) 804.

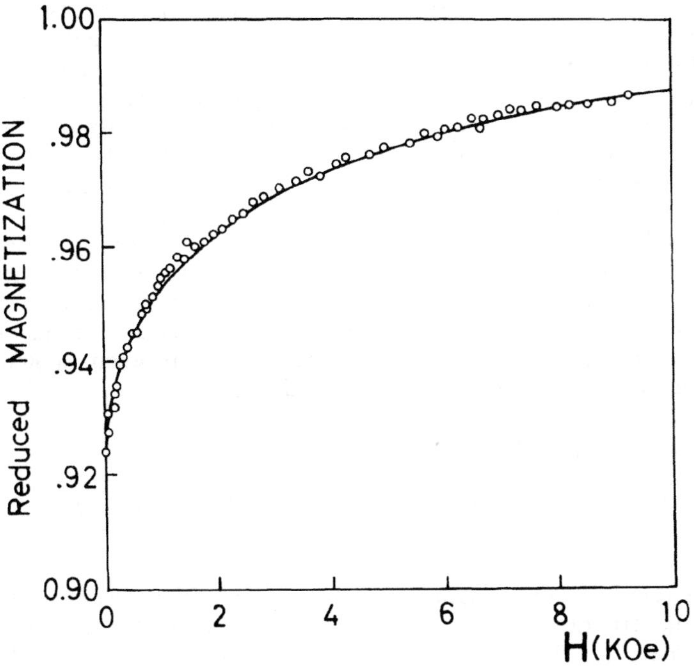


Fig. 13. Reduced magnetization in $K_2CuF_4$ vs the magnetic field at 1.77 K, measured with $^{19}F$ NMR. The solid line represents renormalized spin wave theory. After K. Tsuru and N. Uryû, *J. Phys. Soc. Jpn.* **41** (1976) 804.

absence of a spin wave energy gap leads to substantial values of $n_{\mathbf{k}}$ already at low temperatures. In Figure 12 it is seen that the effects of the spin wave interactions are strong enough to modify the temperature dependence of the magnetization from linear in $T$ (cf. Equation (31)) to approximately proportional to $T^{3/2}$, thus seemingly restoring Bloch's law.

To conclude this section, we mention that spin wave calculations, including renormalization, have been carried out for the nearly 2-d Heisenberg ferromagnets $(C_2H_5NH_3)_2CuCl_4$ [70] and $(C_3H_7NH_3)_2CuCl_4$ [71]. In the former the ferromagnetic layers are coupled antiferromagnetically, while the latter differs from $K_2CuF_4$ by additional exchange anisotropy within the basal plane. Further, in these as well as other members of the series $(C_nH_{2n+1}NH_3)_2CuCl_4$ ($n$ = 1, 2, 3, 4, 8, 10) experimental results for the magnetization from NMR as well as an unrenormalized spin wave analysis have been reported by Dupas *et al.* [72].

## 6. Impurities in Antiferromagnets

In this final section, we discuss the extensions that can be made to the 2-d spin wave theory discussed above in order to treat local thermodynamic properties around isolated impurities, either magnetic or nonmagnetic, in antiferromagnets. Representative as well as instructive examples, for which extensive data and calculations are available, are $K_2MnF_4$ doped with Ni, Zn, and Mg [34], and $K_2NiF_4$ doped with Mn [73], all at concentrations of order 1 at. %. In $K_2MnF_4$ a Ni dopant gives rise to local modes well above the spin wave band, while in the latter cases the impurity-associated modes are low within the band. The theory, developed by Van Luyk *et al.* [34], is quite naturally based on the Green's function formalism, and allows the impurity-associated modifications, themselves worked out up to order $1/2S$, to be calculated to all orders of the perturbation.

We consider a single impurity in a quadratic-layer antiferromagnet on, say, the up lattice. Extreme dilution is implied so that the impurity does not interact with other impurities. The spin Hamiltonian of the pure system $\mathscr{H}_{\text{pure}}$, Equation (1) with $H = 0$, is then augmented with a local perturbation restricted to substitution of the impurity at the origin and modification of the exchange between the impurity site and its first shell of neighbors at displacements $\boldsymbol{\delta}$. That is,

$$\mathscr{H} = \mathscr{H}_{\text{pure}} + V,$$

$$V = \left(2|J_0| \sum_{\boldsymbol{\delta}} \mathbf{S}_0 \cdot \mathbf{S}_{\boldsymbol{\delta}} - g_0\mu_B H_{A0} S_0^z\right) - \left(2|J| \sum_{\boldsymbol{\delta}} \mathbf{S} \cdot \mathbf{S}_{\boldsymbol{\delta}} - g\mu_B H_A S^z\right). \quad (47)$$

Quantities referring to the impurity are labelled with the subscript 0 throughout. We insert the HP transformation in both $\mathscr{H}_{\text{pure}}$ and $V$, and subsequently expand up to first order in $1/2S$, which yields terms containing two or four local spin deviation operators. In view of the local nature of the problem, it is advantageous to decouple with the RPA prior to passage to spin waves, i.e., decoupling scheme

(ii) discussed in Section 3, to obtain the effective two-operator Hamiltonians

$$\mathscr{H}_{\text{pure}} = 2|J|SR \sum_{\langle l,\, m\rangle} [(1 + \alpha/R)(a_l^\dagger a_l + b_m^\dagger b_m) + a_l b_m + a_l^\dagger b_m^\dagger],$$

$$V = 2z|J|SR \left[ \tau a_0^\dagger a_0 + \sigma \sum_{\boldsymbol{\delta}} b_{\boldsymbol{\delta}}^\dagger b_{\boldsymbol{\delta}} + \frac{\lambda}{z} \sum_{\boldsymbol{\delta}} (a_0 b_{\boldsymbol{\delta}} + a_0^\dagger b_{\boldsymbol{\delta}}^\dagger) \right]. \tag{48}$$

Here, the renormalization is contained in the quantity $R = 1 - (\Delta + \mu)/S$, which refers to the pure system; $\Delta = \langle a_l^\dagger a_l \rangle_{\text{pure}} = \langle b_m^\dagger b_m \rangle_{\text{pure}}$ is the number of spin deviations, and $\mu = \langle a_l b_m \rangle_{\text{pure}}$ represents the correlation of the nearest neighbors, with $\langle\ \ \rangle$ denoting the statistical average of the expectation value of the operator within. As before, $\alpha = g\mu_B H_A/2z|J|S$ is the anisotropy parameter, while $z = 4$ is the coordination number. The coefficients $\tau$, $\sigma$, and $\lambda$ are specified in [34]. They are inversely proportional to $R$ and depend further on $S_0$, $J_0$, $\alpha_0$, $\langle a_0^\dagger a_0 \rangle$, $\langle b_{\boldsymbol{\delta}}^\dagger b_{\boldsymbol{\delta}} \rangle$, $\langle a_0 b_{\boldsymbol{\delta}} \rangle$, and the corresponding quantities of the pure system. After passage to spatial Fourier transformations and Bogoliubov transformation to normal modes, $\mathscr{H}_{\text{pure}}$ yields the dispersion relation

$$E_{\mathbf{k}} = 2z|J|SR[(1 + \alpha/R)^2 - \gamma_{\mathbf{k}}^2]^{\frac{1}{2}}, \tag{49}$$

which is to be compared with the result from decoupling method (i), Equation (10), or the 'simple' spin wave approximation derived from it, Equations (17) and (18). For $K_2MnF_4$ and $K_2NiF_4$, $R = 1.0316$ and $1.0790$, respectively, at $T = 0$ [36].

The magnetizations at sites $l$ and $m$ located on the spin-up and spin-down sublattices, respectively, are generally given by

$$\langle S_l^z \rangle = S - \langle a_l^\dagger a_l \rangle, \qquad \langle S_m^z \rangle = -S + \langle b_m^\dagger b_m \rangle. \tag{50}$$

The statistical averages $\langle a_l^\dagger a_l \rangle$ and $\langle b_m^\dagger b_m \rangle$, as well as the ones already introduced, may be calculated from the appropriate double-time Green's functions of the spin deviation operators [74]. These have the general form $G_{AB}(t) = -i\theta(t)\langle [A(t), B(0)] \rangle$, with $\theta(t)$ the unit step function, or, in the energy domain,

$$G_{AB}(E) = \int_{-\infty}^{+\infty} G_{AB}(t)\, e^{iEt}\, \mathrm{d}t, \tag{51}$$

which develops according to the equation of motion

$$EG_{AB}(E) = \langle [A, B] \rangle + G_{[A, \mathscr{H}],\, B}(E). \tag{52}$$

After substituting $\mathscr{H}$, the Hamiltonian of the system, and solving for $G_{AB}(E)$, we recover $\langle a_l^\dagger a_l \rangle$, etc., from

$$\langle BA \rangle = \int_{-\infty}^{+\infty} \mathscr{A}_{BA}(E)\, \mathrm{d}E, \tag{53}$$

where

$$\mathscr{A}_{BA}(E) = -\frac{1}{\pi}\frac{\operatorname{Im} G_{AB}(E)}{e^{\beta E} - 1}, \tag{54}$$

with $\beta = 1/k_B T$, is the spectral intensity. It should be noted at this point that the higher-order Green's functions in Equation (52) reduce to the various $G_{AB}(E)$ due to the two-operator form of both $\mathscr{H}_{\text{pure}}$ and $V$, rendering decoupling of the Green's functions unnecessary.

In solving Equation (52), it is advantageous to deal first with the pure system. We introduce, for convenience, dimensionless energies $\varepsilon_{\mathbf{k}} = E_{\mathbf{k}}/2z|J|SR$ and dimensionless Green's functions $\Gamma_{ij}(\varepsilon) = 2z|J|SRg_j G_{ij}(E)$, with the indices running over all sites (and associated with the operators $a_l$ and $b_m$ in an obvious way), and $g_j = \pm 1$ depending on the sublattice the site $j$ is on. The first $(z+1) \times (z+1)$ part of the Green's function of the pure system, which is easily extensible, then reads

$$\Gamma^0 = \begin{pmatrix} (\varepsilon + 1 + \alpha/R)U(\varepsilon, \mathbf{0}) & V(\varepsilon, \boldsymbol{\delta})\Delta \\ -V(\varepsilon, -\boldsymbol{\delta})\Delta & (\varepsilon - 1 - \alpha/R)U(\varepsilon, \mathbf{p})\Delta' \end{pmatrix}, \tag{55}$$

in which $\Delta$ and $\Delta'$ are $1 \times 4$ and $4 \times 4$ matrices with all elements unity, respectively, and

$$U(\varepsilon, \mathbf{p}) = \frac{1}{N_0}\lim_{s \to 0^+}\sum_{\mathbf{k}} \frac{e^{i\mathbf{k}\cdot\mathbf{p}}}{(\varepsilon + is)^2 - (1 + \alpha/R)^2 + \gamma_{\mathbf{k}}^2},$$

$$V(\varepsilon, \mathbf{p}) = \frac{1}{N_0}\lim_{s \to 0^+}\sum_{\mathbf{k}} \frac{\gamma_{\mathbf{k}}\, e^{i\mathbf{k}\cdot\mathbf{p}}}{(\varepsilon + is)^2 - (1 + \alpha/R)^2 + \gamma_{\mathbf{k}}^2}, \tag{56}$$

with $\mathbf{p} = \mathbf{r}_i - \mathbf{r}_j$. The impurity-associated Green's functions may be subsequently found from Dyson's equation in matrix form

$$\Gamma = (I - \Gamma^0 V)^{-1}\Gamma^0, \tag{57}$$

which is exact to all orders in the perturbation. According to Equation (47), $V$ is a $(z+1) \times (z+1)$ matrix with zeros added. With $V$ and $\Gamma^0$, Equation (57) permits the derivation of algebraic expressions for the elements of $\Gamma$, which, however, turn out to be of considerable complexity, even when restricting the problem to the 13-site cluster of the impurity and its first three shells of neighbors within the layer. These expressions are given in Equations (45) of [34], as is a method to evaluate $U(\varepsilon, \mathbf{p})$ and $V(\varepsilon, \mathbf{p})$ numerically for quadratic-layer systems. The inversion of $I - \Gamma^0 V$ in Equation (57) gives rise to poles of $s$, $p$, and $d$ character, corresponding to impurity modes primarily located on the impurity itself or one of the shells. As to the computer evaluations of the Green's functions, we here only note that their occurrence in the renormalizing factor $R$ and the coefficients $\tau$, $\sigma$, and $\lambda$ necessitates iteration of the numerical process until self-consistency is reached.

With regard to the experimental results, we confine ourselves to a few representative examples. The thermal decrements of the local magnetizations in units of spin, $\Delta S_i(T)$, residing at the first three shells about the nonmagnetic impurities Zn and Mg in $K_2MnF_4$ are, relative to the magnetization of the host at considerable distance, reproduced in Figure 14 [34], i.e., the $\Delta S_i(T)$ are additional to $\Delta S_\infty(T)$ and represent the effect of the substitution. The data points have been extracted from the positions of the weak satellites of the out-of-layer $^{19}F$ NMR that emerge upon the substitution of the impurity ($f_i^{\rm I} = 1, 5, 9$), as well as the NMR of out-of-layer $^{19}F$ nuclei adjacent to the impurity ($f_0^{\rm I}$) and of the in-layer $^{19}F$ nuclei between the impurity and its first Mn neighbors ($f_{01}^{\rm II}$). Care has been taken to eliminate the unknown modifications of the hyperfine parameters due to lattice deformations. The full curves are the results of the Green's function theory outlined above for a 13-site cluster about the impurity. The theory is found to provide a superb description, the more so as all relevant quantities are known. For comparison, Figure 14 also shows the results of the theory with the nearest-neighbor correlations removed by setting $\langle a_0 b_1 \rangle = 0$ and of a 2-d adaptation of a theory by Walker *et al.* [75] based on Green's functions of the spin operators themselves. With the former the effects are clearly overestimated, while in the latter the usage of 'quasi-bosons' describing the decrease of $S - \langle S^z \rangle$ with temperature appears largely to compensate the neglect of the correlations. The most salient conclusion expressed by Figure 14 is probably that such a strong perturbation as the replacement of a Mn spin by a nonmagnetic impurity has effects of a very localized nature and of only moderate size. The effect on the first shell is still significant. At 20 K, the thermal decrement amounts of $\Delta S_5(T) = 0.07$ unit of spin in addition to $\Delta S_\infty(T) = 0.21$ already present in the unperturbed host, i.e., the magnetization on the first shell drops 4/3 times as fast as that on a host site. On shells further out than the first, however, the modifications relative to the host are reduced by at least an order of magnitude. In Ni-doped $K_2MnF_4$ the effects are even smaller [34]. A point to note here is that the analysis allows determination of $J_0$ between Ni and Mn, with the result $J_0/J = 3.0 \pm 0.1$.

In the above Mn-based systems the modifications are due to a removal of spectral density from the spin wave band to modes far above the band, rather than thermal excitation of a local mode. In Mn-doped $K_2NiF_4$, on the other hand, where $J_0/J < 1$, the impurity modes are resonant with the band. It was observed that the reduced magnetization $\langle S_i^z(T) \rangle / \langle S_i^z(0) \rangle$ drops about twice as fast at the impurity ($i = 0$) than it does in the pure system, or at host sites far away from the impurity ($i = \infty$) [73]. The results of the spectral density Im $\Gamma_{ii}(\varepsilon)$ at the impurity and the first few shells, as emerging from the fitting of the above Green's function theory to the measured $\langle S_0^z \rangle$ as a function of the temperature, are displayed in Figure 15. The fit yields $J_0/k_B = -16.2$ K, consistent with the analysis of Ni in $K_2MnF_4$. For comparison, also inserted is Im $\Gamma_{\infty\infty}(\varepsilon)$ of the pure system, to find that Im $\Gamma_{00}(\varepsilon)$ peaks low in the band at the expense of spectral density near the zone boundary. The associated local mode, which of course has an *s* character, is further seen to reside mainly on the impurity. It occurs so low within the band and is broadened to such an extent that it becomes appreciably excited already at

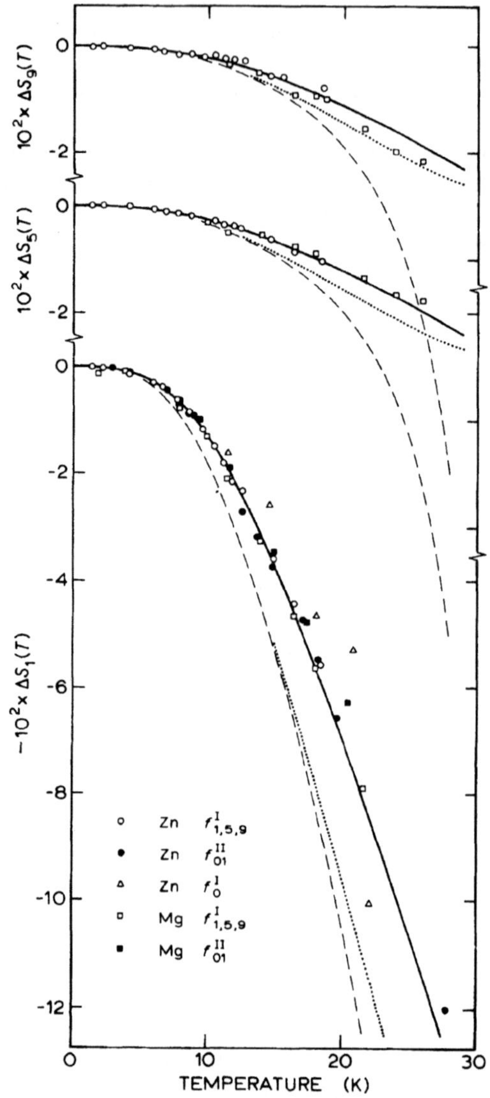


Fig. 14. Thermal spin decrements $\Delta S_i(T)$, relative to the host magnetization, in the first three shells about the nonmagnetic impurities Zn and Mg in $K_2MnF_4$. The data points have been extracted from the weak satellites in the NMR of out-of-layer (open symbols) and in-layer (closed symbols) $^{19}F$ nuclei. The solid curves have been calculated with the Green's function theory applied to a 13-site cluster with Mg or Zn as center. Dashed and dotted curves represent, respectively, the results of calculations with the nearest neighbor correlations set to zero and the results of the quasi-boson spin-operator theory by Walker *et al., Phys. Rev. B* **5** (1972) 1144. After J. A. van Luijk *et al., Phys. Rev. B* **21** (1980) 1963.

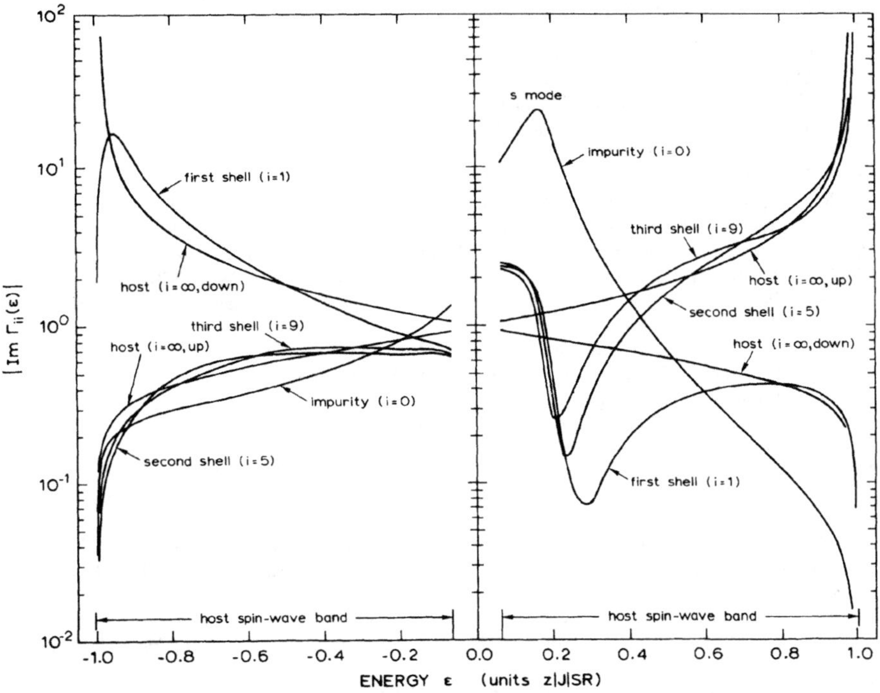


Fig. 15. Spectral density Im $\Gamma_{ii}(\varepsilon)$ at the Mn impurity ($i = 0$) and the Ni sites in the first three neighboring shells ($i = 1, 5, 9$) in Mn-doped $K_2NiF_4$. Also inserted is Im $\Gamma_{\infty\infty}(e)$ in pure $K_2NiF_4$. A strong resonance with $s$ symmetry, primarily residing at the impurity, occurs at $\varepsilon = 0.16$. After H. van der Vlist *et al.*, *Phys. Rev. B* **30** (1984) 5270.

temperatures corresponding to a fraction of the spin wave gap energy ($E_{k=0}/k_B$ = 27.5 K). The simple molecular-field approach, on the other hand, overshoots the energy of the peak by a factor of 2. The broadening corresponds to lifetimes of the local mode below 1 ps. A spatial decomposition of the impurity-associated excitation has finally extended to the case of resonant modes the general conclusion that in 2-d the disturbance by an impurity is limited to a few shells only. This also indicates that in dilute and mixed systems the local variations of the magnetization from the spatial average are primarily determined by the immediate surroundings.

## References

1. F. Bloch, *Z. Physik* **61** (1930) 206.
2. T. Holstein and H. Primakoff, *Phys. Rev.* **58** (1940) 1098.
3. F. J. Dyson, *Phys. Rev.* **102** (1956) 1217.
4. S. V. Maleev, *Zh. Eksp. Teor. Fiz.* **33** (1957) 1010 [*Sov. Phys. — JETP* **6** (1958) 776].
5. F. Keffer, in *Encyclopedia of Physics*, S. Flügge, Ed., Springer-Verlag, Berlin, 1966, Vol. XVIII/2.

6. P. W. Anderson, *Phys. Rev.* **86** (1952) 694.
7. R. Kubo, *Phys. Rev.* **87** (1952) 568.
8. J. M. Ziman, *Proc. Phys. Soc.* **A65** (1952) 540; *ibid.* **65** (1952) 548.
9. T. Oguchi, *Phys. Rev.* **117** (1960) 117.
10. N. D. Mermin and H. Wagner, *Phys. Rev. Lett.* **17** (1966) 1133.
11. G. G. Low, *Proc. Phys. Soc. (London)* **82** (1963) 999.
12. D. J. Breed, *Physica* **37** (1967) 35.
13. L. J. de Jongh and A. R. Miedema, *Adv. Phys.* **23** (1974) 1.
14. R. J. Birgeneau, H. J. Guggenheim, and G. Shirane, *Phys. Rev. B* **8** (1973) 304.
15. R. J. Birgeneau, H. J. Guggenheim, and G. Shirane, *Phys. Rev. B* **1** (1970) 2211.
16. E. Gurewitz, A. Epstein, J. Makovski, and H. Shaked, *Phys. Rev. Lett.* **25** (1970) 1713.
17. M. P. H. Thurlings, E. Frikkee, and H. W. de Wijn, *Phys. Rev. B* **25** (1982) 4750.
18. G. K. Wertheim, H. J. Guggenheim, H. J. Levinstein, D. N. E. Buchanan, and R. C. Sherwood, *Phys. Rev.* **173** (1968) 614.
19. Y. Ito and J. Akimitsu, *J. Phys. Soc. Jpn.* **40** (1976) 1333.
20. M. Hidaka and P. J. Walker, *Solid State Commun.* **31** (1979) 383.
21. D. I. Khomskii and K. I. Kugel, *Solid State Commun.* **13** (1973) 763.
22. K. Hirakawa and H. Ikeda, *J. Phys. Soc. Jpn.* **35** (1973) 1328.
23. I. Yamada, *J. Phys. Soc. Jpn.* **33** (1972) 979.
24. C. Dekker, A. F. M. Arts, and H. W. de Wijn, *Phys. Rev. B* **38** (1988) 11512.
25. D. Vaknin, S. K. Sinha, D. E. Moncton, D. C. Johnston, J. M. Newsam, C. R. Safinya, and H. E. King, Jr., *Phys. Rev. Lett.* **58** (1987) 2802.
26. M. A. Kastner, R. J. Birgeneau, T. R. Thurston, P. J. Picone, H. P. Jenssen, D. R. Gabbe, M. Sato, K. Fukuda, S. Shamoto, Y. Endoh, K. Yamada, and G. Shirane, *Phys. Rev. B* **38** (1988) 6636.
27. J. M. Tranquada, D. E. Cox, W. Kunnmann, H. Moudden, G. Shirane, M. Suenaga, P. Zolliker, D. Vaknin, S. K. Sinha, M. S. Alvarez, A. J. Jacobson, and D. C. Johnston, *Phys. Rev. Lett.* **60** (1988) 156.
28. A. F. M. Arts and H. W. de Wijn, *Phys. Rev. B* **15** (1977) 4348.
29. T. Nakamura, *Prog. Theor. Phys.* **7** (1952) 539.
30. A. Pekalski and W. J. Zietek, *Acta Phys. Pol.* **A39** (1971) 327; A. Pekalski, *Acta Phys. Pol.* **A40** (1971) 189.
31. C. A. M. Mulder, H. W. Capel, and J. H. H. Perk, *Physica* **112B** (1982) 147.
32. P.-A. Lindgard and O. Danielsen, *J. Phys. C* **7** (1974) 1523; P.-A. Lindgard and A. Kowalska, *J. Phys. C* **9** (1976) 2081; J. F. Cooke and P.-A. Lindgard, *Phys. Rev. B* **16** (1977) 408.
33. I. Goldhirsh, *J. Phys. A* **13** (1980) 453.
34. J. A. van Luijk, A. F. M. Arts, and H. W. de Wijn, *Phys. Rev. B* **21** (1980) 1963.
35. H. W. de Wijn, L. R. Walker, S. Geschwind, and H. J. Guggenheim, *Phys. Rev. B* **8** (1973) 299.
36. H. W. de Wijn, L. R. Walker, and R. E. Walstedt, *Phys. Rev. B* **8** (1973) 285; Erratum **9** (1974) 2419.
37. K. Tsuru and N. Uryû, *J. Magn. Magn. Mat.* **31—34** (1983) 1201.
38. C. A. M. Mulder and H. W. Capel, *Physica* **115B** (1983) 310.
39. H. W. de Wijn, R. E. Walstedt, L. R. Walker, and H. J. Guggenheim, *Phys. Rev. Lett.* **27** (1970) 832.
40. H. Mori and K. Kawasaki, *Progr. Theor. Phys.* **27** (1962) 529.
41. J. Skalyo, Jr., G. Shirane, R. J. Birgeneau, and H. J. Guggenheim, *Phys. Rev. Lett.* **23** (1969) 1394.
42. R. A. Cowley, G. Shirane, R. J. Birgeneau, and H. J. Guggenheim, *Phys. Rev. B* **15** (1977) 4292.
43. B. Schröder, V. Wagner, N. Lehner, K. M. Kesharwani, and R. Geick, *Phys. Status Solidi(b)* **97** (1980) 501.
44. R. J. Birgeneau, F. DeRosa, and H. J. Guggenheim, *Solid State Commun.* **8** (1970) 13.
45. K. Nagata and Y. Tomono, *J. Phys. Soc. Jpn.* **36** (1974) 78.
46. K. Strobel and R. Geick, *Physica* **108B** (1981) 951; R. Geick and K. Strobel, in *Reviews on Infrared and Millimeter Waves*, K. J. Button, Ed., Plenum Press, New York, 1983, Vol. 1, p. 249.

47. S. M. Rezende and R. M. White, *Phys. Rev. B* **14** (1976) 2939.
48. C. M. J. van Uijen and H. W. de Wijn, *Phys. Rev. B* **30** (1984) 5265.
49. M. P. H. Thurlings, A. van der Pol, and H. W. de Wijn, *Phys. Rev. B* **25** (1982) 4765.
50. F. Macco, W. Lehmann, W. Breitling, A. E. Slawska-Waniewska, R. Weber, and U. Dürr, *Solid State Commun.* **26** (1978) 429.
51. C. J. Peters, R. J. Birgeneau, M. A. Kastner, H. Yoshizawa, Y. Endoh, J. Tranquada, G. Shirane, Y. Hidaka, M. Oda, M. Suzuki, and T. Marakami, *Phys. Rev. B* **37** (1988) 9761.
52. K. B. Lyons, P. A. Fleury, J. P. Remeika, A. S. Cooper, and T. J. Negran, *Phys. Rev. B* **37** (1988) 2353.
53. S. Funahashi, F. Moussa, and M. Steiner, *Solid State Commun.* **18** (1976) 443.
54. F. Moussa, B. Hennion, J. Mons, and G. Pepy, *Solid State Commun.* **27** (1978) 141.
55. K. Hirakawa, H. Yoshizawa, J. D. Axe, and G. Shirane, *J. Phys. Soc. Jpn.* **52** (1983) 4220.
56. K. Tsuru and N. Uryû, *J. Phys. Soc. Jpn.* **41** (1976) 804.
57. A. S. Borovik-Romanov, N. M. Kreines, R. Laiho, T. Levola, and V. G. Zhotikov, *J. Phys. C* **13** (1980) 879.
58. H. Yamazaki, *J. Phys. Soc. Jpn.* **37** (1974) 667.
59. F. Moussa and J. Villain, *J. Phys. C* **9** (1976) 4433.
60. C. A. M. Mulder, H. L. Stipdonk, P. H. Kes, A. J. van Duyneveldt, and L. J. de Jongh, *Physica* **113B** (1982) 380.
61. M. P. H. Thurlings, B. J. Dikken, A. F. M. Arts, and H. W. de Wijn, *Phys. Rev. B* **25** (1982) 4771.
62. A. F. M. Arts, C. M. J. van Uijen, J. A. van Luijk, H. W. de Wijn, and C. J. Beers, *Solid State Commun.* **21** (1977) 13.
63. C. A. M. Mulder, J. Onderwaater, W. J. Looyestijn, A. J. van Duyneveld, and H. W. Capel, *Phys. Lett.* **87A** (1982) 303.
64. A. F. M. Arts, C. M. J. van Uijen, and H. W. de Wijn, *Phys. Rev. B* **15** (1977) 4360.
65. R. E. Walstedt, H. W. de Wijn, and H. J. Guggenheim, *Phys. Rev. Lett.* **25** (1970) 1119.
66. T. Ishikawa and T. Oguchi, *Progr. Theor. Phys.* **54** (1975) 1282; D. A. Huse, *Phys. Rev. B* **37** (1988) 2380.
67. P. Burlet, C. Vettier, M. J. G. M. Jurgens, J. Y. Henry, J. Rossat-Mignod, H. Noel, M. Potel, P. Gougeon, and J. C. Levet, *Physica C* **153—155** (1988) 1115.
68. H. Kubo, *J. Phys. Soc. Jpn.* **36** (1974) 675.
69. H. Kubo and N. Kaneshima, personal communication to Ref. 56.
70. K. Tsuru and N. Uryû, *Physica* **92B** (1977) 166.
71. K. Tsuru and N. Uryû, *Physica* **92B** (1977) 79.
72. A. Dupas, K. Le Dang, J.-P. Renard, and P. Veillet, *J. Phys. C* **19** (1977) 3399.
73. H. van der Vlist, A. F. M. Arts, and H. W. de Wijn, *Phys. Rev. B* **30** (1984) 5270.
74. D. N. Zubarzv, *Usp. Fiz. Nauk* **71** (1960) 71 [*Sov. Phys. Usp.* **3** (1960) 320].
75. L. R. Walker, B. C. Chambers, D. Hone, and H. Callen, *Phys. Rev. B* **5** (1972) 1144.

# NEUTRON SCATTERING EXPERIMENTS ON TWO-DIMENSIONAL HEISENBERG AND ISING MAGNETS

K. HIRAKAWA[1]* AND H. IKEDA[2]

[1] *Institute for Solid State Physics, The University of Tokyo, Roppongi, Minato-ku, Tokyo, Japan*
[2] *Department of Physics, Faculty of Science, Ochanomizu University, Ohtsuka, Bunkyo-ku, Tokyo, Japan.*

## 1. Introduction

Two-dimensional (2-d) magnetic systems show a variety of properties depending on the symmetry in the spin space, the interaction range, the quantum nature and also the symmetry of the lattice on which the spins are located. No real substance, however, is an ideal system for which the Hamiltonian can be written in terms of a single, simple interaction term. In most cases some additional interaction terms of different symmetry are needed. As will be shown later by examples, the physical property that is characteristic of the symmetry is not only dependent on the substance itself, but also on specified conditions such as temperature, external field, wave number etc. which the observers can vary. At temperatures very close to the critical point, all systems of basically Heisenberg symmetry will behave like an Ising system if the Hamiltonian contains an Ising-like interaction term, no matter how small it is. Such a general property, known as 'universality', is most remarkably seen in 2-d magnetic systems, because the critical region is extremely wide as compared to the analogous 3-d case.

The 2-d systems to be discussed in this article will be classified into three groups. The first group we discuss are the pure compounds typically of the $K_2NiF_4$ structure and with Heisenberg- and Ising-like interactions. We shall not discuss all the compounds but limit ourselves to some typical ones which display the most characteristic features of 2-d systems. Even in such a limited number of substances, we may be able to view the recent progress in this field.

The second subject to be discussed is the recent research on random magnetic systems. The percolation problem, the random field effects and the competing exchange interactions or anisotropies are of current interest and are, for instance, closely connected with the spin glass problem. A new topic may be the real time observation of the evolution of the spin correlations placed in a nonequilibrium state. All of these subjects have been actively investigated in the last few years. For the study of these problems, the Ising-like system is very often used as a basis because of the simplicity of the interactions compared with the Heisenberg or XY case.

The third subject, which is again quite different from the above two, is formed

* Present address: Fukuoka Institute of Technology, Higashi-ku, Fukuoka, Japan.



*L. J. De Jongh (Ed.), Magnetic Properties of Layered Transition Metal Compounds* 231—270.

by the frustrated spin systems in 2-d, as seen e.g. in the odd model treated by Villain. The simplest case is the triangular lattice system with antiferromagnetic interactions among the nearest-neighbour spins only. This system is characterized by the less stable static potentials between the spins due to the cancellation of the Weiss field. Therefore rather minor interactions, such as dipole—dipole interactions or multispin interactions, can drastically alter the properties of the system. Prominent quantum effects which alter the ground state to a completely new type may also be expected. Experimental examples of this new field of study are scarce, but some interesting results have already been obtained, and it seems worthwhile to introduce the reader to it.

## 2. 2-d Systems with Ising and Heisenberg Interactions

### 2.1. $K_2CoF_4$: A 2-d ISING SYSTEM

$K_2CoF_4$ occupies a special position in the $A_2MF_4$ series of compounds because it is undoubtedly the best example of the $S = 1/2$, square lattice, Ising antiferromagnet. The lowest orbital level, split in the cubic crystalline field, has an effective orbital moment of $L = 1$ and it couples with $S = 3/2$ by the $LS$ coupling. The resultant lowest multiplet is well approximated by the effective spin of $S = 1/2$. According to the ESR study on the $Co^{2+}$ ions in $K_2ZnF_4$ [1], $g_a = 3.13$ and $g_c = 6.30$, indicative of a strong Ising-like anisotropy. The Hamiltonian may be written as

$$\mathcal{H} = -2J_{\parallel} \sum_{\langle ij \rangle} S_i^z S_j^z - 2J_{\perp} \sum_{\langle ij \rangle} (S_i^x S_j^x + S_i^y S_j^y) \tag{1}$$

where $J_{\parallel}$ and $J_{\perp}$ are the exchange coupling constants in the 2-d plane, which have been determined by inelastic neutron scattering. The ratio of the interplane to the intraplane exchange constants, $J'/J$, is too small to be determined by neutron inelastic scattering due to the limitation of resolution, but is certairly less than $10^{-3}$. Samuelsen [2] and Ikeda and Hirakawa [3] have shown that the neutron scattering from $Rb_2CoF_4$ and $K_2CoF_4$ is in full agreement with Onsager's exact solution [4]. The explicit expression of the magnetization $M(T)$ is given by

$$\frac{M_S(T)}{M_0} = \left(1 - \sinh^{-4} \frac{J_{\parallel}}{kT}\right)^{1/8} \tag{2}$$

so that near $T = T_N$,

$$M_S(T)/M_0 = [(T_N - T)/T_N]^{\beta} \tag{3}$$

and the inverse correlation length $\kappa$ is given by

$$\kappa a = \ln \coth \left(\frac{J_{\parallel}}{2kT}\right) - \frac{J_{\parallel}}{kT} \tag{4}$$

where $a$ is the lattice constant. The experimental results that have been obtained and the theoretical curves are shown in Figures 1 and 2. The values of the critical indices found are listed in Table I. We should note that these values are derived from the experimental data using Fisher's expression [5] for the correlation function. After the experiment was published, the validity and the limitation of this expression were discussed [6]. A closer examination using a more accurate formula may be worthwhile to undertake in the future. Recently, the scattering at small $q$ was analysed by the Ornstein—Zernike formula rather than with Fisher's formula. (We should note, however, that use of the Ornstein—Zernike formula does not mean that correct answer is $\eta = 0$).

As for the dynamics, propagating spin waves do not exist if the system is purely Ising since the off-diagonal term is absent. The excitation spectrum for the transverse fluctuation would be a dispersionless mode at an excitation energy of

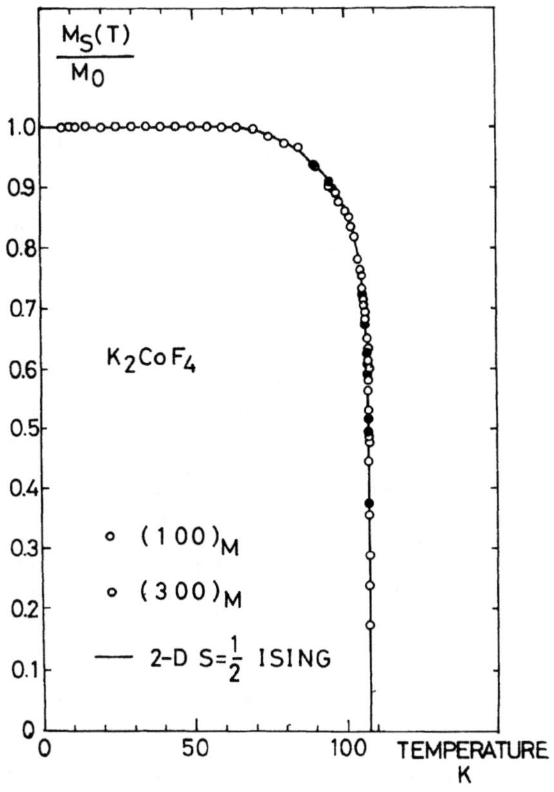


Fig. 1. Temperature variation of the square root of the observed magnetic Bragg scattering from $K_2CoF_4$ showing excellent agreement with the $M_S(T)/M_0$ curve for the 2-d Ising model (Onsager, Yang).

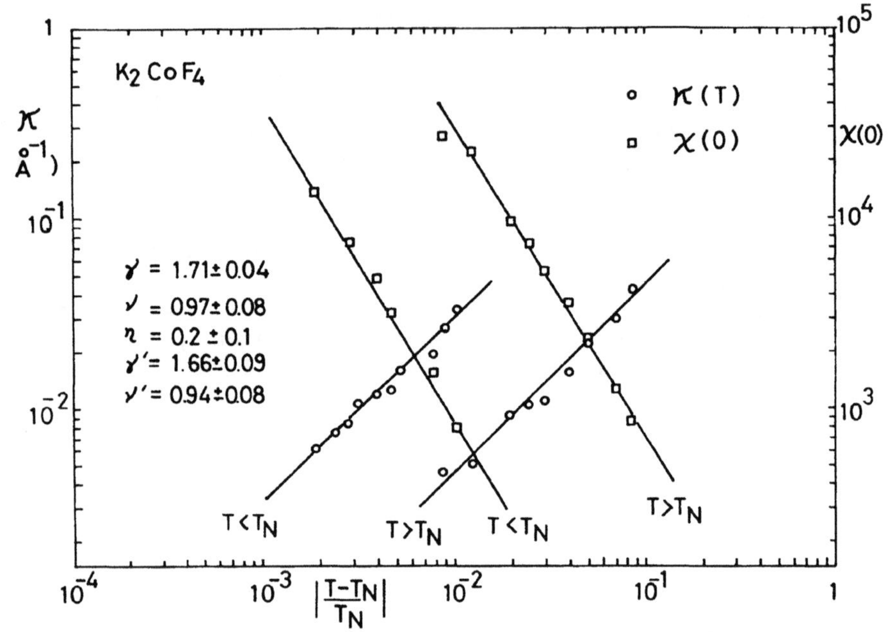


Fig. 2. Temperature variation of the staggered susceptibility $\chi$ ($q = 0$) and the inverse correlation length $\kappa$ for $K_2CoF_4$ showing good agreement with the 2-d, $S = 1/2$, Ising model. Reproduced by courtesy of the Physical Society of Japan.

TABLE I
Critical exponents of 2d Ising-like magnets.

| | $K_2CoF_4$ | $Rb_2CoF_4$ | Theory |
|---|---|---|---|
| $T_N$ | 107.85 K | 102.96 K | |
| $\nu$ | 0.97 ± 0.05 | 0.99 ± 0.04 | 1.0 |
| $\gamma$ | 1.71 ± 0.04 | 1.67 ± 0.09 | 1.75 |
| $\eta$ | 0.2 ± 0.1 | 0.2 ± 0.1 | 0.25 |
| $\beta$ | 0.123 ± 0.008 | 0.115 ± 0.016 | 0.125 |
| $\alpha$ | 0.003 ± 0.039 | 0.016 ± 0.051 | 0 |
| $z$ | | 0.167 ± 0.02 | $(1.75 = \gamma)$ |
| $J_{\parallel}$ | | 89.7 K | |
| $J_{\perp}$ | | 49.6 K | |

$4J$. In $Rb_2CoF_4$, which is essentially identical to $K_2CoF_4$, such a feature is clearly seen but a small fraction of the off-diagonal term still yields a slight dispersion in the excitation spectrum, allowing for the spin wave nature. From the spectrum, Ikeda and Hutchings [7] have determined the parameters in Equation (1). These values are nearly consistent with the parameters deduced from the static measurements [8]. Hutchings *et al.* [9] have also studied dynamical longitudinal fluctuations, and found the dynamics to be governed by $\Gamma_{\parallel} = 0.325t^z$ with $z = 1.69$ and $t = (T - T_N)/T_N$ at $q = 0$, and $\Gamma_{\parallel} = 3.6\ q^z$ with $z = 1.67$ at $T = T_N$ supporting the dynamical scaling relation for $Rb_2CoF_4$. The critical exponent $z$, which is predicted by the conventional theory to be 1.75, is listed in Table I. The above dynamical study proves that the appropriate Hamiltonian for $(K/Rb)_2CoF_4$ is not exactly that of an ideal 2-d Ising system. Nevertheless, all the critical behaviors are in good agreement with Onsager's exact solution. As mentioned, this is consistent with the universality hypothesis.

The experiments on $(K/Rb)_2CoF_4$ have shown that we can in fact have a real substance to which Onsager's theory is directly applicable. With this as a basis, many solid solutions can be studied for the purpose of elaborating the mechanisms of the cooperative phenomena, as will be shown later.

### 2.2. $K_2FeF_4$: A 2-d PLANAR ANTIFERROMAGNET

In contrast to the nearly isotropic Heisenberg systems $K_2NiF_4$ and $K_2MnF_4$, whose properties will be discussed shortly, $K_2FeF_4$ has been considered to be an easy-plane type antiferromagnet. The first elastic neutron scattering study was made by Birgeneau *et al.* [10], who found that below $T_N = 56.3$ K the spins are directed in one direction in the basal $c$-plane. A detailed neutron scattering study has recently been done by Thurlings *et al.* [11], who report the spin structure below $T_N$ as shown in Figure 3. If $K_2FeF_4$ is a truly ideal 2-d planar antiferromagnet, it would be a very interesting example for studying the Kosterlitz–Thouless transition.

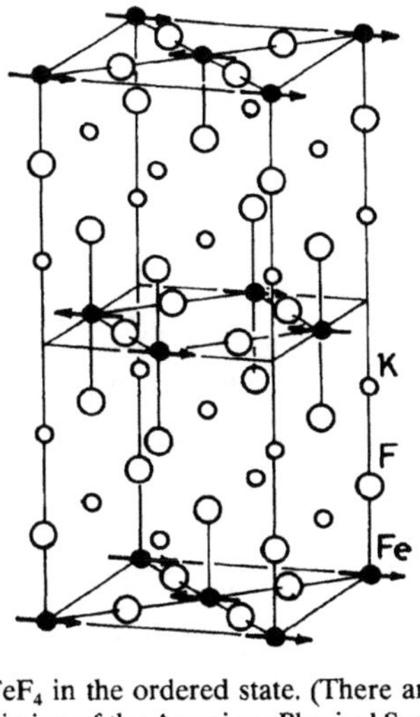


Fig. 3. The spin structure of $K_2FeF_4$ in the ordered state. (There are four domains.) Reproduced by permission of the American Physical Society.

Thurlings *et al.* have shown, however, that the system does not have a continuous symmetry in the basal plane. The anisotropy in the easy plane can not be expected from the quadratic term of the spin Hamiltonian nor from the dipolar interaction because of the 4-fold symmetry. One of the possibilty is the quartic term arising from the crystalline anisotropy. Along this line, but avoiding the mathematical difficulty by decoupling the quartic term of spin operator into the quadratic terms, this term is shown effectively to give rise to an additional anisotropy in the plane of the form $E\ \Sigma_i\ (S_{ix}^2 - S_{iy}^2)$, besides the planar existing anisotropy of the form $D\Sigma_i\ S_{iz}^2$. They treated the system as if it had an orthorhombic symmetry. Physically, the local potential minimum is treated as if an Ising-like anisotropy exists in the plane. In contrast to the case of perfect planar anisotropy, there is an energy gap in the acoustic mode of the spin wave excitation. In this case, the fan-shaped motion of the spin vector in the planar anisotropy system is altered to a cone-shaped motion by this pseudo-Ising-like anisotropy. The following values have been obtained from the spin wave analysis: $J/k = -15.7 \pm 0.3$ K, $D/k = 5.7 \pm 0.1$ K and $E/k = -0.49 \pm 0.03$ K at $T = 0$. The spin wave dispersion is reproduced in Figure 4. Apparently, no dispersion is found along the *c*-axis,

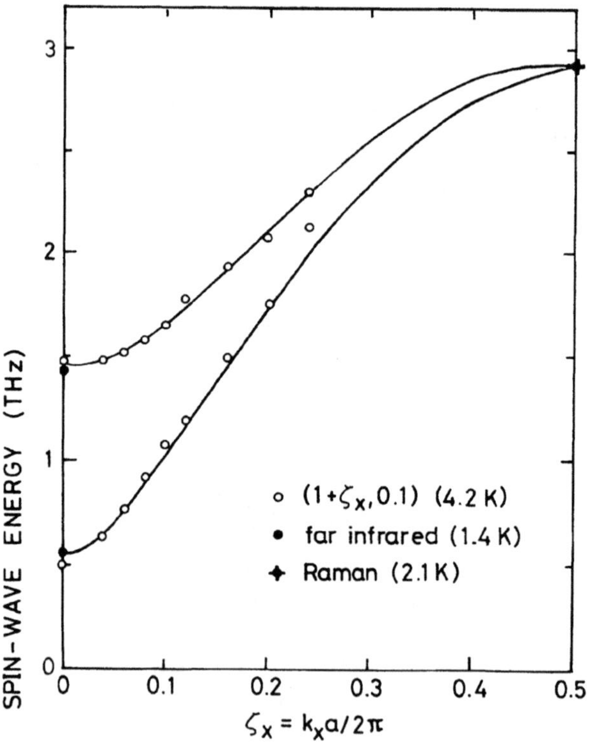


Fig. 4. The spin wave dispersion curve in $K_2FeF_4$. Reproduced by permission of the American Physical Society.

confirming that $J'/J$ is less than $10^{-3}$. The effective $E$ term in $K_2FeF_4$ is unexpectedly large, so large in fact that this compound should be classified within the same Ising universality class as $K_2MnF_4$, although the direction of spin alignment is different.

### 2.3. $K_2MnF_4$ AND $K_2NiF_4$: WEAKLY ANISOTROPIC HEISENBERG MAGNETS

Although these compounds are nearly Heisenberg-like antiferromagnets, a small amount of Ising-like anisotropy is present, the easy axis being the $c$-axis. Most of the thermodynamic properties have already been reviewed in the previous report by de Jongh and Miedema, so that we shall restrict ourselves to the important progress made thereafter. From their neutron scattering studies Shirane *et al.* [12] have shown that the critical exponents for these compounds, when analysed carefully by subtracting the scattering due to the transverse fluctuation, are in good agreement with the values for the pure 2-d Ising system. This means that the critical exponents are due solely to the symmetry of the Hamiltonian and not to the amount of the Ising-like interaction. Since the energy gap due to the uniaxial anisotropy is small for these compounds, the transverse susceptibility, which is inversely proportional to this gap, is very large and cannot be neglected.

The presence of the abovementioned Ising character provides evidence that the transition in these systems is caused primarily by the Ising anisotropy rather than by $J'$. The Ising behavior of $K_2MnF_4$ and $K_2NiF_4$ is, however, not an intrinsic property. For instance, in an external magnetic field, a transition from Ising to XY symmetry can occur. De Jongh *et al.* [13] have reported an interesting transition in $K_2MnF_4$. In a weakly anisotropic 3-d antiferromagnet, a first-order spin-flop transition occurs when the field $H$ applied along the easy axis is increased. But in the 2-d system $K_2MnF_4$, the transition from Ising to XY seems to occur at $H = 60$ kOe over a broad field range. The conventional bicritical point where the first-order spin-flop line bifurcates in the I-P and XY-P boundaries cannot be observed. Instead, the data appear to indicate a bifurcation of I-P into the I-XY and the XY-P curves, at a multicritical point $T^* \simeq 37.5$ K and $H^* \simeq 57.5$ kOe as shown in Figure 4 in the chapter by de Groot and de Jongh. They define an effective anisotropy parameter $\alpha_H = (H_{SF}^2 - H^2)/2H_E^2 = \alpha(1 - H^2/H_{SF}^2)$, where $\alpha = H_A/H_E = H_{SF}^2/2H_E^2$ is the anisotropy parameter at $H = 0$. When $H$ is increased, the effective Ising anisotropy decreases, and one may expect a decrease of $T_N$ with decreasing $\alpha_H$. But in this 2-d system the decrease is expected to be extremely steep, as expressed by $kT_N/|J| \sim \alpha^n$, $n$ being 0.04 as given by Binder *et al.* [14], or even logarithmic as predicted by Nelson *et al.* [15]. The full line in Figure 4 (de Groot and de Jongh's chapter) is for $n = 0.05$, in good agreement with the experimental points. For extremely small $\alpha_H$ the system will approach the 2-d Heisenberg magnet. However, the true Heisenberg behavior can hardly be expected to be observable because even a very small perturbation such as another small anisotropy of $J'$ will cause very drastic changes in $T_N$. For the same reason, the ideal Kosterlitz—Thouless state ($T < T_{KT}$, where $\chi \rightarrow \infty$ but without LRO) may be difficult to observe. The system seems to enter a mixed state, possibly made up of domains in which the ordered moments are not free to rotate but loosely bound in the [110] or [$\bar{1}$10] direction.

### 2.4. $Rb_2CrCl_4$: A PLANAR HEISENBERG FERROMAGNET WITH SMALL ANISOTROPY

The crystal structure of $Rb_2CrCl_4$ is essentially that of $K_2NiF_4$, except for a small modification [16] due to the Jahn—Teller distortion, similar as for $K_2CuF_4$ [17]. It has also in common with $K_2CuF_4$ that both are optically transparent ferromagnets. Hutchings *et al.* [18, 19] have studied the crystallographic as well as the magnetic structures and excitations in detail. According to them, the Curie temperature varies somewhat between 52.4 or 57 K, according to the differences in sample preparation. The lattice parameters $a$ and $b$ of the orthorhombic cell are equal within the experimental errors, so that the simplified pseudo-tetragonal cell will be used here, i.e. $a$ = 5.095 and $c$ = 15.751 Å. In $Rb_2CrCl_4$ the ferromagnetic spin alignment between the nearest-neighbour spins is slightly modified by the different orientation of the local distortion axis of the $Cl_6$-octahedron. Thus, the spins align in the direction of [110] but are slightly canted alternately by an angle of ~ 1° in the basal plane. The spin-wave dispersion has been analysed in detail. The Hamiltonian can be written in a simplified expression by neglecting the small canting angle as

$$\mathscr{H} = -2J \sum_{\langle ij \rangle} \mathbf{S}_i \cdot \mathbf{S}_j + D \sum_i S_i^{x^2} - p \sum_i S_i^{z^2} + \mathscr{H}_{\text{dip}} \tag{5}$$

where the $D$ term expresses the planar anisotropy and the $p$ term expresses the weak anisotropy in the easy plane. The effect of dipole—dipole interaction $\mathscr{H}_{\text{dip}}$ has been proved to be small. The coefficients obtained from the spin-wave analysis are,

$$J/k = 7.56 \pm 0.19\text{K}$$

$$D/k = 1.06 \pm 0.15\text{ K}$$

$$p/k = 0.123 \pm 0.014\text{ K}.$$

The ratio $J'/J$ = 0.0003 is small enough to regard this system as a good 2-d ferromagnet. The finite Curie temperature is mainly due to the small but finite Ising-like in-plane anisotropy, and to a lesser extent to the interplane interaction.

### 2.5. $K_2CuF_4$: A PLANAR HEISENBERG FERROMAGNET

$K_2CuF_4$ is an excellent realization of the 2-d Heisenberg ferromagnet with $S$ = 1/2. The Hamiltonian can be written as

$$\mathscr{H} = -2J \sum_{\langle ij \rangle} \mathbf{S}_i \cdot \mathbf{S}_j + 2J_{\text{A}} \sum_{\langle ij \rangle} S_i^z S_j^z - 2J' \sum_{\langle ik \rangle} \mathbf{S}_i \cdot \mathbf{S}_k , \tag{6}$$

where $J_A$ represents the anisotropic part of the exchange interaction. All the parameters $J/k$ = 11.34 K [20], $J_A/k$ = 0.12 K [21] and $J'/k$ = 0.009 K [22] are known experimentally (here $J'$ is defined with the effective coordination number $z'$

of the adjacent layers taken as 2 instead of 8). Though the system is nearly Heisenberg, the anisotropy $J_A$, which is of the easy plane-type, causes the low-temperature behavior to be XY-like. The reason is similar as for the Ising-type behavior observed for $K_2MnF_4$ and $K_2NiF_4$, as discussed above in connection with the universality concept. The ferromagnetic long-range order occurs at 6.25 K, however, before this the system undergoes successive crossovers from the 2-d Heisenberg (H) regime to the 2-d XY-regime and finally to the 3-d XY regime. The appearance of an Ising(I)-like regime has not been observed in this case. The reason is that the crystalline anisotropy can not be expected in this $S = 1/2$ system and also the dipolar anisotropy vanishes because of the square lattice symmetry. In fact, no in-plane anisotropy (presumably $H_A^{in} < 0.1$ Oe) has been observed even at 30 mK in zero external field [23]. This is important when discussing the transition in $K_2CuF_4$ on the basis of the Kosterlitz—Thouless (KT) transition, slightly modified by the 3-d interaction but not by the Ising-like interaction. For convenience, we shall express the various interactions in the form of effective fields, which have the following values at $T = 0$ K:

$$H_E = 2JzS/g\mu_B = 293 \text{ kOe}$$

$$H_A = 2J_A zS/g\mu_B = 2.3 \text{ kOe}$$

$$H'_E = 2J'z'S/g\mu_B = 126 \text{ Oe}$$

Hirakawa *et al.* [24] have shown that similar crossover phenomena as observed along the temperature ($T$) axis in zero field can also be seen in an increasing external field at a fixed low $T < T_c$. This is shown schematically in Figure 5. For instance, the crossover at 4 K as a function of field can be explained as follows. In a weak applied field $H$, the main mechanisms which bring the system into a 3-d ordered phase are the strong intraplane Weiss field, in combination with the weak interplane Weiss field of ~ 100 Oe. However, the magnetization for fields greater than this $H'_E$ is determined by the strength of $H$ rather than that of $H'_E$. For such field strengths, the response to the external field $H$ is very similar to that of the pure 2-d system with $H'_E = 0$. (We are discussing the case of the field parallel to the easy plane. For the field normal to the plane, i.e. the field non-conjugate to the order parameter, the response is entirely different [24]). The motion of the spins for $H < H_A$ is of a fan-shaped or ellipsoidal form, but it approaches a cone-shaped precession for $H > H_A$. This corresponds to an anisotropy crossover from the 2-d XY to the 2-d H regime.

Similar crossovers can occur in $q$-space as well. One of the simplest ways to understand this crossover is to carefully examine the spin wave dispersion relation which is given by [25]

$$\hbar\omega_q = 2S[\{zJ(1-\gamma(q) + z'J'(1-\gamma'(q))\}\{zJ(1-\gamma(q) + z'J'(1-\gamma'(q)) - zJ_A\gamma(q)\}]^{1/2},$$

$$\gamma(q) = z^{-1}\sum_{\rho}\exp(2\pi q\cdot\rho) \quad \text{and} \quad \gamma'(q) = z'^{-1}\sum_{\rho'}\exp(2\pi q_c\cdot\rho'), \tag{7}$$

$\rho$ and $\rho'$ being the positional vectors to the nearest neighbours.

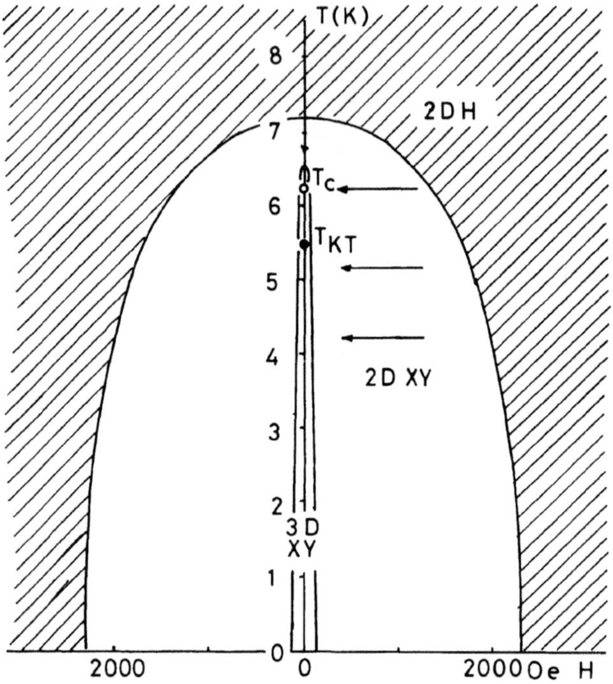


Fig. 5. Schematical phase diagram in $H$—$T$ space in $K_2CuF_4$ explaning the crossover phenomena. Reproduced by courtesy of the Physical Society of Japan.

For large enough $q$ (hereafter $q_a$ will be denoted simply by $q$ unless specially noted), the system behaves like a 2-d H system. For smaller values of $q$, however, the system becomes sensitive to the additional terms in the Hamiltonian which alter the symmetry. In the present case $H_A$ is greater than $H'_E$, so that when $q$ is reduced a 2-d H to 2-d XY anisotropy crossover occurs for $q_2 = 1/\pi(J_A/J)^{1/2} = 0.03$, which is a measure of the transition from the quadratic $q$ dependence for the 2-d Heisenberg ferromagnet to the linear $q$ dependence for the XY ferromagnet. For still smaller $q$ values, the system enters into the 3-d XY regime at $q_1 = 1/\pi(J'/J)^{1/2} = 0.008$. Below this value the dependence of the spin wave energy on $q_c$ can no longer be neglected. So, if we look at larger $q$ values, the Heisenberg nature can be fully seen. A theory for this region has been presented by Moussa and Villain (M.V.) [26], who explain the crossover phenomena qualitatively in the same way. We shall first discuss the Heisenberg regime of the system. The experiments have been done by M.V., following the first study of the dynamics by Funahashi *et al.* [20]. Recently, Hirakawa *et al.* (H.Y.A.S.) [25] performed a more systematic study of the dynamics, extending over much wider $q$ and energy ranges.

In the H regime then, the following simple spin wave dispersion law is found to hold quite accurately:

$$\hbar\omega_q = JzS(1 - \cos 2\pi q), \quad z = 4 \quad \text{and} \quad J/k = 11.93\,\text{K}. \tag{8}$$

The form of the dynamical scattering function has been predicted by M.V. [26]. According to them, one should distinguish three regions on the temperature scale. At $T \sim 0$, a very sharp $\delta$-function-like scattering, independent of the $q$ value is expected. At $T \sim T_c$, the scattering function is predicted to have the form:

$$\mathscr{S}(q, \omega) \propto [1/2 \pm 1/2 \pm n(\hbar\omega_q)] \frac{A}{(\omega \mp \omega q)^{1-\alpha(T)}} \tag{9}$$

where $n$ is the Bose factor and $\alpha(T) = kT/4\pi JS$. For $T > T_c$, the scattering function has a somewhat complicated form, but roughly speaking the spin wave feature is kept. The energy width increases with increasing $T$, as the intralayer correlation length $\xi$ decreases. A noticeable feature is that the width is largest at $q = 1/4$, where the slope of the dispersion is maximum. This means that the damping is caused by the uncertainty in the $q$ value, due to the finite size of $\xi$. The experimental results by M.V. and H.Y.A.S. are in agreement with the theory.

The correlation for smaller $q$ values is somewhat complicated, but interesting because the behavior of the system approaches the characteristic features associated with the KT transition in the 2-d XY system. The magnetic measurements have been performed in detail by Hirakawa *et al.* [24] to determine the susceptibility and magnetization in the H—T diagram. They have shown that in the 2-d XY regime the $\chi$ and $\delta$ (critical exponent for $M \propto H^{1/\delta}$) along the $T$ and the $H$ axes are consistent with Kosterlitz's theory [27] with $T_{KT} = 5.50$ K, which is a little lower than the actual Curie point $T_c = 6.25$ K. The difference is interpreted as being caused by the finite value of $J'$ [28]. The $\xi$, determined by neutron scattering [24], is also consistent with these measurements as well as the Kosterlitz theory, as shown in Figure 6.

One of the most important features of the KT transition is probably the universal jump in the stiffness constant expected at $T_{KT}$. In principle, it may be measured by twisting the magnetic system on a macroscopic scale. But, as we cannot expect a spontaneous magnetization, we cannot perform experiments by applying static twisting fields on the 2-d system and see its response on passing through $T_{KT}$. Moreover, for such a long-wave macroscopic scale, the real system which has interlayer interactions will never show the same response as in a 2-d XY system. The best way is to study the stiffness for the transverse twisting, i.e. the spin wave frequency, for properly small $q$ values. If the system would be an ideal 2-d XY ferromagnet, the $J$ at small $q$ thus measured should retain a constant value up to the very vicinity of $T_{KT}$, and then suddenly reduce to zero at $T_{KT}$. This jump-like decrease is described by the universal value of [27, 29]:

$$J = (1/\pi)kT_{KT} \qquad \text{(setting } S = 1\text{).} \tag{10}$$

In a nonideal system such as $K_2CuF_4$, the position of the transition point is modified a little by $J'$ and the character of the sudden jump will be difficult to see. Still some feature of the steep change in the spin wave frequency should be observable if the $q$ value is chosen small enough to be in the 2-d XY regime, but not so small as to be in the 3-d XY regime. Following this line of thinking, it is worthwhile examining the spin wave energies measured as a function of $T$ at small

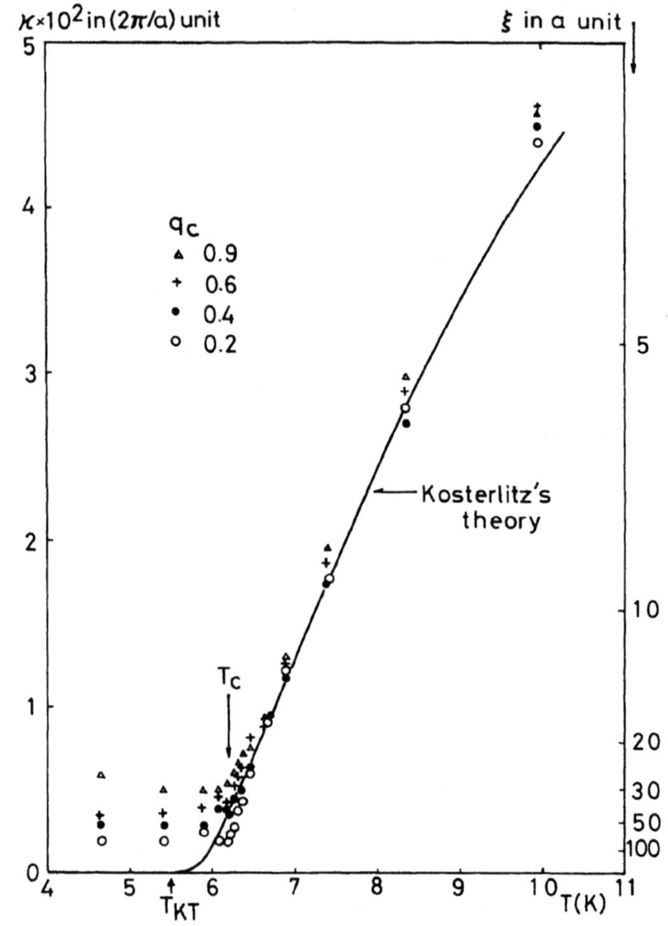


Fig. 6. Temperature variation of correlation length $\xi$ in the 2-d plane in $K_2CuF_4$. Near $T_C$, the apparent $\xi$ becomes $q_C$ dependent showing the 2-d to 3-d crossover. The full line is the theoretical curve given by Kosterlitz with $T_{KT} = 5.5$ K which is deduced from the other experiments. Reproduced by courtesy of the Physical Society of Japan.

$q$ values, as shown in Figure 7. For small $q$ the damping constant does indeed change quite rapidly on passing through $T_c$. Although the details of the frequency change cannot be determined, the sudden decrease is recognized qualitatively.

It should be noted here that the existence of isolated vortices immediately above $T_c$ has not been verified by neutron scattering. If vortices exist in $K_2CuF_4$, they will probably be different from those in a pure XY system because of the Heisenberg-type interaction. Near the centre of the vortex, the spins will curl up out of the easy plane like an instanton rather than a vortex. But except for the very center, the vortex-like features may still be present.

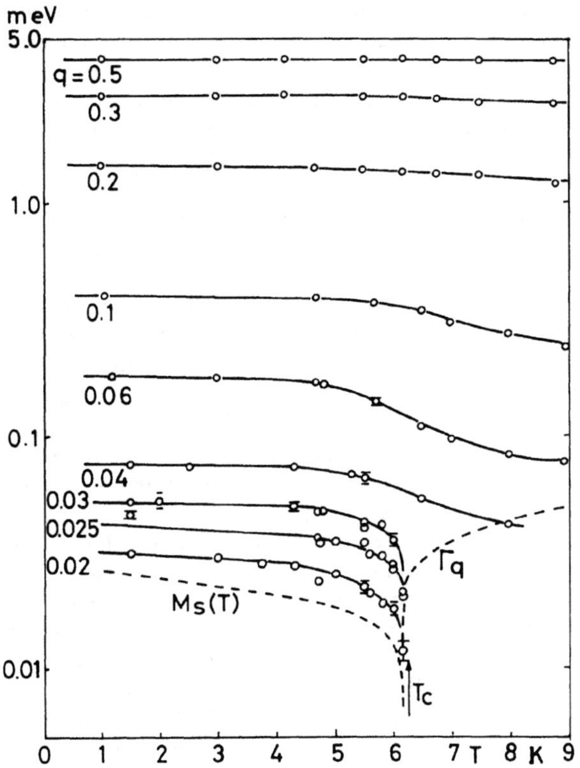


Fig. 7. Temperature variation of the spin wave frequency as a function of $q$ in $K_2CuF_4$, showing steep changes near $T_C$ at small values of $q$.

## 3. 2-d Random Magnetic Systems

The study of disordered insulating magnets is a very rich field for testing theories of phase transitions and excitations in random magnetic systems. Though there is a wide variety of systems available, 2-d systems with the $K_2NiF_4$ structure have been extensively investigated experimentally in recent years, and continue to provide crucial tests of the theories and new challenges for theoretical work. In this section we mainly discuss the neutron scattering measurements on these systems, which reveal the properties of the excitations and phase transitions of disordered systems.

### 3.1. PHASE TRANSITIONS AND CRITICAL PHENOMENA

The nature of the phase transitions in randomly disordered systems has attracted considerable attention. The obvious first question is whether or not a sharp continuous phase transition remains to be present, and if so whether the exponents are the same as for the related pure system. The first attempt to answer this question was made by Harris [30]. In 1974 he suggested, using general arguments, that the sharpness of the phase transition in a system with random quenched impurities is unchanged if the specific heat exponent $\alpha$ in the pure system is negative. If $\alpha$ is positive the asymptotic critical behavior is expected to change as the critical temperature is approached. In later work, using renormalization-group theory and the $\varepsilon$-expansion, this result was confirmed [31, 32]. However, for the 2-d Ising model, which has $\alpha = 0$, it is unclear if any change should be expected to occur or not.

In the initial experiments several 3-d mixed antiferromagnets have been studied, and in many cases a rounded phase transition has been observed [33, 34]. The rounding rather corresponds to macroscopic concentration fluctuations of about a few percent. Below we show some neutron scattering results obtained in the 2-d diluted Ising antiferromagnets $Rb_2Co_cMg_{1-c}F_4$, in which nonmagnetic Mg atoms are randomly substituted on the magnetic sites [35, 36]. In Figure 8 we show the phase diagram of $Rb_2Co_cMg_{1-c}F_4$, which has been determined mainly by observing the critical divergence of the staggered fluctuations in samples with various concentrations. When the magnetic atoms are diluted with nonmagnetic Mg atoms, the antiferromagnetic transition temperature decreases until at a critical concentration (percolation concentration), $c_p = 0.59$, there is no longer any long-range order at $T > 0$. The properties of percolation will also be discussed later. In random magnetic systems, it is very difficult to ensure that the samples are chemically homogeneous. Despite this difficulty, it is possible to measure the properties of the critical fluctuations by analyzing the result in the form of $S^{zz}(Q) = A_T/(q^2 + \kappa^2)$ where $Q = q + \tau$ and $\tau$ is any reciprocal lattice vector of the magnetic lattice. Since in 2-d Ising systems the exchange interactions are highly anisotropic, critical fluctuations are described only by a longitudinal component. In Figure 9 we show the results for the inverse correlation length $\kappa$ for $Rb_2Co_cMg_{1-c}F_4$ ($c = 1$, 0.91, 0.82 and 0.7) as measured by Ikeda [35]. Clearly the critical behavior of all these diluted compounds is very similar. The exponent $\nu$ is equal within experimental error to that predicted by the exact solution of the 2-d Ising model and as found in the pure compound $Rb_2CoF_4$ except that the bare length in diluted systems is enhanced. The bare length in $Rb_2Co_{0.7}Mg_{0.3}F_4$ is enhanced by a factor of 4 with respect to $Rb_2CoF_4$. These results strongly suggest that the critical exponents in 2-d magnets with Ising symmetry are unchanged upon dilution.

Critical properties of the 2-d mixed antiferromagnets $Rb_2Mn_{0.5}Ni_{0.5}F_4$ [37] and $Rb_2Co_cNi_{1-c}F_4$ ($c = 0.8$, 0.65, 0.5) [38] have also been studied by critical neutron scattering and by other techniques. The critical behavior of these materials is again closely in accord with that of the pure 2-d Ising model. All these results are in agreement with the universality predictions.

On the other hand, in recent neutron scattering experiments on a dilute 3-d Ising system $Fe_{0.5}Zn_{0.5}F_2$ [39] it was found that the critical properties are different from those of the pure system $FeF_2$, for which the specific heat exponent $\alpha$ is positive, in agreement with the above-mentioned theoretical predictions by Harris.

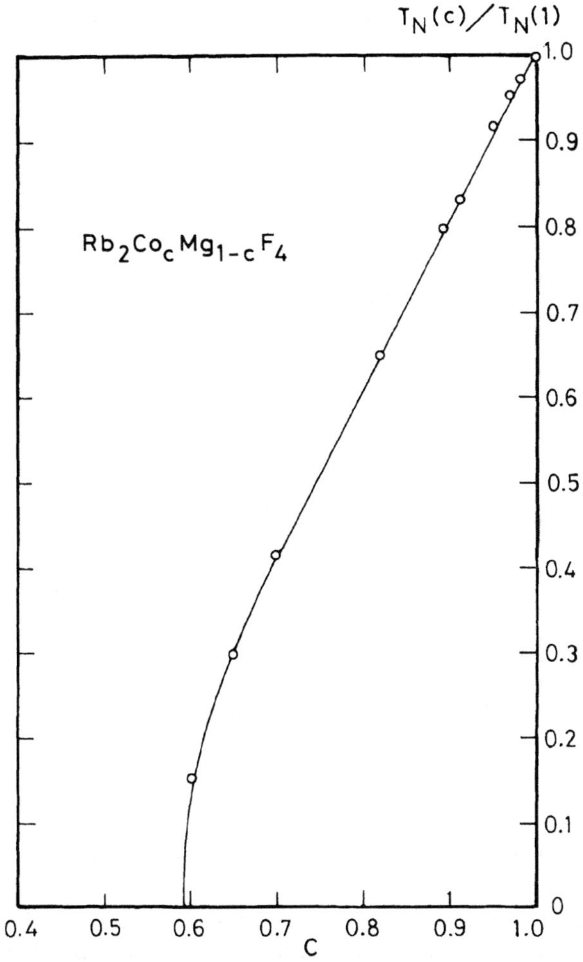


Fig. 8. Concentration dependence of the Néel temperature in $Rb_2Co_cMg_{1-c}F_4$ as measured by neutron scattering. The solid line is a theoretical prediction for the quadratic layer Ising magnet. Reproduced by courtesy of the Physical Society of Japan.

The properties of percolation in $Rb_2Co_cMg_{1-c}F_4$ near the critical concentration $c_p$ (0.59) have been extensively investigated in recent years [40]. Close to the point $c = c_p$ and $T = 0$ K, the long-range order can be destroyed either geometrically by decreasing $c$, or thermally by raising the temperature. This percolation point is therefore an example of a multicritical point [41]. In principle the scattering and consequently the correlation length can be determined as a function of both $c$ and $T$. The properties of percolation, considered as a problem in geometrical disorder, have been studied for many years. It is experimentally very difficult to determine

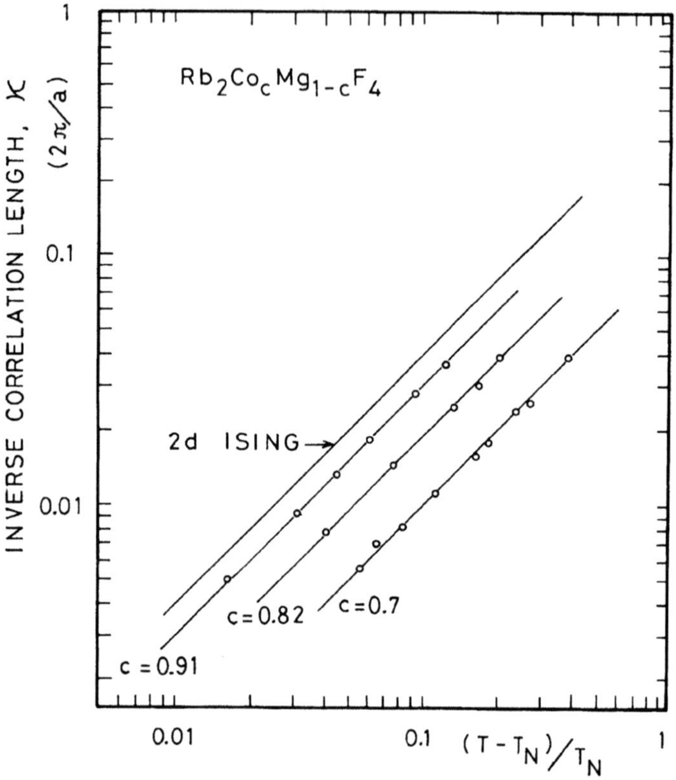


Fig. 9. The inverse correlation length of the longitudinal fluctuations in $Rb_2Co_cMg_{1-c}F_4$. The data agree with an exponent value of $\nu = 1.0$, irrespective of the concentration of magnetic ions. Reproduced by courtesy of the Physical Society of Japan.

the critical properties as a function of concentration $c$ owing to the ambiguity of the concentration used in the experiments. So far all the experiments have been performed so as to determine the behavior of the scattering as a function of temperature for certain fixed concentrations near $c_p$. An essential point of physical importance is that at the percolation point the appropriate scaling field is not the temperature but the inverse correlation length, $\kappa$, of a one-dimensional chain formed by the corresponding magnetic ions [41, 42]. As is well known, in the Ising limit $\kappa_{1d}(T)$ reduces to $\kappa_{1d}(T) = \ln [\tanh(J/kT)] \simeq 2e^{-2J/kT}$ and in the classical Heisenberg limit to $\kappa_{1d}(T) = \ln [\coth(J/kT) - kT/J] \simeq kT/J$. For $c = c_p$ one will obtain thermal critical behavior with exponent $\nu_T = \nu_p/\phi$, $\gamma_T = \gamma_p/\phi$ etc., where $\nu_p$, $\gamma_p$ are geometrical critical exponents ($\xi_G \sim |c_p - c|^{-\nu_p}$, $S(c) \sim |c_p - c|^{-\gamma_p}$ etc.) and $\phi$ is the percolation crossover exponent, while $\kappa(c_p, T) = B(\kappa_{1d}(T)\ a)^{\nu_T}$. Theories for the appropriate geometrical exponents in two dimensions give $\nu_p = 1.355$, $\gamma_p = 2.425$ etc. [43—46], and one finds $\phi = 1$ for an Ising system [47—49].

Indeed, we might argue that, in the Ising system, one may break the connection between two parts of a cluster either by removing a spin or by a thermal excitation (spin flip) at a particular site. Both are local effects which, if equivalent, would suggest $\phi = 1$. However, for the Heisenberg system the thermal excitations are more complicated, which explains why the understanding of the critical behavior near the percolation threshold in a Heisenberg system is more difficult both from the experimental and theoretical side. [50].

We first discuss the experimental results for the 2-d Ising system $Rb_2Co_cMg_{1-c}F_4$ [40]. Four different concentrations were examined with $c = 0.55$, 0.575, 0.583 and 0.595 by Cowley *et al.* In Figure 10 the results for $\kappa(T)$ are plotted against the temperature. The data are consistent with $\nu_T = 1.33$, in agreement with the calculated value of $\nu_p$ and suggesting $\phi = 1$.

Similar measurements had been performed earlier on the 2-d nearly Heisenberg

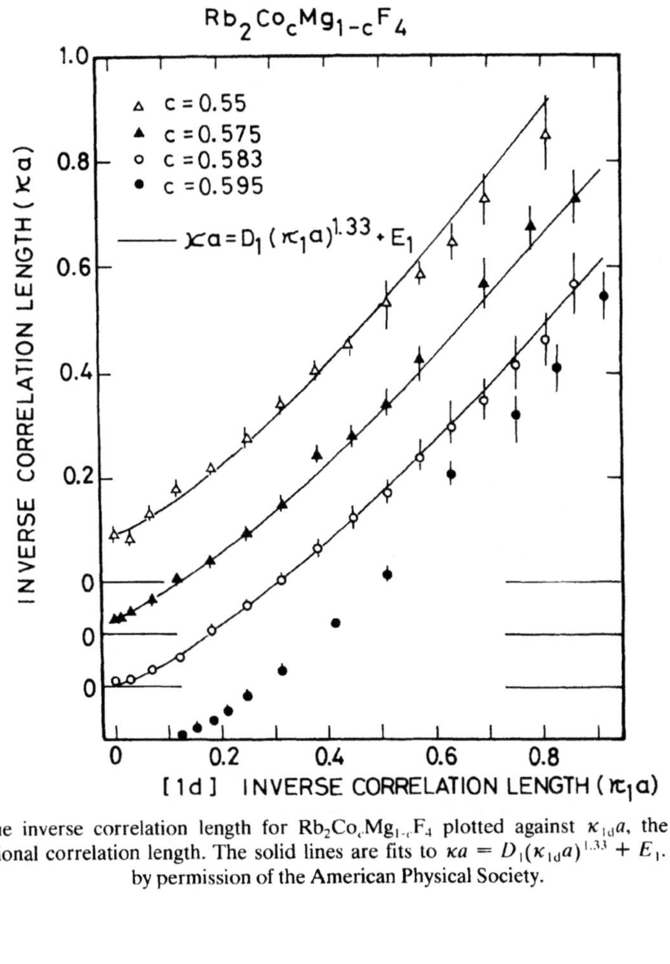


Fig. 10. The inverse correlation length for $Rb_2Co_cMg_{1-c}F_4$ plotted against $\kappa_{1d}a$, the appropriate one-dimensional correlation length. The solid lines are fits to $\kappa a = D_1(\kappa_{1d}a)^{1.33} + E_1$. Reproduced by permission of the American Physical Society.

system $Rb_2Mn_cMg_{1-c}F_4$ [50]. The longitudinal spin fluctuations were fitted to the Lorentzian form and the results suggest that $\nu_T$ for this system is 0.90 ± 0.03, considerably different from the Ising value and suggesting that the crossover exponent for the Heisenberg system is about 1.5. This value is now open to theoretical verification.

The properties near the percolation threshold have also been studied in the 3-d systems $Mn_cZn_{1-c}F_2$ and $KMn_cZn_{1-c}F_3$ [51]. The qualitative picture of the percolation point as a multicritical point, with the temperature scale set by the one-dimensional correlation length, is well established experimentally, although the quantitative agreement between experiment and theoretical description in nearly Heisenberg systems is not always satisfactory.

### 3.2. EXCITATIONS

In the quadratic layer Ising systems in which the magnetic ions are diluted with nonmagnetic ions the excitation spectrum consists of four fairly distinct branches. This behavior was observed in $Rb_2Co_cMg_{1-c}F_4$ with $c = 0.91$, 0.7 [52] and 0.57 [40] by Ikeda and Shirane, and by Cowley *et al.*, respectively. The results for the concentration $c = 0.7$ are shown in Figure 11. In Figure 11(a) scattered neutron distributions at the zone boundary are given, which clearly show the existence of four peaks. The energies of these peaks are in excellent agreement with the Ising

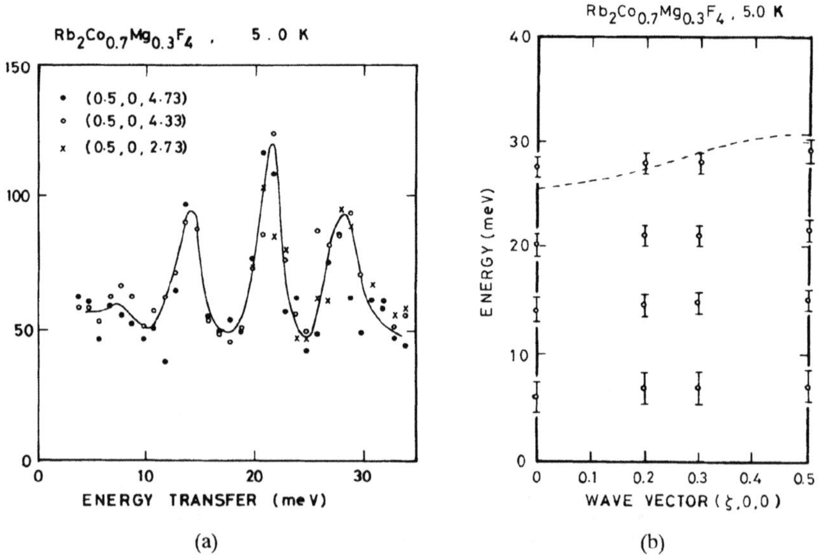


Fig. 11. (a) Zone boundary excitations of $Rb_2Co_{0.7}Mg_{0.3}F_4$ as a function of energy at 5.0 K. The solid line is a guide to the eye. (b) Energies of Ising peaks in $Rb_2Co_{0.7}Mg_{0.3}F_4$ as a function of wave vector at 5.0 K. The broken line is the spin-wave dispersion in pure $Rb_2CoF_4$. Reproduced by courtesy of the Physical Society of Japan.

energies $2zSJ_{\parallel}$ of the Co ions when one takes an effective spin value $S = 1/2$, an Ising exchange $J_{\parallel} = 7.2$ meV and the number of nearest magnetic neighbors $z = 1$, 2, 3 and 4. The value of $J_{\parallel}$ is only slightly smaller than for the pure crystal, $J = 7.73$ meV. Furthermore, the integrated intensities of the peaks are found to be in the ratio of the calculated probability densities of random populations of magnetic ions with 1, 2, 3 and 4 magnetic neighbors, i.e. $4c(1-c)^3$, $6c^2(1-c)^2$, $4c^3(1-c)$ and $c^4$, respectively. The experiments show that a correct theory for the spin wave excitations in random magnetic systems should treat the local environment of each ion in detail, especially in systems with Ising symmetry. Figure 11(b) shows that there is much less dispersion in the excitation frequencies than in the pure material. For the 70% sample, the excitations appear to be no longer propagative, but localized.

The magnetic interactions in $Rb_2Mn_cMg_{1-c}F_4$ are nearly Heisenberg in character and consequently the peak structure over the entire Brillouin zone is less marked, as shown by the results for $Rb_2Mn_{0.54}Mg_{0.46}F_4$ in Figure 12 [53]. Nevertheless, four peaks are also observed in this case, corresponding to the four Ising energies at the zone boundary. Also shown in Figure 12 are the results of computer simulations based on linear spin wave theory by Alben and Thorpe [54].

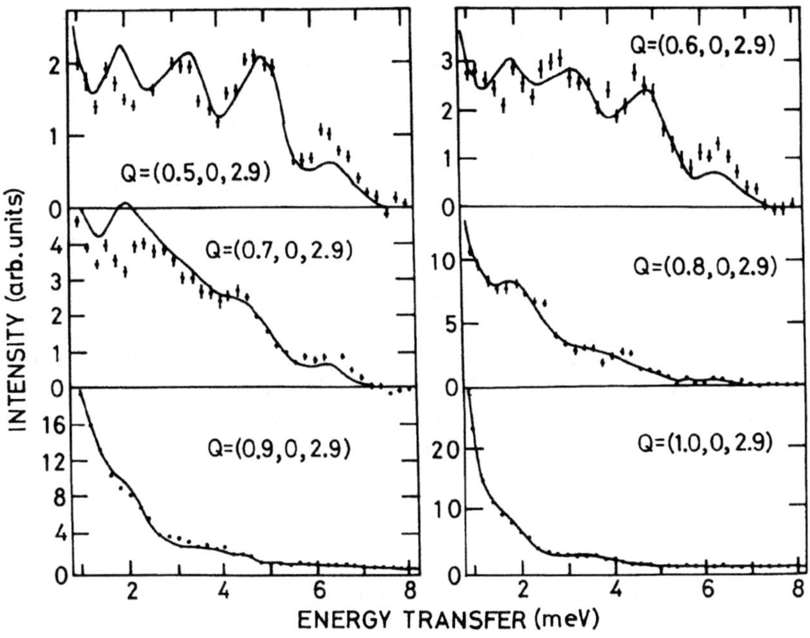


Fig. 12. Scattered neutron distributions in $Rb_2Mn_{0.54}Mg_{0.46}F_4$ at 4.0 K compared with computer simulations obtained using the theory of Alben and Thorpe. Reproduced by permission of the American Physical Society.

The agreement shows that linear spin wave theory is adequate for describing the excitations even in these disordered magnetic systems.

In concentrated systems composed of two different magnetic ions the excitation spectrum consists of two fairly distinct branches. This behavior was first observed in 3-d antiferromagnets and more recently also in 2-d antiferromagnets as $Rb_2Mn_cNi_{1-c}F_4$ [55], $Rb_2Co_cMn_{1-c}F_4$ [56] and $Rb_2Co_cNi_{1-c}F_4$ [57]. Of importance to note here is that the frequency of the magnons at the zone boundary is again given by the simple Ising cluster model, and that the excitation spectra over the whole Brillouin zone can be fitted at low temperatures by linear spin wave theory. The calculations based on linear spin wave theory gives good agreement with the experiments, although the exchange parameters are slightly different from those in the pure compounds, which can be attributed to the change in the lattice constants. It should be remarked that the exchange parameters associated with Co spins can be different from those in the pure compound due to the large orbital contribution to the effective spin Hamiltonian.

### 3.3. RANDOM FIELD EFFECTS

The effect of the random magnetic fields on phase transitions and ordering in magnetic systems has been the subject of considerable experimental and theoretical interest in recent years. Imry and Ma [58] pointed out that in the presence of a random field, the system will be unstable against the formation of domains provided the dimensionality of the system, $d$, is less than the lower critical dimensionality $d_c$, where $d_c = 2$ in Ising systems and $d_c = 3$ for Heisenberg systems. However, it has been argued in a series of papers that the behavior of systems in a random field is analogous to that of the pure system in d − 2 dimensions, suggesting that in Ising systems $d_c$ is 3 rather than 2 [59—62]. Consequently, the lower critical dimensionality of the Ising model in a random field posed an exciting problem. These interesting ideas can be tested experimentally by following the suggestion of Fishman and Aharony [63] that a uniform field applied to a random antiferromagnet produces a random staggered field $H_{RF}$. Application of a uniform field $H_0$ to a disordered antiferromagnet should therefore destroy the long-range order in systems with dimension less than or equal to $d_c$. These suggestions have been tested experimentally by applying a uniform field to the 2-d Ising system $Rb_2Co_{0.7}Mg_{0.3}F_4$ at Brookhaven by Birgeneau *et al.* [64].

The correlation function associated with the thermal spin fluctuations of a $d$-dimensional Ising model is given for $T > T_C$ as

$$\langle S_i S_j \rangle \sim R^{(1-d)/2} e^{-\kappa R}$$

where $R$ is the distance between spins $S_i$ and $S_j$. In the presence of a random field the dimensionality shift predicts that

$$\langle S_i S_j \rangle \sim R^{(3-d)/2} e^{-\kappa R}.$$

The Fourier transform of this expression gives the scattering cross section, and we obtain

$$S(Q) = \frac{A}{\kappa^2 + q^2} + \frac{B}{(\kappa^2 + q^2)^2}\,.$$

For small $\kappa$ and $T < T_c$, the intensity arising from the latter squared-Lorentzian term should go over smoothly into the Bragg peak arising from the true long-range order, so that

$$B \sim \langle M \rangle^2 \kappa^{4-d}$$

where $M$ is the magnetization of the system. Theory predicts that, as $T \to 0$

$$\kappa \sim H_{\mathrm{RF}}^{2/(d_c - 2)} \quad \text{for} \quad d < d_c$$

and

$$\kappa \sim e^{-c/H_{\mathrm{RF}}^2} \quad \text{for} \quad d = d_c$$

and that $S^{zz}(\tau) \sim \kappa^{-d}$. So that one should have $\kappa \sim H_{\mathrm{RF}}^2$ and $S^{zz}(\tau) \sim \kappa^{-2} \sim H_{\mathrm{RF}}^{-4}$ for the 2-d compound $Rb_2Co_{0.7}Mg_{0.3}F_4$.

The principal features of the random-field effects are immediately evident from the qualitative behavior of the Bragg profiles in the applied magnetic field. The field dependence of the scattering in the neighborhood of the magnetic reflection at $T = 5$ K along the transverse direction $Q = (1, \xi, 0)$ is shown in Figure 13. In particular the width becomes progressively broader with increasing field. This increase in the width shows directly that the long-range magnetic order is being destroyed by the application of a magnetic field. The behavior of $\kappa$ and $S^{zz}(\tau)$ as a function of field is shown in Figure 14. It is evident that both exhibit power law dependences on $H$, namely $\kappa \sim H^{1.6}$ and $S^{zz}(\tau) \sim H^{-3.2}$ at low temperatures. Therefore, the destruction of the long-range order by the random field demonstrates explicitly that $d_c \geqslant 2$.

The application of a magnetic field to a dilute antiferromagnet has been used to study the effect of random fields using several techniques other than neutron scattering. One result to be mentioned here is the susceptibility measurement by Ikeda. The diverging static susceptibility at the Néel temperature of $Rb_2Co_{0.7}Mg_{0.3}F_4$ has been found in an extremely weak external field. This divergent susceptibility is strongly suppressed with increasing magnetic field and then the susceptibility shows a remarkable rounding of the transition (see Figure 15). This, and the appearance of the remanent magnetization below the Néel temperature at a constant field, also suggest that the phase transition is destroyed and the system splits into micro-domains, which would be consistent with the neutron scattering observations.

Neutron scattering measurements have also been performed on the 3-d compounds $Co_cZn_{1-c}F_2$ [66], $Fe_cZn_{1-c}F_2$ [67–69] and $Mn_cZn_{1-c}F_2$ [70] by the Brookhaven group. All these materials are antiferromagnets with predominantly exchange interactions between the nearest neighboring antiferromagnetic ions only. Though the interactions between the magnetic ions are quite different in these compounds, they all belong to the 3-d Ising universality class as regards their

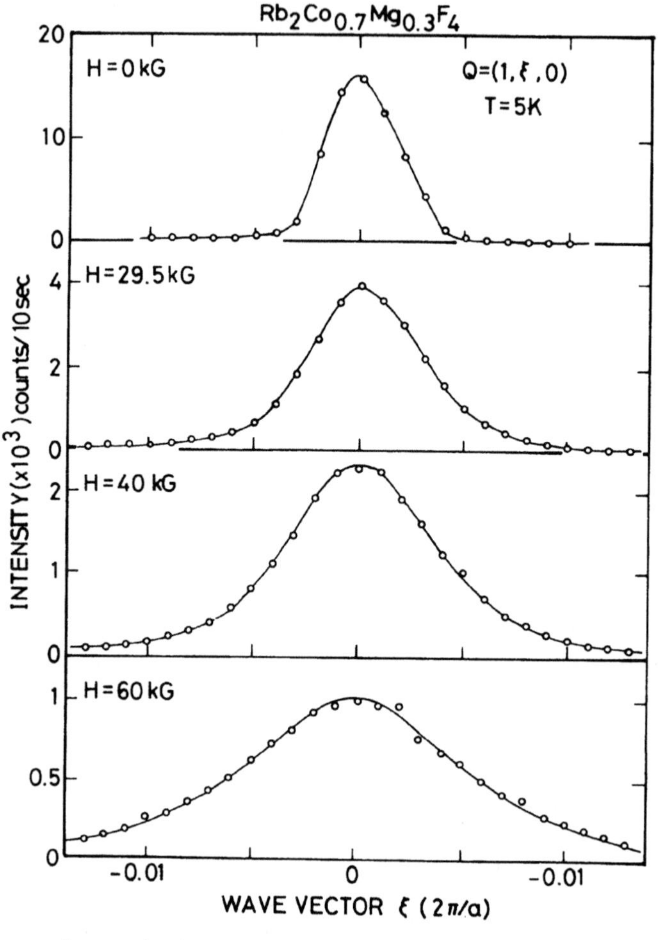


Fig. 13. The magnetic scattering in $Rb_2Co_{0.7}Mg_{0.3}F_4$ at 5 K for four different magnetic fields; scans are along the transverse direction $Q = (1, \xi, 0)$. The solid lines are guides to the eye. Reproduced by permission of the American Physical Society.

critical behavior close to $T_N$. When the Co, Fe and Mn systems are cooled in the presence of a field the scattering in the neighborhood of the magnetic reflections becomes progressively broader with increasing field. This indicates that the long-range order is destroyed by the application of a magnetic field even in the 3-d Ising systems. On the other hand, when the samples are cooled in zero field into the Néel state and then the field is raised, long-range order survives up to the phase boundary. History dependent effects have also been observed using thermal expansion measurements for $Mn_cZn_{1-c}F_2$ [71, 72] and neutron scattering for $Fe_cMg_{1-c}Cl_2$ [73, 74]. Although final conclusions on this interesting problem have

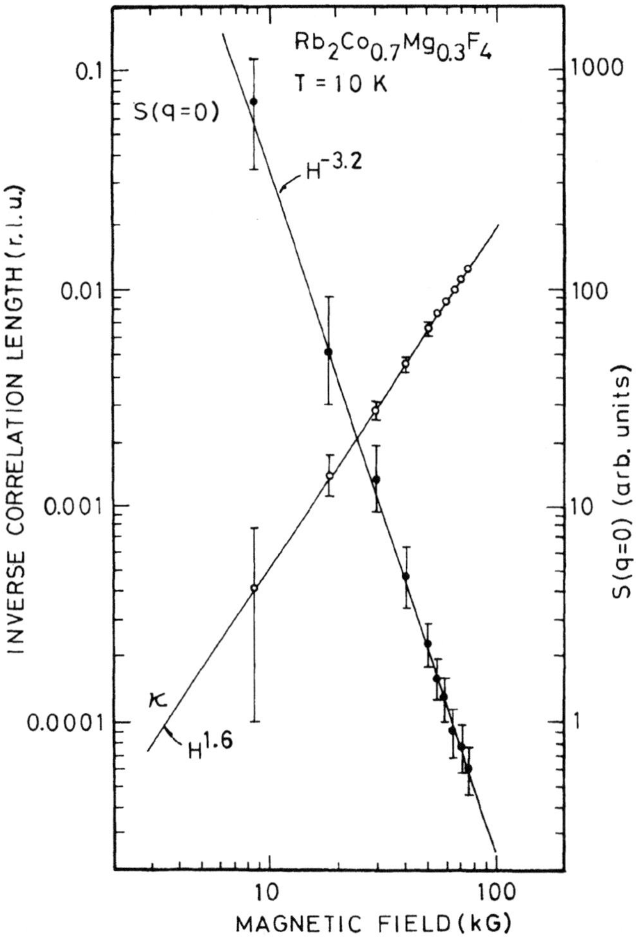


Fig. 14. The dependence on the applied magnetic field at 10 K of the inverse correlation length and the peak intensity of the magnetic scattering from $Rb_2Co_{0.7}Mg_{0.3}F_4$. Reproduced by permission of the American Physical Society.

not yet been reached, many, sometimes conflicting, experimental and theoretical studies are now underway to establish what is the true ground state of the random-field Ising model and, by extension, what $d_c$ is.

All of the abovementioned measurements show that relatively small random fields do indeed produce dramatic effects on the ordering and critical phenomena of systems. Although a detailed understanding of these effects has not yet been achieved, there is no doubt that they are quite significant for our understanding of phase transitions in disordered magnets. The disordering influence of the random fields in competing-anisotropy systems will be briefly discussed in Section 3.5.

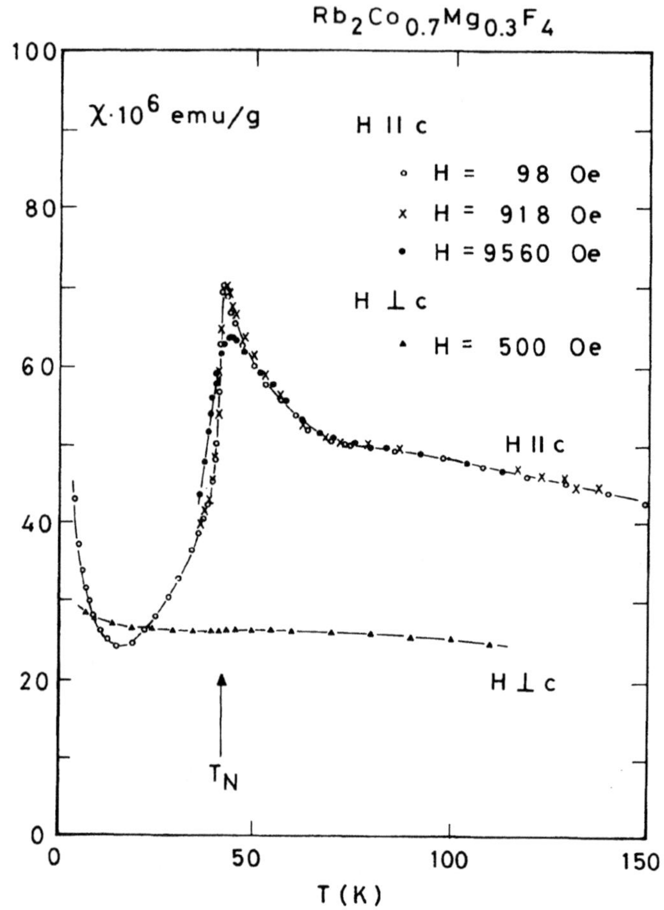


Fig. 15. Parallel and perpendicular susceptibility as a function of temperature in $Rb_2Co_{0.7}Mg_{0.3}F_4$. The solid lines are guides to the eye. Reproduced by permission of IOP Publishing Ltd.

### 3.4. RELAXATION FROM 2-d TO 3-d ORDER

As demonstrated originally by Samuelsen [2], in $Rb_2CoF_4$ the degree of correlations between planes depends on the rate of cooling through the Néel temperature. This effect is enhanced in diluted systems. Even in pure $Rb_2CoF_4$ complete 3-d long-range order is established only when the sample is cooled very slowly through its Néel temperature. On the other hand, if samples of both undiluted and diluted $Rb_2Co_cMg_{1-c}F_4$ are cooled rapidly through $T_N$, the 3-d long-range order is not established and Bragg scattering due to 2-d long-range order is observed below $T_N$ [75]. This is due to the fact the complete 3-d long-range order is unstable against stacking faults along the direction perpendicular to the layers. Then defects occur even in the pure system, and are greatly enhanced in the diluted ones. However, the 3-d long-range ordered state is the ground state because the weak interplanar exchange interactions favor the 3-d long-rage order.

The observation of the time evolution of the interplanar ordering following a quench to low temperatures from the disordered high-temperature state has been performed in diluted $Rb_2Co_{0.7}Mg_{0.3}F_4$ ($T_N$ = 42.8 K) by Ikeda [76]. A characteristic dependence as a function of time was observed on a macroscopic time scale and the observed behavior is in good agreement with recently developed statistical theory for a one-dimensional domain growth model as will be discussed below.

We show in Figure 16 the time dependence of the magnetic intensities at the (100) antiferromagnetic reciprocal lattice position after a quick cooling of the 70% sample from the paramagnetic state (~ 70 K) to the ordered state (5 K) and then raising the temperature to 41.819 ± 0.001 K within a few tens of a second. Thus, at an initial $t$ = 0 the system is in a completely disordered state as regards the inter-planar correlations. It should be noted that the unit of time in Figure 16 is one hour. The measurements of the intensity distribution along the $q_c$ direction (interplanar direction) near the (100) position at several times (A, B, C and D in Figure 16) demonstrates that the width decreases with time, thus evidencing the

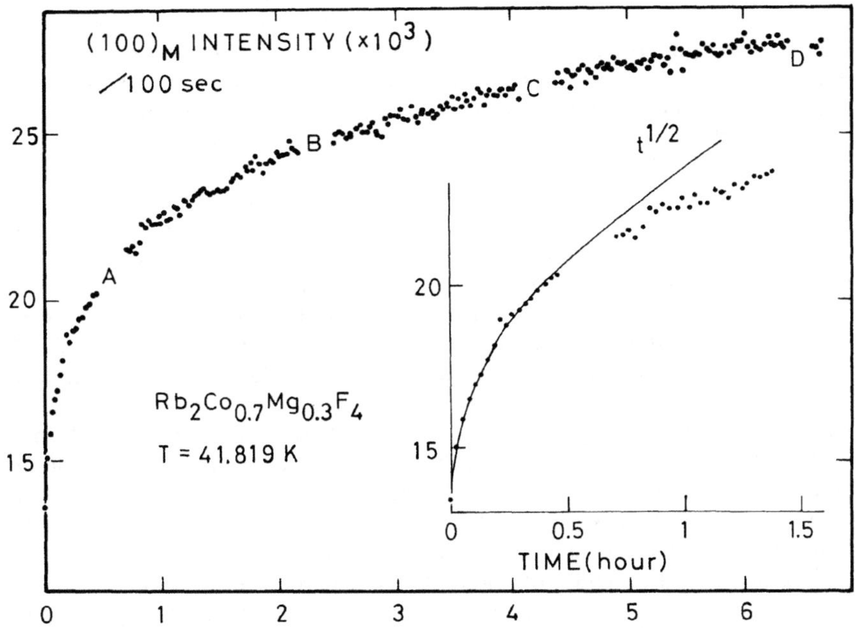


Fig. 16. Intensity of the (100) antiferromagnetic lattice point in $Rb_2Co_{0.7}Mg_{0.3}F_4$ measured at 41.819 K as a function of time. The solid line in the inset represents a $t^{1/2}$ dependence. Reproduced by permission of IOP Publishing Ltd.

growth of 3-d (interplanar) correlations. Therefore, the observed relaxation phenomena can be associated with a 1-d model system, at least phenomenologically. The time dependence of the scattered intensities follows a $t^{1/2}$ dependence up to half an hour from $t = 0$ and thereafter follows a $\log(t)$ dependence. In a 1-d system the scattering function at the reciprocal lattice position, $S(\tau)$, is proportional to the correlation length $\kappa^{-1}$ (domain size) because $S(Q) \sim \kappa/(\kappa^2 + q^2)$. Quite recently the domain growth in one dimension has been examined theoretically by Kawasaki *et al.* [77—79]. They developed a theory based on a time-dependent Ginzburg—Landau model in one dimension, by taking the dynamical motion of kinks and antikinks into consideration, and obtained a qualitative agreement with the experiments. Though the existing theory fails as yet to explain why the relaxation times are so slow, the relaxation is indeed predicted to accelerate as the temperature approaches $T_N$.

### 3.5. COMPETING ANISOTROPIES AND INTERACTIONS

Phase transitions in magnetic systems in which two different magnetic ions with competing spin-anisotropies are randomly distributed have been the subject of considerable interest in recent years. Matsubara and Inawashiro [80], Fishman and Aharony [81] and Lindgard [82] have discussed the composition vs temperature phase diagram for these mixed magnetic systems. It was found that two critical lines are present and correspond to independent ordering of the parallel and perpendicular spin components. The two lines cross at a tetracritical point where both types of spin components order simultaneously. The critical behavior was discussed by Fishman and Aharony, who suggested that the phase boundaries associated with the ordering of the two spin components were decoupled.

There are several examples of systems in which this competition can be studied [83]. The first measurements were made on 2-d antiferromagnets $K_2Fe_cMn_{1-c}F_4$ [84], where the strong crystal-field anisotropy of the Fe ions forces the moments to lie within the magnetic layers perpendicular to the $c$ axis and the weak axial dipolar anisotropy favours alignment of the Mn moments parallel to the $c$ axis. The presence of an intermediate (mixed) phase in this system has been confirmed by Bevaart *et al* [84] in a rather narrow concentration region near $c \sim 0.025$. The resulting phase diagram is shown in Figure 17. It should be noted here that they also examined the phase diagram of $K_2Mn_{0.978}Fe_{0.022}F_4$ in the $H—T$ space and found a new tetracritical point (cf. Figure 18). It has been revealed that, in this $H—T$ phase diagram, four lines of second-order transitions meet at the tetracritical point and each curve divides the $H—T$ diagram into four distinct phases, namely the paramagnetic (P) phase, the axial (A) antiferromagnetic phase, the intermediate (I) antiferromagnetic phase and the planar (PL) phase.

More detailed information on the critical phenomena has been obtained in the quasi-2-d system $Fe_cCo_{1-c}Cl_2$ [85], in which the single-ion anisotropy of the $Fe^{2+}$ ions favors a spin alignment along the $c$ direction, whereas the anisotropic $Co^{2+}$ exchange favors a spin direction perpendicular to the $c$ axis. In Figure 19 we show the phase diagram for $Fe_cCo_{1-c}Cl_2$ as determined by Wong *et al.* from neutron

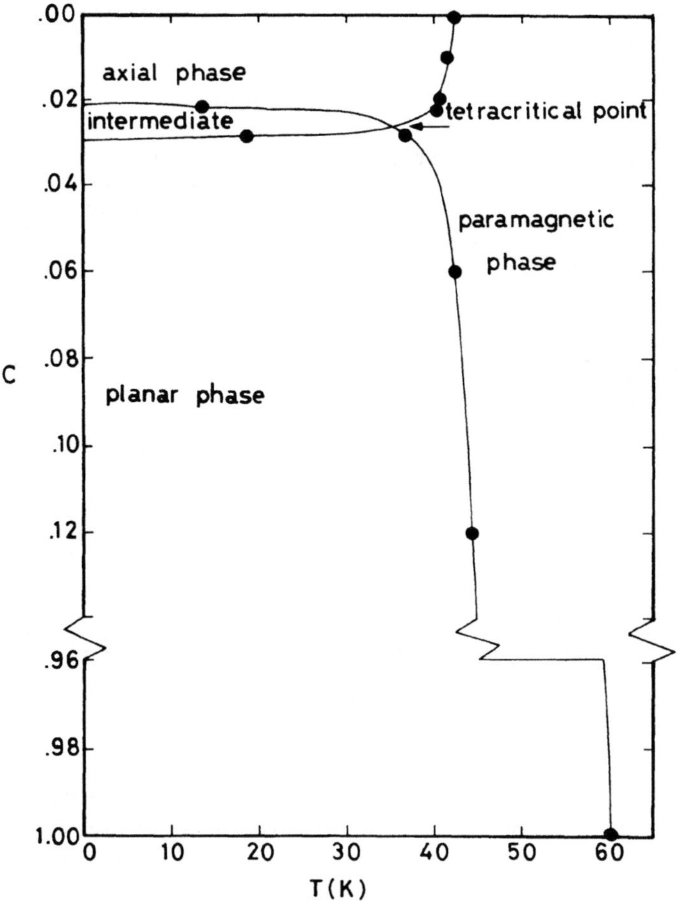


Fig. 17. Phase diagram obtained for $K_2Fe_cMn_{1-c}F_4$. Reproduced by permission of the American Physical Society.

scattering and magnetic susceptibility measurements. The diagram shows the presence of a phase with the spins ordered parallel to the *c* axis, one with the spins perpendicular to the *c* axis, and a mixed phase. The observed phase boundaries are qualitatively in agreement with mean-field predictions. However, detailed critical scattering measurements showed that the transition from the upper phase with one-component ordered to the lower mixed ordered phase is blurred. Therefore, the lower transition lines are either nonexistent or at least drastically altered by the existence of long-range order in the other spin component. It is clear therefore that $S_\parallel$ (parallel component of the spins to the *c* axis) and $S_\perp$ are not decoupled and do not behave independently. An explanation offered is the existence of off-diagonal exchange interactions in the Hamiltonian. As a con-

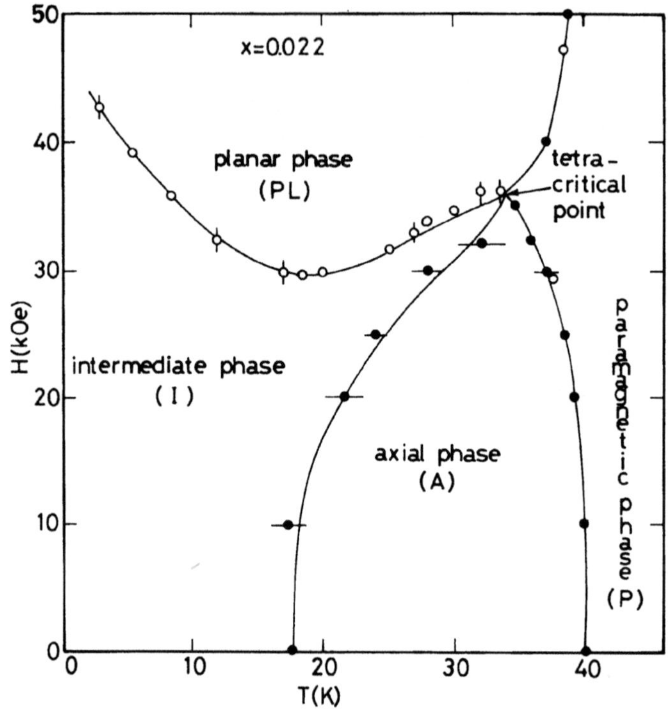


Fig. 18. The $H$—$T$ phase diagram for $K_2Mn_{0.978}Fe_{0.022}F_4$.

sequence of the off-diagonal terms, the lower phase boundaries in the tetracritical phase diagram are destroyed by the random fields generated by the ordered spins. The existence of the random-field-induced domain phase has been strongly suggested in the Mössbauer study performed by Ito *et al.* [86].

When a magnetic field is applied parallel to the spins in the axial phase in these competing-anisotropy systems, the destruction of the antiferromagnetic long-range order can be expected in a similar way as in a diluted Ising-type antiferromagnet. The ordering phenomena of the axial phase in $K_2Co_{0.27}Fe_{0.73}F_4$ in an applied field has been recently examined and turned out to be in agreement with random-field models [87].

The systems with mixtures of ferromagnetic and antiferromagnetic exchange interactions should give insight into the behavior of spin glasses. Two candidates for studying 2-d spin glass are $Rb_2Cr_cMn_{1-c}F_4$ and $K_2Cu_cMn_{1-c}F_4$ where the Mn ions are coupled antiferromagnetically and the Cr and Cu ions ferromagnetically. Although some measurements have been performed on these interesting materials, detailed experiments including neutron scattering have not yet been performed and so we have to await future progress.

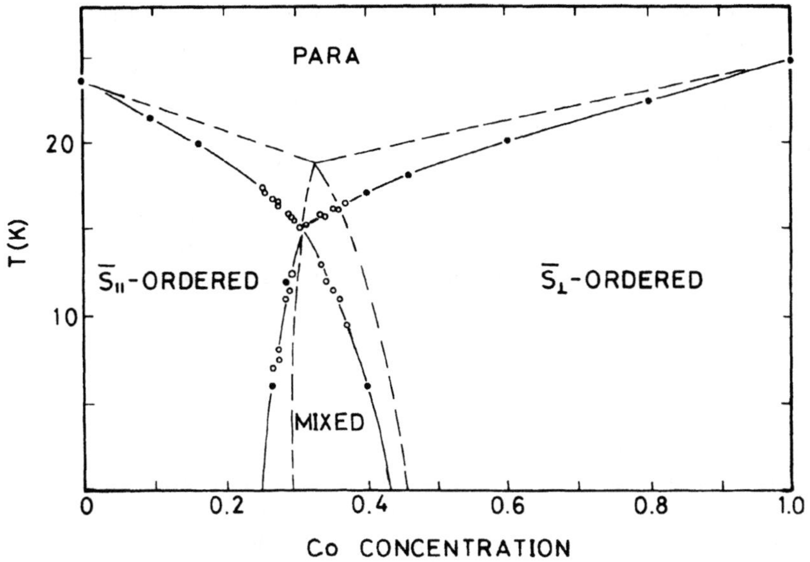


Fig. 19. The phase diagram of $Fe_cCo_{1-c}Cl_2$. The dashed lines are the results of mean field theory and the experimental points are from susceptibility (solid) and neutron scattering (open circles) measurements. Reproduced by permission of the American Physical Society.

## 4. Triangular Lattice Antiferromagnet (TALAF)

The study of frustrated systems is another subject of vivid current interest. Among the many possible frustrated systems such as spin glasses, ANNNI models, etc., we shall focus our attention on a quite simple system: namely the triangular lattice antiferromagnet (TALAF). Even in such a simple system, we can find many interesting problems. However, owing to the difficulty of finding suitable substances, only a few experimental studies have been reported so far. We shall first point out some of the interesting characteristics of the TALAF.

### 4.1. FLUCTUATIONS

In TAL systems the Néel state is not stable if the interaction is antiferromagnetic and between nearest-neighbouring spins only. This is simply a consequence of the competition of the interactions between the spin and its neighbours, and results in strong fluctuations which bring about unexpected and interesting properties.

The absence of stability is most remarkable in the 2-d Ising TALAF, in which no long-range order occurs at any temperature, in strong contrast with the square lattice Ising AF. A similar competion of interactions occurs in the f.c.c. structure but here we shall confine our discussions to the TALAF. The entropy at absolute zero is finite due to a ground state degeneracy that is approximately $2^{0.49N}$-fold, where $N$ is the total number of spins. Theory predicts that at 0 K the correlation function along the row in one of the three sublattices shows a power law decay proportional to $r^{-1/2}$. However this interesting Ising TALAF has unfortunately not yet been studied experimentally.

As is clear from the above, the TALAF has a wide variety of fluctuations in both space and time due to the symmetry properties of the Hamiltonian. For the study of this system the availability of measuring techniques with extremely high $q$-resolution as well as an extremely wide dynamical range will be indispensable. Additional studies by Monte Carlo simulations should also be important.

### 4.2. AN ADDITIONAL DEGREE OF FREEDOM

As mentioned above, the Ising TALAF does not undergo a phase transition. It is then an interesting question whether there is a transition or not if the interaction is Heisenberg (or XY)-like. According to mean-field theory, the ground state of the Heisenberg (or XY) TALAF with classical spins is the 120° structure in which the nearest neighbour spins make an angle of 120° with each other, forming a three sublattice structure. It is well known, however, that for the 2-d system the MF approximation is inadequate at nonzero temperature. In fact, rigorous proof has been given of the absence of LRO in 2-d systems with Heisenberg (or XY) interaction. Therefore at first sight it might seem that the Heisenberg (or XY) TALAF does not undergo a phase transition at nonzero temperature because the frustration effect tends to prevent spin ordering.

However recent theoretical studies have given a strong indication that both the XY and Heisenberg TALAF undergo well-defined phase transitions of their own, which belong to new universality classes. Miyashita and Shiba [90] and Lee *et al.* [91] have pointed out that, in the XY system, the triangular unit cell of spins $\mathbf{S}_1$, $\mathbf{S}_2$ and $\mathbf{S}_3$, has two degenerate states of different chirality, where the chiral vector $\mathbf{K}$ is defined by $\mathbf{K} = (2/3\sqrt{3})\,(\mathbf{S}_1 \times \mathbf{S}_2 + \mathbf{S}_2 \times \mathbf{S}_3 + \mathbf{S}_3 \times \mathbf{S}_1)$. The upward (+) and downward (−) chiral vectors may be taken to characterise the XY model, and the interaction betwee these vectors gives rise to a new type of order similar to that of the 2-d Ising spin system. We should note that in principle the spin ordering has nothing to do with the chiral order [92]. There may even be cases in which the chirality exhibits an order but leaves the spin correlations paramagnetic. Monte Carlo simulation predicts that the transition will occur at $T_c = 1.026\ JS^2$. Slightly lower than this $T_c$, the system then undergoes a KT transition at $T_{KT} = 1.004\ JS^2$ as might be expected for the 2-d XY system.

In the case of the Heisenberg system, chiral vectors of + and − sign as defined above cannot be independent because they can be transfered to one another by a continuous transformation and the $\mathbf{K}$ becomes a three-dimensional vector. In this case, Kawamura and Miyashita [93] found from homotopy group analysis that there is another kind of vortex, called the $Z_2$-vortex, which is a topologically stable point defect. Roughly speaking, this $Z_2$ vortex is a vortex formed by the chiral vectors, and the transition point is driven by the dissociation of paired $Z_2$ vortices. The transition point is given by $T_c = 0.66\ JS^2$ and, surprisingly, the system is still paramagnetic below $T_c$ with an exponentially decaying correlation function. The susceptibility shows only a slight change at $T_c$ and does not become anisotropic below $T_c$.

We emphasize again that the transition in the TALAF is due largely to the additional degree of freedom hidden in this system. Unfortunately it is hardly possible to observe chiral order directly by neutron scattering because fundamentally it involves three spin correlations. We should note however that, as will be shown later, $VX_2$ is most probably an example of this Heisenberg TALAF with the new type of order.

### 4.3. PERTURBATION

The competition of the interactions in the TALAF not only causes strong fluctuations but also renders the spin structure very sensitive to small perturbations of different symmetry, such as the dipole—dipole interaction. As will be shown below, this interaction causes interesting modifications near to the ordering temperature. In fact the establishment of long-range order in the Ising TALAF at a finite $T$ is itself due to the presence of such perturbations as dipolar coupling, next-nearest neighbour interactions, etc.

### 4.4. QUANTUM EFFECT

In the Heisenberg TALAF with $S = 1/2$, the quantum effect may be large enough to alter the ground state appropriate to the case of classical spins. This was first pointed out by Anderson [94]. Based on variational calculations made on simplified models and by extrapolation to the infinite system, he suggested that the ground state of the Heisenbertg TALAF is different from the conventional Néel state of 120° structure. He argues that the triangular lattice might be composed of spin pairs forming singlets which move through the lattice as in a quantum liquid. A more reliable evaluation of the energy has been given by Fazekas and Anderson [95] on an Ising model system with a slight admixture of Heisenberg interactions. They have shown that the degeneracy of the ground state of the Ising TALAF discussed in 4.1 is removed by the off-diagonal term, and that the quantum liquid state is always stable against the Néel state. Unfortunately, a good experimental example of such an Ising—Heisenberg TALAF with $S = 1/2$ is quite rare.

With the above principal features in mind, we shall present some of the experimental work related to the TALAF problem.

#### *$RbFeCl_3$ and $CsFeCl_3$*

Our first example is $RbFeCl_3$, but $CsFeCl_3$ in a strong magnetic field is a similar system. These quasi 1-d compounds are characterised by chains of $Fe^{2+}$ ions, with strongly ferromagnetically coupled spins along the chains and with a strong easy-plane-type anisotropy, the easy plane being normal to the $c$ axis. Although the effective spin associated with each $Fe^{2+}$ ion in these compounds is $S = 1$, each strongly coupled 1-d chain can be regarded as a single, large, classical spin. The coupled system in the basal plane of the hexagonal lattice can then be approximately described by an assembly of classical spins forming a TALAF. Except for these examples, simple TALAFs with large spins have not yet been found. But, even in these nonideal 2-d examples, very interesting magnetic structural transitions have already been discovered. We shall discuss these systems in some detail since, although the systems are not fully 2-d in structure, the physics is essentially 2-d. In $RbFeCl_3$ three transitions occur at $T_{N1} = 2.5$ K, $T_{N2} = 2.3$ K and $T_{N3} = 1.9$ K [96]. In between these temperatures the basic 120° structure which appears for $T < T_{N3}$ is modulated into incommensurate structures, with very long periods of the order of 100 times the fundamental cell constant [97]. When the temperature is lowered from the paramagnetic phase to $T_{N1}$, three satellite peaks appear at the points corresponding to $q_1$, $q_2$ and $q_3$ in Figure 20, replacing of the single Bragg peak at $K = (1/3\ 1/3\ l)$. With further lowering temperature to $T_{N3} < T < T_{N2}$, three additional peaks appear at $q_1'$, $q_2'$ and $q_3'$ in the same figure, and finally the 120° commensurate structure appears. The origin of this ordering in three steps, and of the appearance of the incommensurate phases, has been successfully explained by Shiba and Suzuki [98, 99]. They have demonstrated that in this system the original degeneracy in the spin polarization at the $K$-point where the magnetic Bragg peak appears in the commensurate phase, is easily removed by the weak dipole–dipole interaction or, more precisely, a conical instability occurs at the $K$ point. The newly produced three minima are shown schematically in Fig. 20.

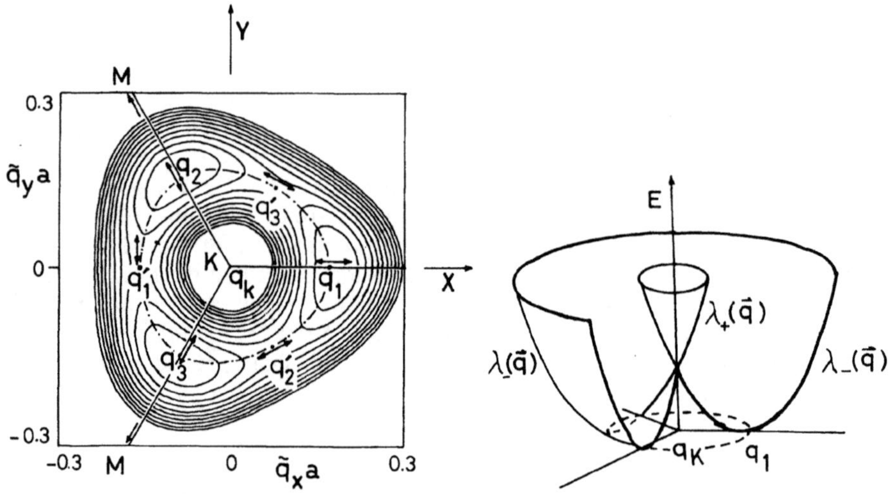


Fig. 20. The energy contour around the $K$-point. $q_1$, $q_2$ and $q_3$ are the minimum points. The direction of the spin polarization is shown by the arrows with two heads. (Taken from Refs. 98, 99.) Reproduced by permission of the American Physical Society.

The wavenumber for the minimum point is given roughly by the ratio $|D/J|$, $D$ being the magnitude of the dipole—dipole interaction. Two possible cases are pointed out by the same authors for the incommensurate phase at $T_{N3} < T < T_{N1}$: (a) the instability occurs at one of the minimum points $q_1$, $q_2$ and $q_3$ and three kinds of domains are formed, and (b) the instability occurs simultaneously at the three minimum points forming a single domain. The physical picture of (a) is as follows. When the temperature is lowered from the paramagnetic phase, the longitudinal component of the spins, as shown in Figure 20 at $q_i$, produces an order below $T_{N1}$, forming an incommensurate structure and leaving the transverse component still paramagnetic. For $T_{N3} < T < T_{N2}$, both the transverse component, as shown by $q_i'$, and the longitudinal components enter into another ordered state. The interesting point is that in this state the periods of modulation, $|q_i|$ and $|q_i'|$, are different. Therefore the spin structure cannot be expressed by a simple rotation of the spin as in a helical structure, but the magnitude of $S$ itself must also be modified. Experiments on $RbFeCl_3$ by Wada *et al.* [97] show the structure to be of the ($a$)-type. The incommensurate structure transforms in the commensurate one through a first-order transition at $T_{N3}$. Steiner *et al.* [100] have found similar structures in $CsFeCl_3$ under the application of strong external fields.

*$VX_2$(X = Cl, Br, I)*

The $VX_2$ compounds are candidates for the Heisenberg TALAF with $S = 3/2$. They have the $CdI_2$ structure. The values for $|J'/J|$ estimated from the spin-wave analysis of the inelastic scattering of neutrons are $5 \times 10^{-3}$ for $VCl_2$ and $2 \times 10^{-2}$ for $VBr_2$ [101]. Therefore these substances are not very ideal 2-d systems. They are, however, excellent realizations of the Heisenberg system as indicated by the $g$-values, $g_c = 1.995$ and $g_a = 1.992$ ($VBr_2$) [102], showing that the contribution of the orbital moment is absent. From the high-temperature analysis of the susceptibility ($\chi$), Niel *et al.* [103] have found the Weiss temperatures $\theta$ and exchange constants $J$ as given in Table II.

TABLE II

Physical properties of $VX_2$(X = Cl, Br and I) and $ACrX_2$(A = Li, Na, K and X = O, S, Se) compounds.

| Compounds | Weiss Temp. $\theta$ | $J$ | $T_N$ | $J'/J$ | Ref. |
|---|---|---|---|---|---|
| $VCl_2$ | −437 K | −23 K | 36 K | $5 \times 10^{-3}$ | 103, 104, 101 |
| $VBr_2$ | −335 | −16 | 29 | $2 \times 10^{-2}$ | 103, 104, 101 |
| $VI_2$ | −143 | −6 | 16 | | 103, 104 |
| $LiCrO_2$ | −700 | | 62 | | 109, 110, 113 |
| $NaCrO_2$ | −290 | | 45 | | 109, 110 |
| $KCrO_2$ | −160 | | 26 | | 109, 110 |
| $LiCrS_2$ | −276 | | 55 | | 112 |
| $LiVS_2$ | | no magnetic order | | | 112 |

Besides the above materials, $VFe_2O_4$($T_N$ = 210 K) shows pronounced 2-d behavior, the spin structure is complex but may correspond to a TALAF [115].

The following results have been obtained from neutron scattering studies [104]. Magnetic Bragg peaks (at least the peaks seem to be a Bragg scattering) appear at ($hhl$) whose $h = n/3$, $n = 3m$ and $l = n/2$, $n = 2m + 1$ below $T_N = 36.0$ K ($VCl_2$), 29.5 K ($VBr_2$) and 16.3 K ($VI_2$). These $T_N$ values are very much lower than the corresponding $\theta$s, in contrast to the usual 3-d antiferromagnets, indicating a strong effect of frustration. These $T_N$ values may possibly not correspond to Néel points of conventional order, because according to the susceptibility measurement no rigid spin structure seems to be formed below them. The $\chi$ is very flat and scarcely depends on temperature. Moreover the appearance of anisotropy in $\chi$ below $T_N$ has not been found even in highly accurate measurements using a SQUID magnetometer ($|\Delta\chi/\chi| = 0.15\%$). It is clear that if the system is an ideal 2-d Heisenberg TALAF, no anisotropy should appear, even for classical spins, but the present system is not ideally 2-d, as stated earlier. The anisotropy is detectable if the above mentioned interlayer coupling is taken into account in the calculation. The origin of this lack of anisotropy is still left open, but it results in very unusual effects on the low-frequency dynamical features as observed by ESR [102] and NMR [106]. The usual type of neutron scattering experiment sees the system in a rapid passage and thus registers a feature of the instantaneous correlation. Following the earlier work on $VX_2$ [104], the spin structure has been studied in more detail using the polarization analysis technique [107]. At present a definite structure has not yet been determined, but the most probable one in $VBr_2$ is that the spins in the basal plane form a 120° structure, that the spins are contained in the *ac* plane, and that the directions of these *ac* planes are distributed homogeneously around the equatorial plane. Whether this structure is indeed rigid or the spins are still fluctuating at lower temperatures is an interesting problem and further work using very slow neutron scattering is desirable.

Two interesting points still need to be discussed. One is that the transition in $VX_2$ might be caused by the mechanism discussed in 4.2. Unfortunately, the order due to chirality cannot be detected by neutron scattering. Theory predicts that the system would be paramagnetic down to 0 K with an exponential decay of the spin correlation function at very long distances. This interesting feature should be checked by neutron scattering. The magnitude of the transition point, the temperature variation of the $\chi$, and the absence of anisotropy below the transition point are in reasonable accord with theory. For $VCl_2$ with $J/k = -23$ K, the theory predicts $T_N = 34$ K which is to be compared with the observed 36 K.

The second point is the quantum effect. One of the means to observe quantum effects on the ground state is the spin reduction. Based on the abovementioned tentative spin structure, the spin reduction can be deduced from scattering experiments. In $VCl_2$ and $VBr_2$, the moment is reduced to $\simeq$ 80% of the full moment value. To what extent this reduction comes from the zero-point spin motions or from the covalency effect is still a problem. Fortunately, however, $VI_2$ has another transition point of 1st order at lower temperature, below which the spin structure changes from the 120° structure to a collinear AF structure. In this structure, a full moment corresponding to $V^{2+}(d^3)$ has been observed. Consequently it seems that the covalency effect is unimportant in $VX_2$. Though the spins in $VX_2$ are $S = 3/2$, the reduction is greater than that found for $K_2NiF_4$, which is a typical 2-d square lattice AF with $S = 1$. In a TALAF one would expect a very large reduction as $S$ changes from 3/2 to 1/2, and possibly a new ground state which can not allow the spin wave approximation, as Anderson points out. We shall present another similar example in the following.

*$AMX_2$($A$ = Li, Na, K; $M$ = 3d metal ion; $X$ = O, S, Se)*

These compounds crystallize into a hexagonal layer structure as shown in Figure 21, and are promising candidates for the TALAF because the $M^{3+}$ ion sheets are well separated by three non magnetic layers of X—A—X ions. There are certainly some indications of the 2-d nature in the powder diffraction patterns, i.e. an asymmetric profile of magnetic diffraction peaks, but a quantitative estimation for $|J'/J|$ has not been reported, presumably because of the difficulty in preparing the single crystals needed to study the spin waves. Some interesting information has, however, been obtained through magnetic as well as neutron scattering measurements. The results for the Cr compounds are summarized in Table II. Among the $ACrX_2$ series the spin structures of $KCrS_2$ and $NaCrS_2$ are different from the others: They are composed of ferromagnetic sheets coupled antiferromagnetically and helically along the $c$ axis, respectively [111]. The other compounds listed are

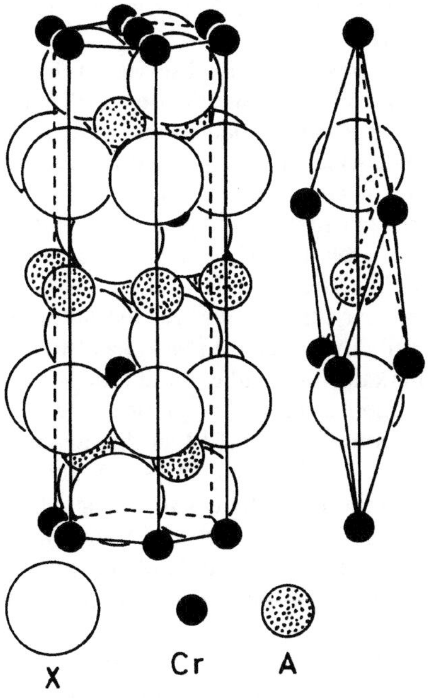


Fig. 21. The crystal structure of the $ACrX_2$ compounds.

TALAFs with noncollinear complex structures. An interesting point to be noted is that the ferromagnetic compounds have moments equal to $3\mu_B$ as expected for the free spin of $Cr^{3+}(d^3)$, whereas the AF ones have moments of about 70 ~ 80% of the above [112, 113]. These results are similar to those in $VX_2$, suggesting that the large reduction is due to the quantum zero-point reduction and not to the covalency effect. Theoretical calculations also support this magnitude of the reduction [114].

As far as the quantum effect is concerned, the systems $AVX_2$, $ATiX_2$ [116] and $ANiX_2$ would be very interesting to study. Van Laar *et al.* [112] have studied $LiVS_2$ by neutron scattering but no evidence for magnetic order has been observed even at 4.2 K. It seems, however, that the $V^{3+}(d^2)$ compounds are not very suitable for study of the quantum effect on the frustrated system, because in these cases the lowest state produced by the LS coupling will be a singlet and so no magnetic order would be expected even for the non-frustrated system.

Hirakawa *et al.* [116—118] have suggested that $LiNiO_2$ would be a promising candidate for the $S = 1/2$, Ising-like AF which is the model of spin quantum liquid proposed by Fazekas and Anderson [95]. The electronic state of $Ni^{3+}$ in $LiNiO_2$ is the same as that of $Co^{2+}$ whose low-lying Kramers doublets are well described by the fictitious spin of $S = 1/2$ with strong Ising-like anisotropic exchange interaction. At the present stage, they have not yet succeeded in growing single crystals and all the measurements have been made on the powder samples. Consequently, a conclusive result has not been drawn. Nevertheless, quite unusual and interesting properties have been found.

To see what special features are to be observed in this quasi-Ising TALAF, they refered to the results of computer simulation [119]. As the simulation for the quantum system is not very easy, calculations were made on the classical spin system. Even for such a classical spin system, some remarkable features have been found. One is the appearance of the two transition points $T_{N1}$ and $T_{N2}$ which becomes a measure of the anisotropy of the exchange interaction. Below the higher transition $T_{N1}$, the three sublattice structure of up-, down- and para-spin configuration is formed. Below the lower transition $T_{N2}$, the system behaves as a ferrimagnet or, more precisely, like an umbrella structure with very big uniform as well as staggered susceptibility. Indeed, $LiNiO_2$ has two transition points at 210 K and 17 K, where remarkable anomalies could be seen in the ESR [117] and magnetic [116] measurements. These transitions were assigned to $T_{N1}$ and $T_{N2}$ respectively, and from this they suggested that the anisotropic AF interactions are $J_{\parallel}/k = 800$ K and $J_{\perp}/k = 200$ K and $g_{\parallel} = 7.1$ and $g_{\perp} = 2.1$. The observed magnetic properties are quite consistent with this simulation. However, a significant disagreement appears in the neutron diffraction. According to the above simulation for the classical spins, well defined three sublattice structure should appear at the temperatures lower than $T_{N1}$ as well as $T_{N2}$. But no indication of sublattice formation could be observed, although they carefully scanned wide range of the $Q$ space and of the temperature range from the room temperature down to 1.4 K. This supports the quantum spin liquid nature proposed by Fazekas and Anderson, being in strong contrast to the conventional AF or ferrimagnets. As the spin moment of the individual $Ni^{3+}$ ion does exist, the scattered intensities summed over the $Q$—$\omega$ space should be conserved. In other words, the magnetic Bragg scattering to be observed must have gone away into the widely spread $Q$—$\omega$ space. One of the inelastic scattering experiments along this line has been performed at BNL [120]. At arbitrarily chosen $Q = 1.4$ Å$^{-1}$ where no Bragg peak was observed, careful analysis of the magnetic inelastic scattering was performed by means of polarized neutrons. Then a broad magnetic diffuse scattering spread over ~20 meV (~$J_\perp$) was observed independently to the change of temperature. Further experiments made by the TOF method at RAL also confirmed the same feature with further extension of the $Q$-space. The existence of such a temperature insensitive scattering, widely spread over both $Q$ and $\omega$ spaces, is difficult to explain if we maintain the classical spin model. What, then, is the nature of the two transitions? This is an interesting problem to be solved in the future.

$NaTiO_2$ is another candidate for the spin quantum liquid [116—118] in which the interaction between the spins $S = 1/2$ is of the Heisenberg-type. This would be the model case which was first proposed by Anderson [94]; an RVB system. As the observed $g = 2.16$ was close to 2, Hirakawa *et al.* assigned that the electronic state is an orbital singlet and the $S = 1/2$ system might be realized. As before, $Ti^{3+}$ ions form triangular lattice sheets and the interaction must be antiferromagnetic because the susceptibility is small, being scarcely dependent on temperature. At low temperature, however, the susceptibility increases, being inversely proportional to the temperature. This increase has been attributed to a small amount of impurity or of imperfections. Again, no magnetic Bragg reflections could be observed in the neutron scattering experiment as low as 1.4 K. The exchange parameter is difficult to estimate, but based on the value of the susceptibility, they suggested that $J/k$ ~ 300 K. Recently, they have found a sharp phase transition at 260 K in a newly prepared sample [121]. Though the X-ray diffraction pattern is nearly the same and no great change is observed above and below the transition, the susceptibility drops down to nearly one half and retains this low value down to the liquid helium temperature. The change of susceptibility is very sharp but continuous. It seems that the system is fairly sensitive to the chemical composition. Further detailed study will be very interesting.

## References


1. V. J. Folen, J. J. Krebs and M. Rubinstein, *Sol. State Comm.* **6** (1968) 865.
2. E. J. Samuelsen, *Phys. Rev. Lett.* 31 (1973) 936; *J. Phys. Chem. Solids* **35** (1974) 785.
3. H. Ikeda and K. Hirakawa, *Sol. State Comm.* **14** (1974) 529.
4. L. Onsager, *Phys. Rev.* **65** (1944) 117; See also C. N. Yang, *Phys. Rev.* **85** (1952) 808 and M. E. Fisher, *Rep. on Progr. Phys.* **30** (1967) 615.
5. M. E. Fisher, *J. Math. Phys.* **5** (1964) 944.
6. C. A. Tracy, *Mag. and Mag. Mater.* AIP, Conf. Proc. No. **29** (1975) 485. C. A. Tracy and B. M. McCoy, *Phys. Rev. Lett.* **31** (1973) 1500.
7. H. Ikeda and M. T. Hutchings, *J. Phys. C.* **11** (1978) L529.
8. D. J. Breed, K. Gilijamse and A. R. Miedema, *Physica* **45** (1969) 205.
9. M. T. Hutchings, H. Ikeda and E. Janke, *Phys. Rev. Lett.* **9** (1982) 386.
10. R. J. Birgeneau, H. J. Guggenheim and G. Shirane, *Phys. Rev. B* **1** (1970) 2211.

11. M. P. H. Thurlings, E. Frikkee and H. W. de Wijn, *Phys. Rev. B* **25** (1982) 4750.
12. G. Shirane and R. J. Birgeneau, *Physica* **86–88B** (1977) 639. J. Als Nielsen, R. J. Birgeneau, H. J. Guggenheim and G. Shirane, *J. Phys.* **C9** (1976) L121.
13. L. J. de Jongh, L. P. Regnault, J. Rossat-Mignod and J. Y. Henry, *J. Appl. Phys.* **53** (1982) 7963.
14. K. Binder and D. P. Landau, *Phys. Rev.* **B13** (1976) 1140.
15. D. R. Nelson and R. A. Percovits, *Phys. Rev.* **B16** (1977) 2191.
16. H. J. Seifert and K. Klatyk, *Z. anorg. allg. Chem.* **334** (1964) 113.
17. R. Haegele and D. Babel, *Z. anorg. allg. Chem.* **409** (1974) 11.
18. M. J. Fair, A. K. Gregson, P. Day and M. T. Hutchings, *Physica* **86–88B** (1977) 657: E. Janke, M. T. Hutchings, P. Day and P. J. Walker, *Harwell Rep.* MPD: NBS: 229 (1983)
19. M. T. Hutchings, J. Als-Nielsen, P. A. Lindgård and P. J. Walker, *J. Phys. C.* **14** (1982) 5327.
20. S. Funahashi, F. Moussa and M. Steiner, *Solid State Comm.* **18** (1976) 433.
21. A. S. Borovik-Romanov, N. M. Kreines, R. Laiho, T. Levola and V. G. Zhotikov, *J. Phys. C.* **13** (1980) 879.
22. H. Yamazaki, *J. Phys. Soc. Jpn.* **37** (1974) 667.
23. M. Fujii and A. Hirai, *J. Mag. Mag. Mater.* **31–34** (1983) and personal communication from Hirai.
24. K. Hirakawa and K. Ubukoshi, *J. Phys. Soc. Jpn.* **50** (1981) 1909: K. Hirakawa, H. Yoshizawa and K. Ubukoshi, *J. Phys. Soc. Jpn.* **51** (1982) 2151: K. Hirakawa, *J. Appl. Phys.* **53** (1982) 1893.
25. K. Hirakawa, H. Yoshizawa, J. D. Axe and G. Shirane, *J. Phys. Soc. Jpn.* **52** (1983) 4220.
26. F. Moussa and J. Villain, *J. Phys.* **C9** (1976) 4433.
27. J. M. Kosterlitz, *J. Phys.* **C7** (1974) 1046.
28. S. Hikami and T. Tsuneto, *Prog. Theor. Phys.* **63** (1980) 387.
29. See for instance, A. P. Young, *Ordering in Strongly Fluctuating Condensed Matter Systems*, Riste, Ed., Plenum Press, 1979 p. 271.
30. A. B. Harris, *J. Phys.* **C7** (1974) 1671.
31. G. Grinstein and A. Luther, *Phys. Rev.* **B13** (1976) 1329.
32. D. E. Khmelnitskii, *Soviet Phys. JETP* **41** (1975) 981.
33. G. M. Meyer and O. W. Dietrich, *J. Phys.* **C11** (1978) 1451.
34. R. A. Cowley and K. Carneiro, *J. Phys.* **C13** (1980) 3281.
35. H. Ikeda, *J. Phys. Soc. Jpn.* **50** (1981) 3215.
36. H. Ikeda, M. Suzuki and M. T. Hutchings, *J. Phys. Soc. Jpn.* **46** (1979) 1153.
37. R. J. Birgeneau, J. Als-Nielsen and G. Shirane, *Phys. Rev.* **B16** (1977) 280.
38. H. Ikeda, T. Abe and I. Hatta, *J. Phys. Soc. Jpn.* **50** (1981) 1488.
39. R. J. Birgeneau, R. A. Cowley, G. Shirane, H. Yoshizawa, D. P. Belanger, A. R. King and V. Jaccarino, *Phys. Rev.* **B27** (1983) 6747.
40. R. A. Cowley, R. J. Birgeneau, G. Shirane, H. J. Guggenheim and H. Ikeda, *Phys. Rev.* **B21** (1980) 4038.
41. D. Stauffer, *Z. Phys.* **B32** (1975) 161.
42. R. J. Birgeneau, R. A. Cowley, G. Shirane and H. J. Guggenheim, *Phys. Rev. Lett.* **37** (1976) 940.
43. P. J. Reynolds, H. E. Stanley and W. Klein, *J. Phys.* **A11** (1978) L199.
44. W. Klein, H. E. Stanley, P. J. Reynolds and A. Coniglio, *Phys. Rev. Lett.* **41** (1978) 1145.
45. J. Blease. J. W. Essam and C. M. Place, *J. Phys.* **C11** (1978) 4009.
46. D. S. Gaunt and H. Rushkin, *J. Phys.* **A11** (1978) 1369.
47. M. F. Thorpe, *J. Phys.* **36** (1975) 1177.
48. M. J. Stephen and G. S. Grest, *Phys. Rev. Lett.* **38** (1977) 567.
49. D. J. Wallace and A. P. Young, *Phys. Rev.* **B17** (1978) 2384.
50. R. J. Birgeneau, R. A. Cowley, G. Shirane, J. A. Tarvin and H. J. Guggenheim, *Phys. Rev.* **B21** (1980) 317.
51. R. A. Cowley, G. Shirane, R. J. Birgeneau, E. C. Svensson and H. J. Guggenheim, *Phys. Rev.* **B22** (1980) 4412.

52. H. Ikeda and G. Shirane, *J. Phys. Soc. Jpn.* **46** (1979) 30.
53. R. A. Cowley, G. Shirane, R. J. Birgeneau and H. J. Guggenheim, *Phys. Rev.* **B15** (1977) 4292.
54. R. Alben and M. F. Thorpe, *J. Phys.* **C8** (1975) L275.
55. J. Als-Nielsen, R. J. Birgeneau, H. J. Guggenheim and G. Shirane, *Phys. Rev.* **B12** (1975) 4963.
56. H. Ikeda, T. Riste and G. Shirane, *J. Phys. Soc. Jpn.* **49** (1980) 504.
57. H. Ikeda, Y. Someya, Y. Endoh, Y. Noda and F. Shibata, *Physica* **120B** (1983) 193.
58. Y. Imry and S. K. Ma, *Phys. Rev. Lett.* **35** (1975) 1399.
59. A. Aharony, Y. Imry and S. K. Ma, *Phys. Rev. Lett.* **37** (1976) 1367.
60. G. Grinstein, *Phys. Rev. Lett.* **37** (1976) 944.
61. A. P. Young, *J. Phys.* **C10** (1977) L257.
62. G. Parisi and N. Sourlas, *Phys. Rev. Lett.* **43** (1979) 744.
63. S. Fishman and A. Aharony, *J. Phys.* **C12** (1979) L729.
64. R. J. Birgeneau, H. Yoshizawa, R. A. Cowley, G. Shirane and H. Ikeda, *Phys. Rev.* **B28** (1983) 1438.
65. H. Ikeda, *J. Phys.* **C16** (1983) L21.
66. M. Hagen, R. A. Cowley, S. K. Satija, H. Yoshizawa, G. Shirane, R. J. Birgeneau and H. J. Guggenheim, *Phys. Rev.* **B28** (1983) 2602.
67. R. A. Cowley, G. Shirane, H. Yoshizawa and R. J. Birgeneau, *Z. Phys.* **B58** (1984) 15.
68. H. Yoshizawa, R. A. Cowley, G. Shirane and R. J. Birgeneau, *Phys. Rev.* **B31** (1985) 4548.
69. D. P. Belanger, A. R. King and V. Jaccarino, *Phys. Rev.* **B31** (1985) 4538.
70. R. A. Cowley, H. Yoshizawa, G. Shirane, M. Hagen and R. J. Birgeneau, *Phys. Rev.* **B30** (1984) 6650.
71. Y. Shapira, *J. Appl. Phys.* **53** (1982) 1931.
72. Y. Shapira and N. F. Oliveira, Jr., *Phys. Rev.* **B27** (1983) 4336.
73. P. Z. Wong, S. von Molnar and P. Dimon, *J. Appl. Phys.* **53** (1982) 7954.
74. P. Z. Wong and J. Cable, *Phys. Rev.* **B28** (1983) 5361.
75. H. Ikeda, M. Suzuki and M. T. Hutchings, *J. Magn. Magn. Mater.* **14** (1979) 163.
76. H. Ikeda, *J. Phys.* **C16** (1983) 3563.
77. K. Kawasaki, T. Ohta and T. Nagai, *J. Phys. Soc. Jpn. Suppl.* **52** (1983) 131.
78. K. Kawasaki and T. Nagai, *Physica* **121A** (1983) 175.
79. T. Nagai and K. Kawasaki, *Physica* **120A** (1983) 587.
80. F. Matsubara and S. Inawashiro, *J. Phys. Soc. Jpn.* **42** (1977) 1529.
81. S. Fishman and A. Aharony, *Phys. Rev.* **B18** (1978) 3507.
82. P. A. Lindgård, *Phys. Rev.* **B14** (1976) 4074; *ibid* **B16** (1978) 2168.
83. See for a recent review K. Katsumata, *J. Mag. Mag. Mater.* **31—34** (1983) 1435.
84. L. Bevaart, E. Frikee and L. J. de Jongh, *Phys. Rev.* **B18** (1978) 3376; *ibid.* **B19** (1979) 4741.
85. P. Wong, P. M. Horn, R. J. Birgeneau and G. Shirane, *Phys. Rev.* **B27** (1983) 428.
86. A. Ito, Y. Someya and K. Katsumata, *Solid State Commun.* **36** (1980) 681.
87. W. A. H. M. Vlak, E. Frikkee, A. F. M. Arts and H. W. de Wijn, *J. Phys.* **C16** (1983) L1025 and *Phys. Rev.* **B33** (1986) 6470.
88. G. H. Wannier, *Phys. Rev.* **79** (1950) 357.
89. J. Stephenson, *J. Math. Phys.* **5** (1964) 1009.
90. S. Miyashita and H. Shiba, *J. Phys. Soc. Jpn.* **53** (1984) 1145.
91. D. H. Lee, J. D. Joannopoulos, J. W. Negele and D. P. Landau, *Phys. Rev. Lett.* **52** (1984) 433.
92. See for instance, J. Villain, *J. Phys.* **C10** (1977) 4793.
93. H. Kawamura and S. Miyashita, *J. Phys. Soc. Jpn.* **53** (1984) 4138.
94. P. W. Anderson, *Mat. Res. Bull.* **8** (1973) 153.
95. P. Fazekas and P. W. Anderson, *Phil. Mag.* **30** (1974) 423.
96. T. Haseda, N. Wada, M. Hata and K. Amaya, *Physica* **108B** (1981) 841.
97. N. Wada, K. Ubukoshi and K. Hirakawa, *J. Phys. Soc. Jpn.* **51** (1982) 2833.
98. H. Shiba, *Sol. State Commun.* **41** (1982) 511.

99. H. Shiba and N. Suzuki, *J. Phys. Soc. Jpn.* **51** (1982) 3488, **52** (1982) 1382.
100. W. Knop, M. Steiner and P. Day, *J. Mag. Mag. Mater.* **31–34** (1983) 1033.
101. H. Kadowaki, K. Ubukoshi and K. Hirakawa, *J. Phys. Soc. Jpn.* **54** (1985) 363.
102. I. Yamada, K. Ubukoshi and K. Hirakawa, *J. Phys. Soc. Jpn.* **53** (1984) 381.
103. M. Niel, C. Cros, G. Le Flem, M. Pouchard and P. Hagenmuller, *Physica* **86–88B** (1977) 702; M. Niel, Thesis.
104. K. Hirakawa, H. Kadowaki and K. Ubukoshi, *J. Phys. Soc. Jpn.* **52** (1983) 1814.
105. K. Hirakawa, H. Ikeda, H. Kadowaki and K. Ubukoshi, *J. Phys. Soc. Jpn.* **52** (1983) 2882.
106. M. Ito and H. Yasuoka, Personal communication.
107. M. Nishi, Y. Ito, H. Kadowaki and K. Hirakawa, *J. Phys. Soc. Jpn.* **53** (1984) 1214.
108. R. J. Birgeneau, H. J. Guggenheim and G. Shirane, *Phys. Rev.* **1** (1970) 2211.
109. J. L. Soubeyroux, D. Fruchart, C. Delmas and G. Le Flem, *J. Mag. Mag. Mater.* **14** (1979) 159.
110. C. Delmas, G. Le Flem, C. Fouassier and P. Hagenmuller, *J. Phys. Chem. Solid* **39** (1978) 55.
111. B. van Laar and F. M. R. Engelsman, *J. Sol. State Chem.* **6** (1973) 384.
112. B. van Laar and D. J. W. IJdo, *J. Sol. State Chem.* **3** (1971) 590.
113. J. L. Soubeyroux, D. Fruch, J. C. Marmeggi, W. J. Fitzgerald, C. Delmas and G. Le Flem, *Phys. Stat. solidi* (a) **67** (1981) 633.
114. T. Oguchi, *J. Phys. Soc. Jpn.* **52** (1983) Suppl. 183.
115. J. Akimitsu, Y. Inada, K. Shiratori, I. Shindo and N. Kimizuka, *Sol. State Commun.* **32** (1979) 1065.
116. K. Hirakawa. H. Kadowaki and K. Ubukoshi, *J. Phys. Soc. Jpn.* **54** (1985) 3526.
117. I. Yamada, K. Ubukoshi and K. Hirakawa, *J. Phys. Soc. Jpn.* **54** (1985) 3571.
118. K. Hirakawa and H. Kadowaki, *Physica* **136B** (1986) 335.
119. S. Miyashita and H. Kawamura, *J. Phys. Soc. Jpn.* **54** (1985) 3385.
120. H. Yoshizawa, G. Shirane and K. Hirakawa, Unpublished.
121. K. Hirakawa *et al.*, to be published.

# PHASE TRANSITIONS IN QUASI TWO-DIMENSIONAL PLANAR MAGNETS

L. P. REGNAULT AND J. ROSSAT-MIGNOD

*Commissariat à l'Energie Atomique, Centre d'Etudes Nucléaires de Grenoble, DRF/SPh/Groupe Magnétisme et Diffraction Neutronique, 85 X, 38041 Grenoble Cedex, France*

## 1. Introduction

In the past ten years a great deal of effort in condensed matter physics has been devoted to theoretical and experimental studies of phase transitions in two-dimensional (2-d) systems (see for example the review article by De Jongh and Miedema [1]). For such systems the thermodynamic behaviour can be described by the general Hamiltonian:

$$\mathscr{H} = -\sum_{i,j} J_{ij}^{\perp}(S_i^x S_j^x + S_i^y S_j^y) + J_{ij}^{\parallel} S_i^z S_j^z \tag{1}$$

where the superscripts $\perp$ and $\parallel$ are defined relative to the $z$-axis (perpendicular to the XY plane) and $J_{ij}^{\perp}$ and $J_{ij}^{\parallel}$ represent the (anisotropic) exchange interaction between spins $i$ and $j$. The Hamiltonian (1) contains the traditional and well-known models, characterized by the particular values of $J_{ij}^{\perp}$ and $J_{ij}^{\parallel}$:

$$\begin{array}{lllll} \text{(a)} & J_{ij}^{\perp}=0 & J_{ij}^{\parallel}=J_{ij}: & \text{Ising system (I)} & n=1 \\ \text{(b)} & J_{ij}^{\perp}=J_{ij}^{\parallel} & =J_{ij}: & \text{Heisenberg system (H)} & n=3 \\ \text{(c)} & J_{ij}^{\parallel}=0 & J_{ij}^{\perp}=J_{ij}: & \text{XY system (XY)} & n=2 \end{array} \tag{2}$$

where $n$ stands for the dimensionality of the spin system.

From the experimental point of view such ideal Hamiltonians rarely describe the actual situation because there are always perturbational terms, which complicate the special cases given in (2). For example for the Heisenberg system a single-ion anisotropy described by $\mathscr{H} = \sum_i D(S_i^z)^2$ or a weak anisotropy in the in-plane interactions $J_{ij}^{\perp}$ ($J_{ij}^x \neq J_{ij}^y$) may induce a more or less marked Ising or planar character. Prototypes of these systems are the well known compounds $K_2NiF_4$, $K_2MnF_4$ (H + $\varepsilon$I) or $K_2CuF_4$ (H + $\varepsilon$XY) [2, 3, 4], the properties of which are described in detail in the Chapter by Hirakawa and Ikeda. Nevertheless the I, H and XY models have attracted a considerable attention because they are representative of systems possessing an Ising, Heisenberg or planar-type anisotropy following the principle of universality.

Although many compounds have been found which are good approximations of



*L. J. De Jongh (Ed.), Magnetic Properties of Layered Transition Metal Compounds* 271—321.

the I or H models [1], very few experimental examples of the 2-d XY model have been studied as yet, because of the lack of good materials with strong XY-anisotropy. In this chapter we shall describe the magnetic properties of the compounds $BaM_2(XO_4)_2$ (M = Co, Ni; X = P, As) which can be regarded as prototypes of the 2-d XY model. The content of this chapter is defined as follows. We start with the description of the main properties of the 2-d XY model, with special focus on the magnetic excitations. We then investigate successively the observed magnetic behaviour of $BaNi_2(PO_4)_2$, $BaCo_2(AsO_4)_2$, $BaNi_2(AsO_4)_2$ and, $BaCo_2(PO_4)_2$. The next section gives a review of other 2-d planar systems investigated up to now, with particular attention to $K_2CuF_4$, $NiCl_2$ and $CoCl_2$ graphite intercalated compounds ($NiCl_2$-GIC and $CoCl_2$-GIC). We end with a comparative analysis of the experimental results in the light of the theory.

## 2. Phase Transition and Excitations in the 2-d XY Model

The aim of this section is to sketch the theoretical background essential for an understanding of the experimental results which will be analysed in the following sections. We mention that an exhaustive discussion of the theory of (quasi) two-dimensional magnets is given in the chapter by Pokrovsky and Uimin. Among the different 2-d models treated in this book, the only one which displays a conventional phase transition is the Ising model. Ever since the work of Onsager [5] it has been well established that there is long-range ordering (LRO) below a finite temperature $T_I$ and that all the quantities in the vicinity of the transition vary as power laws of the variable $t = T/T_I - 1$. For instance $|\langle \mathbf{S} \rangle_T| \sim At^{\beta}$, $\xi \sim Bt^{-\nu}$, $C_m \sim Ct^{-\alpha}$, with the exponents $\beta = 1/8$, $\nu = 1$ and $\alpha \simeq 0$ (in fact a logarithmic divergence) [6—8]. All these properties have been well verified experimentally in systems like $K_2CoF_4$ or $Rb_2CoF_4$, which are certainly the best prototypes of that model [9—11] (more details can be found in [10] and in the chapter by Hirakawa and Ikeda). The advantage of the existence of a phase transition in the pure 2-d Ising model is that perturbations like a small interplane exchange interaction $J'$ are relatively unimportant, and imply only slight modifications of the value of the transition temperature. Qualitatively, the new (3-d) critical temperature can be found by writing that at the transition the interaction energy between correlated segments in adjacent planes is approximatively equal to the thermal energy:

$$T_c \simeq z'|J'|S^2\xi^2(T_c) \tag{3}$$

with $\xi = B/(T/T_I - 1)$. By rearranging terms one obtains: $T_c \simeq T_I + BS(z'|J'|T_I)^{1/2}$, which shows that the 2-d transition temperature is only slightly modified by the effect of interlayer coupling.

As regards the 2-d Heisenberg and XY models the situation is more complicated. Since the paper of Mermin and Wagner [12], it is known rigorously that LRO is absent in such systems down to $T = 0$. However, though it is well established now that the Heisenberg model does not exhibit any phase transition at a finite temperature [13—19], Stanley and Kaplan [20, 21] and, more rigorously, Kosterlitz and Thouless [22, 23] have predicted, for the 2-d XY model, the existence of a phase transition at a *finite* temperature (defined as $T_{KT}$) to a phase without LRO, but with a diverging susceptibility. According to Kosterlitz and Thouless [22] this transition corresponds to an unbinding of vortex—antivortex pairs into a gas of free vortices, which process takes place at a temperature $T_{KT} \simeq 1.8|J|S^2$ in the classical approximation [24, 25]. Computer simulations [25—30] give clear evidence of the topological long-range order (TLRO) of vortex—antivortex pairs. This situation leads to an unusual exponential divergence of the in-plane correlation length and susceptibility when $T$ approachs $T_{KT}$ according to the relations:

$$\begin{aligned} \xi(T) &= A \exp(B(T/T_{KT} - 1)^{-1/2}) \\ \chi(T) &\sim \xi^{(2-\eta)} \end{aligned} \tag{4}$$

with $A \sim 1Å$, $B \sim \pi/2$ and $\eta = 1/4$. The quantities $\xi(T)$ and $\chi(T)$ remain infinite below $T_{KT}$. This behaviour is equivalent to the definition of critical exponents $\beta = \infty$ and $\nu = \infty$ for the spontaneous magnetization and the correlation length. If we now introduce the Josephson relation [31] $\alpha = 2 - d\nu$, which gives a relation between the critical exponents of the specific heat $\alpha$ and of the correlation length $\nu$, one finds that $\alpha = -\infty$, implying that there is no strong singularity of the specific heat at $T_{KT}$ for the *classical* 2-d XY model. This result has been derived more rigorously by Kosterlitz [23] and Berker *et al.* [32] using renormalization group techniques and is also expected from the analysis of high-temperature series (HTS) expansions [33—35]. (See also the chapter by Navarro). Nevertheless, recent results obtained from computer simulations show that the zero-field specific heat of the *classical* 2-d XY model in fact displays a nonnegligible, though finite, peak at a temperature slightly higher than $T_{KT}$ [25, 27]. In the same spirit De Raedt *et al.* [36—39] have recently shown that the *quantum*-mechanical ($S = 1/2$) 2-d XY model must have a phase transition at a finite temperature given approximatively by $T_c \simeq (2.2\text{—}2.3)JS^2$, with a logarithmically diverging specific heat:

$$C_m \sim \ln|T_c - T|. \tag{5}$$

Although this result disagrees in part with the HTS analysis for that model [35, 40] it appears now that the critical behaviour of the specific heat for the 2-d XY model differs from that for the 2-d H model, for which one expects only a Schottky-type anomaly [1, 41] similar to the 1-d model systems [1, 42].

At lower temperatures (i.e. $T \ll T_{KT}$) the thermodynamics of the 2-d XY model is governed by the spin waves (the linear excitations) and by the vortex—antivortex pairs (the nonlinear excitations) of opposite vorticity [18, 22, 43, 44]. Spin waves are expected to have linear dispersion curves for both the ferromagnetic (**F**) and the antiferromagnetic (AF) cases [18, 24]: $\omega(\mathbf{q}) \sim |\mathbf{q}|$, whereas the density of vortex—antivortex pairs $n_v$ is given by [22]:

$$n_v = R \exp(-E_p/T) \tag{6}$$

where $R$ may be a function that is slightly temperature dependent and where $E_p$ (in Kelvins) is the creation energy of a vortex—antivortex pair. Order of magnitude estimates of $E_p$ can be obtained from computer simulations, which give classically $E_p \simeq 13|J|S^2$ for the 2-d XY model [27], and $E_p \simeq 19|J|S^2$ for the planar rotator [26], i.e. when the possibility for the spins to go out of the planes is suppressed. It is easy to show that the spin waves and the vortex—antivortex pairs give contributions to the specific heat which depend on the temperature according to the relation:

$$C_m = AT^2 + B(T)\exp(-E_p/T) \tag{7}$$

where the first term is ascribed to the 2-d spin waves and the second term to the vortex pairs. Such a behaviour is also predicted for superfluid films, for which the exponential contribution arises from the quasi 2-d roton-type excitations [45—48].

Now let us return to the properties of the Kosterlitz—Thouless transition. As shown by Banavar and Jasnow [49] the order parameter is not the staggered magnetization $|\langle \mathbf{S}_k \rangle|$, which is always zero, but more generally what is called the 'helicity modulus'. For superfluid systems the helicity modulus (alias chirality [64]) is equivalent to the superfluid density $\rho_s$, whereas for the magnetic XY systems it is equivalent to the stiffness constant of the hydrodynamic (small $q$) spin waves, following the standard analogy between XY magnetism and superfluidity [50, 51]. Nelson and Kosterlitz [52] have shown that the unbinding of the vortex—antivortex pairs must entail an universal *jump* of the helicity modulus at $T_{KT}$: $\lim \rho_s(T)/\rho_s(0) \simeq \pi^{-1}kT_{KT}/|J|\ S^2 \simeq 0.6$. Such a result is also expected from arguments developed by Pokrovsky and Uimin [53] (with $\rho_s\,(T)/\rho_s(0) \simeq 1/e \simeq 0.37$) and Nelson and Fisher [54]. The existence of this jump has been established experimentally for $^4$He superfluid films [55, 56] and also for superconducting films [57]. Unfortunately for magnetic systems no definite experimental evidence of the KT-transition and of the existence of vortex-type excitations have been reported in the literature, essentially because of the lack of sufficiently good 2-d XY materials, as noted already in [1]. Actually all the experimental 2-d magnetic systems are quasi 2-d systems, because there always exists a small interlayer coupling $|J'/J| \sim 10^{-3}$—$10^{-6}$ [1], which gives rise to a 3-d LRO at a finite temperature $T_c$, when the in-plane correlations are sufficiently well developed. As mentioned at the beginning of this section, $T_c$ is given approximately by Equation (3). Applied to quasi 2-d H systems one finds, for $T_c/|J|\ S^2$, typical values of 2.6, 2.2 and 1.8, when $|J'/J|$ takes the values $10^{-4}$, $10^{-5}$ and $10^{-6}$, respectively. Such a calculation accounts well for the experimental value of $T_c$ observed in typical Heisenberg systems like $K_2CuF_4$, $(C_nH_{2n+1}NH_3)_2$ $CuX_4$ (X = Cl, Br), $K_2NiF_4$ or $K_2MnF_4$ (nonexhaustive list) [1, 41]. Note that for such systems $T_c \simeq 8\pi|J|S^2/\ln(J'/J)$ takes about the same value as $T_{KT}$. Applied to the quasi 2-d XY systems, Equation (3) gives a 3-d long-range ordering at a temperature slightly higher than $T_{KT}$:

$$T_c \simeq T_{KT} + B^2|J|S^2/(\ln(J'/J))^2. \tag{8}$$

This relation has been obtained more rigorously from renormalization group theory by Hikami and Tsuneto [58]. They have shown in particular that the nonlinear vortex-type excitations play a fundamental role for a sufficiently small interlayer coupling. So we could expect to observe a 2-d XY-like behaviour in the vicinity of the transition when $|\langle \mathbf{S}_k \rangle| \simeq 0$, and when the 3-d fluctuations are still weak enough to be outside the 3-d critical regime. The effect of the creation of the vortex—antivortex pairs or vortex rings [58] leads to important consequences: (i) a more rapid decrease of the staggered magnetization in the immediate vicinity of $T_{KT}$, and (ii) a very rapid decrease of the intensity and a broadening of the long wavelength excitations near $T_c$, due to the fact that the vortices are large objects described by an infinite number of Fourier components. This may have a drastic effect on the observation of the jump of $\omega(q)(\sim \sqrt{\rho_s}\, q)$ when $q \to 0$, which may be masked by the rapid damping of the linear excitations. Nevertheless the behaviour of real systems certainly reflects that of pure 2-d systems, if all the undesirable perturbations are minimized. In particular the quasi 2-d planar system must have an in-plane anisotropy $\mathcal{H}_p = \Sigma_i\, h_p \cos(p\phi_i)$ as small as possible with a symmetry $p > 4$ in the notation of José *et al.* [59]. This implies that a small in-plane anisotropy is not relevant to suppress the Kosterlitz—Thouless transition in the case of a hexagonal ($p = 6$) symmetry [53, 59]. If this condition is realized (i.e. for hexagonal or honeycomb layers), the system will first undergo a KT transition at $T_{c2}(h_p) = T_{KT}$, and thereafter a second transition toward an Ising phase at a lower temperature $T_{c1}(h_p)$ [59, 60]. One would hope that this picture would remain valid if in addition a small interlayer coupling would be present in the real system. In that case the lower temperature $T_{c1}$ would coincide with the transition to 3-d LRO. Intuitively one must have $T_{c1} \simeq T_{c2}$, which means that the observation of the KT state is certainly a very difficult problem.

The situation described above, seems to be independent of the sign of the in-plane interaction $J$. In particular one may ask whether there is a modification of the KT picture (i.e. the unbinding of the vortex—antivortex pairs) if we deal now with an antiferromagnetic (AF)-system instead of a ferromagnetic (F)-system, and also whether there is an effect of the lattice symmetry? In fact recent work has shown fundamental differences between F- and AF 2-d XY systems only when the additional problem of *frustration* is present [62—68]. This is the case e.g. for the AF 2-d XY system on the triangular lattice. For this model there is both a continuous as well as a discrete degeneracy, and the transition is ascribed to an order—disorder transition of the chirality [62—64] displaying an Ising-like behaviour. This transition is characterized in particular by a *logarithmic divergence* of the specific heat, as for the 2-d-Ising model [61—63], at a temperature $T_c/|J|S^2 \simeq 1$—$1.1$ [63], i.e. about twice as small as the $T_{KT}$ of the F 2-d XY model. In contrast to the triangular lattice, the AF square lattice exhibits the same properties as the F square lattice. Nothing is known concerning the AF or F honeycomb lattices, but one expects both systems to be equivalent as long as nearest-neighbour interactions are considered. In a similar way the KT picture should be modified if additional in-plane exchange interactions of different signs are introduced, whatever the lattice. This last problem is connected in particular to the critical behaviour of 2-d XY helimagnets, which are predicted to have very peculiar properties [76].

Up to now the discussion has been centered on the static behaviour of the (quasi)2-d XY system. Unfortunately, concerning the dynamics of a magnetic system undergoing a KT-type transition very little is reported in the literature. As indicated above, the dynamical response must reflect the existence of two classes of excitations, namely spin waves and vortices. The spin wave contribution at low temperature, below $T_{KT}$, has been analysed by Berezinski *et al.* [17, 18], Villain [24] and Nelson *et al.* [54], whereas the vortex contribution has been first studied by Huber [69, 70] in the critical regime above $T_{KT}$. Unfortunately the expressions of the dynamical structure factors $S^{\alpha\alpha}(\mathbf{q}, \omega)$ ($\alpha = x, y, z$) are not completely derived in the region $T \gtrsim T_{KT}$, for which the dynamical response of the system must be that of a dilute gas of free vortices. Such expressions are fundamental for the analysis of inelastic neutron scattering or NMR experiments. Nevertheless, central peaks are expected to appear around $q \simeq 0$ and $\omega \simeq 0$ in $S^{xx}(\mathbf{q}, \omega)$ and $S^{zz}(\mathbf{q}, \omega)$, with widths in $\mathbf{q}$ and $\omega$ more or less proportional to the free-vortex density [136]. But from this point of view the dynamics has been much less well investigated theoretically in 2-d systems [141, 142] than in 1-d systems, for which a lot of theoretical work exists [42, 72—76], in particular in connection with soliton problems.

Another important question is what could happen when the system cannot be considered as a classical ($S = \infty$) system at all? This question is fundamental because practically all good prototypes are either spin $S = 1$ or $S = 1/2$ systems. Up to now there exist theoretical predictions for the extreme $S = 1/2$ quantum case only. For that model there appears to be a phase transition, but not necessarily of the same nature as for the classical KT model. In particular, Rogiers *et al.* [35, 77] have found a critical behaviour with power laws with critical exponents $\gamma \simeq 2.5$, $\nu \simeq 1.40$ and $\alpha \simeq -1$ from the HTS expansions of the 2-d XY, $S = 1/2$ model, in disagreement with the exponential behaviour of the classical model. We have mentioned before that De Raedt *et al.* [36, 38] have found, using the generalized Trotter formula, a logarithmic divergence of the specific heat which corresponds rather to $\alpha \simeq 0$. Similar results have been obtained by Suzuki *et al.* [79, 80], who in fact were the first to point out the possibility of reducing a quantum-mechanical problem in $d$ dimensions to a classical problem in $d + 1$ dimensions with the aid of the Trotter formula. But the most important difference can be seen in the nature of the ground state of the 2-d, $S = 1/2$, XY model. As shown by Betts and Kelland [78] the ground state supports both vortex—antivortex pairs (which can exist even at $T = 0$ due to quantum fluctuations) and, more predominantly, spin-up/spin-down pairs which could be responsible of the characteristic features of the phase transition in this model. Further investigations are needed to clarify the exact nature of the elementary excitations of the 2-d, $S = 1/2$, XY model, and the mechanisms which drive the phase transition for this model.

The next paragraphs are devoted to the analysis of the magnetic properties of some typical 2-d planar systems, which can be or have been considered as the best candidates to check the theoretical predictions just described.

## 3. Crystallographic Properties of $BaM_2(XO_4)_2$ Compounds

The isomorphous compounds of chemical formula $BaM_2(XO_4)_2$ (M = Co, Ni; X = P, As) crystallize in a trigonal structure of space groupe $R\bar{3}$, which has been elucidated by Eymond and Durif some years ago [81, 82]. The cell parameters are listed in Table I and the crystallographic structure is drawn in Figure 1. This structure can be described as a stacking of (001) magnetic layers, in which the $M^{2+}$ ions are located on a *honeycomb* lattice. The distance between first neighbours is $d_1 = a/\sqrt{3} \sim 2.8$Å and the magnetic planes are separated by $XO_4$—Ba—$XO_4$ layers. From the large interlayer distance $d_2 = c/3 \sim 7.8$Å, a pronounced two dimensional magnetic character is expected. In particular we note that the ratio $d_2/d_1 \sim 2.70$, is even larger in the $BaM_2(XO_4)_2$ compounds than for example in $K_2NiF_4$ or $K_2CuF_4$, for which $d_2/d_1 \simeq 1.7$ [1]. The exchange interaction within the layers takes place via direct M—M exchange and via M—O—X—O—M superexchange, while the exchange interactions between planes result from long bonds M—O—X—O—O—X—O—M, involving many intermediate ions. Figure 1 gives the definition of the in-plane exchange interactions between first ($J_1$), second ($J_2$) and third ($J_3$) neighbours of the honeycomb lattice and of the interplane exchange interactions $J_1'$, $J_2'$, $J_3'$ and $J_4'$. Note also that the symmetry of the magnetic site is trigonal, and corresponds to $p = 6$, which means that a small in-plane anisotropy must be irrelevant for the formation of vortices at the transition. As we shall see below there exists a strong planar anisotropy in these materials, that is mainly of crystal-field origin, and introduces a strong preference for the moments to lie within the easy plane.

We note that the same honeycomb symmetry also exists in other system as, for example, $KNiAsO_4$ or $NiPS_3$. Also $KNiAsO_4$ [84, 85] has the same type of crystallographic structure as the $BaM_2(XO_4)_2$ compounds, with cell parameters $a \simeq 4.98$Å, $c \simeq 28.50$ and $d_2/d_1 \simeq 3.3$. The compounds $NiPS_3$ [86, 87] crystallizes in a monoclinic structure with cell parameters $a = 5.83$Å, $b = 10.10$Å, $c = 6.63$Å and $\beta \simeq 107.10°$, for which $d_2/d_1 \sim 1.9$.

To summarize, the crystallographic structure of these compounds displays the good characteristics to be used as prototypes to test the 2-d XY model theories, with the restriction that certainly more than one exchange interaction within the

TABLE I
Unit cell parameters of the $BaM_2(XO_4)_2$ compounds at $T = 4.2$ K, the shortest in-plane distance ($d_1$) is compared with the interplane distance ($d_2$).

| Compounds | $a$ (Å) | $c$ (Å) | $d_1$ (Å) | $d_2/d_1$ | Ref. |
|---|---|---|---|---|---|
| $BaNi_2(PO_4)_2$ | 4.81 | 23.22 | 2.78 | 2.78 | [81, 83] |
| $BaNi_2(AsO_4)_2$ | 4.94 | 23.43 | 2.85 | 2.74 | [81, 83] |
| $BaCo_2(AsO_4)_2$ | 5.00 | 23.25 | 2.89 | 2.68 | [81, 83] |
| $BaCo_2(PO_4)_2$ | 4.85 | 23.10 | 2.80 | 2.77 | [83] |
| KNi $AsO_4$ | 4.98 | 28.50 | 2.88 | 3.30 | [84, 85] |

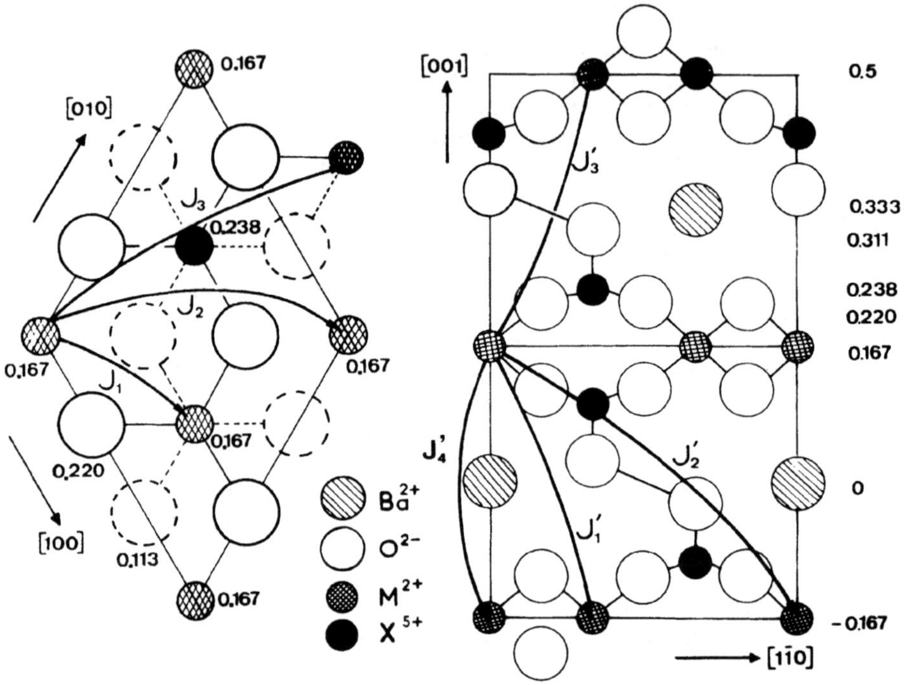


Fig. 1. Crystallographic structure of $BaM_2(XO_4)_2$ compounds. Projections along the [001] and [110] direct axis.

plane must be introduced. This will render the experimental systems slightly more complicated than the actual theoretical models.

## 4. Magnetic Properties of $BaNi_2(PO_4)_2$

In this section we shall give an exhaustive description of the magnetic properties of $BaNi_2(PO_4)_2$, which turns out to be the system for which the physics is easiest to understand.

### 4.1. STATIC PROPERTIES

The planar anisotropic character of this system is evidenced directly from the analysis of the susceptibilities parallel ($\chi_{\parallel}$) and perpendicular ($\chi_{\perp}$) to the $c$ axis. As can be seen in Figure 2, $\chi_{\parallel}$ and $\chi_{\perp}$ exhibit the typical 2-d characteristics, with a broad maximum at $T(\chi_{max}) \simeq 32$ K and an inflexion point ($\delta^2\chi/\delta T^2 = 0$) at $T_N = 24.5 \pm 1$ K which corresponds to the ordering temperature [89, 90]. These provide clearly evidence for antiferromagnetic in-plane interactions and for a strong planar character. The experimental data can be analysed in terms of the

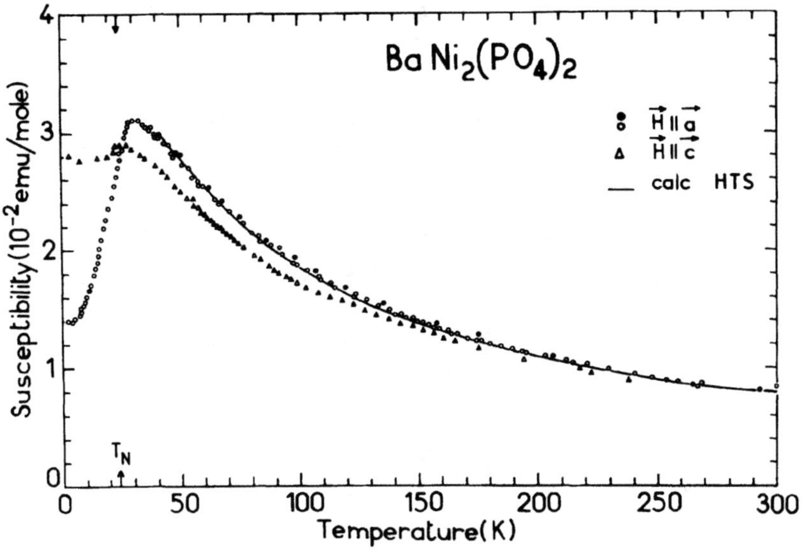


Fig. 2. Temperature dependences of the susceptibilities $\chi_\perp$ and $\chi_\parallel$ of $BaNi_2(PO_4)_2$.

Heisenberg hamiltonian, with a planar anisotropy term added:

$$\mathscr{H} = -\sum_{i,j} J_{ij}\mathbf{S}_i\mathbf{S}_j + D\sum_i (S_i^z)^2 \qquad (10)$$

where $\mathbf{S}_i$ is a spin, $S = 1$, $J_{ij}$ is the exchange interaction between sites $i$ and $j$ and $D(\geqslant 0)$ is the planar anisotropy constant. Unfortunately there exist no high-temperature series expansions for the susceptibilities of the 2-d XY honeycomb lattice to analyse the data properly. Nevertheless, an order of magnitude of the interactions can be obtained from the HTS of the 2-d S = 1 H model [91]. From $\chi_\perp(T)$ one obtains a net effective antiferromagnetic exchange interaction $\tilde{J}/k \sim -8$ K. An order of magnitude of $D$ can be obtained from $\chi(0) \simeq 2N_A(g\mu_B)^2/(2D + 12|\tilde{J}|)$. Taking $g \sim 2.2$ one finds $D/|\tilde{J}| \sim 1$ corresponding to a ratio of the anisotropy field to the exchange field of $H_a/H_e \sim 0.15$. These values give a more quantitative confirmation of the rather strong planar character of the compound $BaNi_2(PO_4)_2$, much stronger anyway than for $K_2CuF_4$ [113—115] ($H_a/H_e \sim 1\%$) or $NiCl_2$-GIC [123] ($H_a/H_e \sim 0.2\%$).

The elastic neutron scattering data confirm the antiferromagnetic ordering below $T_N \simeq 23.5 \pm 0.5$ K, and the localization of the spins within the basal plane. At low temperatures the magnetic structure (Figure 3) is characterized by a wave vector $\mathbf{k} = 0$ and an antiferromagnetic coupling between the two hexagonal sublattices $\phi = \pi$. The temperature dependence of the staggered magnetization has been derived from that of the magnetic peak intensities [92]. In the temperature

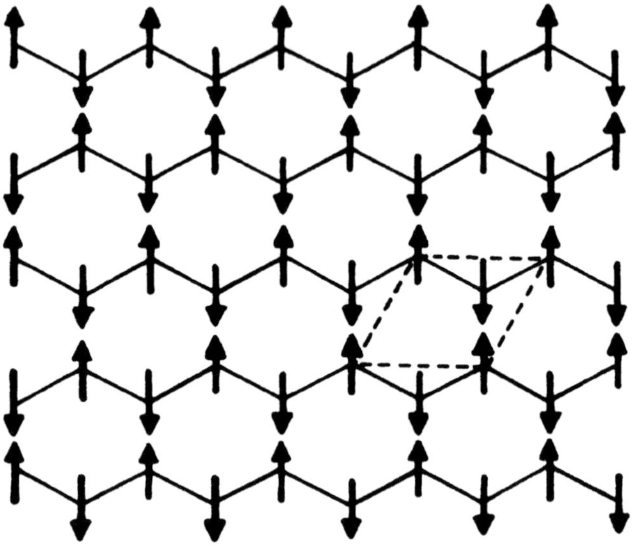

Fig. 3. Antiferromagnetic structure of $BaNi_2(PO_4)_2$.

range $1\text{-}T/T_N \gtrsim 2 \times 10^{-2}$ the order parameter (Figure 4) follows a power law $|\langle \mathbf{S}_k \rangle| \sim (1\text{-}T/T_N)^{\beta}$ with an exponent $\beta \simeq 0.23$ typical of 2-d systems. For temperatures closer to the transition temperature (approximatly when $1\text{-}T/T_N \lesssim 2 \times 10^{-2}$), the critical exponent $\beta$ takes a much larger value $\beta \simeq 0.33$ which is actually very close to that of the 3-d XY model. This change in the slope can be clearly seen in Figure 4 and can be interpreted as a crossover from the 2-d to the 3-d critical regime. But it may also be interpreted as a faster decrease of the staggered magnetization above $T_{KT}$, with $T_{KT} \sim 0.98\ T_N$ in the present case.

The behaviour of the specific heat is quite unusual for a 2-d system with a continuous symmetry [93]. In fact the magnetic specific heat exhibits a $\lambda$-type anomaly at $T_N$ (Figure 5) which contrasts with the Schottky-type anomaly usually observed in 2-d Heisenberg systems, as for instance in $\beta$-Ni(triazole)$_2$(NCS)$_2$ [94] which also has $S = 1$ but a very much weaker anisotropy. Again by lack of high temperature series expansions, the analysis of the high temperature part of the specific heat is not possible. At very low temperatures (when $T \ll T_N$), the total specific heat (i.e. magnetic + lattice) follows a dependence with temperature given by the relation:

$$C = AT^2 + BT^3 \tag{11}$$

with the experimental values $A \simeq 2.8 \pm 0.3$mJ/K$^3$/mole and $B \simeq 0.7 \pm 0.1$mJ/$K^4$/mole. Such a temperature dependence is clearly seen from the plot of $C/T^2$ versus $T$ (Figure 6). In Equation (11), the second term ($\sim T^3$) is ascribed to the contribution of the phonons while the first term ($\sim T^2$) is ascribed to the 2-d spin waves assuming a quasilinear dispersion curve for these excitations. A more

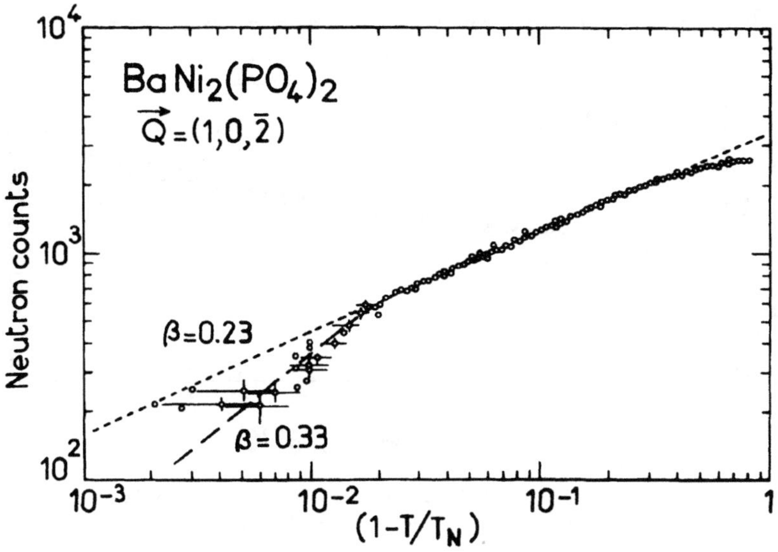


Fig. 4. Temperature dependence of the magnetic Bragg peak [102] of $BaNi_2(PO_4)_2$.

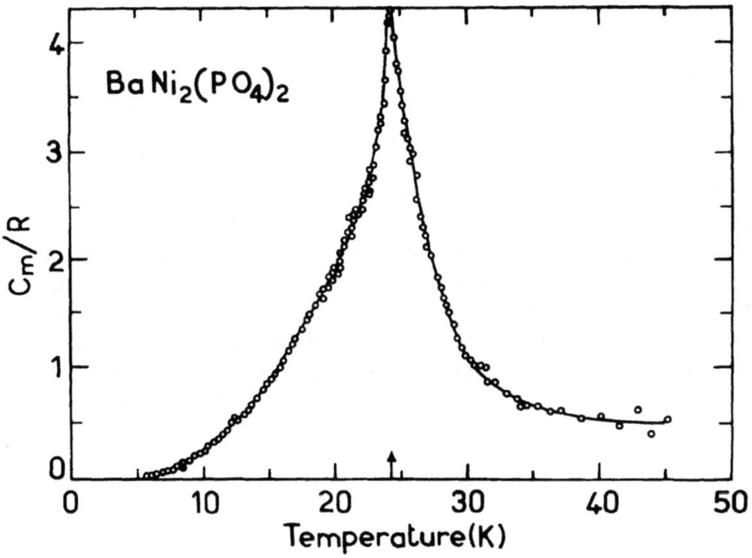


Fig. 5. The specific heat of $BaNi_2(PO_4)_2$ as a function of temperature.

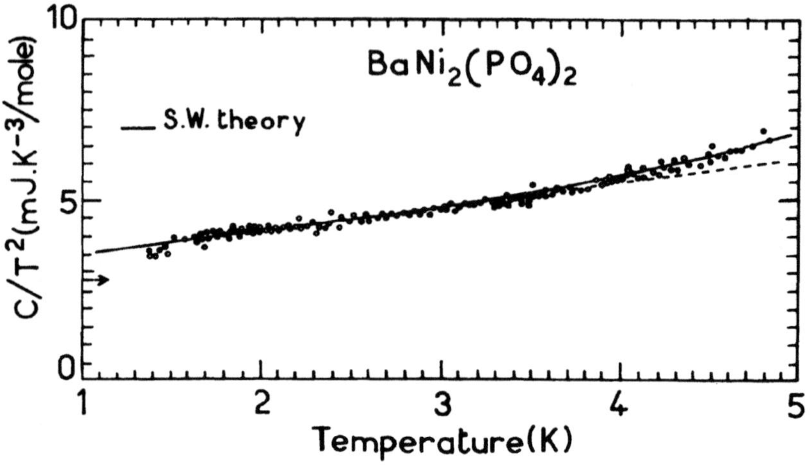


Fig. 6. Low-temperature specific heat of $BaNi_2(PO_4)_2$. The solid line represents the prediction of the linear spin-wave approximation, as described in the text.

quantitative analysis of the low temperature specific heat will be given below, where we shall discuss the problem of the magnetic excitations in $BaNi_2(PO_4)_2$.

### 4.2. DYNAMIC PROPERTIES

A precise knowledge of the exchange and anisotropy parameters is usually deduced from the analysis of the dispersion curves of the magnetic excitations along several directions in the reciprocal lattice. Typical experimental spectra obtained from the inelastic neutron scattering (INS) experiments [95] are shown in Figure 7. Qualitatively, the presence of two branches originates from the fact that two atoms per hexagonal unit cell are needed to build the honeycomb lattice. For $q \to 0$, the lower branch is quasilinear and displays a small gap $\Delta_{IP} \simeq 0.1$ meV $\simeq$ 0.025 THz, reflecting the very weak ($p = 6$) in-plane magnetic anisotropy. The upper branch has a gap $\Delta_{op} \simeq 2.8$ meV $\simeq$ 0.7 THz, due to the strong single-ion planar anisotropy. Quantitatively, the experimental data can be analysed with the Hamiltonian (10) following the standard linear spin wave theory [96—98]. This Hamiltonian is a special case of the more general anisotropic Hamiltonian:

$$\mathcal{H} = -\sum_{i,j} (J_{ij}^{\perp} (S_i^x S_j^x + S_i^y S_j^y) + J_{ij}^{\parallel} S_i^z S_j^z) + D \sum_i (S_i^z)^2 \tag{12}$$

which incorporates single ion anisotropy ($D \neq 0$) as well as exchange anisotropy ($J_{\perp} \neq J_{\parallel}$). In the Holstein—Primakoff approximation one effectively obtains two

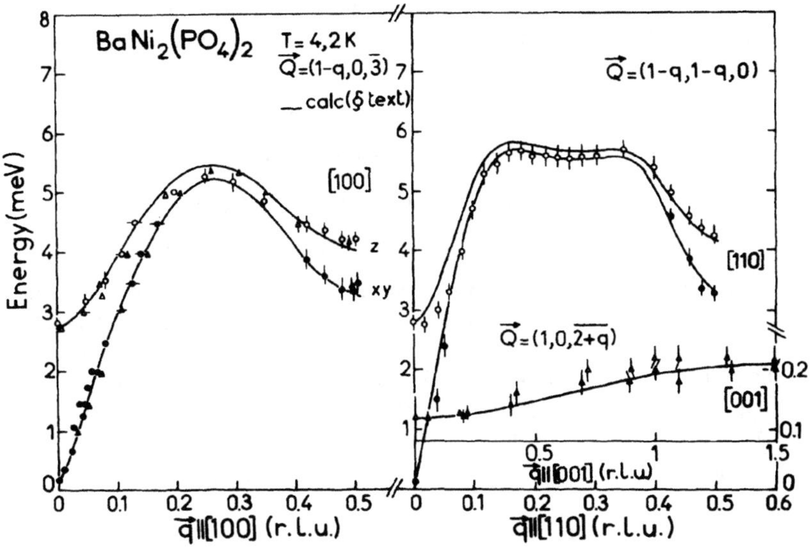


Fig. 7. Excitation spectra of $BaNi_2(PO_4)_2$ along the [100], [110] and [001] directions. The solid lines are fits to the linear spin-wave theory.

modes, with energies [98]:

$$\omega_1(\mathbf{q}) = S(R - Q^{1/2})^{1/2}$$

$$\omega_2(\mathbf{q}) = S(R + Q^{1/2})^{1/2} \tag{13}$$

with

$$R(\mathbf{q}) = |U|^2 - A_1^2 + |B_2|^2 - |A_2|^2$$
$$Q(\mathbf{q}) = 4|UB_2 + A_1A_2|^2 - |A_2B_2^* - A_2^*B_2|^2$$

and

$$A_1 = \mathscr{J}^{\parallel}_{11}(\mathbf{q}) - \tilde{\mathscr{J}}^{\perp}_{11}(\mathbf{q}) - \tilde{D}$$
$$B_1 = \mathscr{J}^{\parallel}_{11}(\mathbf{q}) + \tilde{\mathscr{J}}^{\perp}_{11}(\mathbf{q}) - \tilde{D}$$
$$A_2 = \mathscr{J}^{\parallel}_{12}(\mathbf{q}) - \tilde{\mathscr{J}}^{\perp}_{12}(\mathbf{q}) - \tilde{D}$$
$$B_2 = \mathscr{J}^{\parallel}_{12}(\mathbf{q}) + \tilde{\mathscr{J}}^{\perp}_{12}(\mathbf{q}) \tag{13'}$$
$$U = 2(\mathscr{J}^{\perp}_{11}(\mathbf{k}) + \mathscr{J}^{\perp}_{12}(\mathbf{k})) - B_1$$
$$\tilde{D} = D(1 - 1/2S)$$

In these expressions $\mathbf{q}$ represents the wave vector of the excitation and $\mathbf{k}$ is the wave vector of the ground-state magnetic ordering. For the honeycomb lattice the Fourier transforms $\mathcal{J}^{\alpha}_{11}(\mathbf{q})$ and $\mathcal{J}^{\alpha}_{12}(\mathbf{q})$ ($\alpha = \perp, \parallel$) are given by the following expressions:

$$
\begin{aligned}
\mathcal{J}^{\alpha}_{11}(\mathbf{q}) &= 2J^{\alpha}_{2}(\cos(2\pi q_x) + \cos(2\pi q_y) + \cos(2\pi(q_x + q_y))) \\
\mathcal{J}^{\alpha}_{12}(\mathbf{q}) &= J^{\alpha}_{1}(1 + \exp(-2\pi i\, q_x) + \exp(-2\pi i(q_x + q_y))) + \\
&\quad + J^{\alpha}_{3}(\exp(-2\pi i(2q_x + q_y)) + 2\cos(2\pi q_y)) \qquad (14) \\
\tilde{\mathcal{J}}^{\perp}_{ij}(\mathbf{q}) &= (\mathcal{J}^{\perp}_{ij}(\mathbf{q} + \mathbf{k}) \exp i(\phi_j - \phi_i) + \mathcal{J}^{\perp}_{ij}(\mathbf{q} - \mathbf{k}) \exp(i(\phi_i - \phi_j)))/2
\end{aligned}
$$

where $\phi$ is the phase angle between spins of the two Bravais sublattices at positions (0, 0, 0) and (2/3, 1/3, 0) ($\phi_1 = 0$ and $\phi_2 = \pi$ in $BaNi_2(PO_4)_2$) and where $J_1$, $J_2$ and $J_3$ have been defined in Figure 1. The relations (13) can also be applied when the interlayer interactions are taken into account, namely with the substitution $\mathcal{J}^{\alpha}_{11}(\mathbf{q}) \rightarrow \mathcal{J}^{\alpha}_{11}(q) + j^{\alpha}_{11}(\mathbf{q})$ and $\mathcal{J}^{\alpha}_{12}(\mathbf{q}) \rightarrow \mathcal{J}^{\alpha}_{12}(\mathbf{q}) + j^{\alpha}_{12}(\mathbf{q})$ where the perturbational terms $j^{\alpha}_{11}$ and $j^{\alpha}_{12}$ are given by the relations:

$$
\begin{aligned}
j^{\alpha}_{11}(\mathbf{q}) &= 2J'^{\alpha}_{1}[\cos(2\pi(2q_x + q_y + q_z)/3) + \cos(2\pi(q_x - q_y + q_z)/3) + \\
&\quad + \cos(2\pi(q_x + 2q_y - q_z)/3)] \\
j^{\alpha}_{12}(\mathbf{q}) &= [2J'^{\alpha}_{2}[\cos(2\pi q_x) + \cos(2\pi q_y) + \cos(2\pi(q_x + q_y))] + \\
&\quad + J'^{\alpha}_{4}]\exp(-2\pi i\, q_z/3) + J'^{\alpha}_{3}[\exp(2\pi i(q_x + 2q_y)/3) + \\
&\quad + \exp(2\pi i(q_x - q_y)/3) + \exp(-2\pi i(2q_x + q_y)/3)] \exp(2\pi i q_z/3).
\end{aligned}
$$

In what follows we have assumed for simplicity that $J'_1 \simeq J'_2 \simeq J'_3 \simeq J'_4 \simeq J'$. From the above equations it is easy to deduce an expression of the dispersion law along the $c$ axis (valid only for $BaNi_2(PO_4)_2$-type magnetic structures):

$$
\omega(q_z) \simeq S[24|J_1 + J_3|(10|J'|(1 - \cos(2\pi q_z/3)) + \tilde{D}_{IP})]^{1/2} \qquad (15)
$$

in which we have introduced the effect of an in-plane anisotropy field $H^{IP}_a$ with $D_{IP} = g\mu_B H^{IP}_a S$. The relation (15) allows the determination of $J'$ and $D$. Solid curves, in Figure 7, represent the results of the fit to expressions (13) and (15) with the following values for the exchange and anisotropy parameters:

$$
\begin{aligned}
&J_1/k \simeq -2.2\text{ K}, \quad J_2/k \simeq -0.3\text{ K}, \quad J_3/k \simeq -8.8\text{ K}, \\
&J'/k \simeq -10^{-3}\text{ K}, \quad D \simeq 7.3\text{ K} \quad \text{and} \quad H^{IP}_a \simeq 40\text{ Oe}.
\end{aligned}
$$

So in $BaNi_2(PO_4)_2$ it is necessary to introduce exchange interactions up to at least the third nearest neighbours, and this negative interaction $J_3$ is in fact the strongest one. This makes the comparison with theory more difficult. In the following it will be useful to define an *effective* exchange interaction $\tilde{J}$ between *first* neighbours as $\tilde{J} = -E(\mathbf{k})/3S^2$ where $E(\mathbf{k})$ is the classical energy of the ground-state. $\tilde{J}$ is actually a measure of the magnitude of the interactions present in the system. The application to $BaNi_2(PO_4)_2$ gives $\tilde{J}/k = -(2J_2 + J_1 + J_3) \simeq -11$ K and a ratio $|z'J'/z\tilde{J}| \sim 5 \times 10^{-4}$ which confirms quantitatively the rather good 2-d character of the system. We note that the obtained set of parameters explains quantitatively the susceptibility $\chi_{\|}(0)$, but can only account for the low temperature specific heat if a renormalization of about +10% of the exchange interactions $J_1$, $J_2$ and $J_3$ is applied. The solid curve in Figure 6 represents the result of the calculation using values for $J_1$, $J_2$ and $J_3$ that are 10% larger than that determined from the INS experiments. Such a small discrepancy may be easily ascribed to short-comings in the simple spin wave theory [41]. We may conclude from the strong planar anisotropy ($D/|\tilde{J}| \simeq 0.7$, $H_a/H_e \simeq 0.11$) combined with the small value of $J'$ ($|z'/J'/z\,\tilde{J}| \sim 5 \times 10^{-4}$) and with the weak in-plane anistropy (~40 Oe) that the compound $BaNi_2(PO_4)_2$ certainly is a very good prototype of the 2-d AF-planar model. However, due to the weak frustration, it is expected to behave like the ferromagnetic counterpart.

### 4.3. CRITICAL PROPERTIES

The 2-d nature of the magnetic interactions, which is clearly evidenced by the relatively small value of the ratio $|z'J'/z\,\tilde{J}|$ and by the existence of a $T^2$-contribution in the low-temperature specific heat, is confirmed by the existence of quasi 2-d correlations above $T_N$ up to about $2T_N$ ($\sim \theta$). Information on the temperature dependence of the correlation length $\xi(T)$ can be deduced from the analysis of the widths in $q$ of the scattering law assuming a Lorentzian shape, $S(q) \sim 1/(\Gamma_q^2 + q^2)$, with $\Gamma_q = 1/\xi$. Typical experimental results for various wave lengths and geometries are reported in Figure 8. The correlation length $\xi$ is found to be isotropic in the XY plane. The behaviour of the in-plane correlation length as a function of temperature expected for the 2-d XY model can be fitted to the experimental results. In the temperature range $10^{-2} \lesssim T/T_N - 1 \lesssim 0.5$ the data are well reproduced by the exponential law given by equation (4), with parameters $A \simeq 0.6$ Å, $B \simeq 1.6$ and $T_{KT} \simeq (0.96\text{—}0.97)T_N$, which are very close to the theoretical values. Note that in the same temperature range a fit to a power law $\xi \sim (T/T_N - 1)^{-\nu}$ is also possible, but the obtained exponent $\nu \sim 1.4\text{—}1.5$ has an unusualy large value for a 2-d system ($\nu$(2d-I) = 1 and $\nu$(3d-XY) = 0.66). This anomalously large value is ascribed to the strong 2-d XY character, and to the existence of vortex-type excitations in $BaNi_2(PO_4)_2$. Note also that an order of magnitude of the exponent $\eta$ can be obtained from the scaling relation $2\beta/\nu = d - 2 + \eta$. For the quasi 2-d compound $BaNi_2(PO_4)_2$ one obtains a value $\eta = 2\beta/\nu \simeq 0.31 \pm 0.05$, which is close to the theoretical one $\eta = 1/4$ expected for the 2-d XY model [22], but also agrees quite well with the determination $\eta = 1/\pi \simeq 0.318$ of [53]. Furthermore the shift of 3—4% in $T_N$ with respect to $T_{KT}$ can be accounted for by the interlayer coupling on the basis of the relation (8), slightly modified by taking account of the effective exchange interaction:

$$T_N \simeq T_{KT} + B^2\,\tilde{J}\,S^2(\ell n|J'/\tilde{J}|)^{-2}, \quad \text{with} \quad |J'/\tilde{J}| \sim 10^{-3}\text{—}10^{-4}.$$

This relation gives the good order of magnitude compared with the experimental value of $J'$ deduced from the dispersion of excitations along the $c$ axis. In fact, this value of $T_{KT}$ corresponds approximately to the temperature at which the staggered magnetization starts to drop more rapidly (cf. Figure 4).

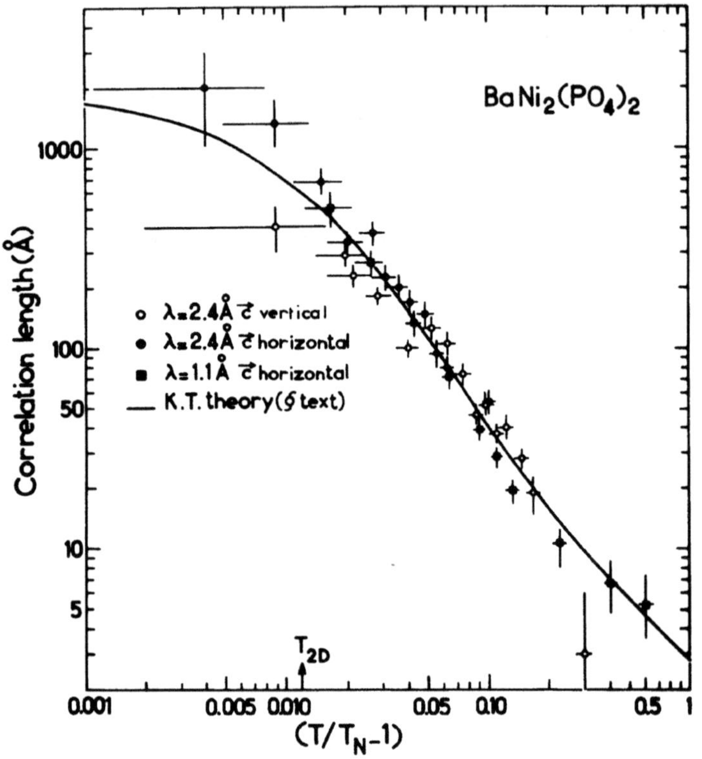


Fig. 8. The temperature dependence of the in-plane correlation length of $BaNi_2(PO_4)_2$. The solid line is a fit to the KT theory. $T_{2D}$ is the temperature above which the interplamar correlations vanish.

To check further the applicability of the KT model, it is interesting to measure the temperature dependence of the long-wavelength excitations. For small $q$-values, the theory predicts a discontinuous decrease of the spin wave energy [52, 53] ($\omega \sim \sqrt{\rho_s}\, q$), but without any information on their intensity and damping. Typical experimental results are given in Figure 9 at $q = 0.03$ r.l.u. in the vicinity of $T_N$. The experimental data, summarized in Figure 10, show that for sufficiently small values of the wave vector **q** the energies of the excitations are strongly renormalized down to $\omega(T \simeq T_N)/\omega(0) \simeq 0.5$, but the renormalization is weakly $q$-dependent. These excitations are observed up to $T_N$, where they disappear suddenly. This is particularly clear from Figure 10, where the excitations disappear within 0.2 K between $T \simeq 23.7$ K and $T_N$ ($\simeq 23.90$ K in the present experiment). A similar situation is also observed in $K_2CuF_4$ [99, 118] and in third-sound measurements on $He^4$ films [100]. However, in $BaNi_2(PO_4)_2$ one finds that the higher $q$ excitations persist above $T_N$, with a renormalization and a damping which increase as $q$ becomes smaller. In our opinion the sudden disappearance of the

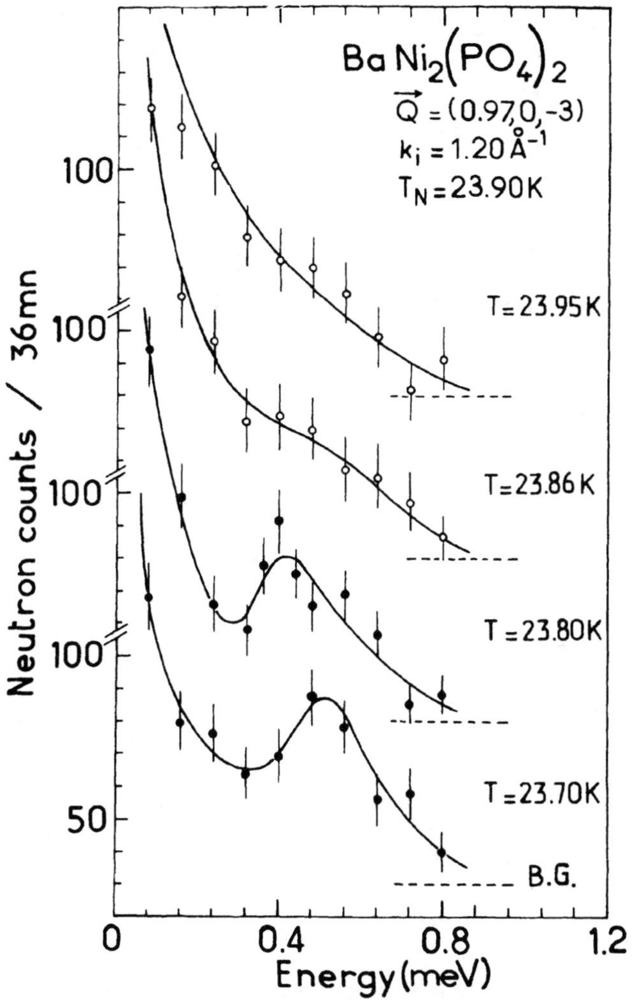


Fig. 9. Scans in energy in the vicinity of $T_N$ in $BaNi_2(PO_4)_2$. The solid lines are guides to the eye.

long wave-length excitations in the quasi 2-d planar system $BaNi_2(PO_4)_2$ is a signature of the existence of vortex-type excitations and of a KT-type transition slightly modified by the very weak interplane interactions. The experimental value of $\rho_s(T)/\rho_s(0)$ as $T \rightarrow T_{KT}$ deduced from the experiments is ~ 0.36, about half as large as the prediction of Nelson *et al.* [52]. Indeed it agrees much better with the prediction of Pokrovsky *et al.* [53], which gives a value $\rho_s(T_{KT})/\rho_s(0) \simeq 0.37$. A more quantitative comparison would necessitate the introduction of dissipation, which is expected to broaden and shift the transition and to mask in part the

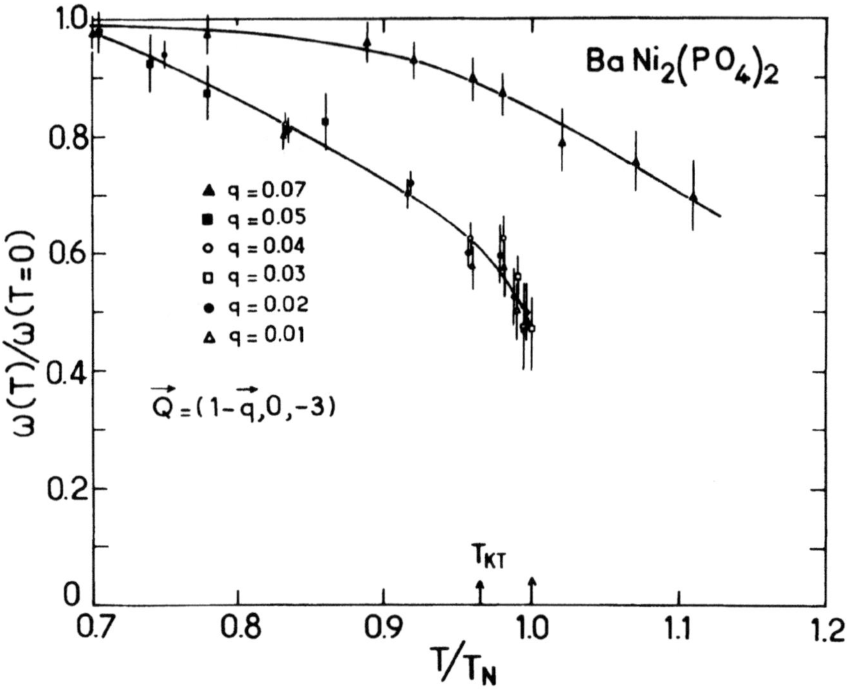


Fig. 10. Temperature and wave-vector dependence of the excitation-energy in $BaNi_2(PO_4)_2$.

universal behaviour [55]. In a same way, the behaviour of the magnetic specific heat shows that for 2-d systems a strong planar anisotropy fundamentally changes the thermodynamic behaviour of the system with respect to the isotropic Heisenberg counterpart. This is also an indication for the existence of such nonlinear excitations, at least in the vicinity of and above $T_N$.

## 5. Magnetic Properties of $BaCo_2(AsO_4)_2$

Although the compound $BaCo_2(AsO_4)_2$ is isomorphous to $BaNi_2(PO_4)_2$ from the crystallographic point of view, the replacement of $Ni^{2+}$ ions by $Co^{2+}$ ions leads to a more complicated physics, as we shall now discuss.

### 5.1. STATIC PROPERTIES

The temperature dependence of the susceptibilities $\chi_\perp$ and $\chi_\parallel$ reveals a very strong planar anisotropy, confining the spins of the $Co^{2+}$ ions in the plane perpendicular to the $c$ axis. The magnitude of the anisotropy is even much larger than in the Ni compound [98, 101], giving rise to a clear XY-like behaviour as shown in Figure

11, where the susceptibilities $\chi_\perp$ and $\chi_\|$ appear to differ by more than an order of magnitude in the low temperature range. The transition temperature is found as $T_c \simeq 5.4 \pm 0.2$ K from the inflexion point, whereas $\chi_\perp$ exhibits a maximum at $T(\chi_{max}) \simeq 6.5$ K. For the quantitative analyses of the experimental data a suitable Hamiltonian can be written in terms of effective spins $s = 1/2$ for the $Co^{2+}$ ions:

$$\mathscr{H} = -\sum_{i,j} [J_{ij}^{\perp} (s_i^x s_j^x + s_i^y s_j^y) + J_{ij}^{\|} (s_i^z s_j^z)] - \mu_B \sum_i [g_\perp (H^x s_i^x + H^y s_i^y) + g_\| H^z s_i^z)] \tag{16}$$

with the parameters $J_\perp/k \simeq 20$—$30$ K, $J_\|/J_\perp \sim 0.4$, $g_\perp \sim 5.0$ and $g_\| \sim 2.5$. These values show in particular that strong positive interactions are present in this compound, and confirm quantitatively the strong planar character, corresponding to a ratio $H_a/H_e = 1 - J_\|/J_\perp \sim 0.6$, i.e. about four times larger than for $BaNi_2(PO_4)_2$. Magnetization measurements confirm the XY-like behaviour and give evidence for two phase transitions at $H_{c1} \sim 2$—$3$ kOe and $H_{c2} \sim 5$—$6$ kOe with an intermediate ferrimagnetic structure. Below we shall give the complete

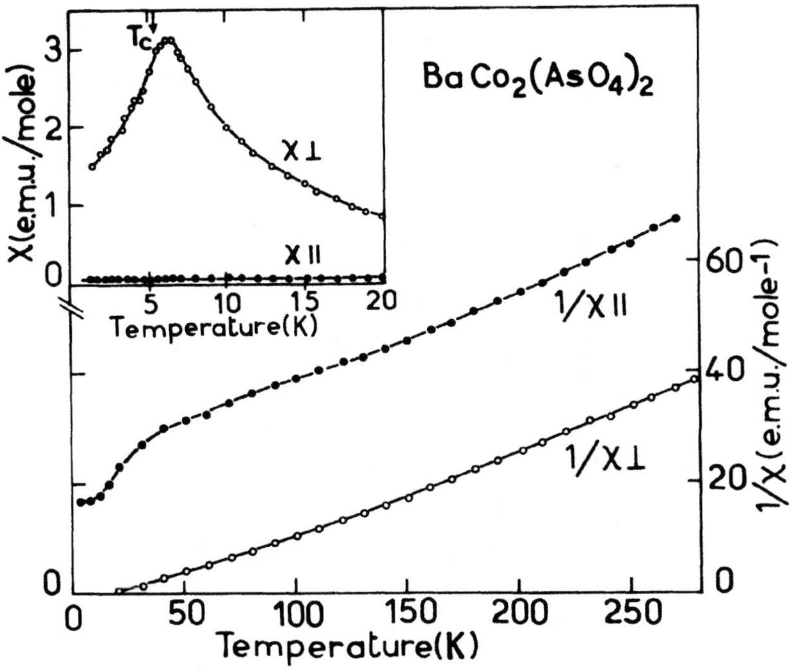


Fig. 11. The temperature dependences of the susceptibilities $\chi_\perp$ and $\chi_\|$ of $BaCo_2(AsO_4)_2$.

phase diagrams in increasing and decreasing fields. In any case, it appears that $BaCo_2(AsO_4)_2$ can be considered as a good prototype of the F 2-d XY model.

On the other hand, the elastic neutron scattering gives evidence for a helimagnetic order at low temperature characterized by a *single* incommensurate wave vector $\mathbf{k} \simeq [0.27, 0, -1.33]$, and a phase angle $\phi \simeq 80 \pm 5°$ between the two hexagonal sublattices. No harmonic of $\mathbf{k}$ is observed. This implies the magnetic structure shown in Figure 12, which can be viewed within a (001) plane as a stacking of quasiferromagnetic chains that are weakly coupled into layers. In the mean field approximation the stability of such a magnetic structure requires the introduction of at least three exchange interactions $J_1$, $J_2$ and $J_3$ [97, 101], and this structure has a classical energy:

$$-E_1(\mathbf{k})/2NS^2 = 2(J_1 \cos(2\pi k_x - \phi) + J_2) + e_c \qquad (17)$$

with $e_c = J_1 \cos(\phi) + 4J_2 \cos(2\pi k_x) + J_3(2\cos(\phi) + \cos(4\pi k_x - \phi))$.

From the experimental values of $k_x$ and $\phi$ one easily observes that the second term of relation (17) is actually much smaller than the first one, in such a way that

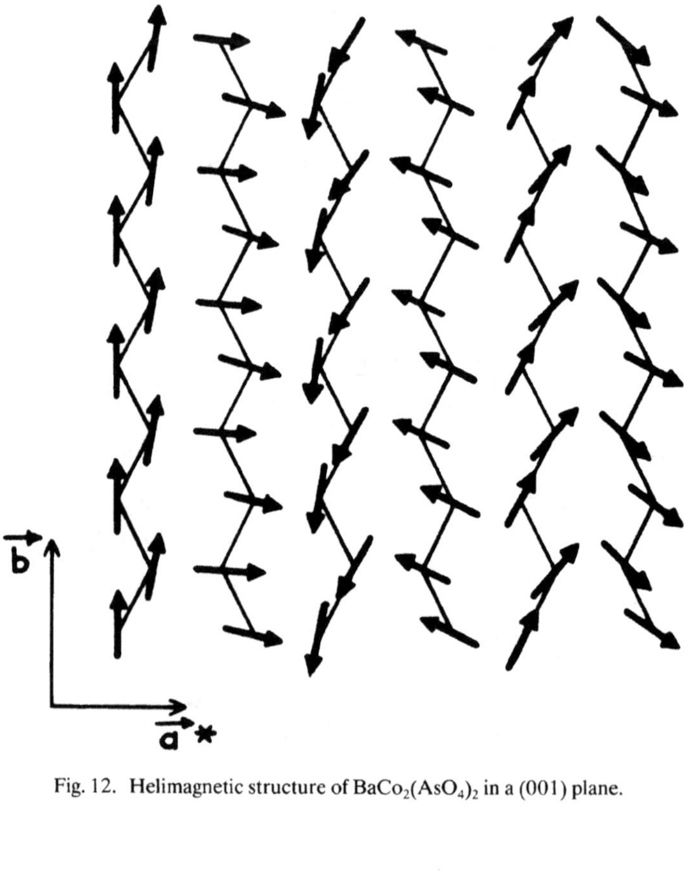


Fig. 12. Helimagnetic structure of $BaCo_2(AsO_4)_2$ in a (001) plane.

strong fluctuations are expected in the [100] reciprocal direction. This helimagnetic ordering disappears at the relatively low temperature $T_c \simeq 5.30$—$5.35$ K, or $T_c/JS^2 \sim 0.75$—1, which is about two to three times smaller than the classical or quantum values. However, the shape of the staggered magnetization curve is rather unusual for a quasi 2-d magnet, as can be seen in Figure 13. The intensities of the magnetic peaks ($\sim |\langle S_k \rangle|^2$) drop quite suddenly between 5 K and $T_c$, a behaviour very similar to that observed in very thin superfluid $^4$He films in which steep decreases are seen in the superfluid density near $T_c$ and are interpreted as the jumps predicted by the theories for the 2-d XY model.

For $BaCo_2(AsO_4)_2$ the quasi elastic neutron scattering data show the existence of strong quasi 2-d magnetic correlations, which persist up to a temperature of about 7—8 $T_c$, a value actually close to the molecular field prediction ($\theta \sim 40$ K). In fact the experimental data give evidence for anisotropic in-plane correlations, depending on whether the neutron wave vector **q** is parallel or perpendicular to the in-plane wave vector **k** of the structure. This means that the in-plane correlation lengths $\xi_{\parallel}$ and $\xi_{\perp}$ along (perpendicular) and perpendicular (along) to the quasichains (wave vector **k**) are different [92]. The temperature dependences of $\xi_{\parallel}$ and $\xi_{\perp}$, as shown in Figure 14, indicate that $\xi_{\perp} \sim (3$—$4)\xi_{\parallel}$, and that $\xi_{\parallel}$ remains

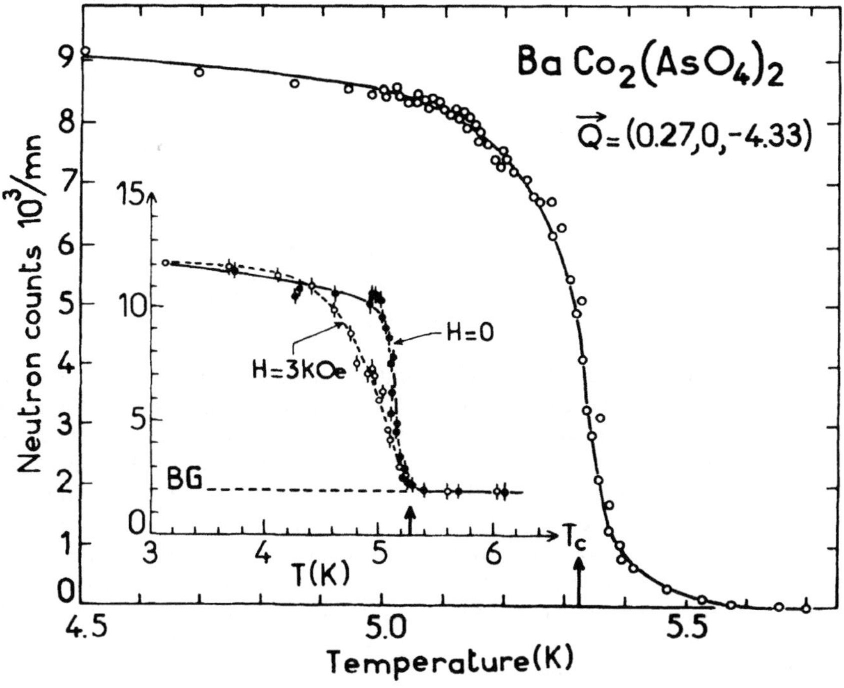


Fig. 13. Temperature dependence of the magnetic peak $(003)^-$ for an applied field $H = 0$ and $H = 2$ kOe. Solid and dashed lines are guides to the eye.

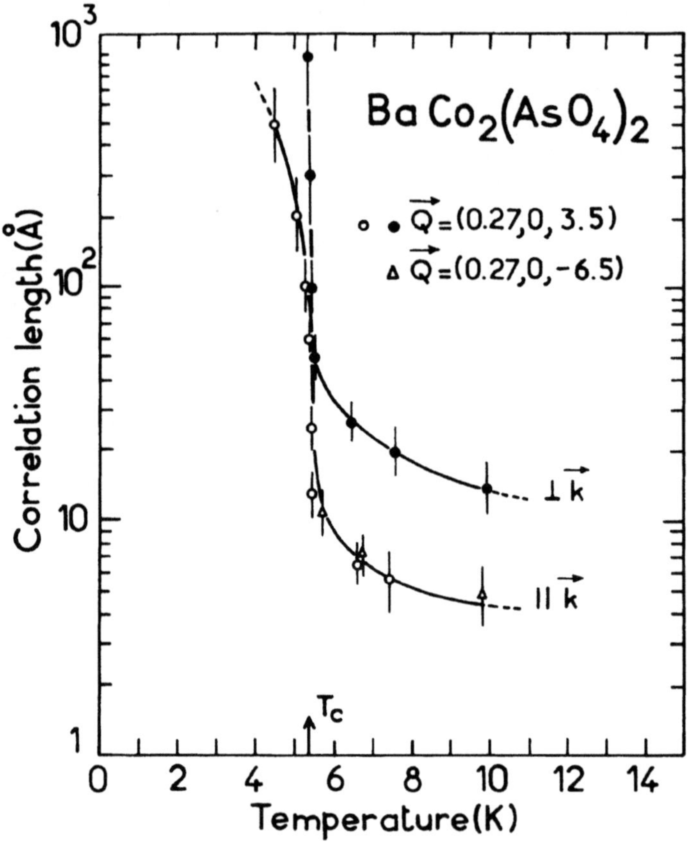


Fig. 14. Temperature dependences of the in-plane correlation lengths $\xi_\perp$ and $\xi_\parallel$. Solid lines are guides for the eye.

finite below $T_c$ whereas $\xi_\perp$ is resolution limited. Nevertheless these results reflect the weak coupling which exists between the quasichains, and show that strong fluctuations do occur even at low temperature. These fluctuations, which are responsible for the abrupt phase transition seen in the temperature dependences of the staggered magnetization and of the correlation lengths, give rise also to a quite unusual behaviour of the magnetic specific heat, as shown in Figure 15. The magnetic specific heat of $BaCo_2(AsO_4)_2$ is characterized by a sharp $\lambda$-type anomaly at $T_c$ and by a large contribution persisting far above $T_c$, reflecting the existence of strong magnetic correlations at high temperatures. The analysis of the specific heat reveals that the total entropy change below $T_c$, $\Delta S(T_c)/2R$ is only 15% of the theoretical limit ln 2 expected for a pseudo spin $S = 1/2$. This confirms the existence of strong magnetic correlations at high temperatures [102]. Such a peak in the specific heat is not currently observed in quasi 2-d systems, except for

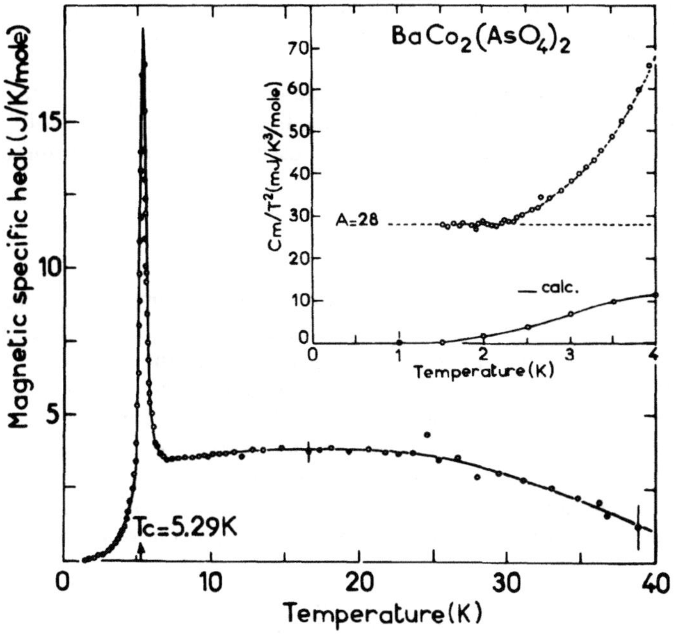


Fig. 15. Magnetic specific heat of $BaCo_2(AsO_4)_2$ as a function of temperature. The insert shows the temperature dependence of $C_m/T^2$.

the 2-d Ising systems, for which the critical specific heat behaves logarithmically. However De Raedt *et al.* [36—39] have predicted such a behaviour for the 2-d, $S = 1/2$, XY model. At the lowest temperatures the experimental specific heat again displays a $T^2$-dependence, $C_m \simeq AT^2$, with $A \sim 27 \pm 2$ mJ/K$^3$/mole, as expected from the existence of 2-d spin waves with a linear dispersion relation. Such a temperature dependence can be seen in the plot of $C_m/T^2$ versus $T$ given in Figure 15. More surprisingly, the difference $C_m - AT^2$ seems to follow an exponential dependence given by the relation:

$$C_m - AT^2 \simeq f(T)\exp(-\Delta/T) \tag{18}$$

where the prefactor $f(T)$ cannot be determined accurately. Assuming a function $f(T) \sim T^{-n}$, with $n = 1, 2, \ldots$, one obtains the order of magnitude estimate $\Delta \simeq 30$ K. A possible interpretation of this exponential term would be the presence of magnetic excitations with an activation energy $\Delta$. According to Equation (7) the excitation of vortex-antivortex pairs would contribute exponentially to the specific heat, so that it is tempting to associate the term with these excitations. This would imply that for $BaCo_2(AsO_4)_2$ one has $\Delta \equiv E_p \sim 30$ K. Qualitatively such excitations at low energy are needed to understand the rather low value of the transition temperature ($T_c/\theta \sim 0.13$). If we assume the *classical* relation $E_p/kT_{KT} \sim 8$ which results from Monte Carlo simulations [27] to be applicable, we find (taking

$T_{KT} \sim T_c$) the estimate $E_p \sim 40$ K, which seems consistent with the above value. In our opinion the low transition temperature and the unusual critical behaviour observed in $BaCo_2(AsO_4)_2$ could originate from the existence of 'vortex–antivortex' pairs which have a relatively small creation energy due to the particular set of in-plane exchange interactions in this compound. We will make this hypothesis more explicit below, after a discussion of the study of the elementary excitations.

## 5.2. MAGNETIC PHASE DIAGRAMS

We first want to comment more extensively on the phase diagrams of $BaCo_2(AsO_4)_2$ in an applied field. As mentioned before, magnetization measurements evidence first-order transitions in a field between the helimagnetic phase, an intermediate ferrimagnetic phase and finally the paramagnetic state. In fact a more accurate determination of the phase diagrams is obtained from the elastic neutron scattering experiments [103]. As can be clearly seen in Figure 16, the phase diagram is quite different in increasing and decreasing field.

In increasing field the neutron data show two phase transitions, namely from the helimagnetic state with wave vector $\mathbf{k}_0 = [0.27, 0, -1.33]$ to a ferrimagnetic state

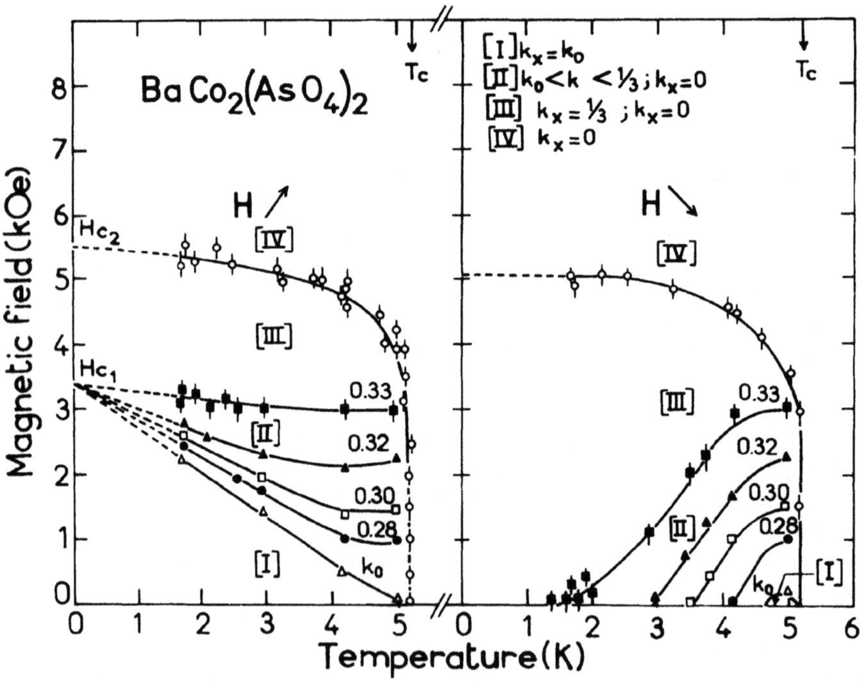


Fig. 16. The H-T phase diagrams of $BaCo_2(AsO_4)_2$ in increasing and decreasing fields.

at a critical field $H_{c1}(T \simeq 0) \simeq 3.3$ kOe and subsequently to the (saturated) paramagnetic state at a critical field $H_{c2}(T \simeq 0) \simeq 5.5$ kOe. These transitions are found to be more or less broadened depending on the temperature. The ferrimagnetic phase in between $H_{c1}$ and $H_{c2}$ can be described in the XY planes by a two wave vector structure in which the magnetic moments are aligned along **H** according to the following Fourier expansion:

$$\mathbf{m}_{ni} = (m_0/3 + (4m/3)\cos(2\pi x_n + \phi_i))\hat{y} \tag{19}$$

where $\hat{y} = \mathbf{H}/H$, $\phi_i$ is the phase angle of the sublattice $i$ ($\phi_1 = -\pi/3$ and $\phi_2 = \pi/3$) and $m_0 = |\mathbf{m}|$. This structure can be viewed as a stacking of ferromagnetic chains alternatively parallel and antiparallel to the field, following the well ordered sequence $++-++-\ldots$ ($\xi_\perp$ and $\xi_\parallel > 1000$ Å). Between two adjacent layers the phases of the sequences are completly *random*. In fact, scans along the $c^*$ direction (Figure 17) show very broadened peaks that are centered around the zero field value $Q_z = \ell_N + k_{0z}$ and can be well accounted for by the existence of stacking faults between the layers. Defining $1\text{-}p$ as the probability of a stacking fault, the experimental data for $H \simeq 4$ kOe are well reproduced by the relation

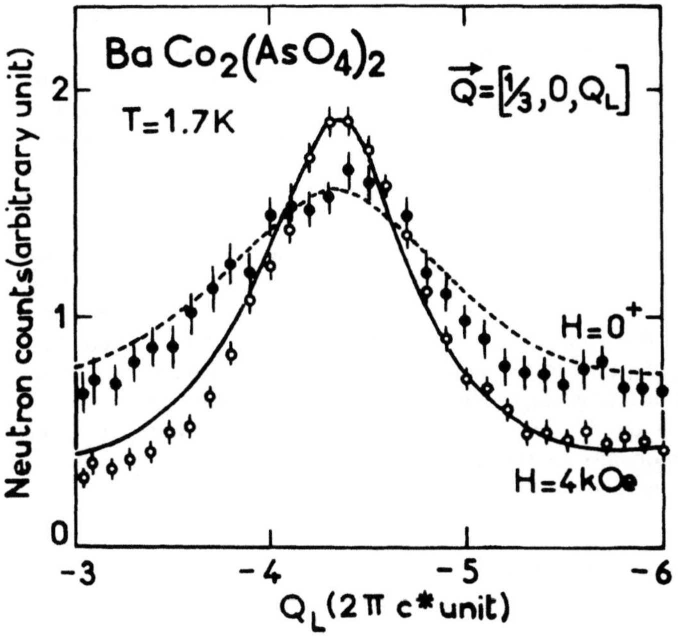


Fig. 17. $q$-scans along the $c^*$-direction for $H = 4$ kOe and $H = 0^+$. Solid and broken lines are fits to the relation (20).

[104]:

$$I^{\pm}(\mathbf{Q}) \sim \delta(Q_x - h_N \pm k_x)\delta(Q_y - k_N \pm k_y)(1-p^2)/(1+p^2 - \\ - 2p\cos(2\pi(Q_z - \ell_N \pm k_z)/3))$$

in which **Q** is the scattering vector of the neutron, $\mathbf{H} = (h_N, k_N, \ell_N)$ is a reciprocal lattice vector and **k** is the wave vector describing the structure, taking $p = 1/3$. This value is justified by the existence of only three possible values for the phase angle in expression (19), and implies that interplane order is absent, since the interplane correlation length $\xi = (c/3)|\ell_N p| \simeq 7$Å, i.e. only *one interlayer* distance long. From the rather good 2-d character it may be concluded that $|J'| \ll g_\perp \mu_B H_{c1} \sim 0.6$ K. In fact a more precise estimate gives $|J'/J| < 10^{-3}$ [98]. As seen in the insert of Figure 13, the ferrimagnetic order at $H \simeq 3$ kOe disappears at about the same temperature as the helimagnetic order at $H = 0$. The transition is again quite sudden, although less abrupt as for $H = 0$. The implication is that the basic 2-d mechanism which drives the phase transition in $BaCo_2(AsO_4)_2$ is not substantially affected by a weak in-plane field (remember that $g\mu_B H_{c1}/kT_c \sim 0.1$). These results are very important because they confirm that the phase transition in this strongly planar system is typically of 2-d character.

In decreasing field the experimental phase diagram is quite different and complicated, characterized as it is by the occurrence of metastable phases and hysteresis. In particular when the field is switched off the system does not return to the initial helimagnetic phase, but to a more complicated phase in which many defects more or less ordered are present, defining an incommensurate wave vector $\mathbf{k} = [k_x, 0]$ with $0.27 \leqslant k_x \leqslant 0.333$. Below approximately 2 K the intermediate ferrimagnetic state remains in zero field. Concerning this metastable state, all the data agree with a magnetic structure that corresponds to a stacking of ferrimagnetic bands with a net magnetization parallel ... $++-++-$ ... and antiparallel ... $--+--+$ ... to the previously applied field direction (with $\xi_\perp > 500$ Å and $\xi_\parallel \sim 100$ Å). Such a structure is probably associated with demagnetizing field effects. The order between the planes is again completely *random* and corresponds to a stacking probability $p = 1/6$ which results now from the six possible sequences of the ferrimagnetic stacking along **c**. In our opinion these results are due in a great part to the 'chain character' of the observed structures and to the possibility for large segments of the chains to rotate more or less easily. For example for the $++-++-$ ... structure at $H = 0$, it can be immediately seen that the moments in the $+$ chains can rotate collectively with respect to the neighbouring chains without cost of exchange energy.

### 5.3. DYNAMIC PROPERTIES

We now turn to the discussion of the excitation spectrum of $BaCo_2(AsO_4)_2$. As in the Ni-compound, two modes are observed for a given field, but the spectrum is quite different [95]. Typical dispersion curves for different values of the magnetic field are reported in Figures 18 and 20. The upper branch is found to be weakly

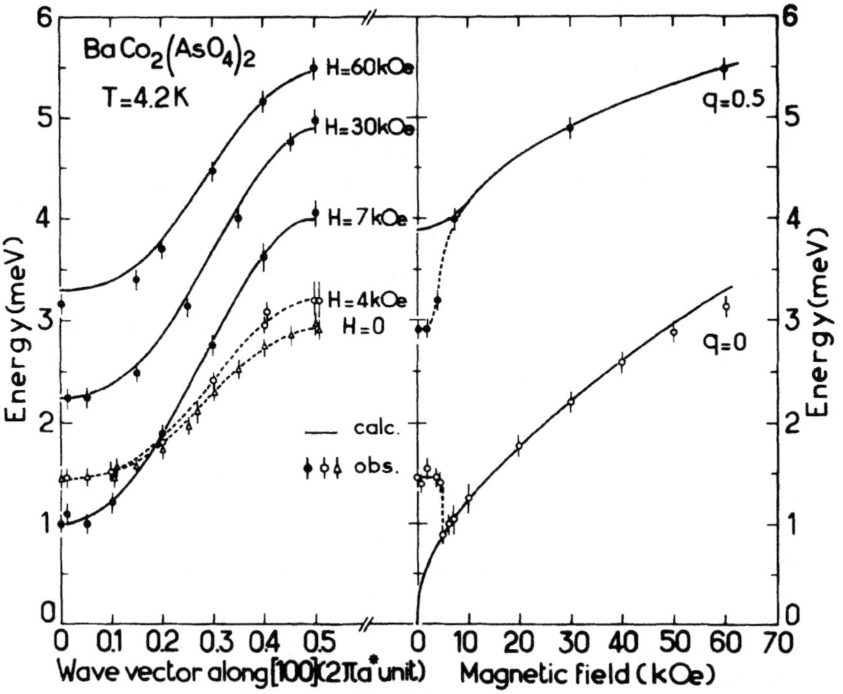


Fig. 18. Magnetic excitation spectra of $BaCo_2(AsO_4)_2$ along a ⟨100⟩ direction as a function of the applied field. The solid lines are fits to the linear spin wave theory. The broken lines are guides to the eye. The energies $\omega$ $(q = 0)$ and $\omega$ $(q = 0.5)$ are plotted as a function of the magnetic field applied within a (001) plane.

$q$-dependent and is approximatively given by the relation $\omega(q, H) \simeq 12.5 + g_\perp \mu_B H$ ($\omega$ in meV and $H$ in kOe), indicating rather strong interactions. The lower energy mode is more difficult to understand, depending on the value of the applied filed. Thus for $H > H_{c2}$ the linear spin wave theory of section 4.2 reproduces the experimental data quite well along the ⟨100⟩, ⟨110⟩ and ⟨001⟩ directions, taking into account the effect of the field by the substitution in equation (13) of $U(\mathbf{q}, 0)$ by $U(\mathbf{q}, H) = U(\mathbf{q}, 0) + g_\perp \mu_B H/S$, with the following values for the exchange and anisotropy parameters:

$$J_1/k = 38 \pm 3\,\mathrm{K}, \quad J_2/k = 1.3 \pm 0.3\,\mathrm{K}, \quad J_3/k = -10 \pm 1\,\mathrm{K}$$

$$J_{\parallel}/J_\perp = 0.4 \pm 0.03, \quad |z'J'/z\tilde{J}| \sim 10^{-3}\text{—}10^{-4}, \quad H_a^{\mathrm{Ip}} \ll 1\,\mathrm{kOe}$$

where $\tilde{J} = -E_1(\mathbf{k})/3S^2 \simeq 30 \pm 3$ K is the effective exchange previously defined. So in $BaCo_2(AsO_4)_2$ the exchange interaction $J_1$ is dominant and ferromagnetic, whereas the interaction $J_3$ is negative with $J_3/J_1 \sim -0.26$. This contrasts with the situation in $BaNi_2(PO_4)_2$, for which $J_{1,2,3}$ are negative and $J_3/J_1 \sim 4$. As remarked above, important frustration effects are to be expected in the Co compound, which are probably much less important in the Ni compound. The estimated value of the in-plane anisotropy field is obtained from the experimental accuracy in energy $\Delta\omega/\omega$ at low field, assuming the relation $\omega(q \simeq 0) \sim \sqrt{H}$ : $H_a \ll \Delta H = 2H_x\Delta\omega/\omega \sim 1$ kOe. It is interesting to remark that from the parameters determined above, the classical ferromagnetic state with energy $-E_3/2NS^2 = 3(J_1 + 2J_2 + J_3)$ is found to be the ground state, in contrast to the experiments which give evidence for a helimagnetic ordering. Nevertheless, this discrepancy is minor because the small difference between $E_3$ and $E_1$ (typically $|E_3\text{-}E_1|/E_1 \sim 4\%$) makes this quantity very sensitive to the accuracy in the determination of $J_1$, $J_2$ and $J_3$ and to quantum corrections, which will certainly be important in this system.

Yet the situation is not so simple for fields $H < H_{c2}$. Because of the helimagnetic nature of the ordering, an inelastic neutron scattering experiment will measure, for a given wave vector $\mathbf{q}$, not only the energy $\omega(\mathbf{q})$, but also the energies $\omega(\mathbf{q} + \mathbf{k}_i)$ and $\omega(\mathbf{q} - \mathbf{k}_i)$, where $\mathbf{k}_i$ is the wave vector associated with the ordering in the domain $i$ [96, 105]. Such effects have been clearly seen in the more classical 3-d helimagnet $NiBr_2$ [106]. For a continuous symmetry and in the case of isotropic exchange interactions, the theory predicts a linear Goldstone mode around $\mathbf{q} = k_i$. Figure 19 summarizes the expected excitation spectrum of $BaCo_2(AsO_4)_2$, using the exchange and anisotropy parameters as previously determined. So along the $\langle 100\rangle$ one must observe the excitations along [100] for the domain $K_1$ with the wave vector $\mathbf{k}_1 = [0.27, 0]$ and the excitations along [010] ([1$\bar{1}$0]) for the domain $K_2(K_3)$ with the wave vector $\mathbf{k}_2 = [0, 0.27]$ ($\mathbf{k}_3 = [0.27, -0.27]$) deduced from $\mathbf{k}_1$ by a rotation of $2\pi/3$ $(-2\pi/3)$. The calculation of the dynamical structure factor [96, 98] indicates that the out-of-plane fluctuations with energy $\omega(\mathbf{q})$ must be unobservable, whereas the in-plane fluctuations are strong around $\mathbf{q} = k_1$ only.

Clearly, the experimental spectrum as shown in Figure 20 differ both in qualitative and in quantitative respect with these predictions. The excitations are found to be only weakly field-dependent, in particular below $q < 0.2$ r.l.u., where the dispersion curves are isotropic when the wave vectors are expressed in Å$^{-1}$ unit. But the most important feature is the existence of a gap $\Delta_0(q = 0, H = 0) \simeq 1.45$ meV $\simeq 0.35$ THz; this value is consistent with the energy $\omega(\mathbf{q} = \mathbf{k}_1)$ obtained from the simple spin wave theory. The scattering intensity is maximum at scattering vectors $\mathbf{Q} = [0, 0, Q_z]$, but not at $\mathbf{Q} = [k_x, 0, Q_z]$, and decreases rapidly when $q$ is increased. This behaviour is observed even in the ferrimagnetic state when the sample is in a single-domain state. We have to conclude that the low-$q$ part of the spectrum is associated with strong XY fluctuations, and corresponds in some sense to excitations centered around $q = 0$ but not around the wave vector of the helimagnetic ($k_x = 0.27$) or ferrimagnetic structure ($k_x = 1/3$). In particular no effect of domains is observed in the helimagnetic state and no linear branch emerging from $\mathbf{q} = \mathbf{k}_i$ has been detected, although the zero-field specific heat gives evidence for a large $T^2$-contribution at low temperature. Assuming linear but anisotropic dispersion curves: $\omega(\mathbf{q}) = \omega(q, \theta) = c(\theta)q$, where $\theta$ is the polar-angle, the specific heat at low temperatures is approximately given by the relation:

$$C_m \simeq 52.3\, R\, \overline{(1/c^2)}(kT)^2 = AT^2 \tag{21}$$

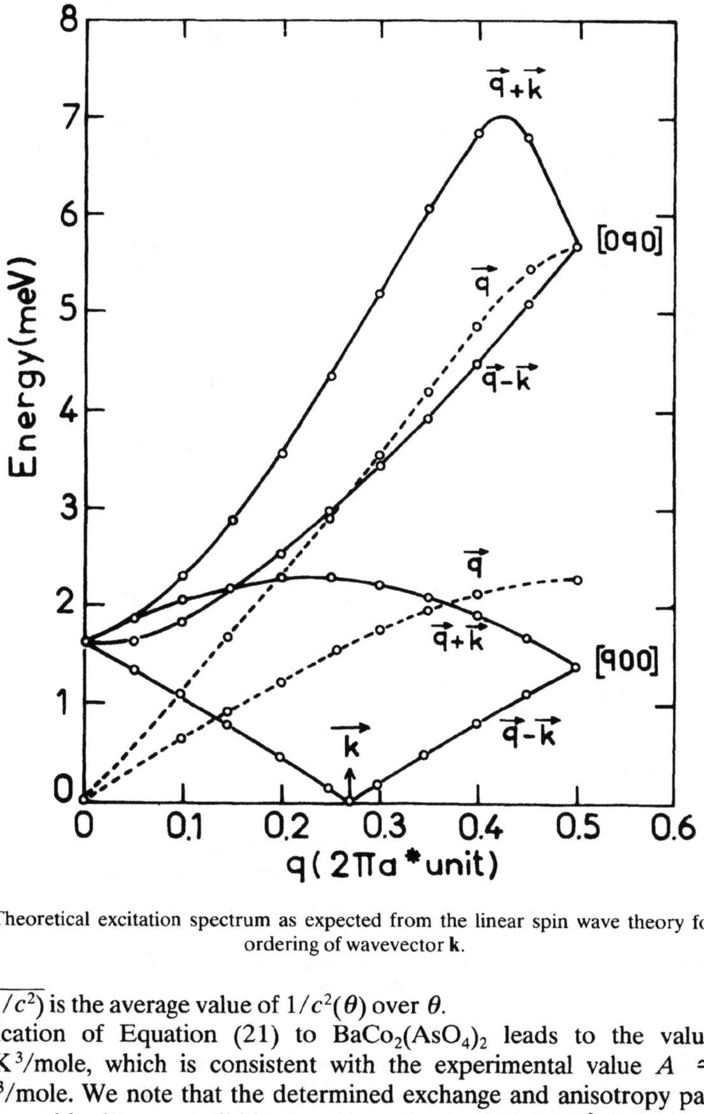


Fig. 19. Theoretical excitation spectrum as expected from the linear spin wave theory for a helical ordering of wavevector **k**.

where $\overline{(1/c^2)}$ is the average value of $1/c^2(\theta)$ over $\theta$.

Application of Equation (21) to $BaCo_2(AsO_4)_2$ leads to the value $A \simeq 36$ mJ/K$^3$/mole, which is consistent with the experimental value $A \simeq 27 \pm 2$ mJ/K$^3$/mole. We note that the determined exchange and anisotropy parameters also agree with the susceptibility result $\chi_\perp(0) \simeq 2N\,(g\mu_B)^2/6(J_1 + 2J_2 + J_3)(1 - J_\parallel/J_\perp)$. So the fact that the linear modes are not seen by the inelastic neutron scattering (INS) experiments may be explained by assuming that these modes become rapidly overdamped.

Whatever may be the explanation, it is clear that the dispersion curves of Figure 20 are not alone sufficient to explain the magnetic specific heat at low temperature. As shown in Figure 15 such a spectrum gives much too small a contribution. Moreover it is worth noting that the experimentally observed gap $\Delta_0$ cannot be

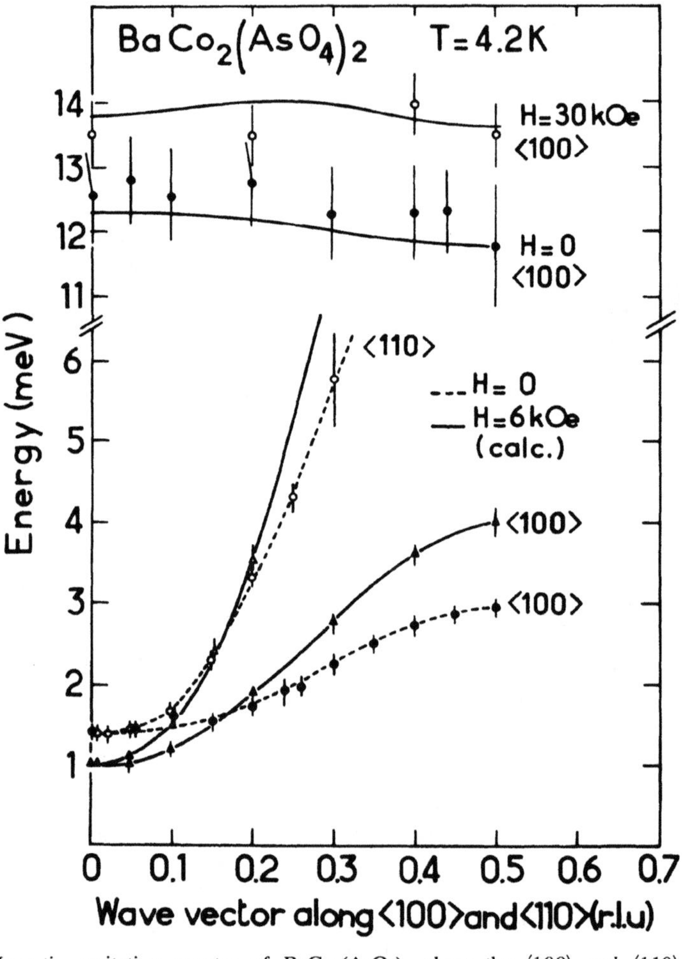


Fig. 20. Magnetic excitation spectra of $BaCo_2(AsO_4)_2$ along the ⟨100⟩ and ⟨110⟩ reciprocal directions.

explained by the presence of in-plane anisotropy, since this would need the introduction of an in-plane anisotropy field $H_a \sim 15$ kOe ($H_a/H_e \sim 5 \times 10^{-2}$), much larger than the upper limit derived previously. Also, such an anisotropy is inconsistent with all the characteristics described before (helimagnetism ordering, large magnetic susceptibility and specific heat, etc.). All these facts suggest the existence of another mode of magnetic excitations with a small energy, or at least indicate that the excitation spectrum is much more complicated than that given by the simple spin wave theory.

In this respect, it is interesting to note that the presence of several exchange interactions of different signs in $BaCo_2(AsO_4)_2$ may in fact lead to the presence of low-energy excitations or defects, which may very well be responsible for the observed strong fluctuations and which may modify the 'simple' mechanism involved in the KT transition. As mentioned above, the energies of the phases observed in the $H$—$T$ phase diagram are given by relations of the type: $-E_i/2NS^2 \simeq 2(J_1 + J_2) + e_{ci}$, where the first term is related to the 'intrachain' energy and $e_{ci}$ is related to the 'interchain' energy. Since $e_{ci}/2(J_1 + J_2)$ is of order of 1/10—1/20, the energy of a vortex, such as that in Figure 21, is considerably reduced, and this vortex can actually be viewed as a stacking of weakly coupled $\pi$-solitons [95, 107]. In the classical approximation such nonlinear excitations have an energy $E_0 \simeq \pi(8(J_1 + 4J_2)e_c)^{1/2}S^2 \sim 30$ K, which is of the same order of magnitude as the gap observed at $q = 0$ in the INS experiments ($\Delta_0 \sim 17$ K), or as the activation energy detected in the specific heat ($\Delta \sim 30$ K). Quantum corrections are generally assumed to reduce the classical value of $E_0$ considerably [108]. One would conclude that the excitation of such types of vortex in $BaCo_2(AsO_4)_2$ is in fact much more favourable than the conventional vortex excitations in 2-d XY systems with nearest-neighbour exchange interactions only. The vortices as sketched in

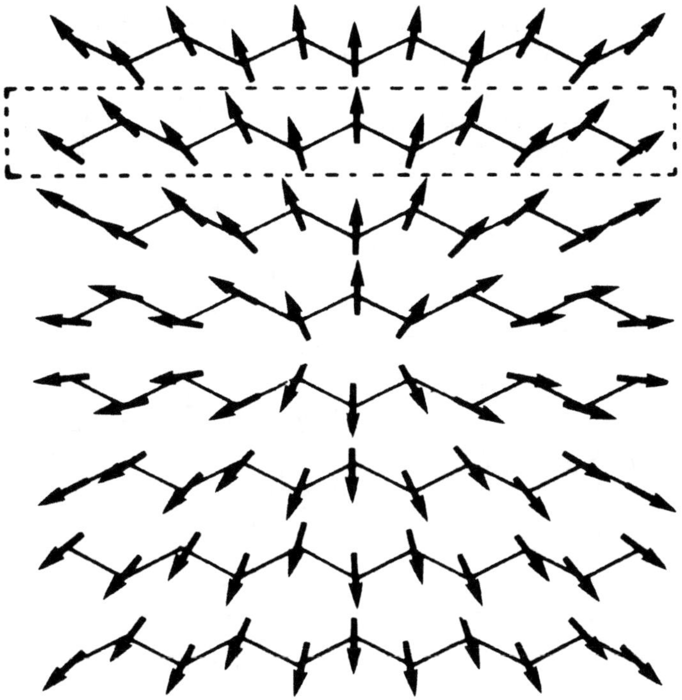

Fig. 21. A possible ferromagnetic vortex in $BaCo_2(AsO_4)_2$. The analogy between a vortex and a $\pi$-soliton is shown.

Figure 21 are anisotropic and they grow preferentially along the direction perpendicular to the ferromagnetic chains, i.e. along the wave vector **k** of the structure. In analogy with the superfluidity of $^4$He [45], they can be viewed as a nucleation of $\pi$-solitons. Note that below $T_c$ the formation of soliton—antisoliton pairs is much more probable than that of isolated solitons, as a consequence of the interchain order that is established.

Further investigations include the measurement of the temperature dependence of the magnetic excitation spectrum. In contrast to $BaNi_2(PO_4)_2$ only a very small renormalization of the energy with temperature occurs, as shown in Figure 22. The main effect of the temperature is seen in the damping and in the behaviour of the intensities of these excitations which reflect the sudden disappearance of the staggered magnetization of the helimagnetic phase. The excitations at $q \simeq 0$ are found to disappear quite rapidly in between 5 K and $T_c \simeq 5.3$ K, without any important shift in their energies up to at least 0.2—0.1 K below $T_c$. Unfortunately

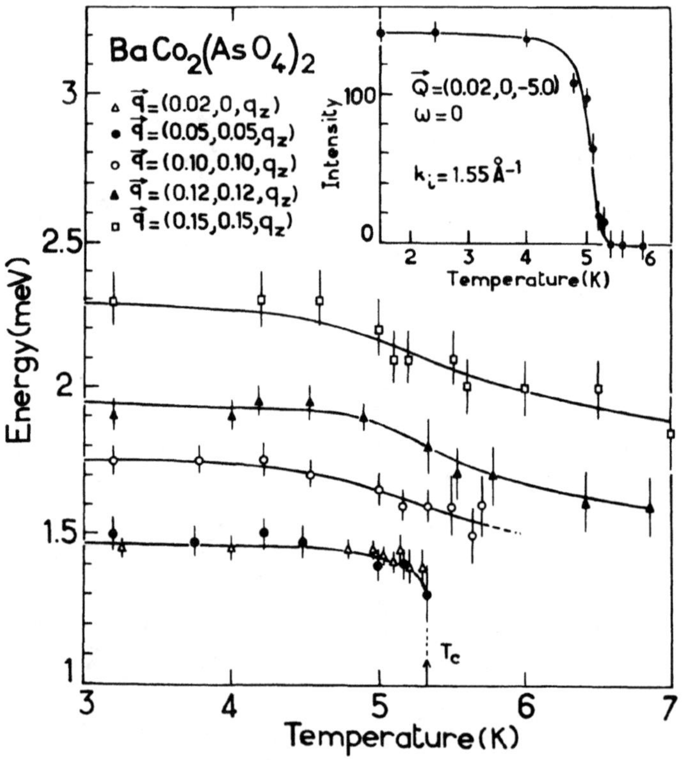


Fig. 22. Energies of magnetic excitations of small wavevectors as a function of temperature for $BaCo_2(AsO_4)_2$. Solid lines are guides to the eye. The intensity maximum is reported in the insert as a function of temperature.

the strong damping near $T_c$ prohibits following the renormalization just below $T_c$ at low wave vector. Nevertheless the observed 'step-like' dependences are qualitatively consistent with the predictions of Nelson and Kosterlitz [52] concerning the discontinuous disappearance of the stiffness constant $\rho_s$ at $T_{KT}$ (recall that in the hydrodynamic limit $\omega(\mathbf{q}) \sim \sqrt{\rho_s}|\mathbf{q}|$). The excitations of larger wave vectors ($q > 0.2$) are found to persist above $T_c$, but are slightly renormalized, with a damping that increases as $T$ increases and $|\mathbf{q}|$ decreases. These results agree with the observed relatively fast decrease of the correlation length above $T_c$.

In parallel with the abrupt disappearance of the 'collective excitations', a quasielastic peak centered around $q = 0$ and $\omega = 0$ is found to grow, with an energy-integrated intensity that is maximum at 6—6.5 K. This corresponds roughly with the temperature of the susceptibility maximum. Figure 23 shows typical scans in energy performed at $T = 2.5$ K and $T = 6$ K. One observes that the central peak (a 2-d basic feature) persists at low temperatures (in the 'ordered' regime) where it has a resolution-limited width in energy. On the other hand it is considerably broadened when $T$ approaches $T_c$. Fits of the experimental data have been obtained assuming a Lorentzian function [138]:

$$S(\omega) \sim \frac{\omega}{1 - \exp(-\omega/T)} \frac{\Gamma_\omega}{\Gamma_\omega^2 + \omega^2} \tag{22}$$

where the first term is the balance factor. These fits provide quantitative information on the temperature dependence of $\Gamma_\omega$ (see the insert of Figure 23). Typical

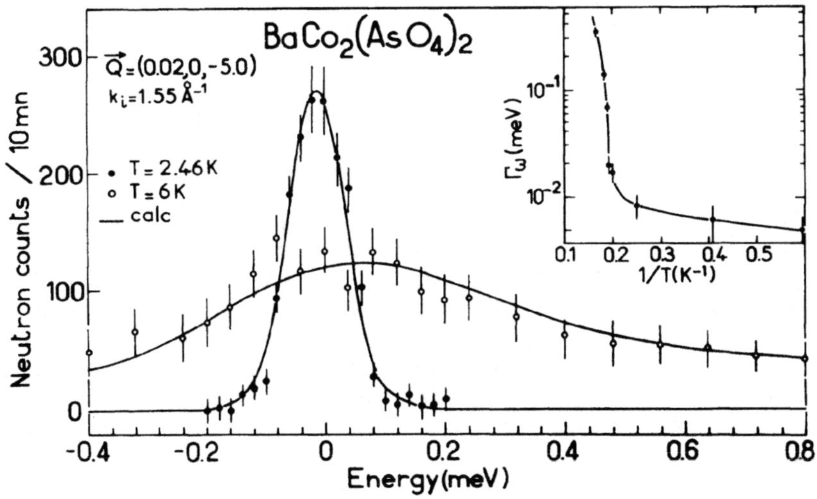


Fig. 23. Scans in energy at $T = 2.5$ K and $T = 6$ K at $\mathbf{Q} = [0.20, 0, -5]$. The solid lines are fits to the Lorentzian function (22). The energy width $\Gamma_\omega$ deduced from the fits is reported in insert as a function of the inverse of the temperature.

values are $\Gamma_\omega \simeq 0.08$ meV $\simeq 0.1$ K at $T = 2.5$ K and $\Gamma_\omega \simeq 0.35$ meV $\simeq 4.2$ K at $T = 6$ K. The main effect of the temperature is to induce a fast increase of $\Gamma_\omega$ above $T_c$. We note that such effects are not observed around $\mathbf{q} = \mathbf{k}$, where the scattering always remains resolution-limited in energy. Anyway these results suggest the existence of fast XY-type fluctuations around $q = 0$ above $T_c$ which are not well explained by the present dynamical theories [69, 70]. Nevertheless it is suggested that $\Gamma_\omega$ could be connected to the density of free vortices $n_v$ and to the in-plane correlation lengths $\xi_\perp$ and $\xi_\parallel$ following the relations:

$$\Gamma_\omega \sim \bar{v}\, n_v$$

$$n_v \sim 1/\xi_\perp \xi_\parallel$$

where $\bar{v}$ represents a characteristic thermal velocity which varies qualitatively as the inverse of the creation energy of the vortices. This is another indication for the existence of low-energy excitations in $BaCo_2(AsO_4)_2$. Recent calculations of the dynamic structure factors for the 2-d easy plane ferromagnet [136] suggest a squared Lorentzian profile instead of Equation (22):

$$S^{xx}(\mathbf{q}, \omega) \sim \gamma^3 \xi^2/[\omega^2 + \gamma^2(1 + (\xi q)^2)]^2$$

where $\gamma \simeq \sqrt{\pi}\bar{v}/\xi$, with $\xi$ the correlation length and $\bar{v}$ the thermal average velocity of a vortex gas.

Such a relation explains quantitatively the scattering observed experimentally above $T_c$. It seems that the proximity of the ferromagnetic and paramagnetic phases (Lifchitz point) strongly modify the behaviour of the system [139]. Yet elastic neutron scattering experiments under pressure up to 20 kbars [140] have shown no modification in the phase transition in $BaCo_2(AsO_4)_2$: we observe in particular the same wave vector $\mathbf{k}$, the same $T_c$ and the same step-like behaviour.

## 6. Magnetic Properties of $BaNi_2(AsO_4)_2$

### 6.1. STATIC PROPERTIES

Similar to the former two compounds the temperature dependence of the susceptibilities $\chi_\perp$ and $\chi_\parallel$ reported in Figure 24 give evidence for a strong planar anisotropy [109], intermediate between that of $BaCo_2(AsO_4)_2$ and $BaNi_2(PO_4)_2$. The transition temperature is found at $T_N = 19 \pm 2$ K from the inflexion point in the $\chi - T$ curve, whereas $T(\chi_{max}) \simeq 32$ K, as in $BaNi_2(PO_4)_2$. Proceeding in a similar way as before, order of magnitude estimates of the AF exchange interactions and of the planar anisotropy can be obtained. From $\chi_\perp(T)$ and $\chi_\parallel(0)$ one finds $\tilde{J}/k \sim -(6-7)$K and $|D/\tilde{J}| \sim 1$, values which are comparable to those of $BaNi_2(PO_4)_2$.

As in the former compounds the magnetic specific heat (reported in Figure 25) exhibits a sharp $\lambda$-type anomaly at $T_N \simeq 19.2 \pm 0.5$ K and the usual 2-d behaviour above $T_N$, characteristic of strong correlations. At much lower temperature ($T < 6$ K) the total (i.e. magnetic and lattice) specific heat can be adjusted to

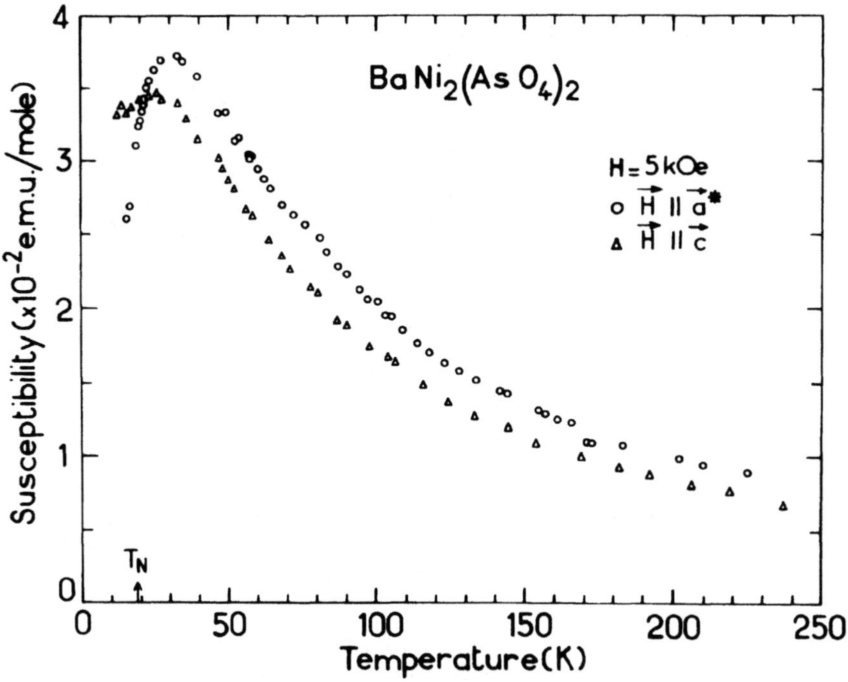


Fig. 24. Susceptibilities perpendicular ($\chi_\perp$) and parallel ($\chi_\parallel$) to the $c$-axis as a function of temperature for $BaNi_2(AsO_4)_2$.

the relation (Figure 26):

$$C = AT^2 + BT^3 + DT^4 \tag{23}$$

with $A \simeq$ 2.1 mJ/K$^3$/mole, $B \simeq$ 0.7 mJ/K$^4$/mole and $D \simeq$ 0.4 mJ/K$^5$/mole. In this expression the $T^3$-term arises as usual from the lattice contribution and has about the same value as in $BaNi_2(PO_4)_2$. But the physical meaning of the term in $T^4$ is questionable. More realistic fits can be obtained using functions of the type:

$$C = AT^2 + BT^3 + D(T)\exp(-\Delta/T) \tag{24}$$

where $B$ represents as before the magnitude of the lattice contribution and where the exponential term could arise from an 'optical' mode with a dispersion curve $\omega(q) = \Delta + \delta\, q^2$, for example. The experimental data give the values $A \simeq 2.8 \pm 0.3$ mJ/K$^3$/mole and $\Delta \sim$ 14 K. This corresponds to about the same value for the prefactor of the quadratic term.

The elastic neutron scattering experiments confirm an antiferromagnetic ordering below $T_N \simeq 18.7 \pm 0.3$ K, with a magnetic structure characterized by the wave vector $\mathbf{k} = [1/2, 0, 1/2]$. The coupling between the two sublattices is antiferromagnetic ($\phi = \pi$) and the magnetic moments ($m_0 \simeq 2 \pm 0.2\mu_B$) are parallel to the $b$ axis (i.e. parallel to the 'chains'), which is probably due to the dipolar

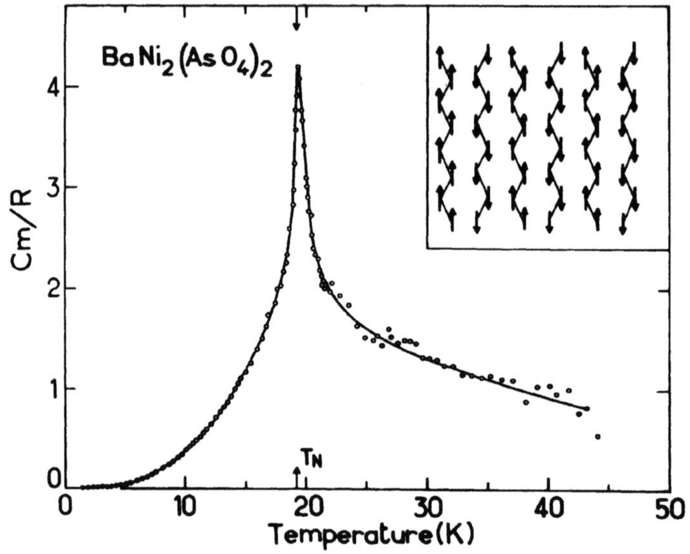


Fig. 25. The magnetic specific heat of $BaNi_2(AsO_4)_2$ as a function of temperature. The insert shows the antiferromagnetic structure within the (001) plane.

anisotropy, estimated to the about 0.4 K. This magnetic structure (Figure 25) can be considered as a stacking of ferromagnetic chains, coupled antifierromagnetically, a situation quite similar to that found in $BaCo_2(AsO_4)_2$, but different from that of $BaNi_2(PO_4)_2$ in the sense that some frustration exists in this AF-structure, which has the classical energy $-E/2NS^2 = 2J_2 - J_1 - 3J_3$.

The temperature dependence of the staggered magnetization (Figure 27) shows a different behaviour. For temperatures such that $1 - T/T_N \gtrsim 10^{-2}$, $|\langle \mathbf{S}_k \rangle|$ follows a power law $|\langle \mathbf{S}_k \rangle| \sim (1 - T/T_N)^{\beta}$ with an exponent $\beta \simeq 0.135$ which is very close to the critical behaviour of the 2-d Ising model, which has $\beta = 1/8$. This Ising-like critical behaviour is also obtained from the in-plane correlation length (Figure 28) which varies as $\xi(T) \sim (T/T_N - 1)^{-\nu}$, with $\nu \simeq 1$, and from the value of the exponent $\eta = 2\beta/\nu \simeq 0.27 \pm 0.05$ ($\nu = 1$ and $\eta = 1/4$ for the 2-d Ising model). So, quite surprisingly, the critical regime of $BaNi_2(AsO_4)_2$ seems to be essentially of the 2-d Ising-type. On the other hand the KT-expression (4) for the in-plane correlation length also fits the experimental data quite well, with parameters $A \sim 0.8$ Å, $B \sim 1.5$ and $T_{KT} \sim 0.95\ T_N$ which are consistent with the theoretical values as in $BaNi_2(PO_4)_2$. It is therefore difficult to say whether the 2-d Ising-like character is intrinsic or fortuitous. In particular, this Ising-like character is in contradiction with the $T^2$-contribution observed in the low-temperature magnetic specific heat, but nevertheless is actually predicted for the frustrated 2-d, XY, AF triangular model [63].

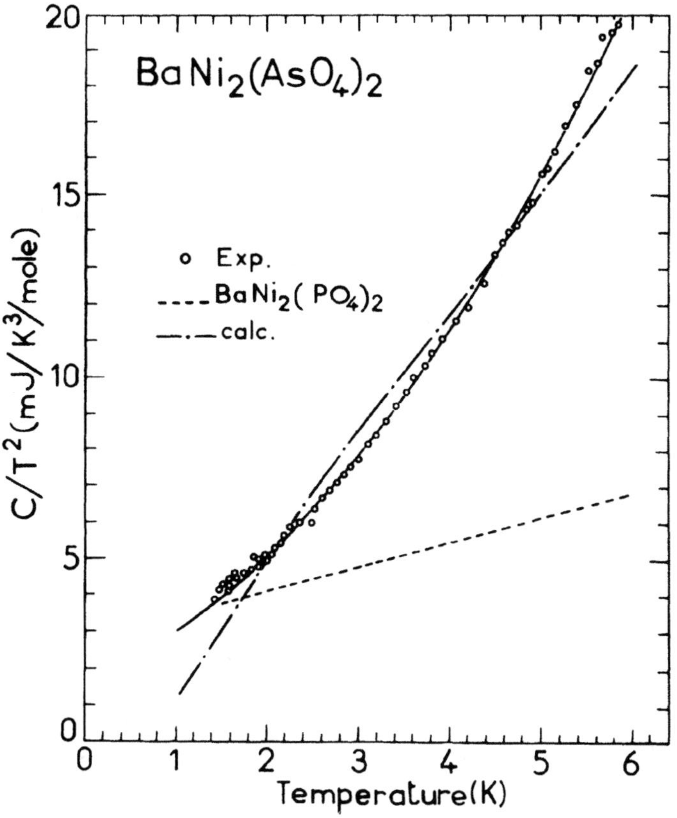


Fig. 26. $C/T^2$ as a function of temperature. The solid line corresponds to a fit to the relation (23). The dot-dashed line is the spin wave contribution and the broken line is the $C/T^2$ contribution for the $BaNi_2(PO_4)_2$ compound.

## 6.2. DYNAMIC PROPERTIES

As mentioned above, the determination of the excitation spectrum is fundamental to an understanding of the physical properties of magnetic materials because it gives a direct determination of the exchange and anisotropy parameters. We have also seen in Section 4 that such a determination can be quite complicated in case of magnetic structures characterized by nonzero wave vectors **k**, or by the existence of multidomain structures. This is typically the case of $BaNi_2(AsO_4)_2$, for which the dispersion curves are shown in Figure 29. We must bear in mind that in an inelastic neutron scattering experiment one observes for a given wave vector **q** simultaneously the excitation $\omega(\mathbf{q})$ and $\omega(\mathbf{q} \pm \mathbf{k})$ for each domain. In $BaNi_2(AsO_4)_2$, four curves must be observed, as indicated in Figure 29, which result from the simple spin wave theory of section 4, using the following parameter

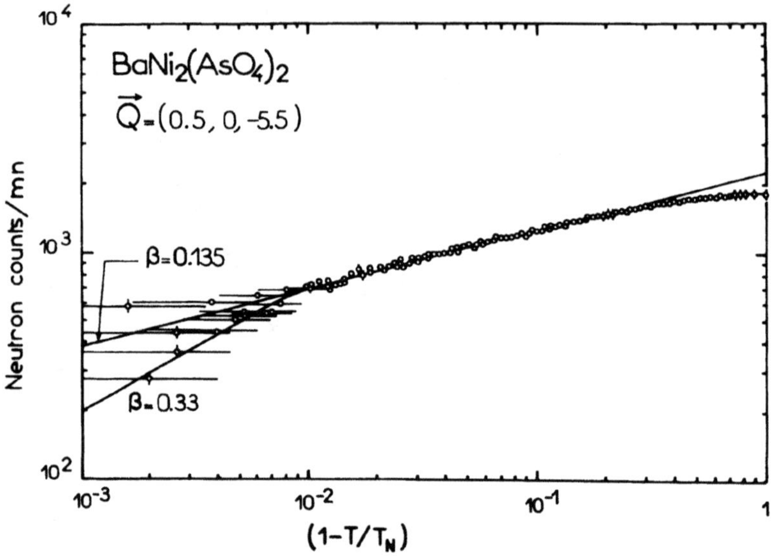


Fig. 27. Intensity of the magnetic superlattice peak $[006]^+$ as a function of temperature for $BaNi_2(AsO_4)_2$.

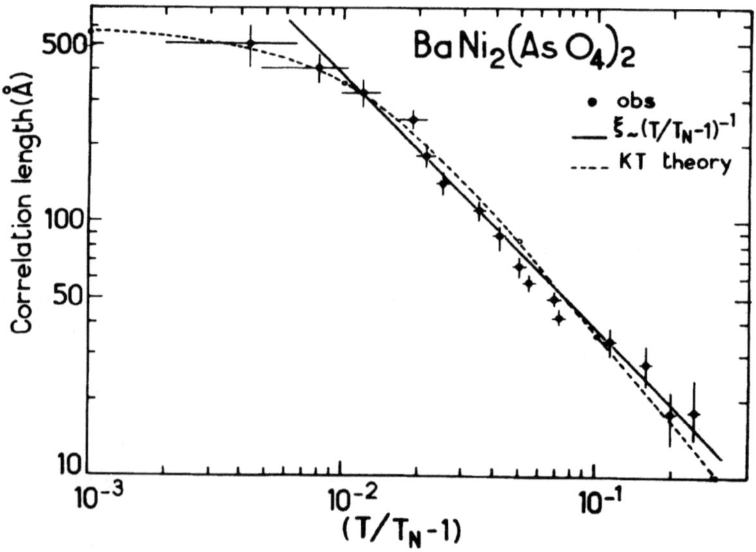


Fig. 28. In-plane correlation length as a function of temperature for $BaNi_2(AsO_4)_2$.

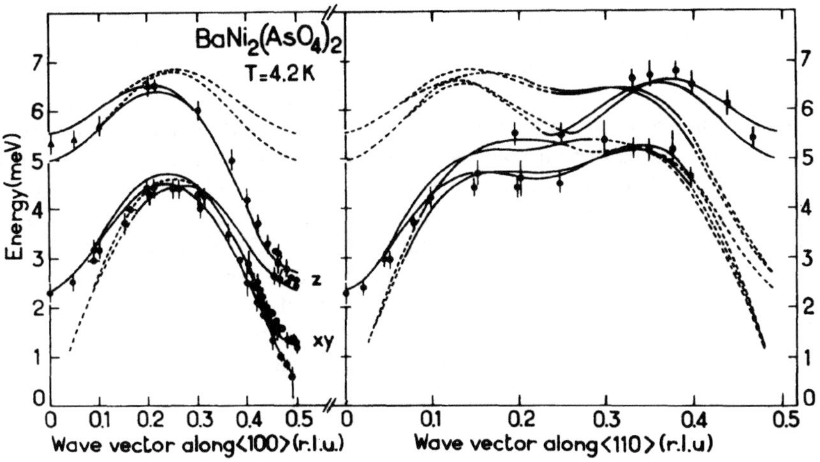


Fig. 29. Magnetic excitation spectra of $BaNi_2(AsO_4)_2$ along the $\langle 100\rangle$ and $\langle 110\rangle$ directions. The solid and broken lines are fits to the linear spin wave theory.

values:

$$J_1 \simeq 3.4\,\mathrm{K}, \quad J_2 \simeq 0.2\,\mathrm{K}, \quad J_3 \simeq -10\,\mathrm{K}, \quad |z'J'/z\tilde{J}| \leqslant 10^{-3},$$
$$D \simeq 7.0\,\mathrm{K}$$

and an effective exchange $\tilde{J}/k \simeq -9$ K. The agreement with the experiment is quite good, except around $\mathbf{q} = [1/2, 0]$ where the experimental spectrum displays gaps at 0.6 meV ($\simeq 7$ K) and 1.2 meV ($\simeq 14$ K). The first one $[\omega(0, 0)]$ is associated with fluctuations parallel to the $a^*$-axis and is attributed to the existence of a dipolar anisotropy term, simulated in the fits by the Hamiltonian $\mathscr{H}_{\mathrm{IP}} \simeq -D_{\mathrm{IP}}\Sigma_j(S_j^y)^2$ with $D_{\mathrm{IP}} \simeq 0.2$—$0.25$ K ($H_{\mathrm{a}}^{\mathrm{IP}} \simeq 3$—$4$ kOe). The gap at 1.2 meV $[\omega(1/2, 1/2)]$ is associated with in-plane fluctuations and is in fact more difficult to understand. From the simple spin wave theory one expects to observe the XY-type fluctuations of domains $K_2$ and $K_3$ at $\mathbf{q} = [1/2, 0]$ with about the same energy as for the domain $K_1$. A more appropriate treatment of the dipolar term is certainly necessary to explain the difference between the two observed values. Nevertheless the specific heat computed from this spectrum is in qualitative agreement with the experimental determination (Figure 26). Note that no $T^2$ term is predicted in the specific heat at low teperature and that one expects effectively an exponential term with an energy of activation of the order of 8—14 K.

Moreover it is worth noting that the actual reduced transition temperature of $BaNi_2(AsO_4)_2$, $T_N/|\tilde{J}|S^2 \simeq 2.0$, is even smaller than for $BaNi_2(PO_4)_2$, $T_N|\tilde{J}|S^2 \simeq 2.3$. This is partly in contradiction with the existence of a strong in-plane anisotropy, or at least suggests that the dipolar anisotropy is not too relevant for the fluctuations which occur in the vicinity of the transition (remember that $H_a/H_e \sim 10^{-4}$ in $BaNi_2(PO_4)_2$, whereas $H_a/H_e \sim 10^{-2}$ in $BaNi_2(AsO_4)_2$). The situation is very different from that observed in $CoCl_2 \cdot 6\,H_2O$ for example, for which an in-plane anisotropy as large as $H_a/H_e \sim 4 \times 10^{-2}$ induces a phase transition at $T_N/|J|S^2 \sim 4$ [1]. It is therefore suggested that $BaNi_2(AsO_4)_2$ could also be a good prototype of the 2-d XY model for temperatures $T \gtrsim T_N$. Unfortunately no information on the temperature dependence of the excitations has been obtained as yet. We mention that the almost isomorphous compound $KNiAsO_4$ displays properties very similar to that of $BaNi_2(AsO_4)_2$. In particular the antiferromagnetic ordering at low temperature (wave vector $\mathbf{k} = [1/2,0,1]$ and phase $\phi = \pi$) and the transition temperature ($T_N \simeq 19.5$ K) are almost the same [85], in spite of a probably much smaller (negative) value of $J'$, which results from the larger $d_2/d_1$ value for this compound ($d_2/d_1 \sim 3.3$). This demonstrates again that the interlayer coupling is of minor importance for the phase transition in the XY plane. The same kind of magnetic structure is also found in the planar system $NiPS_3$ below $T_N \simeq 160$ K [86]. Unfortunately, no precise determination of the exchange and anisotropy parameters has been obtained for that compound, so that a comparison with the present data is not possible.

## 7. Magnetic Properties of $BaCo_2(PO_4)_2$

Although the room temperature structure of $BaCo_2(PO_4)_2$ is isomorphous to that of $BaCo_2(AsO_4)_2$, it exhibits a crystallographic phase transition at about 760°C from the low-temperature $BaMg_2(AsO_4)_2$-type trigonal phase, to a hexagonal $Ba(ZnPO_4)_2$-type phase, with cell parameters $a \simeq 5.208$ Å and $c \simeq 8.22$ Å [83], and in which the $Co^{2+}$ ions are now in tetrahedral sites. Upon cooling this hexagonal phase becomes unstable at room temperature and transforms into an unknown metastable monoclinic phase, which eventually returns to the initial phase after appropriate thermal treatment [83]. The existence of these phase transitions prohibits the growth of large single crystals of $BaCo_2(AsO_4)_2$. Consequently all the present studies have been performed on polycrystalline samples [83].

The susceptibility shows a behaviour very similar to that of $BaCo_2(AsO_4)_2$, also with probably a strong planar character. Although no quantitative informations about the strength of the anisotropy can be obtained, one may estimate the exchange and anisotropy parameters to be not far from those of $BaCo_2(AsO_4)_2$. In fact the neutron scattering study at low temperature reveals the presence of superlattice peaks that can be indexed by the wave vectors $\mathbf{k}_1 = [0.25_2, 0, 0.13]$ and $\mathbf{k}_2 = [1/2, 0, k_z]$ ($k_z = 0$ and 1), which are the same as observed in $BaCo_2(AsO_4)_2$ and $BaNi_2(AsO_4)_2$, respectively. There is no clear explanation for such a coexistence, but it requires that the energies of both structures are close. Assuming $J_2/J_1 \sim 0$, this yields the ratio $J_3/J_1 \sim -0.4$. For the structure associated with $\mathbf{k}_1$ strong fluctuations are expected, due to the very weak coupling which exists between the ferromagnetic chains. These stronger fluctuations in the helical structure probably explain why the transition temperature $T_{c1} \simeq 3.8$ K has a smaller value than in the antiferromagnetic phase $T_{c2} \simeq 7$ K. However the ratio $T_{c1}/T_{c2} \sim 0.54$ is consistent with the ratio of $t = T_c/|J|S^2$ values obtained for $BaCo_2(AsO_4)_2$ and $BaNi_2(AsO_4)_2$, which is of the order of 0.4. An estimate of $J_1$ can be obtained by assuming that the relation $t_{c2} \simeq 2.0$ remains valid. One gets $J_1 \sim 20$ K, which is about half as large as for $BaCo_2(AsO_4)_2$.

The magnetic specific heat does not exhibit the same behaviour as the previous compounds. The most remarkable feature is the absence of a $\lambda$-type anomaly. As shown in Figure 30, the specific heat shows a bump around $T(C_{max}) \simeq 5.5$ K $(\simeq (T_{c1} + T_{c2})/2)$, preceded on the high-temperature side by a slowly varying contribution characteristic of strong high-temperature correlations. Note that strong 2-d correlations are also seen up to 40 K in the quasielastic neutron scattering. Below $T \simeq T_{c1}$ the specific heat is found to decrease very rapidly, much faster than in $BaCo_2(AsO_4)_2$. Its dependence on temperature is described by relation (18), with $A \simeq (33 \pm 2)$mJ/K$^3$/mole and $\Delta \simeq 25$—27 K. A comparison of the experimental parameters of both cobalt-compounds is reported in Table II. It is difficult to account for the coefficient $A$ because the proportion of phases 1 and 2 is not very well known. Nevertheless the above estimated parameters give a good order of magnitude estimate for the activation energy calculated from the energy of a $\pi$-soliton to be $\Delta \simeq \pi\sqrt{8}\sqrt{J_1 e_c}\, S^2 \sim 25$ K. As in $BaCo_2(AsO_4)_2$, these strong fluctuations at low energies are thought to be responsible for the relatively small value of the reduced transition temperature $t_{c1} \sim 1$, which is about half as large as the classical 2-d planar value. They also explain the qualitatively more rapid disappearance of magnetic peaks in the helical structure than in the anti-

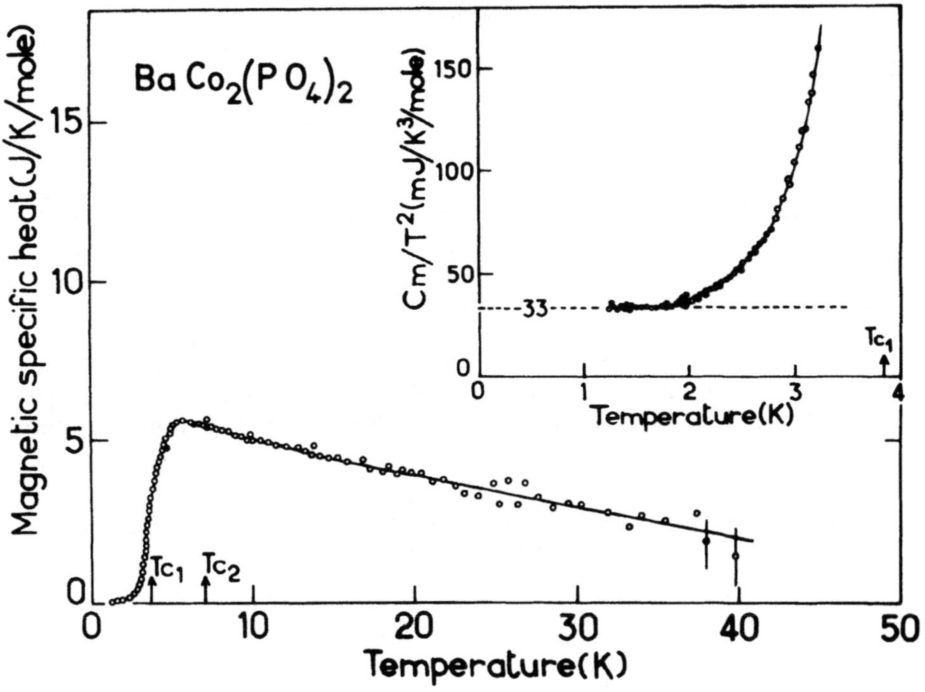


Fig. 30. The magnetic specific heat of $BaCo_2(PO_4)_2$ as a function of temperature. $C_m/T^2$ is given in insert as a function of temperature. The solid lines are guides to the eye.

TABLE II
Experimental parameters of the specific heat assuming a magnetic contribution $C_m = AT^2 + BT^{-n}\exp(-\Delta/T)$.

| Compound | $A$ (mJ/K$^3$/mole) | $B$ (mJ K$^{n-1}$/mole) | | $\Delta$ (K) | |
|---|---|---|---|---|---|
| | | $n = 1$ | $n = 2$ | $n = 1$ | $n = 2$ |
| $BaCo_2(PO_4)_2$ | 33 ± 2 | 56 | 250 | 25 | 27 |
| $BaCo_2(AsO_4)_2$ | 27 ± 2 | 19 | 160 | 27 | 31 |

ferromagnetic structure. Unfortunately, studies on polycrystalline samples do not in general allow a precise analysis of the temperature dependence of the data in the vicinity of the transition. In particular it is not possible to say if the order parameter $|\langle \mathbf{S}_k \rangle|$ decreases as steeply as in $BaCo_2(AsO_4)_2$, and no information exists about the spin dynamics in $BaCo_2(PO_4)_2$.

## 8. Other Experimental Realizations of the 2d-Planar Model

In this section we shall discuss some other magnetic systems which have been considered as possible experimental examples of the 2-d XY model. As pointed out already, in most cases either the 2-d character or the planar anisotropy are not pronounced enough to make them really good examples.

### 8.1. $K_2CuF_4$

In many respects $K_2CuF_4$ can be considered as an excellent realization of the 2-d, $S = 1/2$ Heisenberg ferromagnet, rather than as an example of the 2-d XY model. The planar anisotropy in this compound is only 1% and the symmetry of the (weak) in-plane anisotropy corresponds to $p = 4$ (tetragonal system) [110—115]. An extensive description of the properties of $K_2CuF_4$ is given in the chapter by Hirakawa and Ikeda in this book, to which we refer the reader for additional information. In Table III we have collected some relevant parameters for comparison with the former compounds and with the theoretical predictions. In spite of the strong Heisenberg character, which is clearly evident from the zero-field

TABLE III
Characteristics of some quasi 2-d planar magnets: $K_2CuF_4$ and the $NiCl_2$, $CoCl_2$ stage-2 graphite intercalated compounds.

| Compound | $T_c$(K) | $J$(K) | $S$ | $H_a^{ip}/H_e$ | $H_a^{op}/H_e$ | $T_c/\|J\|S^2$ | $\|J'/J\|$ | $p$ |
|---|---|---|---|---|---|---|---|---|
| $K_2CuF_4$ | 6.25 | 11.3 | 1/2 | $\sim 10^{-5}$ | $1.2 \times 10^{-2}$ | 2.3 | $10^{-3}$—$10^{-4}$ | 4 |
| $NiCl_2$-GIC (2) | 18—20 | 9—10 | 1 | $\sim 10^{-5}$ | $2 \times 10^{-3}$ | 2.1 | $10^{-4}$—$10^{-5}$ | 6 |
| $CoCl_2$-GIC (2) | 8—9 | ~13 | 1/2 | $\sim 10^{-5}$ | ~0.6 | 2.5 | $\sim 10^{-4}$ | 6 |

specific heat [119] by the existence of a large Schottky-type bump, $K_2CuF_4$ exhibits critical properties which are consistent with the KT picture when the in-plane correlations are sufficiently large [113—115]. In particular the susceptibility and the in-plane correlation length agree quite well with the expression (4) with $T_{KT} \simeq 5.5$ K, which is a little bit smaller than the actual transition temperature $T_c \simeq 6.25$ K (with $T_{KT} \simeq 0.88 T_c$). Quite interestingly the spin dynamics of $K_2CuF_4$ has been studied by the inelastic neutron scattering technique as a function of temperature and wave vector [116—118]. We have mentioned that theoretical or experimental work on the dynamics of 2-d magnets is scarce. For $K_2CuF_4$ it has been shown that if the wave vector is chosen in the range $(J'/J)^{1/2} \lesssim \pi q \lesssim (H_a/H_e)^{1/2}$, which allow a proper definition of the 2-d XY regime, then the magnon frequency undergoes a steep decrease at the transition [118], which is attributed to the jump predicted by Nelson and Kosterlitz. In fact these results are similar to those observed in $BaNi_2(PO_4)_2$, with the difference that in the latter compound the 2-d XY regime is better defined and more extended. Note that the higher $q$ excitations ($q \gg 0.03$) persist above $T_c$, more or less renormalized and damped depending on the relative values of $q$ and $T$. But the system is typically 2-d Heisenberg in this range. At low temperature (that is $T/J < 0.3$) $K_2CuF_4$ behaves like a perfect Heisenberg ferromagnetic system, with a magnetic specific heat linearly dependent on temperature and well accounted for by the linear spin wave theory [41]. The characteristic properties observed in $BaCo_2(AsO_4)_2$ are not seen in $K_2CuF_4$ which behaves in fact quite similar as $BaNi_2(PO_4)_2$ from the point of view of the transition. To our opinion the close analogy between the magnetic behaviour of these two compounds is independent of the ferromagnetic or antiferromagnetic character of the interactions but arises mainly from the fact that, in both compounds, the ordering wave vector ($k = 0$) induces no frustration.

## 8.2. $NiCl_2$ AND $CoCl_2$ GRAPHITE INTERCALATED COMPOUNDS

Graphite intercalation compounds (GIC) provide very interesting systems for the experimental study of the phase transition of the 2-d magnets, since in principle one may control the 2-d character by intercalating more and more graphite layers between adjacent magnetic layers. In practice the situation is not so ideal because the size of crystallographically well correlated 2-d segments is difficult to control. As shown by Flandrois *et al.* [120] a magnetic layer is in fact composed of small islands of about 100—200 Å in diameter, the size of which is very difficult to increase. This poses a serious limitation to a study of the critical behaviour of such systems. In particular the boundary effects are expected to be very important and to change the thermodynamic behaviour dramatically. Also, it is very difficult to obtain single crystals of good quality of GIC compounds, and most experiments are performed on highly oriented pyrolitic graphite (HOPG) samples. Finally, we mention that the estimation of the number of intercalated layers is a difficult problem. This is very important for the analysis of the specific heat measurements, for which an accurate determination of the graphite lattice contribution is needed to get the magnetic one. Nevertheless a lot of experimental studies have been undertaken on GIC compounds, among which the most characteristic for our purpose are the stage 2 $NiCl_2$ and $CoCl_2$-GIC compounds.

$NiCl_2$-GIC

Stage 2 $NiCl_2$-GIC can be regarded as a very good 2-d ferromagnet with spins $S = 1$ in which the $NiCl_2$ are separated by two graphite layers. The characteristic Ni—Ni distances within ($d_1$) and between the layers ($d_2$) are 3.5 Å and 12.5 Å respectively ($d_2/d_1 \simeq 3.6$). But the planar anisotropy is only 0.2% so that the character is mainly of the Heisenberg type. The in-plane anisotropy is also very weak (see Table III) with an hexagonal symmetry corresponding to $p = 6$. In $NiCl_2$-GIC the most remarkable feature is certainly the discovery of two phase transitions from magnetization and specific heat measurements [121—123], namely at $T_{c1} \simeq 20$ K and $T_{c2} \simeq 18$ K. The former transition has been interpreted as a transition from the paramagnetic state to the XY-like phase and the latter as a transition from the XY-like phase to the ferromagnetic phase. This is exactly the prediction of José *et al.* [59]. Moreover Suzuki and Ikeda [123] have reported susceptibility measurements versus field and temperature which are interpreted in terms of Equation (4) assuming $T_{KT} \simeq T_{c2}$ and with experimental parameters consistent with those derived by Tobochnik and Chester [25] from Monte Carlo simulations. Unfortunately the exponential behaviour is observed at $T/T_{c2} - 1 \gtrsim 0.15$, which is certainly outside the critical regime. Indeed, the interpretation in terms of the KT transition is incompatible with recent precise magnetic measurements in very low fields by Suzuki *et al.* [124], who have observed a finite value of the spontaneous magnetization between $T_{c2}$ and $T_{c1}$, in disagreement with the absence of long-range ordering required for the 2-d XY model. This indicates strongly that the state in between $T_{c2}$ and $T_{c1}$ has nothing to do with the topological order of vortex—antivortex pairs. In our opinion the observed phenomena could arise from the utilization of samples for which the coherence length is not large enough. We recall that actually a similar coexistence is also observed in $BaCo_2(PO_4)_2$ for which it has been proved, by elastic neutron scattering measurements, that both transitions are related to long-range ordering.

$CoCl_2$-GIC

Stage 2 $CoCl_2$-GIC is isomorphous to the Ni-compound. But from the pure $CoCl_2$ compound [129], a much stronger planar anisotropy is expected. This compound should in fact approach much closer the 2-d XY model, apart from the already mentioned difficulties (see Table III). As in $NiCl_2$-GIC, a double-peak structure has been detected in the susceptibility measurements [121, 126], with $T_{c1} \simeq 9$ K and $T_{c2} \simeq 8$ K. As in the former Ni-compound, the intermediate state between $T_{c2}$ and $T_{c1}$ was suspected to be the state predicted by Kosterlitz and Thouless. Moreover Elahy and Dresselhaus [130, 131] have reported susceptibility measurements versus temperature which could be also interpreted in terms of the equation $\chi = \chi_0 + \chi_1 \exp[B'/(T/T_{KT} - 1)^{1/2}]$ with $B' = B(2 - \eta) \simeq 1.4$ and $T_{KT} \simeq 8.3$ K in the range $0.14 \lesssim T/T_{KT} - 1 \lesssim 2.5$. But quite similarly to the Ni-compound the magnetic susceptibility and spontaneous magnetization data of Murakami *et al.* [126] give evidence for a finite magnetization below 10 K, in disagreement with such an explanation. In addition the elastic neutron scattering data unambigously show the appearance of long-range ordering below $T_{c2}$ characterized by a wave vector $\mathbf{k} = [0, 0, 1/2]$ [125, 127]. This implies the existence of a relatively well ordered structure made of ferromagnetic layers coupled antiferromagnetically, and definitively excludes the explanation in terms of an intermediate state of the K—T type.

The magnetic specific heat is found to display a broad peak at $T_{c2}$ [127—128, 132], with a shape that seems qualitatively consistent with the Monte Carlo simulations based on the 2-d XY model [132]. On the other hand, strong metastable states could exist, which could result in the appearance of a more or less important $\lambda$-type peak between $T_{c2}$ *and* $T_{c1}$, in addition to the broad maximum [128]. We think that specific heat measurements on samples of much better quality are needed to verify more precisely if the specific heat of the 2-d XY $S = 1/2$ model indeed shows the logarithmic behaviour predicted by De Raedt *et al.* [38]. Nevertheless the order of magnitude of $T_c/JS^2$ found experimentally (about 2.3) is consistent with the theoretical predictions of that model [38], which is probably more adequate to describe the properties of $CoCl_2$-GIC systems than the classical model. Note that the HTS expansions predict a power law for the critical behaviour with $\gamma \simeq 2.5$ and $\nu = 1.4$ for the 2-d XY $S = 1/2$ model [35, 77]. Elahy and Dresselhaus [130] have been successful in fitting their data for $\chi_\perp$ to a function of the form $\chi \sim (T/T_c - 1)^{-\gamma}$, with $\gamma \simeq 2$ for stage 2 $CoCl_2$-GIC and $\gamma \simeq 2.16$ for stage 4 $CoCl_2$-GIC. Neutron scattering experiments are hardly needed to verify if the critical exponents $\beta$ and $\nu$ are anomalous or not. Few is known concerning the excitations in $CoCl_2$-GIC as well as in $NiCl_2$-GIC compounds [143—145]. Finally we mention that recent calculations by D. G. Rancourt [137] explain quantitatively the behaviour of the susceptibilities of these compounds, assuming 2-d superferromagnetism, intraplane and interplane supermoment interactions, and plane-to-plane variations of the intercalant concentration, without any relation to the K—T transition.

## 9. Concluding Remarks

The first aim in writing this chapter was to fill the blank pointed out by De Jongh and Miedema [1] regarding the absence of good magnetic examples of the 2-d XY model. We hope to have shown that the compounds of formula $BaM_2(XO_4)_2$ (M = Co, Ni; X = P, As) can be considered as very good prototypes of the 2-d XY model. In particular the strong planar character ($H_a/H_e \sim 0.10$—$0.6$), the weak interlayer coupling ($|J'/J| \sim 10^{-3}$—$10^{-4}$) and the weak hexagonal in-plane anisotropy are well established from magnetization, specific heat and neutron scattering experiments. In the same spirit the compounds $KNiAsO_4$ and $NiPS_3$ have certainly physical properties very close to the former compounds. Nevertheless these systems are certainly more complicated than the simple (quasi) 2-d XY model with exchange interactions between nearest neighbouring spins only. In the compounds $BaM_2(XO_4)_2$ we have proved that at least three different in-plane exchange interactions have to be taken into account, introducing then some difficulties in the quantitative comparison of the experimental data with the actual theories. We hope that extended theories will be developed in the future.

The compound which seems best to fit the theoretical predictions of the 2-d XY model is certainly $BaNi_2(PO_4)_2$ (see Table IV the principal characteristics of these compounds). For this system we have shown that the correlation length follows the

TABLE IV
Exchange and anisotropy parameters of the compounds $BaM_2(XO_4)_2 \cdot \tilde{J}$ represents an effective exchange between first neighbours defined as $\tilde{J} = -E(\mathbf{k})/3S^2$, where $E(\mathbf{k})$ is the energy of the classical ground state.

| Compound | $S$ | $J_1$(K) | $J_2$(K) | $J_3$(K) | $J_{\|}/J_{\perp}$ | $D_{op}$(K) | $\tilde{J}$ (K) | $\|z'J'/z\tilde{J}\|$ | $D_{lp}(K)$ | $T_c/\|\tilde{J}\|S^2$ |
|---|---|---|---|---|---|---|---|---|---|---|
| $BaNi_2(PO_4)_2$ | 1 | −2.2 | −0.3 | −8.8 | 1 | 7.3 | −11 | $5 \times 10^{-4}$ | 0.005 | 2.3 |
| $BaNi_2(AsO_4)_2$ | 1 | 3.5 | 0.2 | −10 | 1 | 7 | −9 | $10^{-3}$—$10^{-4}$ | 0.2 | 2.0 |
| $BaCo_2(AsO_4)_2$ | 1/2 | 38 | 1.5 | −10 | 0.4 | — | 30 | $10^{3}$—$10^{-4}$ | $\ll 0.15$ | 0.75 |
| $BaCo_2(PO_4)_2$ | 1/2 | 20 | 0 | −8 | 0.4 | — | 15 | $10^{-3}$—$10^{-4}$ | — | 1.0 |

exponential prediction of Kosterlitz and Thouless in a range of temperature consistent with the value of the ratio $J'/J$ and with the existence of a well defined 2-d planar regime above $T \simeq 1.01\,T_N$. In the antiferromagnet $BaNi_2(PO_4)_2$ the interactions between the two sublattices are negative ($J_1 + J_3 < 0$) and no frustration is present in such a manner that the predictions of the ferromagnetic 2-d XY model apply. The situation is more complicated for the other compounds because the interactions $J_1$ and $J_3$ have the same order of magnitude and strong frustration effects are expected to occur in these systems. The most characteristic example is the 2-d helimagnet $BaCo_2(AsO_4)_2$, in which the main interaction $J_1$ is ferromagnetic ($J_1 \sim 38$ K, $J_3/J_1 \sim -0.26$). This compound exhibits magnetic properties which are explained qualitatively by the KT-theory and by the existence of vortex-type excitations. In particular one observes anomalies in the behaviour of thermodynamic quantities in the vicinity of the critical temperature that are quite similar to the predictions of Nelson and Kosterlitz and to the experimental results on $^4$He superfluid films. However the vortex-type excitations can have small energies in $BaCo_2(AsO_4)_2$ due to the particular set of $J_1$, $J_2$ and $J_3$ values. This could explain the low value of the phase transition temperature $T_c/JS^2 \sim 0.75$, i.e. about three times smaller than expected for the $S = 1/2$, 2-d XY model. We have seen that such phenomena are not present in $CoCl_2$-GIC which displays a more conventional ferromagnetic phase.

The most unexpected characteristic concerns the existence of strong anomalies in the specific heats of these compounds, the sharpness of which increases starting from $BaNi_2(PO_4)_2$ towards $BaNi_2(AsO_4)_2$, $BaCo_2(AsO_4)_2$ and finally disappears in $BaCo_2(PO_4)_2$. In particular one observes strong $\lambda$-type anomalies, except for

$BaCo_2(PO_4)_2$, in contrast to the specific heat of Heisenberg-like magnets which are typically composed of a broad Schottky-type bump followed by a very weak $\lambda$-type anomaly [1, 33, 94]. We suggest these pronounced peaks to be due to the creation of vortices or vortex—antivortex pairs in the vicinity of the transition. To our opinion the strong planar character must drastically change the thermodynamics of the system. Such effects are actually expected from recent theoretical work, e.g. from Monte Carlo simulations for the 2-d XY classical model and more analytically for the 2-d XY, $S = 1/2$ model which exhibits a logarithmic specific heat $C \sim \ln|T\text{-}T_c|$ as in the 2-d Ising model. Nevertheless this behaviour is not observed in either $BaCo_2(PO_4)_2$ or in $CoCl_2$-GIC, which in fact exhibit a very similar specific heat.

In the same way the strong damping of the magnetic excitations which occurs at the transition, and the rapid decrease of the correlation lengths above the transition can also be interpreted in terms of the crucial role played by the vortex-type excitations. As mentioned at the beginning of this chapter, the damping can be interpreted as a broadening of the excitations resulting from the description of a vortex by a large number of Fourier components: $\mathbf{S}(r) = \Sigma_k \mathbf{S}_k \exp(2\pi i\, \mathbf{k} \cdot \mathbf{r}_i)$. The resulting structure factor of the excitation is then the superposition of many modes emerging from the various **k**-vectors with an appropriate weight related to $|\mathbf{S}_k|$. Similarly the vortices and the vortex-antivortex pairs will contribute to the dynamical structure factors $S^{xx}(\mathbf{q}, \omega)$, $S^{yy}(\mathbf{q}, \omega)$, which are more or less complicated functions of the densities of vortex pairs and free vortices [136]. Unfortunately there exists no complete calculation of these quantities for the 2-d XY model [70]. We emphasize that a knowledge of these dynamical structure factors $S^{xx}(\mathbf{q}, \omega)$ and $S^{zz}(\mathbf{q}, \omega)$ will be very useful for the understanding of the temperature dependence of the central peak observed in $BaCo_2(AsO_4)_2$, which corresponds to relatively rapid fluctuations above $T_c$.

Another unexpected result concerns the existence of excitation modes, at low energy, which cannot be explained by simple spin wave theory. These excitations have been detected in $BaCo_2(AsO_4)_2$ around $\mathbf{q} = 0$ ($\omega_q \simeq 17$ K) and exist certainly also in $BaCo_2(PO_4)_2$. Yet they cannot be interpreted by any in-plane anisotropy (single ion or dipolar) because the magnitude of the equivalent anisotropy field which is needed ($\sim$ 15 kOe) disagrees completely with the observation of a strong quadratic contribution in the low-temperature specific heat of these compounds. Such a $T^2$-law is expected for conventional spin-waves with a linear dispersion curve, propagating in a 2-d magnet. We have shown also that these terms are well accounted for by the exchange interactions determined from the experimental dispersion curves. The main problem is that we do not observe any linear mode emerging from $\mathbf{q} \sim \mathbf{k}$, the wave vector of the ground state. We are thus lead to the assumption that these modes are overdamped even at low temperature, or that the structure factor has a quite small value. Unfortunately the existing theories are not very informative as concerns the damping of spin waves [116] or the effects of the magnon—vortex interaction. In fact, such calculations would be very important because the experimental results seem to show that it is the *damping* of the excitations which increases anomalously near $T_c$ rather than the energy that indeed rapidly decreases (jumps?) to zero, as shown in $BaNi_2(PO_4)_2$, $BaCo_2(AsO_4)_2$ and $K_2CuF_4$.

In most of the theoretical calculations the classical spin approximation is used instead of the quantum-mechanical approach, which however would certainly be more adequate in describing the physics of spin $S = 1$ or $S = 1/2$ systems like $BaM_2(XO_4)_2$ (M = Co, Ni; X = P, As), $KNiAsO_4$, $NiPS_3$, $CoCl_2$-GIC, etc. The quantum character would modify severely the spectrum of elementary excitations of the 2-d XY model. Such effects have been more or less predicted by Betts and Kelland [78] who have shown numerically the relevance of spin-up spin-down pairs which may be responsible for the pecularities of the transition in the $S = 1/2$, 2-d XY model. It is perhaps not too mistaken to assume that these pairs are the quantum version of the $\pi$-solitons or pairs of $\pi$-solitons discussed earlier as a possible explanation for the unusual excitations observed in $BaCo_2(AsO_4)_2$ and certainly also in $BaCo_2(PO_4)_2$. In a certain sense they can be viewed as the first level of quantized vortices, or more probably vortex—antivortex pairs, by analogy with superfluidity [46, 47]. But this remains to be proved. Finally whatever may be the interpretation of these unusual results, we hope they will stimulate theoretical work especially concerning the spin dynamics and the quantum effects in 2-d systems. To paraphrase J. Villain [134], “this will not be a simple exercise”.

## Acknowledgements

We are grateful to J. Y. Henry for his help in the single crystal growth and to L. J. de Jongh and J. Villain for very stimulating discussions.

## References

1. L. J. de Jongh and A. R. Miedema, *Adv. Phys.* **23** (1974) 1.
2. R. J. Birgeneau, H. J. Guggenheim and G. Shirane, *Phys. Rev.* **B1** (1970) 2211; **B8** (1973) 304.
3. H. Ikeda and K. Hirakawa, *J. Phys. Soc. Jpn.* **33** (1972) 393.
4. K. Hirakawa and H. Ikeda, *J. Phys. Soc. Jpn.* **35** (1973) 1328; **35** (1973) 1608.
5. L. Onsager, *Phys. Rev.* **65** (1944) 117; **76** (1949) 1244.
6. C. N. Yang, *Phys. Rev.* **85**(1952) 808.
7. R. B. Griffiths, *Phys. Rev. Lett.* **24** (1970) 1479.
8. M. E. Fisher, *Rep. Prog. Phys.* **30** (1967) 615.
9. D. J. Breed, *Physica* **37** (1967) 35.
10. H. Ikeda and K. Hirakawa, *Solid State Comm.* **14** (1974) 529.
11. F. Samuelsen, *Phys. Rev. Lett.* **31** (1973) 936.
12. N. B. Mermin and H. Wagner, *Phys. Rev. Lett.* **17** (1966) 1133.
13. G. S. Rushbrooke, G. A. Baker and P. J. Wood in *Phase Transition and Critical phenomena*, Domb and Green Eds., Vol. 3, Academic Press 1974, p. 245.
14. A. A. Migdal, *Zh. Eksp. Teor. Fiz.* **69** (1975) 1457; Sov. Phys. JETP **42** (1976) 743.
15. A. M. Polyakov, *Phys. Lett.* **59B** (1975) 79.
16. E. Brezin and J. Zinn-Justin, *Phys. Rev. Lett.* **36** (1976) 691.
17. V. L. Berezinski, *Sov. Phys. JETP* **32** (1971) 493; **34** (1972) 610.
18. V. L. Berezinski and A. Y. Blank, *Sov. Phys. JETP* **37** (1973) 369.
19. V. L. Pokrovsky and G. V. Uimin, *Adv. Phys.* **28** (1979) 595.

20. H. E. Stanley and T. A. Kaplan, *Phys. Rev. Lett.* **17** (1966) 913.
21. H. E. Stanley, *Phys. Rev. Lett.* **20** (1968) 589.
22. J. M. Kosterlitz and D. J. Thouless, *J. Phys.* **C6** (1973) 1181.
23. J. M. Kosterlitz, *J. Phys.* **C7** (1974) 1047.
24. J. Villain, *J. Physique* **35** (1974) 27; **36** (1975) 581.
25. J. Tobochnik and G. V. Chester, *Phys. Rev.* **B20** (1979) 3761.
26. D. P. Landau and K. Binder, *Phys. Rev.* **B24** (1981) 1391.
27. R. W. Gerling and D. P. Landau in *Magnetic Excitations and Fluctuations* Lovesey *et al.* Eds., Springer series in Solid State Sciences, Vol. 54, Springer, 1984, p. 61.
28. C. Kawabata and K. Binder, *Solid State Comm.* **22** (1977) 705.
29. M. Suzuki, S. Miyashita, A. Kuroda and C. Kawabata, *Phys. Lett.* **60A** (1977) 478.
30. C. Kawabata and A. R. Bishop, *J. Phys. Soc. Jpn.* **52** (1983) 27.
31. B. D. Josephson, *Physics Lett.* **21** (1966) 608.
32. A. N. Berker and D. R. Nelson, *Phys. Rev.* **B19** (1979) 2388.
33. D. D. Betts in *Phase Transition and Critical Phenomena*, Domb and Green Eds., Vol. 4, Academic Press, 1974, p. 569.
34. H. E. Stanley, *Phys. Rev. Lett.* **20** (1968) 589.
35. J. Rogiers, T. Lookmann, D. D. Betts and C. J. Elliot, *Can. J. Phys.* **56** (1978) 409.
36. H. de Raedt, B. de Raedt, J. Fivez and A. Lagendijk, *Physics Lett.* **104** (1984) 430.
37. A. Lagendijk and H. de Raedt, *Phys. Rev. Lett.* **49** (1982) 602.
38. H. de Raedt, B. de Raedt and A. Lagendijk, *Z. Phys.* **B57** (1984) 209.
39. H. de Raedt and A. Lagendijk, *Helvetica Physica Acta* **56** (1983) 593.
40. D. D. Betts, *Physica* **86—88B** (1977) 556.
41. P. Bloembergen, *Physica* **85B** (1977) 51; H. A. Algra, L. J. de Jongh and R. L. Carlin, *Physica* **93B** (1978) 24.
42. M. Steiner, J. Villain and G. Windsor, *Adv. Physics* **25** (1976) 87.
43. R. Savit, *Phys. Rev.* **B17** (1978) 1340.
44. R. Savit, *Rev. Mod. phys.* **52** (1980) 453.
45. W. F. Vinen in *Progress in Low Temperature Physics* C. J. Gorter Ed., Vol. 3, North Holland Publishing Company, Amsterdam, 1955.
46. J. de Boer in the *Proceeding of the International School of Physics Enrio Fermi, Liquid Helium*, G. Garezi Ed., 1967.
47. T. C. Padmore, *Phys. Rev. Lett.* **32** (1974) 826.
48. J. E. Rutledge, W. L. McMillan, J. M. Mucheland and T. E. Washburn, *Phys. Rev.* **B18** (1978) 2155.
49. J. R. Banavar and D. Jasnow, *J. Phys.* **All** (1978) 1361.
50. T. Matsubara and H. Matsuda, *Prog. Theor. Phys.* **16** (1956) 569; **17** (1957) 19.
51. B. I. Halperin and P. C. Hohenberg, *Phys. Rev.* **188** (1969) 898.
52. D. R. Nelson and J. M. Kosterlitz, *Phys. Rev. Lett.* **39** (1977) 1201.
53. V. L. Pokrovsky and G. V. Uimin, *Physics Lett.* **45A** (1973) 467; *Sov. Phys. JETP* **38** (1974) 847; *Adv. Phys.* **28** (1979) 595.
54. D. R. Nelson and P. S. Fisher, *Phys. Rev.* **B16** (1977) 4945.
55. D. J. Bishop and J. D. Reppy, *Pays. Rev. Lett.* **40** (1978) 1727: B. A. Huberman, R. J. Myerson and S. Doniach, *Phys. Rev. Lett.* **40** (1978) 780: V. Ambegaokar, B. I. Halperin, D. R. Nelson and E. D. Siggia, *Phys. Rev. Lett.* **40** (1978) 783.
56. I. Rudnick, *Phys. Rev. Lett.* **40** (1978) 1454.
57. A. F. Hebard and A. T. Fiory, *Phys. Rev. Lett.* **44** (1980) 291.
58. S. Hikami and T. Tsuneto, *Prog. Theor. Phys.* **63** (1980) 387.
59. J. V. José, L. P. Kadanoff, S. Kirkpatrick, D. R. Nelson, *Phys. Rev.* **16** (1977) 1217.
60. D. R. Nelson in *Phase Transitions and Critcal Phenomena*, Vol. 7, Domb and Lebowitz Eds., Academic Press, 1983.
61. S. Teitel and C. Jayapraskash, *Phys. Rev.* **1983**) 598.
62. D. H. Lee, J. D. Joannapoulos, J. W. Negele and D. P. Landau, *Phys. Rev. Lett.* **52** (1984) 433.

63. S. Miyashita and H. Shiba, *J. Phys. Soc. Jpn.* **53** (1984) 1145; H. Kawamura and S. Miyashita, *J. Phys. Soc.* **53** (1984) 4138.
64. J. Villain, *J. Physique* **38** (1977) 385.
65. J. V. José, *Phys. Rev.* **B20** (1979) 2167.
66. S. Miyashita, *J. Phys. Soc. Jpn.* **52** (1983) 780; **53** (1984) 44.
67. S. Miyashita, *Prog. Theor. Phys.* **69** (1983) 714.
68. H. Kawamura and S. Miyashita, *J. Phys. Soc. Jpn.* **53** (1984) 9.
69. D. L. Huber, *Phys. Lett.* **76A** (1980) 406; **68A** (1978) 125.
70. D. L. Huber, *Phys. Rev.* **B26** (1982) 3758.
71. H. J. Mikeska, *J. Phys.* **C11** (1978) L29; **C13** (1980) 2913.
72. J. Villain, *Physica* **B79** (1975) 1.
73. K. Maki, *Phys. Rev.* **B24** (1981) 335.
74. G. Müller, H. Thomas, H. Beck and J. C. Bonner, *Phys. Rev.* **B24** (1981) 1429.
75. J. C. Bonner and G. Müller, *Phys. Rev.* **B29** (1984) 5216.
76. T. Garel and S. Doniach, *J. Phys.* **C13** (1980) L887: D. H. Lee and G. Grinstein, *Phys. Rev. Lett.* **55** (1985) 541: D. H. Lee, G. Grinstein and J. Toner, *Phys. Rev. Lett.* **56** (1986) 2318.
77. J. Rogiers, E. W. Grundke and D. D. Betts, *Can. J. Phys.* **57** (1979) 1719.
78. D. D. Betts and S. B. Kelland, *J. Phys. Soc. Jpn.* **52** (1983) Suppl. 11.
79. M. Suzuki, *Prog. Theor. Phys.* **56** (1976) 1454.
80. M. Suzuki, S. Miyashita and A. Kuroda, *Prog. Theor. Phys.* **58** (1877) 1377.
81. S. Eymond and A. Durif, *Mat. Res. Bull.* **4** (1969) 595.
82. S. Eymond, Thesis, University of Grenoble (1968), unpublished.
83. J. Y. Henry, Thesis, University of Grenoble (1976), unpublished.
84. G. Ladwig and B. Ziemer, *Z. anorg. allg. Chem* **457** (1979) 143.
85. L. P. Regnault, unpublished results.
86. A. Wiedenmann, D. Delacote, J. Rossat-Mignod, R. Buder, P. Segransan, C. Berthier, G. Ouvrard and R. Brec, Proceedings of VII International Conference on Solid Compounds of Transition Elements, Grenoble, 21—25 June, 1982.
87. H. Hahn and W. Klingen, *Natürwiss* **52** (1968) 494.
88. R. Brec, G. Ouvrard, A. Louisy and J. Rouxel, *J. Ann. Chem. Fr.* **5** (1980) 499.
89. L. P. Regnault, J. Rossat-Mignod, J. Y. Henry and L. J. de Jongh, *J. Mag. Mag. Mat.* **31—34** (1983) 1205.
90. L. P. Regnault, J. Rossat-Mignod, L. J. de Jongh and J. Y. Henry, to be published.
91. G. S. Rushbrooke and P. J. Wood, *J. Mol. Phys.* **1** (1958 (1958) 257; **6** (1963) 409.
92. L. P. Regnault, J. Rossat-Mignod and J. Y. Henry, *J. Phys. Soc. Jpn.* **52** (1983) Suppl. 1.
93. L. P. Regnault, J. Y. Henry, J. Rossat-Mignod and A. de Combarieu, *J. Mag. Mag. Mat.* **15—18** (1980) 1021.
94. D. W. Engelfriet, W. L. Groeneveld and G. M. Nap, *Z. Naturforsch.* **35a** (1980) 1382.
95. L. P. Regnault, J. Rossat-Mignod, J. Y. Henry, R. Pynn and D. Petitgrand in *Magnetic Excitations and Fluctuations* Lovesey *et al.* Eds., Springer Series in Solid State Sciences 54 (1984) p. 201.
96. P. A. Lindgard and A. Kowalska, *J. Phys.* **C9** (1976) 2081.
97. E. Rastelli, A. Tassi and L. Reatto, *Physica* **97B** (1979) 1.
98. L. P. Regnault, Thesis, University of Grenoble (1981), unpublished.
99. K. Hirakawa, H. Yoshizawa, J. D. Axe and G. Shirane, *J. Phys. Soc. Jpn.* **52** (1983) Suppl. 19.
100. K. R. Atkins and I. Rudnick in *Progress in Low Temperature Physics* C. J. Gorter Ed., Vol. VI, Chap. 2, North Holland, Amsterdam, 1970.
101. L. P. Regnault, P. Burlet and J. Rossat-Mignod, *Physica* **86—88B** (1977) 660.
102. L. P. Regnault, J. Rossat-Mignod, J. Villain and A. DeCombarieu, *J. Physique* **39** (1978) C6—759.
103. L. P. Regnault and J. Rossat-Mignod, *J. Mag. Mag. Mat.* **14** (1979) 194.
104. A. Guinier, *Théorie et technique de la Crystallographie*, Dunod, Paris, 1956.
105. J. Villain, Thesis, unpublished.

106. L. P. Regnault, J. Rossat-Mignod, A. Adam, D. Billerey and C. Terrier, *J. Physique* **43** (1982) 1283.
107. J. Villain, personal communication.
108. K. Maki in *Physics in One Dimension* Bernasconi and Schneider Eds., Springer Series in Solid State Sciences 23 1981, p. 63; *Phys. Rev.* **B24** (1981) 3991.
109. L. P. Regnault, J. Rossat-Mignod, L. J. de Jongh and J. Y. Henry, to be published.
110. S. Funahashi, F. Moussa and M. Steiner, *Solid State Comm.* **18** (1976) 433.
111. S. Borovik-Romanov, N. M. Kreines, R. Laiho, T. Levola and V. G. Zhotikov, *J. Phys. C* **13** (1980) 879.
112. H. Yamazaki, *J. Phys. Soc. Jpn.* **37** (1974) 667.
113. K. Hirakawa and K. Ubukoshi, *J. Phys. Soc. Jpn.* **50** (1981) 1909.
114. K. Hirakawa, H. Yoshizawa and K. Ubukoshi, *J. Phys. Soc. Jpn* **51** (1982) 2151.
115. K. Hirakawa, *J. Appl. Phys.* **53** (1982) 1893.
116. F. Moussa and J. Villain, *J. Phys.* **C9** (1976) 4433.
117. F. Moussa, B. Hennion, J. Mons and G. Peppy, *Solid State Comm.* **27** (1978) 141.
118. K. Hirakawa, H. Yoshizawa, J. D. Axe and G. Shirane, *J. Phys. Soc. Jpn.* **52** (1983) 4220.
119. I. Yamada, *J. Phys. Soc. Jpn.* **33** (1972) 979.
120. S. Flandrois, J. Masson, J. C. Rouillon, J. Gaultier and C. Hauw, *Synthetic Metals* **3** (1981) 1.
121. Yu. S. Karimov, *Sov. Phys. JETP* **38** (1974) 129; **39** (1974) 547; **41** (1976) 772.
122. Yu. S. Karimov and Yu. N. Noviko, *JETP Lett.* **19** (1974) 159.
123. M. Suzuki and H. Ikeda, *J. Phys. C* **14** (1981) L923.
124. M. Suzuki, H. Ikeda, Y. Murakami, M. Matsuura, H. Suematsu, R. Nishitani and R. Yoshizaki, *J. Mag. Mag. Mat.* **31—34** (1983) 1173.
125. C. Simon, F. Batallan, I. Rosenman, J. Schweitzer, H. Lauter and R. Vangelisti, *Synthetic Metals* **8** (1983) 53.
126. Y. Murakami, M. Matsuura, M. Suzuki and H. Ikeda, *J. Mag. Mag. Mat.* **31—34** (1983) 1171.
127. M. Suzuki, H. Ikeda and Y. Endoh, *Synthetic Metals* **8** (1983) 43.
128. C. Ayache, C. Simon, F. Batallan and I. Rosenman, to be published.
129. M. T. Hutchings, *J. Phys.* **C6** (1973) 3143.
130. M. Elahy and G. Dresselhauss, *Phys. Rev.* **B30** (1984) 7225.
131. M. Elahy and G. Dresselhauss in *Intercalated Graphite*, M. S. Dresselhauss, G. Dresselhauss, J. E. Fisher and M. J. Moran Eds. Elsevier/North Horth Holland, New York, 1983, vol. 20. p. 207.
132. M. Shayegan, L. Salamanca-Riba, J. Heremans, G. Dresselhauss and J. P. Issi ibid., p. 213.
133. L. J. de Jongh in *Recent Developments in Condensed Matter Physics*, J. T. Devreese Ed., (Plenum Publishing Corporation, 1981, Vol. 1, p. 343.
134. J. Villain, *Proceeding of Inst. Phys. Conf.* Ser. n° 64: Section 4 (1982) 239.
135. E. Loh, D. J. Scalapino and P. M. Grant, *PHYS. Rev.* **B31** (1985) 4712; *Phys. Rev.* **B33** (1986) 5104.
136. F. G. Mertens, A. R. Bishop, G. M. Wysin and C. Kawabata, *Phys. Rev. Lett.* **59** (1987) 117.
137. D. G. Rancourt, *J. Mag. Mag. Mat.* **51** (1985) 133.
138. L. P. Regnault, J. P. Boucher, J. Rossat-Mignod, J. Bouillot, R. Pynn, J. Y. Henry and J. P. Renard, *Physica* **136B** (1986) 329.
139. J. Villain, *Physica* **86—88B** (1977) 631.
140. L. P. Regnault and T. Chattopadhay, unpublished results.
141. R. Loft and T. A. DeGrand, *Phys. Rev.* **B35** (1987) 8528.
142. R. Coté and A. Griffin, *Phys. Rev.* **B34** (1986) 6240.
143. H. Zabel and S. M. Shapiro, *Phys. Rev.* **B36** (1988) 7292.
144. D. G. Wiesler, H. Zabel and S. M. Shapiro, *Physica* **B156** & **157** (1989) 292.
145. J. Rogerie, Ch. Simon, I. Rosenman, J. Schweitzer, Ch. Vettier, R. Vangelisti and P. Pernod, *J. de Physique* **C8** (1988) 1437.

# SPIN DYNAMICS IN THE PARAMAGNETIC REGIME: NMR AND EPR IN TWO-DIMENSIONAL MAGNETS

H. BENNER

*Institut für Festkörperphysik, Technische Hochschule Darmstadt, D 6100 Darmstadt, Federal Republic of Germany*

and

J. P. BOUCHER

*Département de Recherche Fondamentale, Section de Résonance Magnétique, Centre d'Etudes Nucléaires de Grenoble, F 38041 Grenoble Cedex, France*

## 1. Introduction

The study of spin fluctuations in magnetic systems raises challenging questions which, since the pioneering work of Kubo and Tomita [1] and of Moriya [2], have been widely investigated. In the last fifteen years, the discovery of quasi one-dimensional (1-d) and two-dimensional (2-d) magnetic compounds [3] has renewed the interest in this field. Some very specific properties which exist in any dimension become magnified in low-dimensional systems.

As a starting point let us consider the simplified model of a set of electronic spins with strong mutual interactions described by the Heisenberg exchange Hamiltonian

$$\mathscr{E} = -J \sum_{i \neq j} \mathbf{S}_i \cdot \mathbf{S}_j \tag{1.1}$$

The scalar form $\mathbf{S}_i \cdot \mathbf{S}_j$ makes this coupling completely isotropic. This specific property will be seen to play an essential role in the spin dynamics. In the high-temperature limit ($T \gg JS(S+1)$), to which the first part of the present discussion will be confined, the spin system exhibits a paramagnetic behaviour. In particular, the susceptibility (in units of $g^2\mu_B^2/k_B$) follows the simple Curie law:

$$\chi^\alpha = \frac{1}{T} \sum_{i,j} \langle S_i^\alpha S_j^\alpha \rangle = \frac{1}{3} S(S+1)\, N/T, \qquad \alpha = x, y, z,$$

showing that there is no static correlation between spins located at different sites. In terms of the correlation functions we have:

$$g_{ij}^\alpha = \langle S_i^\alpha S_j^\alpha \rangle \begin{cases} = 0 & \text{for} \quad i \neq j \\ = \langle (S_i^\alpha)^2 \rangle & \text{for} \quad i = j, \end{cases}$$

as if there were no coupling between the spins.

*L. J. De Jongh (Ed.), Magnetic Properties of Layered Transition Metal Compounds* 323—378.

### 1.1. DYNAMICS OF THE 2-SPIN CORRELATION FUNCTIONS

In the high-temperature limit, the effect of the coupling between spins can be seen by studying the dynamic behaviour of the local fluctuations. As an example, the time evolution of the 2-spin correlation functions

$$g_{ij}^{a}(t) = \langle S_i^a(t) S_j^a \rangle \tag{1.2}$$

is shown in Figure 1a for the case of three-dimensional (3-d) exchange coupling.

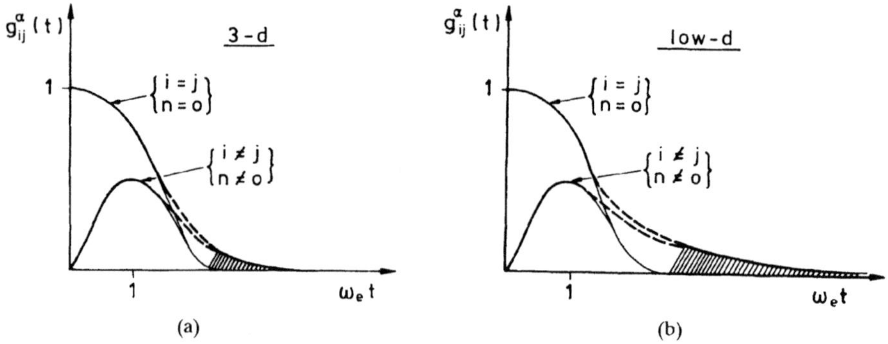


(a) (b)

Fig. 1. 2-spin correlation functions in 3-d (a) and low-d (b) magnetic systems. The dashed areas correspond to the diffusive part. The time scale is defined by the exchange frequency $\omega_e$.

For any operator $\mathcal{O}$, the time dependence is defined by:

$$\mathcal{O}(t) = e^{i\mathcal{H}t}\,\mathcal{O}\,e^{-i\mathcal{H}t} \tag{1.3}$$

where $\mathcal{H}$ is the full Hamiltonian of the electronic spin system. The self-correlation function ($i = j$) decreases smoothly with respect to the characteristic frequency $\omega_e = J[8zS(S+1)/3]^{1/2}$, where $z$ is the number of nearest neighbours. The cross-correlation functions ($i \neq j$) start from zero at $t = 0$ (no static correlation), and pass through a maximum at $\omega_e t \approx 1$. In Figure 1b a schematic description is given for the case of low-d systems. Compared to the 3-d case, an important difference occurs for $\omega_e t \gg 1$: the functions $g_{ij}^{a}(t)$ decrease much more slowly. This behaviour results from diffusive processes which in low dimensions retain the correlations for much longer times.

Experimentally, information on the spin dynamics can be obtained by probing the frequency behaviour of the correlation functions, which are described by the spectral densities:

$$f_{ij}^{a}(\omega) = \frac{1}{2\pi}\int_{-\infty}^{+\infty} \mathrm{d}t\, e^{-i\omega t}\, g_{ij}^{a}(t). \tag{1.4}$$

The long time behaviour ($t \to \infty$) of $g_{ij}^{a}(t)$ is essentially reflected in the low-frequency behaviour ($\omega \to 0$) of $f_{ij}^{a}(\omega)$.

Two experimental magnetic resonance methods, namely NMR and EPR, will be discussed here, which are well suited to probing the very low frequency part — or the long time behaviour — of the local fluctuations defined by Equations (1.2) and (1.4).

### 1.2. NUCLEAR MAGNETIC RESONANCE (NMR)

In magnetic compounds, the relaxation of nuclear spins is essentially governed by the fluctuations which take place in the electronic spin system. The nuclear relaxation rates can be expressed explicitly in terms of the spectral densities $f_{ij}^{\alpha}(\omega)$. In the following, the formulation will be given more often for 1-d systems, since it is the simplest. For instance, the correlation between spins can be defined with only one index $|i-j| = n$. Extension to 2-d systems usually is straightforward. For the longitudinal relaxation rate in the laboratory frame, we can write

$$\frac{1}{T_1} = \sum_n \{A_n^z f_n^z(\omega_N) + A_n^+ f_n^+(\omega_N)\},$$

where $\omega_N$ is the nuclear Larmor frequency. The coefficients $A_n^{\alpha}$ are the usual 'geometrical' factors associated with the hyperfine couplings between nuclear and electronic spins. The $z$-axis refers to the external magnetic field needed in magnetic resonance experiments. The transverse spin operators are defined by $S_j^{\pm} = S_j^x \pm iS_j^y$. In the infinite temperature limit, a slightly different formulation can be given for $1/T_1$:

$$\frac{1}{T_1} = \sum_n \{A_n^z \tilde{f}_n^z(\omega_N) + A_n^+ \tilde{f}_n^+(\omega_0)\} \tag{1.5}$$

where both the nuclear and electronic Larmor frequencies, $\omega_N$ and $\omega_0$, appear explicitly. $\tilde{f}_n(\omega)$ is the spectral density of the correlation function $\tilde{g}_n^{\alpha}(t)$ defined by:

$$g_n^{\alpha}(t) = \tilde{g}_n^{\alpha}(t)\, e^{i\alpha\omega_0 t}. \tag{1.6}$$

When $\alpha$ is used as a coefficient, it takes the values 0, $\pm 1$ corresponding to the index $\alpha = z, \pm$, respectively. Note that for $\alpha = z$, $\tilde{g}_n^z(t) = g_n^z(t)$. The function $\tilde{g}_n^{\alpha}(t)$ describes the fluctuations induced by the mutual interactions between the spins. For instance, in our simplified model, the full Hamiltonian to be considered in (1.3) is:

$$\mathcal{H}_0 = \mathcal{Z} + \mathcal{E} \tag{1.7}$$

with the Zeeman term $\mathcal{Z} = \omega_0 S^z$, where $S^z$ is the $z$-component of the total spin operator

$$\mathbf{S} = \sum_i \mathbf{S}_i. \tag{1.8}$$

The isotropic character of $\mathcal{E}$ yields the fundamental commutation rule:

$$[\mathcal{E}, S^{\alpha}] = 0 \tag{1.9}$$

and consequently, for the function $\tilde{g}_n^a(t)$ one obtains

$$\tilde{g}_n^a(t) = \langle e^{i\mathscr{E}t} S_i^a e^{-i\mathscr{E}t} (S_{i+n}^a)^\dagger \rangle. \quad (1.10)$$

The time modulation is seen to result only from the Heisenberg Hamiltonian $\mathscr{E}$, Equation (1.1). The functions represented in Figure 1 actually correspond to Equation (1.10).

The expression (1.5) shows that by $T_1$ measurements the magnetic fluctuations are probed not only at $\omega_N$, but also at $\omega_0 \pm \omega_N \simeq \omega_0$. These Larmor frequencies are usually much smaller than the characteristic exchange frequency $\omega_e$. For instance, for protons and free electrons in an external field $H = 3$ kOe we evaluate $\omega_N = 8 \times 10^7$ rad/s and $\omega_0 = 5 \times 10^{10}$ rad/s, while for a $J$ of a few Kelvins, $\omega_e$ is in the range of $10^{12}$ rad/s. Expressions similar to (1.5) can be written for the transverse relaxation time $T_2$ and for the relaxation time in the rotating frame $T_{1\rho}$ [4].

Finally, it is worth noting that, since $\omega_N$ and $\omega_0$ are both proportional to the applied field, an experimental investigation of $1/T_1$ as a function of $H$ is equivalent to a frequency analysis of the local magnetic fluctuations.

### 1.3. ELECTRON PARAMAGNETIC RESONANCE (EPR)

The electron paramagnetic resonance gives a way to probe the electronic spin system directly. Assuming for simplicity a circular polarization of the microwave, the absorption part of the EPR line is given by the spectral density $F^+(\omega)$ of the total spin correlation function

$$G^+(t) = \langle S^+(t)\, S^-(0) \rangle / \langle |S^+|^2 \rangle \quad (1.11)$$

where $S^+$ is the transverse component of the total spin operator, Equation (1.8).

Within the simplified model described by the Hamiltonian (1.7), the function $\tilde{G}^+(t)$ defined through

$$G^+(t) = \tilde{G}^+(t)\, e^{i\omega_0 t}$$

is evaluated as constant in time:

$$\tilde{G}^+(t) = \langle e^{i\mathscr{E}t} S^+ e^{-i\mathscr{E}t} S^- \rangle / \langle |S^+|^2 \rangle = 1.$$

The self-correlation function of the total spin operator turns out to be insensitive to the exchange interaction $\mathscr{E}$. Again it seems to behave as if there were no coupling between the spins, as in liquids.

The spectral density $F^+(\omega)$ corresponding to Equation (1.11) is consequently described by a Dirac delta function centered at $\omega_0$:

$$F^+(\omega) = \delta(\omega - \omega_0).$$

However, this function can also be written as the sum of the local spectral densities $f_n^+(\omega)$, normalized to $\langle |S_i^+|^2 \rangle$:

$$F^+(\omega) = \sum_n f_n^+(\omega) / \langle |S_i^+|^2 \rangle.$$

In contrast to $F^+(\omega)$, the $f_n^+(\omega)$ are rather broad functions, their characteristic widths being of the order of $\omega_e$, as shown in Figure 2 for the self-correlation functions. Therefore, one is led to the following conclusion: except for the specific value $\omega = \omega_0$, we have exactly

$$\sum_{n(\neq 0)} f_n^+(\omega) = -f_{n=0}^+(\omega) \quad \text{for} \quad \omega \neq \omega_0$$

or, equivalently,

$$\sum_{n(\neq 0)} \tilde{f}_n^+(\omega) = -\tilde{f}_{n=0}^+(\omega) \quad \text{for} \quad \omega \neq 0.$$

In case of a pure Heisenberg system the spectral contributions (at $\omega \neq 0$) of all the cross-correlations exactly cancel the contribution of the self-correlation. This gives an illustration of the very specific role played by the microscopic fluctuations in magnetic systems [5]. For instance, in liquids, the situation is totally different (there is no cross-correlation), although the exchange and 'motional' narrowing phenomena are usually developed within the same formalism.

In solids, the exact compensation between the cross-correlation and the self-correlation functions is a consequence of the isotropic property of the interaction Hamiltonian $\mathscr{E}$, Equation (1.1). However, in real systems one has to take into account the dipolar interactions between the spins which introduce a non-isotropic term $\mathscr{D}$ in the total Hamiltonian:

$$\mathscr{H} = \mathscr{Z} + \mathscr{E} + \mathscr{D}. \tag{1.12}$$

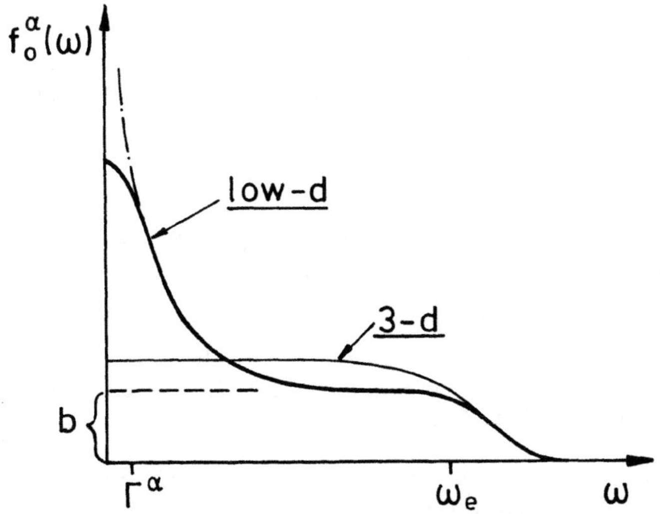


Fig. 2. Spectral density of local fluctuations in 3-d and low-d magnetic systems. $\Gamma^\alpha$ is the cut-off for the low-d diffusive behaviour. $b$ determines the high frequency contribution to the spectrum.

The coupling $\mathscr{D}$ can be very small compared to $\mathscr{E}$, but has the property that

$$[\mathscr{D}, S^{+}] \neq 0. \tag{1.13}$$

A primary consequency of this property is that now the isotropic symmetry of the spin Hamiltonian is broken. As a result one expects drastic changes in the behaviour of the local magnetic fluctuations at low frequency. For instance, the sum rule given above for a pure Heisenberg system is no longer fulfilled, and this leads to a broadening of the EPR line. The corresponding half-width at half maximum $\Delta H$ can be expressed in terms of spectral densities that also describe local fluctuations. One can write approximately:

$$\Delta H \approx \omega_d{}^2\{F_0{}^2\,\tilde{\Gamma}(\omega=0)+10\,F_1{}^2\,\tilde{\Gamma}(\omega_0)+F_2{}^2\,\tilde{\Gamma}(2\omega_0)\}, \tag{1.14}$$

where $\omega_d$ and $F_m$ are the characteristic frequency and the geometrical coefficients associated with the dipolar couplings. $\tilde{\Gamma}(\omega)$ is the spectral density of correlation functions involving four spin operators. These functions and the 2-spin correlation functions have a very similar time dependence. Both describe local fluctuations. Through EPR measurements, these fluctuations are seen to be probed at low frequency, namely at $\omega = 0$, $\omega_0$ and $2\omega_0$.

## 2. General Formalism

The purpose of this section is to outline the basic features associated with the dynamics of a strongly interacting spin system. Emphasis will be made on the specific behaviour related to the low lattice dimensionality for the description of the microscopic correlation functions.

### 2.1. DIFFUSION AND DIMENSIONALITY

The shape of the correlation functions defined in Equation (1.10) is sketched in Figure 1. As an example we now consider the case of the 2-spin self-correlation ($n = 0$) function.

For $\omega_e t \ll 1$, a short time expansion can be used

$$\tilde{g}_0^{\alpha}(t) \sim 1 - M_2^0 t^2/2 \tag{2.1}$$

where $M_2^0 = 2\omega_e^2$ is the second moment of $\tilde{g}_0^{\alpha}(t)$. Equation (2.1) can also be considered as the expansion of the Gaussian function

$$\tilde{g}_0^{\alpha}(t) \sim \exp(-\omega_e^2 t^2). \tag{2.2}$$

It turns out that Equation (2.2) gives a reasonably good approximation in the range $0 \leqslant \omega_e t \lesssim 1$. The knowledge of higher order moments ($M_{2n}^0$) can be used to improve the description of $\tilde{g}_0^{\alpha}(t)$ to longer times.

The dimensionality plays only a minor role at short times. The main change occurs in the value of the characteristic frequency $\omega_e$ which depends explicitly on the number $z$ of nearest neighbours.

For $\omega_e t \gg 1$ a hydrodynamic description can be used. The appropriateness of such a description for the transport of spin excitations in Heisenberg magnetic insulators was first pointed out by Bloembergen and van Hove [6]. The idea is that for a spin system obeying a Heisenberg Hamiltonian, Equation (1.1), the total spin operator, according to Equation (1.9), is a *constant of motion*. Therefore, in the high temperature limit, the microscopic spin fluctuations are governed at long times by a diffusion equation:

$$D\nabla^2 \tilde{g}^\alpha(\mathbf{r}, t) = \frac{\partial}{\partial t}\, \tilde{g}^\alpha(\mathbf{r}, t), \tag{2.3}$$

where $D$ is the diffusion coefficient. The solution of Equation (2.3) depends strongly on the dimensionality $d$ of the magnetic lattice:

$$\tilde{g}^\alpha(\mathbf{r}, t) = (4\pi Dt)^{-d/2} \exp(-\mathbf{r}^2/4Dt). \tag{2.4}$$

The diffusive behaviour completes the description of $\tilde{g}_0^\alpha(t)$ for $\omega_e t \gg 1$. In three dimensions it adds only a small tail which decreases rapidly with time as $t^{-3/2}$. In 1-d and 2-d systems, on the other hand, the tail persists at much longer times according to

$$\tilde{g}_n^\alpha(t) \sim (4\pi Dt)^{-1/2} \quad \text{for} \quad d = 1.$$

$$\tilde{g}_n^\alpha(t) \sim (4\pi Dt)^{-1} \quad \text{for} \quad d = 2.$$

In terms of spectral densities, such a diffusive behaviour leads to dramatic effects in low-d systems, since it implies a divergence of the functions $\tilde{f}_n^\alpha(\omega)$ at $\omega = 0$. One obtains

$$\tilde{f}_n^\alpha(\omega) \sim \frac{1}{2\pi}\,(2D\omega)^{-1/2} \qquad \text{for} \quad d = 1$$

$$\tilde{f}_n^\alpha(\omega) \sim \frac{1}{4\pi D}\ln(4\pi D/\omega) \quad \text{for} \quad d = 2.$$

The predicted divergences have their origin in two implicit conditions:

(i) the spin arrangements correspond to a perfect 1-d or 2-d lattice,
(ii) the interaction between the spins is described by the fully isotropic Heisenberg Hamiltonian ($\mathscr{E}$).

In real systems one expects the divergences to be truncated for two reasons:

(i) In crystals there exist small magnetic couplings between chains or between planes, giving a 3-d character to the magnetic interaction. Consequently the divergences are suppressed.
(ii) Anisotropic terms are always present as perturbations in the Hamiltonian, such as the intrachain or intraplane dipolar interactions $\mathscr{D}$. The total spin operator is then no longer a constant of motion:

$$[\mathscr{D}, S^\alpha] \neq 0 \tag{2.5}$$

which, as discussed above, was the basic argument for justifying a diffusive behaviour. In the following, we only consider the intrachain or intraplane truncation effects induced by dipolar interactions.

### 2.2. CUT-OFF AND EPR LINEWIDTH

For describing the truncation effects it is easier to consider the correlation functions in reciprocal space. Defining the spin operators by

$$S_{\mathbf{q}}^{a} = N^{-1/2} \sum_{j=1}^{N} S_j^a \, e^{i\mathbf{q}\cdot\mathbf{r}_j}, \tag{2.6}$$

one can write

$$g_n^a(t) = \sum_{\mathbf{q}} g_{\mathbf{q}}^a(t)\, e^{-i\mathbf{q}\cdot\mathbf{r}_n}$$

where $g_{\mathbf{q}}^a(t)$ represents the normal mode associated with the wave vector $\mathbf{q}$:

$$g_{\mathbf{q}}^a(t) = \langle S_{\mathbf{q}}^a(t)\,(S_{\mathbf{q}}^a)^{\dagger}\rangle. \tag{2.7}$$

The corresponding spectral density $f_{\mathbf{q}}^a(\omega)$ is obtained by the usual Fourier transform. In reciprocal space the hydrodynamic regime corresponds to the limit $q \to 0$. From Equation (2.3) one obtains

$$\tilde{f}_{\mathbf{q}}^a(\omega) = \frac{1}{\pi}\langle |S_{\mathbf{q}}^a|^2\rangle \frac{Dq^2}{\omega^2 + (Dq^2)^2}\,. \tag{2.8}$$

The evaluation of the diffusion coefficient $D$ will not be reported here. It does not raise any specific feature associated with the dimensionality [7]. It is a function of $\omega_e$ and, therefore, depends on the coordination number $z$ and on the spin value [8].

For a pure Heisenberg system ($\mathcal{H} = \mathcal{E}$) the width of the normal mode $f_{\mathbf{q}}^a(\omega)$ is zero at $q = 0$, according to

$$\Gamma_{\mathbf{q}} = Dq^2.$$

This explains the divergence of the spectral densities $f_n^a(\omega)$ at $\omega = 0$ in 1-d and 2-d systems.

Now, the presence of an anisotropic term in the Hamiltonian ($\mathcal{H} = \mathcal{E} + \mathcal{D}$) yields an extra contribution to the linewidth:

$$\Gamma_{\mathbf{q}}^a = Dq^2 + \Gamma^a.$$

$\Gamma^a$, which is independent of $q$, defines the cut-off frequency of the 2-spin correlation functions $\tilde{g}_n^a(t)$. It is given by the width of the $q = 0$ mode. The divergence of the spectral density $\tilde{f}_{\mathbf{q}=0}^a(\omega)$ is truncated. For 2-d systems the low frequency part of $\tilde{f}_{\mathbf{q}=0}^a(\omega)$ can be described by

$$\tilde{f}_{\mathbf{q}=0}^a(\omega) = \langle |S_i|^2\rangle \frac{1}{4\pi D} \ln\left\{\frac{4\pi D}{\sqrt{\omega^2 + (\Gamma^a)^2}}\right\} + b \tag{2.9}$$

which takes a finite value at $\omega = 0$. As shown in Figure 2, the high-frequency fluctuations yield a contribution which is represented by the term $b$ in Equation (2.9).

For the nuclear relaxation rate in the limit $\omega_N, \omega_0 \ll \omega_e$ the following expression can be derived from Equation (1.5):

$$\frac{1}{T_1} = \frac{A^z}{4\pi D} \ln \left\{ \frac{4\pi D}{\sqrt{\omega_N^2 + (\Gamma^z)^2}} \right\} + \\ + \frac{A^+}{4\pi D} \ln \left\{ \frac{4\pi D}{\sqrt{\omega_0^2 + (\Gamma^+)^2}} \right\} + B. \tag{2.10}$$

The coefficients $A^z$ and $A^+$ depend on the hyperfine coupling between nuclei and electrons. The coefficient $B$ depends on both the hyperfine coupling and the matching procedure used to describe the intermediate regime between the short and long time parts of the correlation functions (the dashed lines in Figures 1a and 1b). For each compound a specific procedure must be defined.

The cut-off $\Gamma^+$ in Equation (2.10) is associated with the fluctuations of the transverse spin operator $S^+$. It is worth noting that the uniform mode $q = 0$ is simply the EPR line, compare Equations (1.11) and (2.7). The width of the EPR line (in units of $\gamma_0 = g\mu_B/\hbar$) gives a direct measure of the cut-off frequency $\Gamma^+$:

$$\Delta H = \Gamma^+. \tag{2.11}$$

The description of the small $q$ modes and of the EPR line is a difficult problem in low-d systems. The difficulty arises from the fact that the 4-spin correlation functions, which describe the width $\Delta H$ (or $\Gamma^+$), also follow a diffusive behaviour. A drastic dependence on frequency is therefore expected for the corresponding spectral densities as $\omega$ goes to zero, yielding a non-Markovian character to these modes. Because of this specific feature, a new discussion of the basic work of Kubo and Tomita [1] has been started in the last ten years [9—11].

## 3. EPR Spectrum

In this section we discuss the problem of the shape of the EPR line. Its absorption part $F^+(\omega)$ is the spectral density of (1.11), where the time dependence results from the full Hamiltonian

$$\mathcal{H} = \mathcal{Z} + \mathcal{E} + \mathcal{D}.$$

A very general form for $F^+(\omega)$ is the following [12]:

$$F^+(\omega) = \frac{1}{\pi} \frac{\operatorname{Re} M^+(\omega)}{[\omega - \omega_0 + \operatorname{Im} M^+(\omega)]^2 + [\operatorname{Re} M^+(\omega)]^2} \tag{3.1}$$

where Re $M^+(\omega)$ and Im $M^+(\omega)$ are the real and imaginary parts of a complex function $M^+(\omega)$ which depends explicitly on frequency. Re $M^+(\omega)$ and Im $M^+(\omega)$

are related by the Kramers—Kronig (KK) relations. In order to help in understanding their relative behaviour we remind here that the KK transformation is qualitatively similar to a derivation procedure. This means that, if Re $M^+(\omega)$ is strongly frequency dependent, Im $M^+(\omega)$ takes large values. On the other hand, if Re $M^+(\omega)$ is constant in frequency, one expects Im $M^+(\omega)$ to be zero. In the latter case $F^+(\omega)$ has a simple Lorentzian shape which corresponds to the Markovian limit. A non-Lorentzian shape of $F^+(\omega)$ can therefore be attributed — and vice versa — to the frequency dependence of Re $M^+(\omega)$, which is a signature of memory effects ('non-Markovian behaviour').

In evaluating the 'memory spectrum' $M^+(\omega)$ we shall consider separately the 'high' and 'low' frequency regions defined by the conditions $|\omega - \omega_0| \gg |M^+(\omega)|$ and $|\omega - \omega_0| \lesssim |M^+(\omega)|$, respectively. Physically, Re $M^+(\omega)$ represents the width of the EPR line while Im $M^+(\omega)$ describes a shift of the resonance frequency. Therefore, the high frequency region corresponds to the wings of the EPR line ($|\omega - \omega_0| \gg \Delta H$) while the low frequency region describes its central part ($|\omega - \omega_0| \lesssim \Delta H$).

In the high frequency region the following approximation is easily deduced from the general form (3.1):

$$\mathrm{Re}\, M^+(\omega) \approx \pi(\omega - \omega_0)^2 F^+(\omega). \tag{3.2}$$

On the other hand, it can be shown from the definition (1.4) that the function

$$T^+(\omega) = (\omega - \omega_0)^2 F^+(\omega) \tag{3.3}$$

represents the spectral density of

$$T^+(t) = \langle e^{i\mathscr{H}t}[\mathscr{D}, S^+]\, e^{-i\mathscr{H}t}[S^-, \mathscr{D}]\rangle / \langle |S^+|^2\rangle. \tag{3.4}$$

This function is the starting point for a description of the width of the EPR line or, equivalently, the cut-off of the spectral density $f_n^+(\omega)$. It is worth noting that the operator involved in $T^+(t)$ is the commutator $[\mathscr{D}, S^+]$, which was considered in Equation (1.13) to account for the broadening of the EPR line, and in Equation (2.5) to account for the cut-off of the 2-spin correlation functions $g_{ij}^+(t) = \langle S_i^+(t)\, S_j^-\rangle$. As a matter of fact, this is a general feature: for describing the cut-off of a correlation function $\langle A_n(t)\, A_{n'}^\dagger\rangle$, one has to consider the new correlation function:

$$\left\langle \left[\mathscr{D}, \sum_n A_n\right](t) \left[\sum_n A_n^\dagger, \mathscr{D}\right]\right\rangle, \tag{3.5}$$

where $\Sigma_n A_n$ is the 'total' operator corresponding to the wave vector $q = 0$.

The dipolar term is the sum of secular ($\mathscr{D}_0$) and nonsecular ($\mathscr{D}_{\pm 1}, \mathscr{D}_{\pm 2}$) terms. For magnetic chains these terms can be expressed in a rather simple way, since all vectors $\mathbf{r}_n$ joining two electron spins on the same chain have the same direction:

$$\mathscr{D}_0 = \frac{1}{4}(1 - 3\cos^2\theta)\,\hbar\gamma_0{}^2/c^3 \sum_i \sum_{n(\neq 0)} \left|\frac{1}{n}\right|^3 (3S_i^z S_{i+n}^z - \mathbf{S}_i\mathbf{S}_{i+n})$$

$$\mathscr{D}_{\pm 1} = -\frac{3}{4}\sin\theta\cos\theta\, e^{\mp i\phi}\,\hbar\gamma_0{}^2/c^3 \sum_i \sum_{n(\neq 0)} \left|\frac{1}{n}\right|^3 (S_i^z S_{i+n}^{\pm} + S_n^{\pm} S_{i+n}^z)$$

$$\mathscr{D}_{\pm 2} = -\frac{3}{8}\sin^2\theta\, e^{\mp 2i\phi}\,\hbar\gamma_0{}^2/c^3 \sum_i \sum_{n(\neq 0)} \left|\frac{1}{n}\right|^3 S_i^{\pm} S_{i+n}^{\pm}\,. \tag{3.6}$$

Here, $\theta$ and $\phi$ are the azimuthal and polar angles defining the chain direction $\mathbf{c}$ with respect to the magnetic field. In these equations $\gamma_0$ is the gyromagnetic ratio and $c$ is the distance between two neighbouring spins along a chain. In 2-d or 3-d spin lattices, on the other hand, one has to use the general expressions for $\mathscr{D}_m$ where $n\mathbf{c}$ is replaced by lattice vectors $\mathbf{r}_n$ of different orientations, and the angular dependent terms of Equation (3.6) now depend on the individual directions $\theta_n$ and $\phi_n$, and have to be included in the summation, compare Equation (3.16) and (3.17) below.

Under the experimental condition of high field, $\omega_0 \gg \Delta H$, the function $T^+(t)$ can be written as the sum of four contributions

$$T^+(t) = \sum_{m=-2}^{+1} \langle e^{i\mathscr{H}t}[\mathscr{D}_m, S^+]\, e^{-i\mathscr{H}t}[S^-, \mathscr{D}_{-m}]\rangle / \langle |S^+|^2\rangle. \tag{3.7}$$

Usually, an extra approximation is made: the dipolar term $\mathscr{D}$ is neglected in the time modulation Hamiltonian $\mathscr{H}$, which is therefore reduced to the Hamiltonian $\mathscr{H}_0 = \mathscr{E} + \mathscr{Z}$ of our simplified model. Equation (3.7) can then be rewritten as

$$T^+(t) \approx \sum_{m=-2}^{+1} \langle e^{i\mathscr{E}t}[\mathscr{D}_m, S^+]\, e^{-i\mathscr{E}t}[S^-, \mathscr{D}_{-m}]\rangle\, e^{i(m+1)\omega_0 t} / \langle |S^+|^2\rangle. \tag{3.8}$$

This means that in the wings of the EPR line the function $T^+(\omega)$ describes the fluctuations of a pure Heisenberg system. This basic feature arises in the Kubo-Tomita formalism [1]. Equation (3.8) has been shown to be adequate for 3-d systems, where it can be used to describe not only the wings but also the central part of the EPR line [13]. The essential reason is that in 3-d systems Re $M^+(\omega)$ does not exhibit any drastic behaviour at low frequencies. In particular, around $\omega \approx \omega_0$, Re $M^+(\omega)$ is nearly frequency independent. The EPR line has a Lorentzian shape, and one can write

$$\Delta H = \mathrm{Re}\, M^+(\omega_0) \approx \pi\, T^+(\omega_0). \tag{3.9}$$

In low-d systems, however, the same description leads to dramatic effects because of the diffusive behaviour of the 4-spin correlation functions involved in $T^+(t)$.

### 3.1. DIFFUSION OF 4-SPIN CORRELATION FUNCTIONS

Defining the bilinear spin operators

$$\begin{aligned}
A_n^0 &= \sum_i (S_i^z S_{i+n}^+ + S_i^+ S_{i+n}^z) \\
A_n^{+1} &= \sum_i S_i^+ S_{i+n}^+ \\
A_n^{-1} &= \sum_i (S_i^z S_{i+n}^z - \frac{1}{3} \mathbf{S}_i \mathbf{S}_{i+n}) \\
A_n^{-2} &= \sum_i (S_i^z S_{i+n}^- + S_i^- S_{i+n}^z)
\end{aligned} \tag{3.10}$$

and considering only the simpler case of magnetic chains, Equation (3.7) can be expressed as

$$T^+(t) = \omega_d^2 \{F_0^2 \Gamma^0(t) + 4F_1^2 \Gamma^{+1}(t) + + 6F_1^2 \Gamma^{-1}(t) + F_2^2 \Gamma^{-2}(t)\},$$

where

$$\begin{aligned}
\omega_d^2 &= \tfrac{3}{4} S(S+1) (\hbar \gamma_0^2 / c^3)^2 \\
F_0 &= 3 \cos^2 \theta - 1 \\
F_1 &= \sin \theta \cos \theta \\
F_2 &= \sin^2 \theta,
\end{aligned}$$

as already introduced in Equation (1.14), and

$$\Gamma^m(t) = \sum_{\substack{n, n' \\ (\neq 0)}} \left| \frac{1}{n} \right|^3 \left| \frac{1}{n'} \right|^3 \phi_{n,n'}^m(t). \tag{3.11}$$

Here, the 4-spin correlation functions are defined by

$$\phi_{n,n'}^m(t) = \langle e^{i\mathcal{H}t} A_n^m e^{-i\mathcal{H}t} (A_{n'}^m)^\dagger \rangle / \langle |A_n^m|^2 \rangle. \tag{3.12}$$

If we neglect $\mathcal{D}$ in the time modulation Hamiltonian $\mathcal{H} = \mathcal{Z} + \mathcal{E} + \mathcal{D}$, as in the 3-d case, the functions $\phi_{n,n'}^m(t)$ show specific rotationally invariant properties [14]. We can write

$$\phi_{n,n'}^m(t) \approx \tilde{\phi}_{n,n'}(t)\, e^{i(m+1)\omega_0 t}$$

where the function

$$\tilde{\phi}_{n,n'}(t) = \langle e^{i\mathscr{E}t} A_n^m e^{-i\mathscr{E}t} (A_{n'}^m)^\dagger \rangle / \langle |A_n^m|^2 \rangle \qquad (3.13)$$

is in fact independent of $m$. Using Equation (3.9) one obtains the expression for $\Delta H$ given in (1.14), where $\tilde{\Gamma}(\omega)$ is the spectral density of the function

$$\tilde{\Gamma}(t) = \sum_{\substack{n,n'\\(\neq 0)}} \left|\frac{1}{n}\right|^3 \left|\frac{1}{n'}\right|^3 \tilde{\phi}_{n,n'}(t).$$

In this formalism the spin dynamics are described by the 4-spin correlation functions $\tilde{\phi}_{n,n'}(t)$. To evaluate their time dependence, we can follow the procedure given previously for the 2-spin correlation functions. For instance, for describing the long time behaviour it is worth noting that the 'total' operators $\Sigma_n A_n^m$ remain constants of motion with respect to the Heisenberg Hamiltonian

$$\left[\mathscr{E}, \sum_n A_n^m\right] = 0.$$

This suggests that a hydrodynamic description can also be used for $\tilde{\phi}_{n,n'}(t)$ at long times ($\omega_e t \gg 1$) yielding the following diffusion law:

$$\tilde{\phi}_{n,n'}(t) \sim (\hat{D}t)^{-d/2}. \qquad (3.14)$$

Other descriptions relying on decoupling procedures [14] give the same result and predict for the diffusion coefficient

$$\hat{D} = 2D,$$

where $D$ was defined for the 2-spin correlation functions.

The 2-spin and 4-spin correlation functions $\tilde{g}_n^a(t)$ and $\tilde{\phi}_{n,n'}(t)$ are seen to behave in a very similar way. One purpose of EPR experiments in low-d systems was to check the validity of the diffusive assumption for the 4-spin correlation functions. However, a modification of the formalism must be introduced to remove the unrealistic divergence of $\Gamma(\omega)$ at $\omega = 0$, which would result in an infinite value of the EPR linewidth $\Delta H$, see Equation (1.14).

It has been suggested that Equation (3.12) should be used instead of (3.13), or equivalently (3.7) instead of (3.8) [10, 11, 16]. In other words, it would be sufficient to retain the dipolar term $\mathscr{D}$ in the time modulation Hamiltonian of $T^+(t)$, the memory spectrum Re $M^+(\omega)$ remaining well described by $T^+(\omega)$:

$$\text{Re}\, M^+(\omega) = \pi\, T^+(\omega). \qquad (3.15)$$

This point, however, needs to be carefully discussed.

Before going into further detail, we still have to generalize Equation (3.11) for a 2-d (or 3-d) magnetic lattice. Owing to the more complicated dipolar coefficients, we have to write

$$T^+(t) = \omega_d^2 \left\{ \sum_{n,n'} F_n^0 F_{n'}^0 \, \phi_{n,n'}^0(t) + 4F_n^{+1}F_{n'}^{-1} \, \phi_{n,n'}^{+1}(t) \right.$$
$$\left. + 6F_n^{+1}F_{n'}^{-1} \, \phi_{n,n'}^{-1}(t) + F_n^{+2}F_{n'}^{-2} \, \phi_{n,n'}^{-2}(t) \right\}, \tag{3.16}$$

where $\phi_{n,n'}^m(t)$ and $\omega_d^2$ have the same meaning as in Equation (3.11) and (3.12), apart from normalizing the dipolar coefficients to the lattice parameter $a$ of a square spin lattice. The geometrical coefficients $F_n^m$ are defined as

$$\begin{aligned} F_n^0 &= (3\cos^2\theta_n - 1)(a/r_n)^3 \\ F_n^{\pm 1} &= \sin\theta_n \cos\theta_n \, e^{\mp i\phi_n} (a/r_n)^3 \\ F_n^{\pm 2} &= \sin^2\theta_n \, e^{\mp 2i\phi_n} (a/r_n)^3. \end{aligned} \tag{3.17}$$

Here, $\theta_n$ and $\phi_n$ refer to the directions of the individual lattice vectors $\mathbf{r}_n = \mathbf{r}_i - \mathbf{r}_j$ joining two spins at different sites, and the summation has to be performed over all pairs of interacting spins. Apparently, this generalization will not affect any of the arguments justifying the hydrodynamic description of $\tilde{\phi}_{n,n'}(t)$.

In the following sections we will mainly discuss the long-time behaviour of $T^+(t)$. In this limit, following Equation (3.14), the $\tilde{\phi}_{n,n'}(t)$ become independent of $n$ and $n'$, and the geometrical coefficients can be written as $|\Sigma_n F_n^m|^2$. Evaluating these lattice sums for a square spin lattice, the same angular dependence with respect to the direction of $\mathbf{H}$ is obtained as in the 1-d case [15]:

$$\left| \sum_n F_n^m(\theta) \right|^2 \sim (F_m(\theta))^2.$$

We will use this simpler form in the following discussion of characteristic long time effects in 2-d magnetic lattices, but we have to bear in mind that different coefficients must be used for short or intermediate times (compare Section 4.2.1).

### 3.2. SECULAR CONTRIBUTION $\mathscr{D}_0$

When the magnetic field $\mathbf{H}$ is parallel to the chain axis or perpendicular to the magnetic plane ($\theta = 0$), only the secular term $\mathscr{D}_0$ contributes to the EPR linewidth. Equation (3.7) then becomes

$$T^+(t) = \langle e^{i(\mathscr{E}+\mathscr{D}_0)t} [\mathscr{D}_0, S^+] \, e^{-i(\mathscr{E}+\mathscr{D}_0)t} [S^-, \mathscr{D}_0] \rangle \, e^{i\omega_0 t} / \langle |S^+|^2 \rangle,$$

which involves the 4-spin correlation functions

$$\phi_{n,n'}^0(t) = \langle e^{i(\mathscr{E}+\mathscr{D}_0)t} A_n^0 \, e^{-i(\mathscr{E}+\mathscr{D}_0)t} (A_{n'}^0)^\dagger \rangle \, e^{i\omega_0 t} / \langle |A_n^0|^2 \rangle \tag{3.18}$$

with $A_n^0$ defined in Equation (3.10). At this stage it is essential to reconsider the exact Equation (3.3) which reminds us that the spectral representation $T^+(\omega)$ of $T^+(t)$ is equal to zero at $\omega = \omega_0$: $T^+(\omega_0) = 0$. This point is illustrated in Figure 3, where the functions $F^+(\omega)$, $T^+(\omega)$ and $\operatorname{Re} M^+(\omega)$ are schematically represented. In the central part of the EPR line $T^+(\omega)$ and $\operatorname{Re} M^+(\omega)$ have completely different

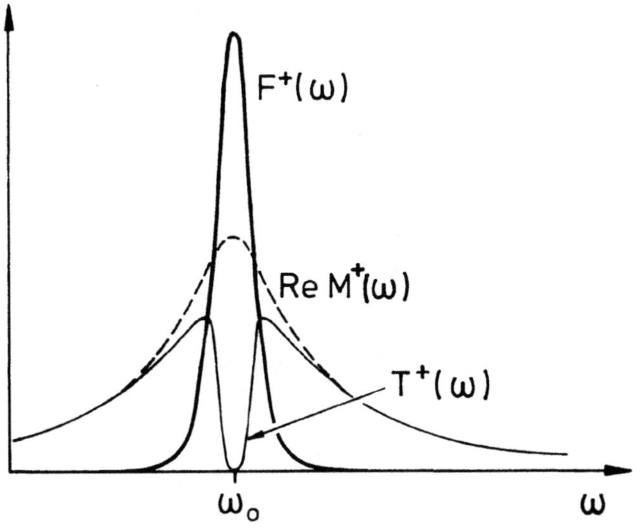


Fig. 3. EPR spectrum $F^+(\omega)$, memory spectrum Re $M^+(\omega)$ and $\pi T^+(\omega)$ for the secular contribution $\mathscr{D}_0$.

behaviours. For $|\omega - \omega_0| \lesssim \Delta H$, Equation (3.15) is certainly not correct, but it is a good approximation in the wings of the line where the fluctuations are actually dominated by the Heisenberg Hamiltonian. For $|\omega - \omega_0| \gg \Delta H$, $\mathscr{D}_0$ can definitely be neglected in the time modulation Hamiltonian of Equation (3.15), which means that the dynamics is governed by diffusive processes. In the absence of a better theory, only a phenomenological description can be given for the central part of the EPR line. It relies on the cut-off concept. We may write for a 2-d system

$$\operatorname{Re} M^+(\omega) = F_0{}^2 \frac{\hat{A}\omega_d{}^2}{4\pi\hat{D}} \ln \left\{ \frac{4\pi\hat{D}}{\sqrt{(\omega - \omega_0)^2 + (\Gamma_0^+)^2}} \right\} + \hat{B}_0, \tag{3.19}$$

where $\Gamma_0^+$ is a 'phenomenological' cut-off. In this equation $\hat{A}$ is a coefficient which depends on the 2-d spin lattice and $\hat{B}_0$ takes into account a possible contribution from the high frequency fluctuations of the memory spectrum Re $M^+(\omega)$.

To evaluate $\Gamma_0^+$, we present the heuristic and simple argument that the cut-off should remain comparable to the EPR linewidth itself [17]. We write simply

$$\Gamma_0^+ \approx \Delta H.$$

Equation (3.17) shows that around $\omega \approx \omega_0$ the memory spectrum Re $M^+(\omega)$ depends appreciably on the frequency $\omega - \omega_0$. Therefore, one expects the EPR line to have a non-Lorentzian shape characteristic of the dimensionality. From the definition $\Delta H$ = Re $M^+(\omega_0)$ we obtain the self-consistent equation for the linewidth

$$\Delta H = F_0^2 \frac{\hat{A}\omega_d^2}{4\pi\hat{D}} \ln\left\{\frac{4\pi\hat{D}}{\Delta H}\right\} + \hat{B}_0. \tag{3.20}$$

### 3.3. NONSECULAR CONTRIBUTIONS

For another special direction of **H**, i.e. at the 'magic' angle where $3\cos^2\theta = 1$, the secular term $\mathscr{D}_0$ vanishes. In that case only the nonsecular terms $\mathscr{D}_m$, with $m = \pm 1, -2$ contribute to the memory spectrum.

In section 3.1 we have seen that by neglecting the dipolar interaction in the time modulation of $T^+(t)$, Equation (3.7), the local fluctuations can be described by a unique function $\tilde{\phi}_{n,n'}(t)$ which is independent of $m$. The same frequency spectrum $\Gamma(\omega)$ is therefore obtained for the different functions $\Gamma_m(t)$, except that they are shifted from the central value $\omega_0$ by $m\omega_0$. These shifts are approximately maintained when the full Hamiltonian is considered in $T^+(t)$, Equation (3.7).

For the specific term $\mathscr{D}_{+1}$, the functions $T^+(\omega)$, Re $M^+(\omega)$ and the resulting EPR line $F^+(\omega)$ are represented in Figure 4. Again, except for the central part where it fails completely, the relation (3.15) gives a good description of the EPR line.

In contrast to the previous case, the description of the memory spectrum in the wings of the EPR line is better obtained from Equation (3.7) than from (3.8) which would lead to a divergence of $M^+(\omega)$ at $\omega = 2\omega_0$. For the 4-spin correlation function $\phi^{+1}_{n,n'}(t)$, Equation (3.12), the dipolar term in the time modulation Hamiltonian introduces a cut-off in a very formal way.

For the central part of the line there is no good theoretical expression to work with. As in the 3-d case, we simply assume that Re $M^+(\omega)$ does not differ very much from the behaviour predicted by $T^+(\omega)$ on each side of the EPR line (this is equivalent to neglecting $\mathscr{D}$ in the time modulation Hamiltonian and using (3.8)). In low-d systems this behaviour is expected to be diffusive. For 2-d systems the following expression can be used:

$$\text{Re } M^+_{+1}(\omega) = 4F_1^2 \frac{\hat{A}\omega_d^2}{4\pi\hat{D}} \ln\left\{\frac{4\pi\hat{D}}{\sqrt{(\omega - 2\omega_0)^2 + (\Gamma^+_{+1})^2}}\right\} + \hat{B}_{+1} \tag{3.21}$$

where $\Gamma^+_{+1}$ is the cut-off of the correlation function $\phi^{+1}_{n,n'}(t)$. In the central part of the line ($\omega \approx \omega_0$) Re $M^+(\omega)$ is practically frequency independent. At the magic

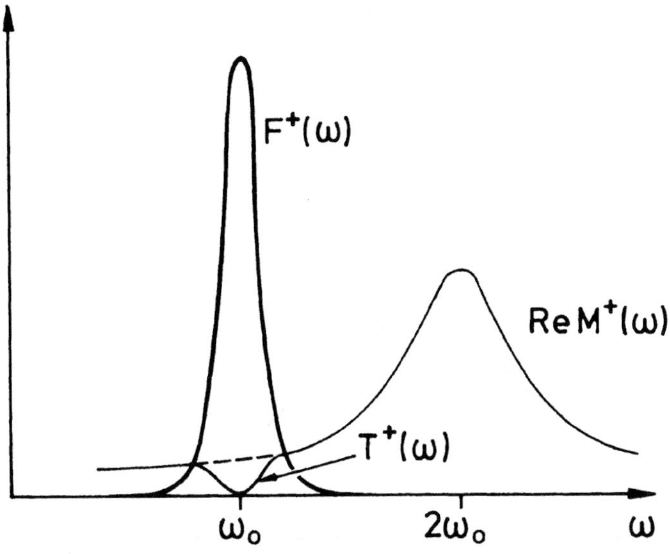


Fig. 4. EPR spectrum $F^{+}(\omega)$, memory spectrum Re $M^{+}(\omega)$ and $\pi T^{+}(\omega)$ for the nonsecular contribution $\mathscr{D}_{+1}$.

angle the EPR line is therefore expected to show a Lorentzian shape. Since $\omega_0 \gg \Gamma^{+}_{+1}$, the contribution of $\mathscr{D}_{+1}$ to the EPR linewidth is evaluated as

$$4F_1^{\,2}\,\frac{\hat{A}\omega_d^{\,2}}{4\pi\hat{D}}\ln\left\{\frac{4\pi\hat{D}}{\omega_0}\right\} + \hat{B}_{+1}.$$

A similar description can be given for the two other nonsecular terms $\mathscr{D}_{-1}$ and $\mathscr{D}_{-2}$. The shape of the resulting memory spectrum at the magic angle is given in Figure 5. In the central part of the line all these contributions follow a diffusive law. In $\Delta H$ the diffusion is probed at $\omega = \omega_0$ and $2\omega_0$, and the expression for the linewidth can be completed as

$$\Delta H = \frac{\hat{A}\omega_d^{\,2}}{4\pi\hat{D}}\left[10F_1^{\,2}\ln\left\{\frac{4\pi\hat{D}}{\omega_0}\right\} + F_2^{\,2}\ln\left\{\frac{4\pi\hat{D}}{2\omega_0}\right\}\right] + \hat{B}. \qquad (3.22)$$

So, at the magic angle, a study of $\Delta H$ as a function of $\omega_0 \sim H$ offers a way of probing the diffusive behaviour of the correlation function $\tilde{\phi}_{n,n'}(t)$.

In the general case, for any direction of the magnetic field, the secular ($m = 0$) and nonsecular ($m = \pm 1, -2$) terms together contribute to the EPR linewidth. The memory spectrum consists of four contributions as shown in Figure 6 for low-d systems. The discussion developed for the two specific cases, $\theta = 0$ and

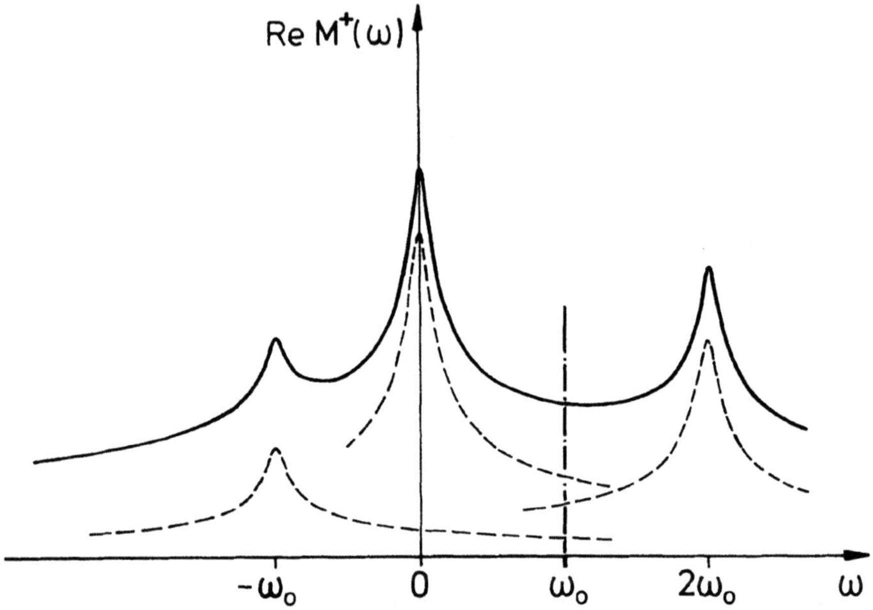


Fig. 5. Memory spectrum at the 'magic angle', where $3\cos^2\theta - 1 = 0$ (solid line). The EPR linewidth probes the sum of all nonsecular contributions (dashed) at $\omega_0$.

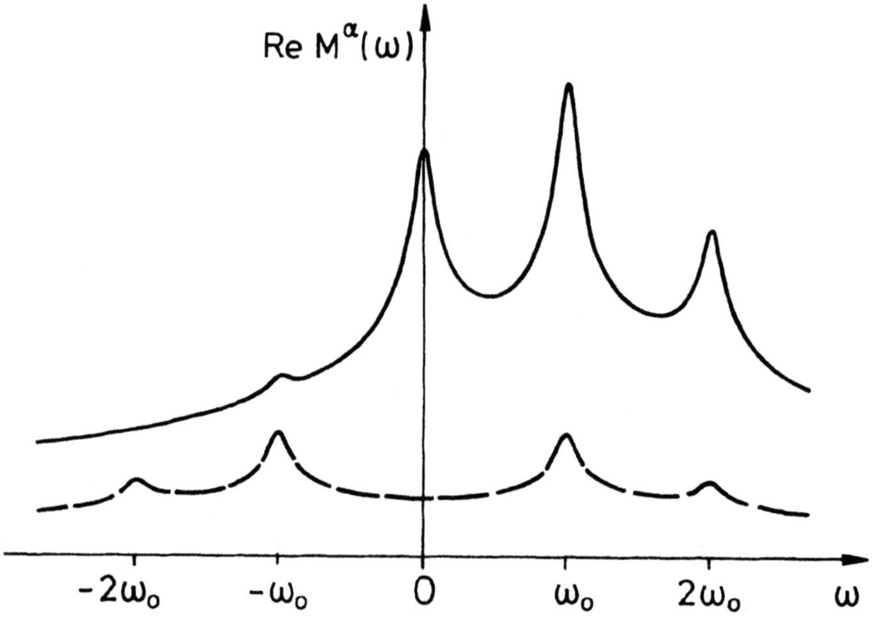


Fig. 6. Memory spectrum of the spin components $S^+$ (solid line) and $S^z$ (dashed line) for arbitrary orientation.

$3\cos^2\theta = 1$, can be retained. For 2-d systems $\Delta H$ is given by

$$\Delta H = \frac{\hat{A}\omega_d^2}{4\pi\hat{D}}\left[(F_0(\theta))^2 \ln\left\{\frac{4\pi\hat{D}}{\Delta H}\right\} + 10(F_1(\theta))^2 \ln\left\{\frac{4\pi\hat{D}}{\omega_0}\right\} + \right.$$

$$\left. + (F_2(\theta))^2 \ln\left\{\frac{4\pi\hat{D}}{2\omega_0}\right\}\right] + \hat{B}(\theta). \tag{3.23}$$

Since $\Delta H \ll \omega_0$, the width is dominated by the secular contribution. This is another interesting feature of low-d systems where the EPR line is characterized by a specific anisotropy [9, 15].

### 3.4. SATELLITE LINE

As shown for the general case in Figure 6, the memory spectrum of low-d systems exhibits maxima far in the wings of the EPR line. The problem is to study the effect of such peaks in the EPR spectrum. Let us consider, for instance, the peak originating from the nonsecular term $\mathscr{D}_{+1}$ which occurs at $\omega = 2\omega_0$. Around this frequency, $F^+(\omega)$, Equation (3.2), is well approximated by:

$$F^+(\omega \approx 2\omega_0) \approx \frac{1}{\pi}\frac{\text{Re}\, M^+(\omega)}{(\omega - \omega_0)^2} \approx \frac{1}{\pi}\frac{\text{Re}\, M^+(\omega)}{\omega_0^2}. \tag{3.24}$$

Corresponding to the peak of Re $M^+(\omega)$ at $2\omega_0$, an extra line — a satellite line — will occur in the EPR spectrum. This is a general feature of spin resonance lines. For instance, the so-called 'forbidden' lines commonly observed in NMR can be explained in the same way. In that case the memory spectrum is a narrow function, as narrow as the NMR line itself. For this reason such satellite lines can easily be observed.

In low-d systems the peaks result from the diffusive behaviour of the 4-spin correlation functions. For the term $\mathscr{D}_{+1}$, the function to be considered is $\phi^{+1}_{n,n'}(t) = \langle A^{+1}_n(t)\, A^{-1}_{n'}\rangle / \langle |A^{+1}_n|^2\rangle$. The corresponding memory spectrum has been described in Section 3.3. It is explicitly given by Equation (3.21). The width of the satellite line gives a direct measurement of the cut-off $\Gamma^+_{+1}$. For the evaluation of $\Gamma^+_{+1}$ we use the argument outlined in Equation (3.5). One has to consider the commutator:

$$\left[\mathscr{D}, \sum_n A^{+1}_n\right] = [\mathscr{D}, (S^+)^2]$$

and to build the correlation function:

$$\hat{T}^+(t) \sim \langle e^{i\mathscr{H}t}[\mathscr{D}, (S^+)^2]\, e^{-i\mathscr{H}t}[(S^-)^2, \mathscr{D}]\rangle.$$

The spectral representation of $\hat{T}^+(t)$ defines, in fact, a 'second-order' memory spectrum. The same discussion presented for Re $M^+(\omega)$ can be repeated here. If we neglect, for instance, $\mathscr{D}$ in the time modulation Hamiltonian $\mathscr{H}$ we obtain

$$\hat{T}^+(t) \sim \sum_{m=-2}^{+1} \langle e^{i\mathscr{E}t}[\mathscr{D}_m, (S^+)^2]\, e^{-i\mathscr{E}t}[(S^-)^2, \mathscr{D}_{-m}]\rangle\, e^{i(m+2)\omega_0 t}.$$

The corresponding spectral density therefore is the sum of four contributions centered at the frequencies $(m + 2)\omega_0$. Moreover, the correlation functions in $\hat{T}^+(t)$ involve six spin operators given by $\hat{A}_n^m = [A_n^m, (S^+)^2]$. The important point is that the total operators $\Sigma_n \hat{A}_n^m$ remain constants of motion with respect to $\mathscr{E}$:

$$\left[\sum_n \hat{A}_n^m, \mathscr{E}\right] = 0.$$

A diffusive behaviour can be reasonably expected for the 6-spin correlation functions. The description for $\Gamma_{+1}^+$ and $\Delta H$ are completely similar. The dominant contribution to $\Gamma_{+1}^+$ is associated with the secular term $\mathscr{D}_0$. It involves diffusion and cut-off processes. The nonsecular terms yield a purely diffusive contribution probed at the frequencies $\pm\omega_0$ and $-2\omega_0$. The similarity is so strong that, in practice, it is sufficient to write

$$\Gamma_{+1}^+ \approx 2\Delta H \tag{3.25}$$

which completes the self-consistent character of the description of the EPR linewidth in low-d systems. The factor 2 in Equation (3.25) results from the fact that the initial commutator contributes two equivalent terms

$$[\mathscr{D}, (S^+)^2] = S^+[\mathscr{D}, S^+] + [\mathscr{D}, S^+]S^+.$$

Equation (3.25) also shows that both the main EPR line and the satellite line follow the same anisotropy.

To conclude, we note that the 2-spin and 4-spin correlation functions, which explicitely occur in the formulation of the nuclear relaxation rate $1/T_1$ and of the EPR linewidth $\Delta H$, behave diffusively at long times. In low-d systems such a behaviour leads to a divergence of the corresponding spectral densities. In real systems this divergence is limited by cut-off processes associated with the anisotropic part of the Hamiltonian. For magnetic resonance experiments the evaluation of the cut-off frequency becomes the central point of the description, since it determines the low frequency behaviour of the local fluctuations. For the transverse 2-spin correlation functions $g_n^+(t) = \langle S_i^+(t)\, S_{i+n}^-\rangle$, the cut-off $\Gamma^+$ is given by the EPR linewidth $\Delta H$.

For the cut-off $\Gamma^z$ of the longitudinal 2-spin correlation functions $g_n^z(t) = \langle S_i^z(t)\, S_{i+n}^z\rangle$ the same analysis can be developed. According to Equation (3.5), one has to consider the correlation functions of the commutator

$$[\mathscr{D}, S^z] = \sum_{m=\pm1,\pm2} [\mathscr{D}_m, S^z].$$

Since $[\mathscr{D}_0, S^z] = 0$, only the nonsecular terms of the dipolar Hamiltonian are expected to play a role in the corresponding memory spectrum $M^z(\omega)$, as shown in Figure 5. Their contributions are essentially diffusive. For $\Gamma^z$, which is given by Re $M^z(\omega = 0)$, one obtains

$$\Gamma^z = \frac{\hat{A}\omega_D^{\ 2}}{4\pi\hat{D}} \left[ 4(F_1(\theta))^2 \ln \left\{ \frac{4\pi\hat{D}}{\omega_0} \right\} + 4(F_2(\theta))^2 \ln \left\{ \frac{4\pi\hat{D}}{2\omega_0} \right\} \right] + \hat{B}. \tag{3.26}$$

An $m = 2$ satellite line is also expected to show up in the longitudinal EPR spectrum at $\omega \approx 2\omega_0$:

$$F^z(\omega \approx 2\omega_0) \approx \frac{1}{\pi} \frac{\text{Re}\, M^z(\omega)}{4\omega_0^{\ 2}}. \tag{3.27}$$

The 4-spin correlation functions describing the transverse memory spectrum in Equation (3.24) and the longitudinal one in Equation (3.27) are the same. The width $\Gamma^z_{+2}$ of this longitudinal forbidden line therefore is described by Equation (3.25):

$$\Gamma^z_{+2} = \Gamma^+_{+1} \approx 2\Delta H. \tag{3.28}$$

Equations (3.23) and (3.26) give the cut-off frequencies of the longitudinal and of the transverse 2-spin correlation functions $g^z_n(t)$ and $g^+_n(t)$. They complete the description of the nuclear relaxation rate for 2-d systems given explicitly in Equation (2.10).

## 4. Experiments on Quasi 2-d Heisenberg Magnets

The best 2-d Heisenberg magnets known at present are layer-type perovskites with magnetic $S$-state ions, such as $K_2MnF_4$ or $(C_nH_{2n+1}NH_3)_2MnCl_4$. In these compounds the $Mn^{2+}$ ions couple antiferromagnetically within the layers. As a matter of fact, even in nearly ideal 2-d Heisenberg magnets the expected anomalies are less pronounced than in most linear chains, since the influence of diffusive modes diminishes with increasing dimensionality of the spin lattice. So, for probing the low frequency singularities discussed in the preceding sections, a very high experimental sensitivity is generally required.

For non-$S$-state ions the low-d anomalies will be more or less hidden by competing relaxation mechanisms resulting from spin-orbit coupling. A lot of EPR and NMR experiments have been performed on the isomorphous $Cu^{2+}$ compounds which, however, turn out to be less appropriate for such investigations: the antisymmetric exchange occuring in these compounds results in an additional spin-lattice relaxation, which generally exceeds the spin-spin relaxation of interest. Moreover, the anisotropic $g$-factor tends to broaden the line inhomogeneously. $Ni^{2+}$ and $Co^{2+}$ in most of the layer-type compounds show crystal field splittings which either introduce a large anisotropy or even change the type of spin-spin interaction from isotropic to planar or to Ising-like.

For all the above reasons we want to confine the following discussion of experimental results exemplarily to manganese layer-type compounds, and mainly to $(C_2H_5NH_3)_2MnCl_4$. For a more complete review the reader is referred to the literature [3, 18].

### 4.1. NMR EXPERIMENTS

The nuclear spin—lattice relaxation time $T_1$ was shown to directly probe the spectral density of local electron 2-spin correlation functions at low frequencies. So it should be possible to study the diffusive spin dynamics by means of the characteristic frequency dependence of $T_1$ and to evaluate the relevant diffusion constant $D$, provided that the hyperfine coupling coefficients are known with sufficient accuracy.

Although in 1-d systems this technique has been successfully applied in very detailed and elaborate studies of the electron spin dynamics [17, 19], there are only very few results yet on 2-d systems. Ferrieu and Boucher [20] used proton NMR to study the 2-d character of orthorhombic $Cu(NH_3)_4SO_4 \cdot H_2O$. Evidence for 2-d spin diffusion was found by Rappold [21] on $(C_2H_5NH_3)_2MnCl_4$. Nishihara *et al.* [22] also found a weak logarithmic frequency dependence of $1/T_1$ in $(CH_3NH_3)_2CuCl_4$.

For practical use we may simplify Equation (2.10) in the following way:

$$1/T_1 = a(\theta) - b(\theta) \ln \sqrt{\omega_N^2 + (\Gamma^z)^2} - c(\theta) \ln \omega_N. \tag{4.1}$$

In the first term we have included both the short-time contributions (term $B$ in Equation (2.10)) and part of the long-time contributions ($\sim$ ln $4\pi D$). $\Gamma^z$ denotes the cut-off frequency of the longitudinal spin fluctuations resulting, for instance, from the dipolar coupling of the electron spins or from weak interplane exchange coupling. According to its physical origin $\Gamma^z$ may also vary with orientation. The cut-off $\Gamma^+$ of the transverse spin fluctuations generally turns out to be negligible with respect to the much larger electronic Larmor frequency $\omega_0 = \omega_N \gamma_0/\gamma_N$, and one can map the logarithmic dependence on $\omega_0$ to a dependence on the measuring frequency $\omega_N$ simply by splitting off the constant contribution $\ln (\gamma_0/\gamma_N)$. Within the frequency range of a conventional pulsed NMR spectrometer and in the absence of other competing relaxation mechanisms, the logarithmic frequency dependence of the $c$-term should show up rather distinctly. Whether the $b$-term yields an additional logarithmic contribution to $1/T_1$ or not depends on the magnitude of $\Gamma^z$. Usually, electronic damping rates will be much larger than the nuclear Larmor frequency, and the $b$-term will be independent of frequency and will only diminish the $a$-term.

In Figure 7 we have presented $T_1$ data on $(C_2H_5NH_3)_2MnCl_4$ for two different orientations vs. a logarithmic frequency scale. For either orientation the data follow a straight line proving the expected diffusive dynamics. A sharp bend below 20 MHz indicates the changeover in the $b$-term at the cut-off frequency $\Gamma^z$, which fortuitously can be directly observed in this system. The value of $\Gamma^z$ is consistent with the observed EPR linewidth. This confirms that the cut-off originates from in-plane dipolar interactions rather than from a residual isotropic exchange coupling between the planes (which does not contribute to the EPR linewidth).

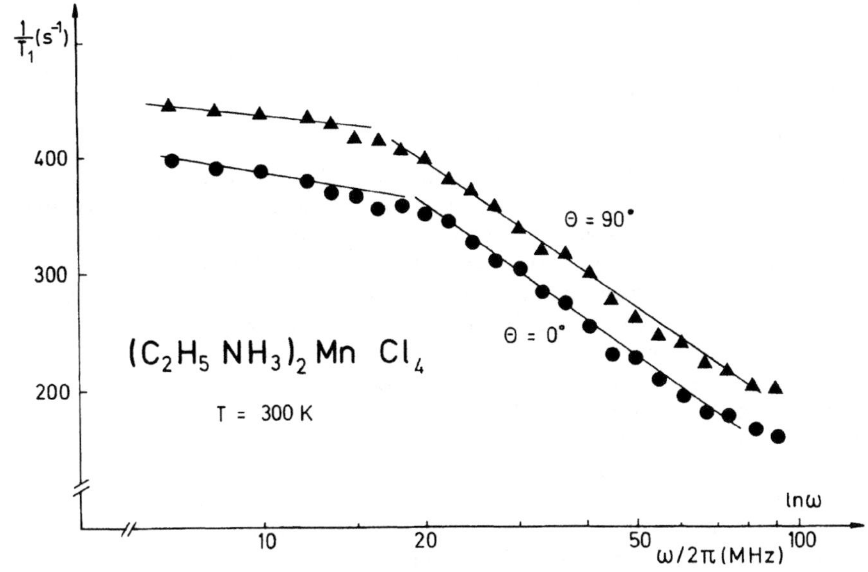


Fig. 7. Frequency dependence of the proton spin-lattice relaxation rate $1/T_1$ in $(C_2H_5NH_3)_2MnCl_4$ for the magnetic field parallel (▲) and perpendicular (●) to the layers.

The evaluation of the dipolar hyperfine coupling coefficients reveals two principal difficulties of proton NMR: the large number of protons in the magnetic unit cell and the uncertainty of their exact positions due to thermal rotations of the $CH_3$ or $NH_3$ groups. In $(C_2H_5NH_3)_2MnCl_4$ there are 64 inequivalent protons. Their mean positions are known from X-ray diffraction [23]. The observed NMR signal then represents the superposition of 64 single proton contributions which — due to their different strengths of dipolar coupling — may differ both in relaxation times and resonance fields. At room temperature the static dipolar shift remains sufficiently small, and the observed exponential decay of magnetization can be described by a 'mean relaxation time' resulting from the effective coupling coefficients:

$$A^{\alpha} = N_p^{-1} \sum_p A_p^{\alpha} .$$

Here $N_p$ denotes the total number of protons and the $A_p^{\alpha}$ denote their individual couplings. Using these effective coefficients and the experimental slopes of Figure 7, the diffusion constant is evaluated to amount to $D/k_B = (12 \pm 2)$ K (normalized to the lattice unit $a$). However, it should be emphasized that the error bar refers only to the experimental error and not to the uncertainty of the calculated hyperfine coefficients. Using $J/k_B = -5$ K, theoretical values of $D$ for the high temperature limit come close to this result [15, 24].

## 4.2. EPR EXPERIMENTS

EPR offers a variety of different possibilities to study directly or indirectly the diffusive decay of 4-spin correlation functions.

### 4.2.1. *Angular Dependence of Linewidth*

Conventional EPR investigations deal mainly with the resonance field and with the linewidth which — in our notation — is given by the real part of the transverse memory spectrum at resonance, Re $M^{+}(\omega_0)$, Equation (3.9). The simplest and most obvious '2-d anomaly' of the linewidth shows up in the characteristic angular dependence resulting from the dominating secular contribution. A typical example is given in Figure 8a. Increasing the angle $\theta$, which is the angle between the $c$-axis (perpendicular to the planes) and the magnetic field, from 0° to 90°, the experimental peak-to-peak linewidth rapidly decreases, passes a shallow minimum and increases again. A similar linewidth anisotropy has been observed in many other layer compounds [17, 25, 26]. This angular dependence is in contrast to the 3-d case where the linewidth shows a monotonic decrease proportional to $1 + \cos^2 \theta$ [1, 27]. The characteristic minimum occurring close to the magic angle $\theta = 55°$ is a pecularity of low-d systems. It corresponds to the vanishing secular contribution, which dominates the linewidth at other orientations.

It was first pointed out by Richards and Salamon [15] that, in contrast to the 1-d case, the short-time and the long-time part of $T^{+}(t)$ in a 2-d spin lattice contribute to the EPR linewidth with different angular dependencies. As an example, we consider here only the secular part. In the long-time limit, where the 4-spin correlation function $\tilde{\phi}_{n,n'}(t)$ is independent of $n$ and $n'$, the dipolar coupling coefficients for a square lattice, according to Equation (3.16), are given by

$$\left|\sum_n F_n^0\right|^2 = \frac{1}{4}(3\cos^2\theta - 1)^2 \left[\sum_n (a/r_n)^3\right]^2. \tag{4.2}$$

In the short-time limit, where $\tilde{\phi}_{n,n'}(t)$ is proportional to $\delta_{n,n'}$, one finds a quite different dependence given by

$$\sum_n \left|F_n^0\right|^2 = \left\{\frac{1}{4}(3\cos^2\theta - 1)^2 + \frac{9}{8}\sin^4\theta\right\} \sum_n (a/r_n)^6 + $$
$$+ \frac{9}{8}\sin^4\theta\cos 4\alpha \sum_n e^{4i\phi_n}(a/r_n)^6 . \tag{4.3}$$

Here, $r_n$ denotes the spacing between two interacting spins and $\phi_n$ denotes the direction of the corresponding lattice vector with respect to the $a$-axis. $\alpha$, finally, describes an additional rotation of the whole crystal around its $c$-axis. It is important to note that in a square lattice $(\Sigma_n r_n^{-3})^2$ exceeds $\Sigma_n r_n^{-6}$ by nearly a factor of twenty. As a consequence, the term $B(\theta)$ in Equation (3.23) depends more sensitively on the appropriate weighting of $\tilde{\phi}_{n,n'}(t)$ for intermediate times than on

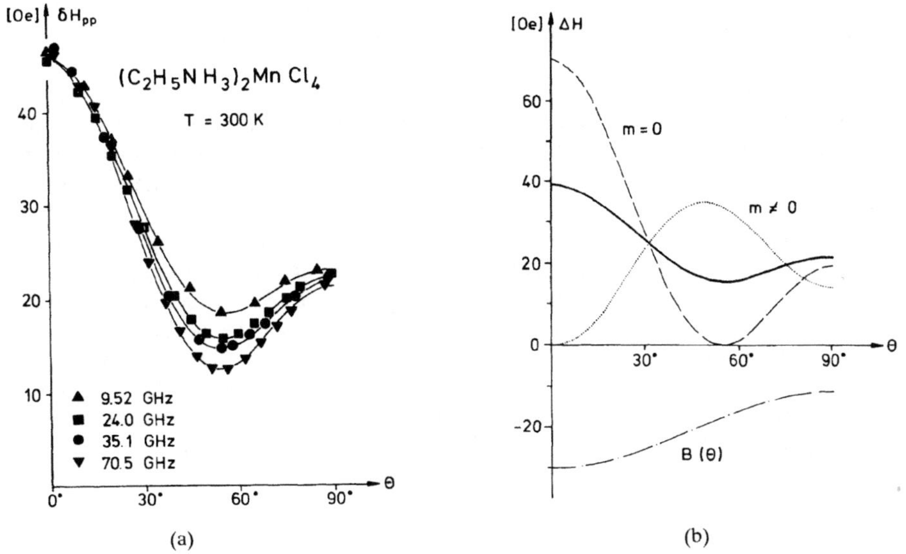


Fig. 8. (a) Angular dependence of the experimental peak-to-peak linewidth $\delta H_{pp}$ of $(C_2H_5NH_3)MnCl_4$ for different frequencies (solid line: theory). (b) Theoretical angular dependence of the EPR linewidth $\Delta H$ at 9.5 GHz (solid line). The secular (dashed) and nonsecular (dotted) contributions and the correcting term $B$ (dash-dotted) are given separately.

the short-time contribution at all. In other words, when drawing any quantitative conclusions from the observed angular dependence of linewidth, it is very important to treat the dependence on $n$, $n'$ — or equivalently in reciprocal space, the dependence on $q$, $q'$ — of these coefficients with sufficient accuracy.

For the case of $(C_2H_5NH_3)_2MnCl_4$ we have presented in Fig. 8b the angular dependence of the different contributions to $\Delta H$ according to Equation (3.23). For intermediate times a realistic matching was obtained by an exact numerical calculation of the dipolar coefficient while approximating $\tilde{\phi}_{n,n'}(t)$ by a purely diffusive decay (by analogy with Equation (2.4), replacing $\mathbf{r}$ by $\mathbf{r}_n - \mathbf{r}_{n'}$). This way, the correction term $B(\theta)$ in Equation (3.23) essentially compensates the overestimation of short- and intermediate-time contributions resulting from the improper coefficients (4.2). $B(\theta)$ therefore comes out negative.

To get a direct comparison with our experimental data, we have also calculated the peak-to-peak linewidth $\delta H_{pp}$ by taking into account the non-Lorentzian lineshape (see Section 4.2.4). It gives rise to a relative increase of $\delta H_{pp}$ at $\theta = 0°$ with respect to the half width at half maximum $\Delta H$. As shown in Figure 8a, the resulting theoretical angular dependence (solid line) describes our data very well. The prevailing $(3\cos^2\theta - 1)^2$ dependence supports the conclusion that the important contributions to the EPR linewidth mainly result from the low frequency fluctuations of the secular part. Due to competing fluctuations of higher frequency, the characteristic minimum at the magic angle is, in fact, less pronounced than in typical 1-d systems like TMMC — where such fluctuations turn out to be negligible — but it still differs markedly from the common 3-d behaviour.

### 4.2.2. *Frequency Dependence of Magic Angle Linewidth*

More conclusive than the preceding indirect proof of diffusive fluctuations are experiments which directly aim at the frequency or field dependence of the memory spectrum. Such information can, for instance, be obtained from the frequency dependence of the linewidth.

This technique was first used by Lagendijk and Siegel for studying the diffusive spin dynamics of 1-d systems [28] and could also succesfully be applied to the analysis of the weak logarithmic frequency dependence of non-secular resonances in 2-d systems [29, 30]. It is limited by the condition that the observed linewidth has to be much smaller than the resonance field. For $(C_2H_5NH_3)_2MnCl_4$ in X-band, for example, the ratio of $\omega_0/\Delta H$ is of the order of 100, so the condition is well satisfied. Due to the vanishing secular contribution, this analysis is most simply performed at the magic angle $\theta = 55°$, where the lineshape is Lorentzian. For this special orientation the peak-to-peak linewidth is given by

$$\delta H_{pp} = 1.15 \text{ Re } M^+(\omega_0) = \text{const} - 32.2\ \omega_d^2/\hat{D} \ln \omega. \tag{4.4}$$

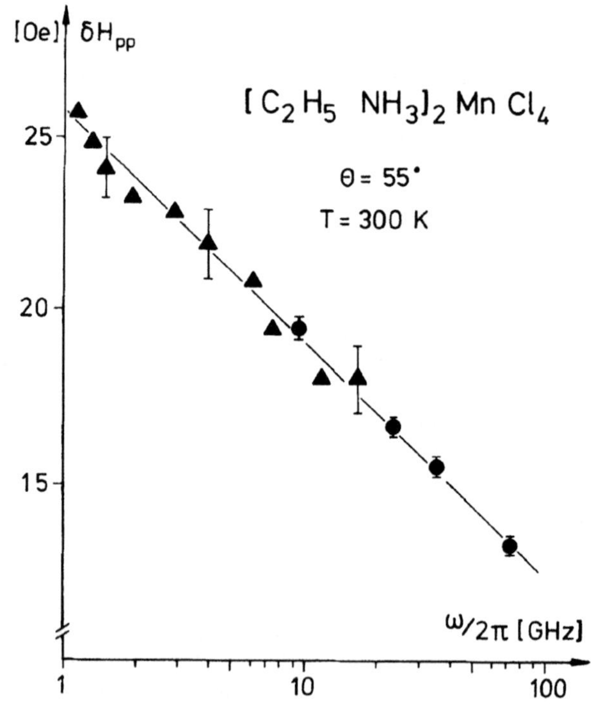


Fig. 9. Frequency dependence of linewidth at the 'magic angle' $\theta = 55°$. (Data from refs. [29] ● and [30] ▲.)

Here, the factor 1.15 describes the ratio of $\delta H_{pp}/\Delta H$ for a Lorentzian line, and the coefficient $\hat{A}$ in Equation (3.22) amounts to 80.6 for a square lattice. In Figure 9 the magic angle linewidth of $(C_2H_5NH_3)_2MnCl_4$ has been plotted against a logarithmic frequency scale. Within the limits of experimental error the data fit a straight line well, which has to be taken as direct experimental evidence for the diffusion of the 4-spin correlation functions. From the slope one can directly evaluate the diffusion constant. Normalized to the lattice unit $a$, $\hat{D}/k_B$ amounts to $(26 \pm 4)$ K. This is about twice the value of the 2-spin diffusion coefficient $D/k_B$ obtained from NMR experiments.

#### 4.2.3. *Dynamic Shift*

In principle, the same kind of information can be obtained from the dynamic shift of the resonance field which is given by the imaginary part of the memory spectrum Im $M^+(\omega_0)$, Equation (3.1). Compared to the frequency dependence of the linewidth, however, this technique suffers from two severe disadvantages: as a general matter of fact, the imaginary contributions of the different resonances of $M^+(\omega)$ partly cancel each other owing to different sign. Moreover, in the case of 2-d spin diffusion, the dynamic shift resulting from each nonsecular resonance is proportional to

$$\mathrm{Im}\,\ln(im\omega_0 + \Gamma_m^+) = \arctan(m\omega_0/\Gamma_m^+), \qquad m = \pm 1, -2.$$

So, for a small cut-off $\Gamma_m^+$, Im $M^+(\omega_0)$ nearly behaves as a step function and shows an even weaker frequency dependence than the real part.

Recently a different kind of dynamic shift resulting from the low-symmetric spin distribution of 1-d and 2-d lattices was observed and claimed to present a typical effect of low dimensionality [31]. A critical examination of the underlying theory, however, attributes this shift to special apparatus conditions (linear polarization of the microwave) rather than to low-d spin dynamics [32]. So, due to other competing shift mechanisms, which may be even larger than the interesting one, and due to experimental uncertainly of the $g$-factor, it would be very difficult to probe the diffusive dynamics by analyzing the dynamic shift. Therefore, in contrast to the 1-d case [28], experimental results on 2-d systems are not yet available.

#### 4.2.4. *Lineshape of the Main Line*

It is also possible to single out particular peaks of $M^+(\omega)$ directly from the EPR spectrum. Starting from the general expression (3.1) we have to keep in mind that for technical reasons the EPR usually records the field dependence of $F^+$ rather than its frequency dependence. The constant value of $\omega$ then will define the resonance field $H_0 = \omega/\gamma_0$. Considering a 3-d system, the constant memory spectrum results in a Lorentzian line profile. In 1-d or 2-d systems, close to the resonance field $H_0$, the field dependence of the secular resonance of Re $M^+(H)$ competes with the much stronger resonance behaviour of the denominator of $F^+(H)$, thus leading to a more or less pronounced deviation from the Lorentzian lineshape. The deviation occurs most distinctly at $\theta = 0°$, where the secular contribution shows a maximum, and vanishes at the magic angle.

The non-Lorentzian lineshape of 2-d systems was first observed in $K_2MnF_4$ by

Richards and Salamon [15] and was analysed by plotting the inverse signal intensity versus $(H - H_0)^2$. Within this diagram a Lorentzian is given by a straight line while the expected 2-d lineshapes are represented by slightly curved lines, where the deviation from the Lorentzian depends on orientation. Due to the weak logarithmic field dependence the observed deviations from a Lorentzian shape [15, 26] occurred less pronounced than in 1-d systems.

A more transparent quantitiative interpretation can be obtained from directly recording the field dependence of the memory spectrum. This was achieved from very sensitive measurements of both the absorption and dispersion spectra [33], which we will denote here by Re $F(H)$ rather than $F^+(H)$ and by Im $F(H)$, respectively. From the definition (3.1) and using the Kramers–Kronig relations, one may derive

$$\mathrm{Re}\, M^+(H) = \frac{1}{\pi} \frac{\mathrm{Re}\, F(H)}{[\mathrm{Re}\, F(H)]^2 + [\mathrm{Im}\, F(H)]^2} \tag{4.5a}$$

$$\mathrm{Im}\, M^+(H) = \frac{1}{\pi} \frac{-\,\mathrm{Im}\, F(H)}{[\mathrm{Re}\, F(H)]^2 + [\mathrm{Im}\, F(H)]^2} + H - H_0. \tag{4.5b}$$

In order to substantiate only a slight field dependence of $M^+(H)$ one has to detect the EPR signal far out on the wings and, thereby, to resolve an intensity range of three or four orders of magnitude. Re $F(H)$ and Im $F(H)$ are obtained by numerical integration of the experimental derivative spectra combined with a very sensitive base-line fit. Figure 10 shows the experimental memory spectrum

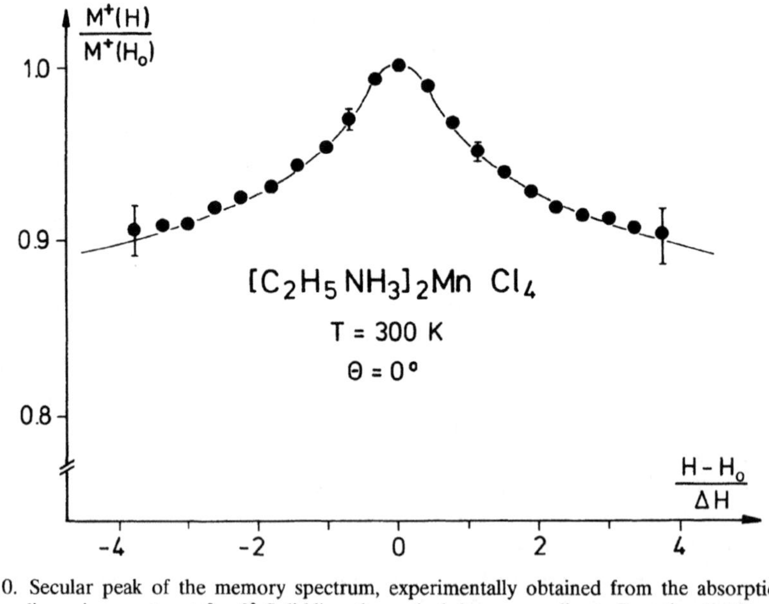


Fig. 10. Secular peak of the memory spectrum, experimentally obtained from the absorption and dispersion spectra at $\theta = 0°$. Solid line: theoretical shape according to Equation (3.19).

observed in $(C_2H_5NH_3)_2MnCl_4$. The peak at $H_0$ representing the secular resonance occurs most distinctly at $\theta = 0°$ and vanishes at the magic angle. The diffusive behaviour of the 4-spin correlation functions shows up in the logarithmic shape of its wings, as expected from Equation (3.19).

4.2.5. *Satellite Lines at Half Resonance Field*

Far off from resonance, where the denominator of $F^+(H)$ varies only slightly with field, the EPR signal may directly result from the field dependence of the memory spectrum. Then, in the common case of transverse microwave excitation ($H_1 \perp H$), the nonsecular resonance at $2\omega_0$ appears as a satellite line at half resonance field (see Equation 3.24)

$$F^+(H \approx H_0/2) \approx \frac{4}{\pi} \operatorname{Re} M^+(H)/H_0^2. \tag{4.6}$$

A similar satellite line also occurs in the case of longitudinal excitation ($H_1 \parallel H$, see Equation 3.27),

$$F^z(H \approx H_0/2) \approx \frac{1}{\pi} \operatorname{Re} M^z(H)/H_0^2, \tag{4.7}$$

which is even more remarkable, since usually no longitudinal EPR signal occurs at all. ($F^z$ does not depend on field in the Markovian limit.)

Half-field transitions have been observed by Nagata *et al.* and by Yokozawa and Tanimoto [34] in pure and impurity-doped $K_2MnF_4$, and by Benner [33] in $(C_2H_5NH_3)_2MnCl_4$. Though for the latter example the maximum intensity of these transitions amounts less than $10^{-5}$ of the main line's (compare Figure 11), the

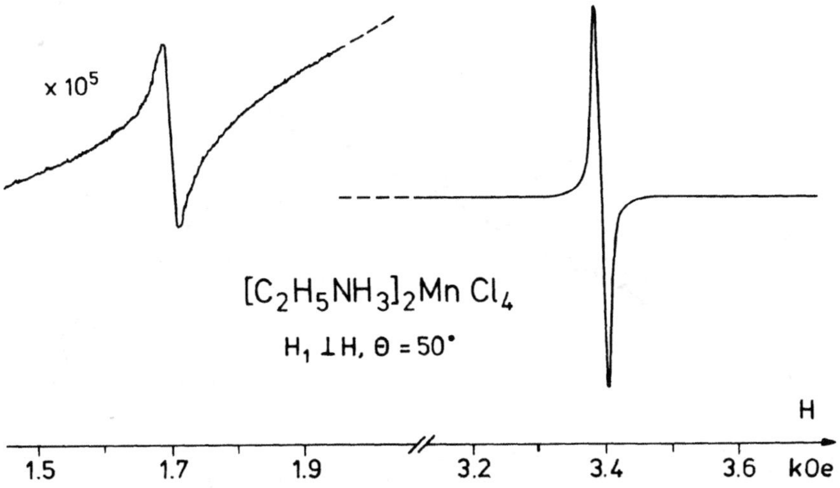


Fig. 11. Satellite line at half resonance field in the derivative EPR spectrum.

experimental sensitivity was still sufficient for a quantitative analysis of both shape and width.

Considering the experimental derivative spectrum, the satellite lines are expected to show the following characteristic properties:

(i) a strong anisotropy of intensity reflecting the dipolar coupling, which for transverse excitation is proportional to $\sin^2\theta\,\cos^2\theta/\delta H_{pp}(\theta)$ and for longitudinal excitation proportional to $\sin^4\theta/\delta H_{pp}(\theta)$,
(ii) a characteristic lineshape based on the logarithmic form of the peaks in the memory spectrum,
(iii) an anisotropic linewidth which is expected to scale with the width of the main line.

All of these properties could be well confirmed by experiments on $(C_2H_5NH_3)_2MnCl_4$.

Let us, for instance, discuss the satellite lineshape in detail. In the transverse case, having subtracted the wings of the main line, the logarithmic shape of the $2\omega_0$-peak, Equation (3.19), results in the following (derivative) EPR spectrum at half resonance field:

$$I'_{sat}(H) \sim \frac{H - H_0/2}{(H - H_0/2)^2 + (\Delta H)^2}, \tag{4.8}$$

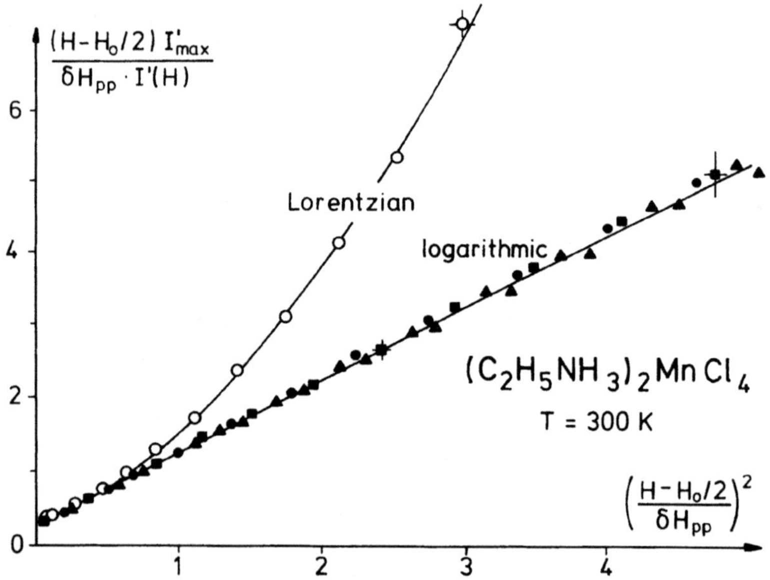


Fig. 12. Shape analysis of the satellite line (▲ 9.5 GHz, $H_1 \perp H$, $\theta = 50°$; ■ 9.5 GHz, $H_1 \| H$, $\theta = 90°$; ● 35 GHz, $H_1 \perp H$, $\theta = 50°$). For comparison, in the latter case the shape of the main line has also been plotted (○), changing $H_0/2$ to $H_0$.

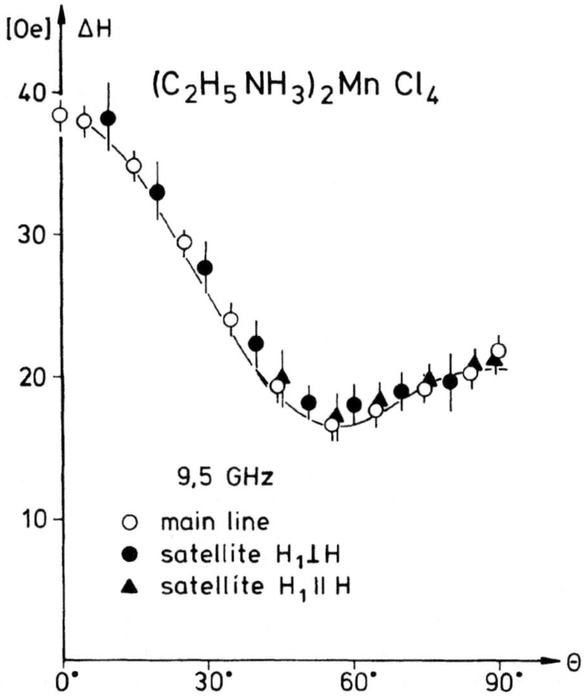


Fig. 13. Comparison of the experimental widths of main and satellite lines. According to the different shapes $\Delta H$ is given by $\sqrt{3}\ \delta H_{pp}/2$ (main line) and by $\delta H_{pp}/2$ (satellite), respectively. Solid line: theory.

and the same shape is obtained for the longitudinal satellite line. Note that the dependence on frequency and on field are no longer equivalent: there occurs a factor 1/2 when converting the scales. For this reason, only half the cut-off frequencies $\Gamma^{+}_{+1}$ or $\Gamma^{z}_{+2}$ enter the expression for the field dependent lineshape. Then, according to Equation (3.28), the cut-off in a field scale is given directly by $\Delta H$.

In Figure 12 we have compared both the longitudinal and transverse satellite lineshapes with the shape of the main line. In the transverse EPR spectrum the wings of the main line had to be subtracted very carefully. Plotting $(H - H_0/2)/I'_{sat}$ vs. $(H - H_0/2)^2$ a straight line is expected from Equation (4.8). The observed lineshapes were found to be in very good agreement with theory, and to differ strongly from the Lorentzian shape of the main line. So, not only the occurrence of half-field transitions, but especially their shape yields a direct quantitative evidence for spin diffusion in 2-d systems.

A very interesting result finally concerns the angular dependence of the experimental linewidths. There occurs a rather pronounced variation with $\theta$ in both the main line and the satellite lines. For a quantitative comparison of the observed peak-to-peak widths we have to take into account the different lineshapes. While for the nearly Lorentzian main line $\delta H_{pp}$ is given by 1.15 $\Delta H$, it follows from the foregoing lineshape analysis that the peak-to-peak width of the satellite lines exactly equals the cut-off parameter $\Delta H$. In Figure 13 the widths of both the transverse and longitudinal satellite lines are compared with the width of the main line for different sample orientations. There is not only qualitative agreement consistent with the arguments of Section 3.4, but — within the limits of experimental error — also an excellent quantitative agreement. It is interesting to note that this result has also been predicted by a self-consistent theory based on the decoupling of the 4-spin correlation functions [11, 33].

## 5. Critical Dynamics

We now want to turn to another interesting topic, where again a lower lattice dimensionality will give rise to specific dynamic properties, namely the critical behaviour at the magnetic phase transition.

Considering an ideal 2-d Heisenberg magnet, rigorous arguments have been presented that no phase transition to long-range order can occur above zero temperature [35]. The same conclusion also holds for the 2-d XY magnet. Instead, more complicated types of transitions were predicted for the latter [36, 37]. In real magnetic compounds, on the other hand, there always exist weak anisotropic and interplane interactions which, at sufficiently low temperatures, result in a 3-d ordering transition. The question is whether the critical behaviour observed close to this transition merely reflects the 3-d character of real layer-type systems, or whether it still represents specific 2-d properties. Since the static aspects of this problem have been extensively discussed in the preceding articles of this book, we confine ourselves here to the dynamic aspects only.

While at high temperature the sign of the exchange integral $J$ was of no importance for the spin dynamics, it is decisive for the behaviour close to the phase transition, In a ferromagnetic (F) layer, approaching $T_c$ from above, the long-wave ($q \approx 0$) fluctuations which already dominate at high temperature, become critical and are singled out even more distinctly. In an antiferromagnetic (AF) layer, however, the influence of these long-wave fluctuations will diminish, and fluctuations with $q$-values close to the ordering vector $\mathbf{q}_0$ will increasingly dominate the NMR and EPR linewidths.

Analyzing the temperature dependence of linewidths or relaxation rates near the phase transition $T_c$, one generally obtains a power-law behaviour

$$\Delta H(T) = \Delta H_{\text{non-crit}} + C\,\varepsilon^{-p}$$

with respect to the reduced temperature

$$\varepsilon = |T - T_c|/T_c.$$

$\Delta H_{\text{non-crit}}$ and $C$ may be regarded as temperature independent. The critical exponent $p$ turns out to represent a rather universal property depending mainly on type and sign of the interaction and on the dimension and symmetry of the spin lattice considered. We will discuss the critical behaviour of the linewidth in terms of this exponent with respect to a possible crossover both in lattice dimensionality (from 2-d to 3-d) and in spin dimensionality (from Heisenberg to XY or Ising behaviour). We will also discuss changes of the angular and frequency dependences accompanying this crossover. For convenience, the theoretical description will consequently be given in reciprocal space.

## 5.1. CRITICAL BEHAVIOUR OF THE NMR LINE

Most of the NMR experiments in the critical regime have been performed with continuous-wave spectrometers. So, instead of $T_1$ we will rather discuss the linewidth which corresponds to the transverse nuclear relaxation rate $1/T_2$. Restricting our discussion to the paramagnetic regime and neglecting the nuclear Larmor frequency in a time scale set by the electronic spin fluctuations, we can work with an expression which is very similar to Equation (1.5):

$$\Delta H_{\mathrm{NMR}} = \sum_n \left(A_n^z \tilde{f}_n^z(\omega \approx 0) + A_n^+ \tilde{f}_n^+(\omega_0)\right). \tag{5.1}$$

So, in practice, the linewidth probes the same electronic 2-spin correlation functions, and $1/T_1$ and $1/T_2$ only differ by the respective hyperfine coupling coefficients (here this seems not worth distinguishing in the notation). Since we prefer a formulation in reciprocal space, we write instead:

$$\Delta H_{\mathrm{NMR}} = \sum_{\mathbf{q}} \left(A_{\mathbf{q}}^z \tilde{f}_{\mathbf{q}}^z(\omega \approx 0) + A_{\mathbf{q}}^+ \tilde{f}_{\mathbf{q}}^+(\omega_0)\right), \tag{5.2}$$

where the spatial Fourier transforms are defined by analogy with Equation (2.6). We will use this expression for the linewidth as the starting point for the following discussion on the critical behaviour.

### 5.1.1. *Isotropic Regime*

Let us first consider the F case. In order to describe the critical behaviour of the NMR linewidth in terms of an exponent $p$, we have to separate the noncritical **q**-contributions — which are roughly given by the high temperature linewidth — from the contributions with **q** close to the ordering vector. The aim of the following derivation is to trace this critical part of the linewidth back to well-known static and dynamic properties of the 2-spin correlation functions $\tilde{f}_{\mathbf{q}\approx 0}^{\alpha}(\omega)$, and to relate $p$ to the respective critical exponents $\eta$, $\nu$ and $z$. The procedure can be simplified by the following assumptions:

(i) Restricting o[...]del to temperatures not too close to $T_c$, we assume isotropic spin fluctuatio[...]h decay exponentially in time with a rate $\Gamma_{\mathbf{q}}$:

$$\langle S_{\mathbf{q}}^{\alpha}(t) S_{\mathbf{q}}^{\alpha}\rangle = T\chi_{\mathbf{q}}\, e^{i\alpha\omega_0 t - \Gamma_{\mathbf{q}} t}, \tag{5.3}$$

where we have put $\langle |S_{\mathbf{q}}^{\alpha}|^2\rangle = T\chi_{\mathbf{q}}$ for $\alpha = x, y, z$.

(ii) The critical increase of the static susceptibility can be described by a modified Ornstein-Zernike form [38]:

$$T\chi_{\mathbf{q}} \sim (q^2 + \kappa^2)^{-1+\eta/2} \tag{5.4}$$

Here, $\kappa$ denotes the inverse correlation length. The Fisher exponent $\eta$, which is generally rather small, describes the spatial decay of static correlations according to

$$\langle S_i^\alpha S_j^\alpha \rangle \sim |\mathbf{r}_i - \mathbf{r}_j|^{-d+2-\eta}.$$

(iii) The $q$-dependence of the coupling coefficients $A_{\mathbf{q}}^\alpha$ may be neglected for $q \lesssim \kappa$ (excluding certain nuclear sites of high symmetry, where $A_{\mathbf{q}}^\alpha \sim q^2$).

(iv) According to the dynamic scaling hypothesis [39], the critical decay rate can be expressed by

$$\Gamma_{\mathbf{q}} = \kappa^z P(q/\kappa). \tag{5.5}$$

The scaling exponent $z$ characterizes the critical dynamic behaviour in a very general way. The function $P(x)$ only depends on the wave number $q$ scaled by the inverse correlation length.

We apply these assumptions to Equation (5.2). The $\tilde{f}_{\mathbf{q}}^\alpha(\omega)$ can be written as

$$\tilde{f}_{\mathbf{q}}^\alpha(\omega) = \frac{1}{\pi} T\chi_{\mathbf{q}} \operatorname{Re} (\Gamma_{\mathbf{q}} + i\omega)^{-1}.$$

Depending on the lattice-dimensionality, the sum over critical $q$-modes in Equation (5.2) is converted to a $d$-dimensional integral. With respect to Equations (5.4) and (5.5) it is convenient to scale $q$ with the inverse correlation length

$$\sum_{q \lesssim \kappa}{}' \ldots = \kappa^d \int_{x \lesssim 1} \mathrm{d}x\, x^{d-1} \ldots, \qquad x = q/\kappa.$$

Neglecting the frequency dependence for a moment and picking up all powers of $\kappa$, we arrive at a critical linewidth contribution

$$\Delta H_{\mathrm{NMR}}^{\mathrm{crit}} = (A_{\mathbf{q}=0}^z + 2A_{\mathbf{q}=0}^+)\, T \sum_{q \lesssim \kappa} \chi_{\mathbf{q}}/\Gamma_{\mathbf{q}},$$

which is proportional to

$$\Delta H_{\mathrm{NMR}}^{\mathrm{crit}} \sim \kappa^{d-z-2+\eta} \int_{x \lesssim 1} \mathrm{d}x\, x^{d-1} (1 + x^2)^{-1+\eta/2}/P(x). \tag{5.6}$$

To answer the important question whether the integral remains finite or not, one has to go into further detail. The F order parameter $\mathbf{S}_{\mathbf{q}=0}$, which is the total spin, represents a constant of motion, and therefore the critical modes behave diffusively: $\Gamma_{\mathbf{q}} = Dq^2$. (Following Equation (5.5), the diffusion constant will slow down according to $D \sim \kappa^{z-2}$.) The diffusive behaviour means that $P(x)$ in Equation

(5.6) vanishes as $x^2$ for small $x$. Then, in three dimensions the integral will remain finite, but in two dimensions there occurs exactly the same complication as in the high temperature regime: the integrand shows a pole at $q = 0$ and the integral will diverge.

As a physical consequence, in a 2-d Heisenberg ferromagnet (2-d HF) one cannot describe the critical increase of linewidth simply by some dynamical exponent $p$, but has to take into account all those mechanisms which will remove the pole at $q = 0$. Considering again the Larmor modulation of $\tilde{f}_q^+(\omega)$ and the dipolar cut-off of $\tilde{f}_q^z(\omega)$ – which formally means to replace $P(x)$ by $x^2 + (ia\omega_0 + \Gamma^a)/D\kappa^2$ – the critical behaviour of the linewidth depends in a more complicated way both on correlation length and frequency:

$$\Delta H_{\mathrm{NMR}}^{\mathrm{crit}} \sim \kappa^{-z+\eta}\{A_{\mathbf{q}=0}^z \ln(D\kappa^2/\Gamma^z) + 2A_{\mathbf{q}=0}^+ \ln(D\kappa^2/\omega_0)\}. \qquad (5.7)$$

The above considerations still hold for the AF case as well, replacing $A_{\mathbf{0}}$ by $A_{\mathbf{q}_0}$ and $\mathbf{q}$ by $\mathbf{q}^* = \mathbf{q} - \mathbf{q}_0$. An important difference occurs, however: the AF order parameter $\mathbf{S}_{\mathbf{q}_0}$ is *not* a constant of motion, which means that the damping of the critical mode remains nonzero: $\Gamma_{\mathbf{q}_0} \neq 0$. Therefore, no divergence will arise at $\mathbf{q}^* = 0$ and the asymptotic behaviour of (5.6) will hold in three as well as in two dimensions.

Both the dynamic exponent $z$ and the Fisher exponent $\eta$ depend on the dimensionality $d$ of the spin lattice and on the type and sign of the interaction. Mode coupling calculations [40] and the dynamic renormalization group theory [39] predict $z$ to amount to $(d + 2 - \eta)/2$ for the HF and to $d/2$ for the HAF. In 3-d systems $\eta$ turns out to be negligible. For a direct comparison with experiments we still have to consider the dependence on the reduced temperature $\varepsilon$. In 3-d systems, the correlation length is expected to diverge according to

$$\frac{1}{\kappa} \sim \varepsilon^{-\nu}$$

with a critical exponent $\nu \approx 2/3$. Then, from Equation (5.6), the asymptotic behaviour of the NMR linewidth in the 3-d F case is given by $\varepsilon^{-1}$ and in the 3-d AF case by $\varepsilon^{-1/3}$ [41, 42].

Unfortunately, the theoretical results for $z$ cited above were limited to a dimensionality $d > 2$, as a consequence of the absence of a long-range ordered state for $T > 0$ in 1-d or 2-d Heisenberg magnets. So, in a strict sense, these results could not be applied to the case of magnetic layers. However, assuming a phase transition to occur at nonzero temperature, Richards could prove by self-consistent mode coupling that the AF result $z = d/2$ still remains valid for the limit $d = 2$ [42]. With some precautions concerning the non-Markovian behaviour of the critical modes, a similar argumentation should also hold for the F case. The resulting $\kappa$-dependence of the critical part of the linewidth in 2-d and 3-d H-F's and H-AF's has been compared in Table I.

The absence of a conventional type of phase transition gives rise to further complications concerning the static properties of the 2-d Heisenberg magnet (and also of the XY magnet). Although vanishing in 3-d systems, the Fisher exponent $\eta$

TABLE I
Theoretical linewidth exponents of 3-d and 2-d magnets.

| | $\Delta H_{\mathrm{NMR}}$ | | $\Delta H_{\mathrm{EPR}}$ | |
|---|---|---|---|---|
| 3-d HF | $\kappa^{-3/2+3\eta/2}$ | $\sim \varepsilon^{-1}$ | $\kappa^{-3/2+3\eta/2}$ | $\sim \varepsilon^{-1}$ |
| 3-d IF | $\kappa^{-3+2\eta}$ | $\sim \varepsilon^{-1.9}$ | [a] $\kappa^{-5+3\eta}$<br>[b] $\kappa^{-1+3\eta}$ | $\sim \varepsilon^{-3.1}$<br>$- \varepsilon^{-0.6}$ |
| 2-d HF | $\kappa^{-2+3\eta/2} F(\omega, \kappa)$ | | $\kappa^{-2+3\eta/2} F(\omega, \kappa)$ | |
| 2-d IF | $\kappa^{-4+2\eta} F(\omega, \kappa)$ | $\sim \varepsilon^{-3.5} G(\omega, \varepsilon)$ | [a] $\kappa^{-6+3\eta} F(\omega, \kappa)$<br>[b] $\kappa^{-2+3\eta}$ | $\sim \varepsilon^{-5.3} G(\omega, \varepsilon)$<br>$\sim \varepsilon^{-1.3}$ |
| 3-d HAF | $\kappa^{-1/2+\eta}$ | $\sim \varepsilon^{-0.3}$ | $\kappa^{-5/2+2\eta}$ | $- \varepsilon^{-1.7}$ |
| 3-d IAF | $\kappa^{-1+2\eta}$ | $\sim \varepsilon^{-0.6}$ | [a] $\kappa^{-3+3\eta}$<br>[b] const | $\sim \varepsilon^{-1.8}$ |
| 2-d HAF | $\kappa^{-1+\eta}$ | $\sim \varepsilon^{-0.8}$ | $\kappa^{-3+2\eta}$ | $\sim \varepsilon^{-2.6}$ |
| 2-d IAF | $\kappa^{-2+2\eta}$ | $\sim \varepsilon^{-1.5}$ | [a] $\kappa^{-4+3\eta}$<br>[b] const | $\sim \varepsilon^{-3.3}$ |

[a] $F_{\mathbf{0}}^{+1}$ or $F_{\mathbf{q}_0}^{+1} \neq 0$. [b] $F_{\mathbf{0}}^{+1}$ or $F_{\mathbf{q}_0}^{+1}$.
The $\kappa$-dependence was obtained by means of the dynamic exponents $z = (d + 2 - \eta)/2$ (HF); $z = 4 - \eta$ (IF); $z = d/2$ (HAF); $z = 2 - \eta$ (IAF). For calculating $p$ the following static exponents $\eta$ and $\nu$ were used: 0.03/0.71 (3-d H); 0.03/0.63 (3-d I); 0.25/1 (2-d I) and — from experiment — 0.2/1 (2-d HAF).

becomes important in 2-d systems. For the 2-d Ising model, which has been solved exactly and which, indeed, shows a phase transition to long-range order at $T_c > 0$, $\eta$ amounts to 1/4 and $\nu$ equals 1. For the 2-d Heisenberg model, however, both are not known exactly and could only be estimated rather crudely from series expansions [43]. For $\eta$ a value of 0.2 was consistently obtained, but the results for $\nu$ vary from 0.9 to 1.6 [44, 45]. Moreover, doubts have also been expressed as to whether the correlation length follows a power law at all [46].

Regardless of the rather large theoretical uncertainty in both exponents, Table I offers the possibility of deriving a 'semi-empirical' exponent $p$ by using experimental values of $\eta$ and $\nu$ as, for instance, are obtained from neutron scattering experiments [47]. We will compare these values for $p$ with the experimental ones observed by magnetic resonance.

Finally, we mention a peculiarity occurring in the AF case. As a consequence of mode coupling, the critical modes also dominate the dynamics of the noncritical $q \approx 0$ fluctuations. Both $\Gamma_{\mathbf{q} \approx \mathbf{q}_0}$ and $\Gamma_{\mathbf{q} \approx 0} = Dq^2$ follow the same 'extended scaling' law [39]. Hence, within the regime of isotropic critical dynamics the diffusion constant $D$ will *speed up* as $\kappa^{z-2}$ ($z$ is smaller in the AF case than in the F case). That means that $D$ is proportional to $\kappa^{-1/2}$ for 3-d AF's and to $\kappa^{-1}$ for 2-d AF's. Though unimportant in three dimensions, the $q \approx 0$ fluctuations must still be considered in two dimensions, as they strongly influence the noncritical part of the linewidth near $T_N$. Therefore, in low-d systems one cannot simply evaluate this noncritical part from the high temperature linewidth.

5.1.2. *Anisotropic Regime*

Very close to the phase transition real spin systems can no longer be regarded as ideally isotropic. Even a very weak anisotropic interaction will introduce a preferential direction for the spins to line up. In the highly isotropic manganese layers, for example, the AF exchange and the dipolar coupling favour a spin orientation perpendicular to the layers (Ising type). In F layers, on the other hand, the dipolar coupling favours a spin orientation parallel to the layers (XY type), which, still closer to $T_c$ and in the presence of a slight orthorhombic perturbation, may change over to a spin alignment along an easy axis within the layers (Ising type). It is known from neutron scattering that this anisotropy is manifested in $\kappa$ and $\chi_{\mathbf{q}}$ [47]. In case of Ising type anisotropy, below a certain temperature $T_a > T_c$, the spin system can be characterized by two different correlation lengths $1/\kappa^{\parallel}$ and $1/\kappa^{\perp}$, describing the correlation of spin components parallel and perpendicular to the easy direction. While the parallel correlation length was found to diverge with a power law according to $\kappa^{\parallel} \sim \varepsilon^{\nu}$, the perpendicular correlation length remains finite, reaching a value $q_a$ at $T_c$. We furthermore assume that the anisotropy in $\chi_{\mathbf{q}}$ can be attributed to the anisotropy of $\kappa$, and write

$$T\chi_{\mathbf{q}}^{\parallel} \sim (\kappa^2 + q^2)^{-1+\eta/2} \tag{5.8a}$$

$$T\chi_{\mathbf{q}}^{\perp} \sim (\kappa^2 + q^2 + q_a^2)^{-1+\eta/2}, \tag{5.8b}$$

where we have put $\kappa^{\parallel} = \kappa$ and $(\kappa^{\perp})^2 = \kappa^2 + q_a^2$, in this way still retaining a unique correlation length $\kappa$. By means of dynamic scaling [48] and by mode coupling calculations [49], it could be shown that in the 'anisotropic regime' ($\kappa < q_a$) the critical dynamics will change over from Heisenberg to Ising-like. That means that the dynamic exponent for $\Gamma_{\mathbf{q}}^{\parallel}$ takes the values of the 'conventional theory' [50] $z = 4 - \eta$ in the F case, and $z = 2 - \eta$ in the AF case, whereas $\Gamma_{\mathbf{q}}^{\perp}$ remains finite.

Making use of these results, we find that in the anisotropic regime the critical broadening of the NMR linewidth is dominated by spin fluctuations along the easy axis (which we have assumed to be the $c$-axis):

$$\Delta H_{\mathrm{NMR}}^{\mathrm{crit}} \sim \begin{cases} A_{\mathbf{q}=0}^{c}\, T \sum\limits_{q \lesssim \kappa} \chi_{\mathbf{q}}^{\parallel}/\Gamma_{\mathbf{q}}^{\parallel} \sim \kappa^{d-6+2\eta} & \text{(FM)} \\ A_{\mathbf{q}_0}^{c}\, T \sum\limits_{q^* \lesssim \kappa} \chi_{\mathbf{q}^*}^{\parallel}/\Gamma_{\mathbf{q}^*}^{\parallel} \sim \kappa^{d-4+2\eta} & \text{(AFM),} \end{cases} \tag{5.9}$$

where the spin fluctuations perpendicular to the easy axis remain constant and only contribute to $\Delta H_{\text{non-crit}}$. Considering a 2-d F, one again has to pay attention to the convergence of the $q$-integral. We introduce the same cut-off mechanisms for small $q$ as discussed above. Describing the static anisotropic behaviour by the Ising exponents $\nu \approx 2/3$ and $\eta \approx 0$ (3-d) and by $\nu = 1$ and $\eta = 1/4$ (2-d), respectively, we finally arrive at the linewidth exponents p given in Table I. These exponents are larger than in the isotropic regime by roughly a factor of 2.

5.1.3. *Experiments*

Existing NMR investigations in the critical regime of 2-d F's [51] refer mainly to static properties below $T_c$ or to details of the hyperfine coupling mechanism, which we do not intend to discuss here. Kimishima and Nishihara [52] report the critical broadening of the proton spin echo width in $(CH_3NH_3)_2CuCl_4$, but do not present a quantitative evaluation. Moreover, one might question whether the true critical behaviour is probed in such resonance experiments, because of the presence of the external magnetic field which tends to suppress the critical fluctuations of the F order parameter [53]. To date, little is known about the intrinsic influence of $H$ on 2-d correlation functions.

The critical dynamics of 2-d AF's have been widely studied by Maarshall *et al.* [54] and by Bucci *et al.* [55] through $^{19}$F NMR on several compounds of the $K_2NiF_4$ type. In these systems the magnetic ions are surrounded by corner shared fluorine octahedra with two magnetically inequivalent fluorine sites, which are located out of ($F_I$) and within ($F_{II}$) the magnetic layers. Both the $F_I$ and $F_{II}$ resonances can be observed separately. Approaching $T_N$ from above, only the $F_I$ resonance shows the expected critical increase of linewidth, whereas the $F_{II}$ linewidth exhibits a decrease (compare Figure 14). These opposite behaviours can be understood to result from the different symmetries of the fluorine sites: Regardless of whether the coupling originates from transferred or dipolar hyperfine interaction, at the axial fluorine site I the hyperfine coefficient $A_{\mathbf{q}_0}$ of the critical spin fluctuations is nonzero — as postulated in the theoretical derivation — whereas for the planar $F_{II}$ nuclei, which are located symmetrically between two neighbouring magnetic ions belonging to opposite sublattices, $A_{\mathbf{q}_0}$ will vanish and the critical fluctuations can scarcely affect the linewidth.

In the nearly isotropic system $K_2MnF_4$ the diverging width of the $F_I$ resonance could be described by an exponent $p = 1.5$ within the temperature range $1.03\, T_N \leqslant T \leqslant 1.5\, T_N$. This value is far larger than expected from the 2-d Heisenberg model and rather agrees with the predictions of the 2-d Ising model. A similar exponent has been obtained in the highly anisotropic systems $K_2CoF_4$ and $Rb_2CoF_4$ [56]. It remains an open question how the very small anisotropy ($H_A/H_E = 10^{-4}$) present in $K_2MnF_4$ can affect the critical fluctuations over a broad temperature range in such a drastic way.

The narrowing of the $F_{II}$ resonance in $K_2MnF_4$, on the other hand, reflects the temperature dependence of the noncritical $\mathbf{q} \approx 0$ fluctuations. Both the continuous decrease of $T\chi_{\mathbf{q}=0}$ and the speeding up of the diffusive modes near $T_N$ — as discussed in Section 5.1.1 — diminish the $F_{II}$ linewidth by a factor 4 compared to its room-temperature value. Indeed, a decrease of the linewidth proportional to $1/D \sim \kappa \sim \varepsilon^{\nu}$ with $\nu = 0.9$ can be observed in the log—log plot of Figure 14. Only very close to the phase transition does the linewidth pass through a minimum and increases again, giving rise to a distinct cusp at $T_N$. Since even 3-d ordering could not account for this effect because of the high symmetry of the $F_{II}$ site, its

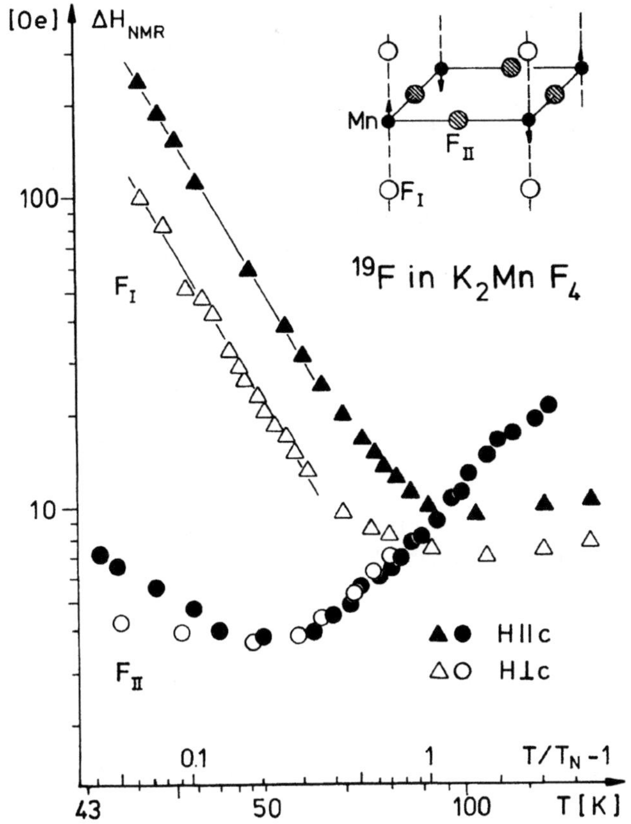


Fig. 14. Critical behaviour of the NMR linewidth for both $^{19}$F resonances in $K_2MnF_4$. (Data from Ref. [55].)

origin is not very clear. Bucci explained the cusp anomaly as a result of a subsequent slowing down of the diffusive modes.

## 5.2. CRITICAL BEHAVIOUR OF THE EPR LINEWIDTH

Similar asymptotic expressions describing the critical behaviour have also been derived for the EPR linewidth. As a starting point we may use Equation (1.14), but have to take into account the more complicated dipolar coefficients $F_{\mathbf{q}}^{m}$ of Equations (3.16) and (3.17):

$$\Delta H_{\mathrm{EPR}} \approx \omega_{\mathrm{d}}^{2} \sum_{\mathbf{q},\mathbf{q}'} \{ F_{\mathbf{q}}^{0} F_{-\mathbf{q}'}^{0} \tilde{\phi}_{\mathbf{q},\mathbf{q}'}(\omega = 0) + 10 F_{\mathbf{q}}^{+1} F_{-\mathbf{q}'}^{-1} \tilde{\phi}_{\mathbf{q},\mathbf{q}'}(\omega_0) + $$

$$+ F_{\mathbf{q}}^{+2} F_{-\mathbf{q}'}^{-2} \tilde{\phi}_{\mathbf{q},\mathbf{q}'}(2\omega_0) \}.$$

Little is known, however, about the critical behaviour of the involved 4-spin correlation functions $\phi^m_{\mathbf{q},\mathbf{q}'}(t)$, defined by Equation (3.12), except that they should retain their diffusive behaviour in the F case for the general reasons presented above. Usually, the only practical way to treat these functions in the critical regime is a decoupling into products of 2-spin correlation functions [42]

$$\langle S^\alpha_q(t) S^\beta_{-\mathbf{q}}(t)(S^\alpha_{\mathbf{q}'} S^\beta_{-\mathbf{q}'})^\dagger\rangle = \delta(\mathbf{q}-\mathbf{q}')\langle S^\alpha_{\mathbf{q}}(t)(S^\alpha_{\mathbf{q}})^\dagger\rangle\langle S^\beta_{\mathbf{q}}(t)(S^\beta_{\mathbf{q}})^\dagger\rangle,$$

which we also want to follow here. Within this approximation, the EPR linewidth can be treated in a similar way as the NMR linewidth, except that $\Delta H_{\text{EPR}}$ now represents a more complicated combination of 2-spin correlation functions:

$$\begin{aligned}\Delta H_{\text{EPR}} \sim \langle|S^+_{\mathbf{0}}|^2\rangle^{-1}\sum_{\mathbf{q}}\int \mathrm{d}t\, e^{i\omega_0 t}\{&|F^0_{\mathbf{q}}|^2\langle S^+_{\mathbf{q}}(t)S^-_{-\mathbf{q}}\rangle\langle S^z_{\mathbf{q}}(t)S^z_{-\mathbf{q}}\rangle+\\ &+|F^1_{\mathbf{q}}|^2(3|\langle S^+_{\mathbf{q}}(t)S^-_{-\mathbf{q}}\rangle|^2+8|\langle S^z_{\mathbf{q}}(t)S^z_{-\mathbf{q}}\rangle|^2)+\\ &+|F^2_{\mathbf{q}}|^2\langle S^-_{\mathbf{q}}(t)S^+_{-\mathbf{q}}\rangle\langle S^z_{\mathbf{q}}(t)S^z_{-\mathbf{q}}\rangle\}.\end{aligned} \tag{5.10}$$

Note that there occurs a factor of $\sqrt{2}$ in the normalization of $S^\pm$, which makes the coefficients differ from Equation (1.14).

### 5.2.1. *Ferromagnets*

Let us first discuss the F case and start again with the isotropic regime. Restricting to the same conditions as in Section 5.1.1, the critical part of the EPR linewidth is given by

$$\begin{aligned}\Delta H^{\text{crit}}_{\text{EPR}} &\sim (|F^0_{\mathbf{q}=\mathbf{0}}|^2+10|F^1_{\mathbf{q}=\mathbf{0}}|^2+|F^2_{\mathbf{q}=\mathbf{0}}|^2)\,T(\chi_{\mathbf{q}=\mathbf{0}})^{-1}\sum_{q\lesssim\kappa}\chi_{\mathbf{q}}{}^2/2\Gamma_{\mathbf{q}}\\ &\sim \kappa^{d-z-2+\eta}\int_{x\lesssim 1}\mathrm{d}x\, x^{d-1}(1+x^2)^{-2+\eta}/P(x)\end{aligned} \tag{5.11}$$

where $z=(d+2-\eta)/2$. The integral will essentially behave in the same way as discussed for the NMR linewidth, so only in three dimensions we expect the critical broadening to be independent of frequency and to follow a power law [42, 57]

$$\Delta H^{\text{crit}}_{\text{EPR}} \sim \kappa^{-3/2} \sim \varepsilon^{-1} \qquad \text{(3-d HF)}. \tag{5.12}$$

It should be mentioned that in cubic spin lattices, where all dipolar coefficients vanish at $q=0$, the ESR linewidth remains unaffected by the critical fluctuations.

For a 2-d spin lattice, taking into account the Larmor modulation of non-secular linewidth contributions and the dipolar cut-off of the secular part, we obtain instead

$$\begin{aligned}\Delta H^{\text{crit}}_{\text{EPR}} \sim \kappa^{-2+3\eta/2}\{&|F^0_{\mathbf{q}=\mathbf{0}}|^2\ln(D\kappa^2/\Delta H)+10|F^1_{\mathbf{q}=\mathbf{0}}|^2\ln(D\kappa^2/\omega_0)+\\ &+|F^2_{\mathbf{q}=\mathbf{0}}|^2\ln(D\kappa^2/2\omega_0)\}\qquad\text{(2-d HF)}.\end{aligned} \tag{5.13}$$

According to Equation (3.20), this cut-off is given directly by the total EPR linewidth $\Delta H$. As long as $\Delta H$ remains much smaller than $\omega_0$ — which is generally the case at higher temperature for typical experimental conditions — the secular part of the linewidth still remains the dominating one and gives rise to the characteristic angular dependence discussed in section 4.2.1. The nonsecular contributions, especially for magic angle orientation, again introduce a logarithmic frequency dependence. Compared to the behaviour at high temperature, these 2-d features are strongly enhanced by the critical slowing down of the diffusive modes.

Provided that the isotropic dynamics persist very close to $T_c$, the extreme line broadening will also manifest as a simultaneous increase of the cut-off frequency, and one expects the Larmor modulation of the nonsecular linewidth contributions to become less efficient than the respective dipolar cut-offs. The resulting decrease of frequency dependence should be accompanied by a changeover in angular dependence from $(3\cos^2\theta - 1)^2$ to roughly $1 + \cos^2\theta$. This effect, however, will scarcely be observable in the very broad line.

Owing to the complex structure of the EPR linewidth, its anisotropic critical behaviour turns out to be more complicated than that of the NMR linewidth, and depends sensitively on the direction and strength of the anisotropy with respect to the magnetic field. We mentioned already that in most F layer compounds an easy axis is found within the planes. Let us, for simplicity, confine to a sample orientation where this axis is parallel to $\mathbf{H}$ (i.e. $\theta = 90°$). Then, in Equation (5.10) only the term containing parallel spin components will further increase according to

$$\sum_{\mathbf{q}} |F_{\mathbf{q}}^1|^2 \chi_{\mathbf{q}}^{\|2}/(2\Gamma_{\mathbf{q}}^{\|} + i\omega_0) \sim |F_{\mathbf{q}=\mathbf{0}}^1|^2 \kappa^{d-8+\eta} F(\omega_0, \kappa), \tag{5.14}$$

whereas the other terms containing also transverse components remain constant. In 2-d systems the diffusive behaviour of $\Gamma_{\mathbf{q}}^{\|}$ for $q \lesssim \kappa$ again results in a pole at $q = 0$. So, the Larmor modulation occurring in this term becomes important and gives rise to a frequency dependence. Provided that $F_{\mathbf{q}=\mathbf{0}}^1$ does not vanish, the linewidth is expected to diverge as $\kappa^{-6+3\eta}$ in two dimensions and like $\kappa^{-5}$ in three dimensions (this result corresponds to case a in Table I).

For a uniaxial spin lattice and for the chosen orientation, the dipolar coefficient $F_{\mathbf{q}}^1$, however, vanishes exactly at $q = 0$, so the EPR linewidth probes the critical parallel fluctuations only through the weak coupling $F_{\mathbf{q}}^1 \sim q^2$ at nonzero $q$. This implies that instead of (5.14) one has to consider the contribution

$$\sum_{\mathbf{q}} q^4 \chi_{\mathbf{q}}^{\|2}/(2\Gamma_{\mathbf{q}}^{\|} + i\omega_0) \sim \kappa^{d-4+3\eta} \tag{5.15}$$

which is still diverging but with a lower power of $\kappa$. There occurs no longer a frequency dependence of linewidth in two dimensions, since the pole at $q = 0$ now has been removed by the $q$-dependence of the coupling coefficient. Using again the critical exponents of the 2-d and 3-d Ising model, we finally arrive at the exponents $p$ given in Table I (case b). It is clear from the above discussion that a slight change of sample orientation ($\theta \neq 0°$ or $90°$), or a decrease of the lattice symmetry, should drastically change the critical broadening of the EPR linewidth.

Convincing EPR data proving the isotropic critical behaviour of F layer compounds have not yet been obtained for several reasons: firstly, in contrast to the AF case, no F layer-type compounds with $S$-state ions are known so far. So the observed relaxation phenomena generally do not result from spin-spin interactions only, as we have assumed. Secondly, in our theoretical analysis we have not yet considered the intrinsic influence of the applied magnetic field, which tends to suppress the critical spin fluctuations. Moreover, since most single crystals cleave readily along the layers and do not allow the preparation of sufficiently small spheres, EPR experiments have generally been performed on naturally grown platelets of rectangular or hexagonal shape. As a consequence of the resulting inhomogeneous demagnetizing field, there usually occurs a variety of magnetostatic modes close to $T_c$ [58], hiding the true paramagnetic signal and impeding a detailed analysis.

A fairly good example of a Heisenberg 2-d F is represented by the layer-type chromium compound $NaCrS_2$ ($J' = 1/6\ J$). Its high-temperature EPR has been studied in detail by Richards *et al.* [59] proving that the dominating source of anisotropy is the dipolar interaction, as we have assumed in our model. With decrasing temperature the EPR line shows the continuous broadening expected for 2-d F's. At $T_N = 17.0$ K there occurs 3-d AF long-range order where the F planes line up with alternating spin direction. Elliston [60] has investigated the critical increases of the linewidths, applying $H$ both parallel and perpendicular to the $c$-axis. In a wide temperature range, $1.7 \leqslant T/T_c \leqslant 4.5$, he obtained for either orientation a power-law behaviour, with exponent $p = 1.25 \pm 0.2$. This value might be consistent with isotropic behaviour, taking into account the theoretical uncertainty of $\nu$, but also agrees with our result for the 2-d F Ising model. Below $1.7\ T_c$ a crossover takes place and the divergence of the linewidth becomes less pronounced. For $1.06 \leqslant T/T_c \leqslant 1.7$ he obtained $p = 0.45 \pm 0.1$ ($H \perp c$) and $p = 0.55 \pm 0.2$ ($H \parallel c$). An explanation of these values based on 3-d AF fluctuations will be given in the next section.

Recently, a very interesting EPR study on another 2-d F, $K_2CuF_4$, has been presented by Yamada *et al.* [61], confirming that the non-Markovian character of spin relaxation is strongly enhanced in the critical regime. Besides analysing the temperature dependence of the linewidth, they also studied the angular dependence of the linewidth, the lineshape, and the intensity of the satellite lines.

Below 100 K a critical broadening of the linewidth was observed, occurring most distinctly for $\theta = 0°$ (Figure 15a). Even for this orientation, where only a frequency-independent secular contribution is expected, the linewidth was strongly affected by the magnetic field, indicating an intrinsic influence of $H$ on the critical spin fluctuations. So, an interpretation in terms of Equations (5.13) or (5.15) seems not to be meaningful. In parallel with this broadening, the characteristic angular dependence of long-wave fluctuations, discussed in Equation (4.1), turned out to be far more pronounced than at room temperature (Figure 15b). This indicates the increasing importance of the diffusive modes. The slowing down of

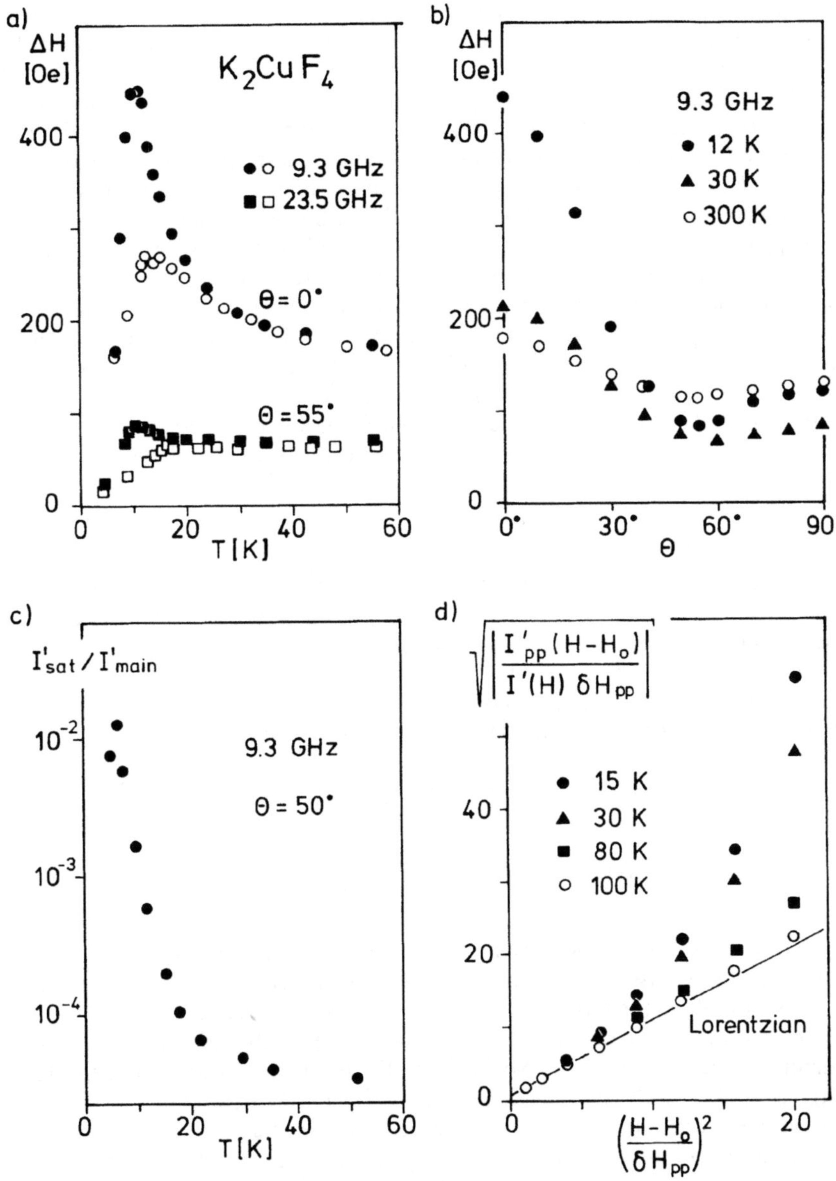


Fig. 15. Critical behaviour of the EPR spectrum in $K_2CuF_4$. (a) Temperature dependence of linewidth for different frequencies and orientations. (b) Angular dependence of linewidth for different temperatures. (c) Temperature dependence of the relative intensity of the $2\omega_0$ satellite line. (d) Lineshape analysis for different temperatures. Solid line: Lorentzian shape. (Data from Ref. /61/.)

the critical spin fluctuations also becomes manifest in the lineshape. At room temperature, the lineshape for $\theta = 0°$ deviates only slightly from a Lorentzian. This changes very rapidly below 50 K, and at 15 K the observed non-Lorentzian character is found to be nearly as pronounced as in typical 1-d systems at high temperatures (Figure 15c). The most dramatic effect can be observed in the intensity of the satellite lines. Decreasing the temperature from 30 K to 10 K, the intensity of the $2\omega_0$ satellite increases by more than two orders of magnitude compared to that of the main line (Figure 15d). Since the intensity ratio of both lines is proportional to $(\omega_d/D)^2$, this increase directly measures the critical slowing down of the diffusive modes. Moreover, below 10 K an additional satellite line at $3\omega_0$ occurs, which at room temperature, has only been observed so far in 1-d compounds [62] or in impurity-doped 2-d compounds [34] (compare Section 6.). This line, being of higher order in $\omega_d/D$, is expected to be weaker in intensity than the $2\omega_0$ satellite even by several orders of magnitude.

Altogether, these effects are characteristic for the slowing down of $q \approx 0$ fluctuations and can, for instance, not be observed in the critical dynamics of a 2-d AF.

### 5.2.2. *Antiferromagnets*

The critical behaviour of the EPR linewidth in layered AFs can be characterized by the range of dominating $q$-contributions changing over from $\mathbf{q} \approx 0$ to $\mathbf{q}_0$. So, in contrast to the F case, EPR no longer directly probes the fluctuations of the order parameter. For this reason, comparing the F and the AF case, some asymmetry occurs in the corresponding relations.

In the isotropic regime the critical increase of the linewidth is given by

$$\Delta H_{\mathrm{EPR}}^{\mathrm{crit}} \sim (|F^0_{\mathbf{q}_0}|^2 + 10|F^1_{\mathbf{q}_0}|^2 + |F^2_{\mathbf{q}_0}|^2)\, T(\chi_{\mathbf{q}=0})^{-1} \sum_{q^* \lesssim \kappa} \chi^2_{\mathbf{q}^*}/2\Gamma_{\mathbf{q}^*}$$

$$\sim \kappa^{d-z-4+2\eta} \int_{x \lesssim 1} \mathrm{d}x\, x^{d-1} (1+x^2)^{-2+\eta}/P(x). \qquad (5.16)$$

In addition to the dynamic exponent $z = d/2$, the asymptotic behaviour also differs from the F case by a factor $\kappa^{-2+\eta}$. It results from the fact that the uniform static susceptibility, appearing as a prefactor in both Equations (5.11) and (5.16), remains uncritical and no longer compensates one of the critical susceptibilities occurring in the $q$-summation. So in 2-d AF's the EPR linewidth is expected to diverge as $\kappa^{-3+2\eta}$ [42] and in 3-d AFs as $\kappa^{-5/2}$ [41, 63]. Again one has to exclude cubic spin lattices (all $F^m_{\mathbf{q}_0} = 0$), where critical spin fluctuations cannot couple strongly enough to the uniform mode to broaden its width. As the order parameter $\mathbf{S}_{\mathbf{q}_0}$ is no longer a constant of motion, $\Gamma_{\mathbf{q}_0}$ remains nonzero, and we may neglect any frequency dependence or additional cut-off mechanism at $q^* = 0$. For a 2-d magnetic lattice of uniaxial symmetry, the equal weight of secular and non-secular linewidth contributions close to $T_N$ will change the angular dependence from rougly $(3\cos^2\theta - 1)^2$ to $1 + \cos^2\theta$.

We have mentioned already that in most AF layers close to $T_N$ the dipolar interaction favours the spins to line up perpendicularly to the layers. Then, in the anisotropic regime, only the spin fluctuations in the $c$-direction will determine the critical behaviour. Accordingly, only the term containing the parallel spin fluctuations:

$$\sum_{\mathbf{q}^*} |F^1_{\mathbf{q}^*}|^2 \chi^{\parallel 2}_{\mathbf{q}^*}/2\Gamma^{\parallel}_{\mathbf{q}^*} \sim |F^1_{\mathbf{q}_0}|^2 \kappa^{d-6+3\eta} \tag{5.17}$$

will diverge, whereas all the other terms will remain constant. The resulting EPR linewidth increases asymptotically as $\kappa^{-3}$ in 3-d and as $\kappa^{-4+3\eta}$ in 2-d systems. Confining ourselves to the orientations $H \parallel c$ and $H \perp c$, however, this critical contribution will just vanish for a uniaxial spin lattice, even when including the $q$-dependence of $F^1_{q^*}$. Only in orthorhombic spin lattices would one expect to observe the above anisotropic critical behaviour for $H \perp c$.

The critical broadening of the EPR linewidth in some manganese layer-type compounds has been studied by several authors [64—68]. As an example, let us discuss the results obtained on $K_2MnF_4$. Lowering $T$ from room temperature towards the AF transition, the linewidth initially decreases as expected for a 2-d AF [15], it passes through a minimum and again increases rapidly near $T_N = 45$ K (Figure 16a). Simultaneously, a change of angular dependence can be observed, such that the flat minimum at the magic angle vanishes, and $\Delta H(55°)$ becomes larger than $\Delta H(90°)$ (Figure 16b). We may regard this as an indication that the range of dominating $q$-modes is changing over from $\mathbf{q} \approx 0$ to $\mathbf{q} \approx \mathbf{q}_0$. At $T = 1.2\ T_N$, finally, the linewidth shows exactly the expected $1 + \cos^2\theta$ dependence. Analyzing the critical increase of linewidth within the range $1.1 \leqslant T/T_N \leqslant 1.5$, de Wijn *et al.* [64] obtained a critical exponent $p = 2.5 \pm 0.2$. Using the static critical exponents $\nu = 0.9 \pm 0.1$ and $\eta = 0.2 \pm 0.05$, as obtained from neutron scattering data [47], the experimental value of $p$ agrees exactly with our theoretical result for the 2-d isotropic regime. This is in some contrast with the NMR data of Bucci [55], implying a 2-d anisotropic behaviour for $1.06 \leqslant T/T_N \leqslant 1.5$. An explanation might be that Bucci's data on $K_2MnF_4$ were partly taken closer to the transition point, and that he used a different value for $T_N$ (41.5 K instead of 42.1 K). Nearly the same EPR exponents have been obtained in several other 2-d AF's, which all consist of $S$-state ions and therefore are expected to be suitable canditates for 2-d isotropic behaviour. We have listed them in Table II.

It is interesting to note a 40% reduction of the EPR exponent, when doping $K_2MnF_4$ with a small amount of Fe impurities [68]. This effect was attributed to an additional spin—phonon coupling of the (non-$S$-state) $Fe^{2+}$ ions, which, sufficiently close to $T_N$ should impede the slowing down of the critical fluctuations. Namely, when the correlation length $1/\kappa$ has reached the average distance of the impurities, the damping rate $\Gamma_{\mathbf{q}_0}$ will no longer result from the Heisenberg dynamics, but from the spin—lattice relaxation rate of the $Fe^{2+}$ ions and will no longer depend on $\kappa$. Thus, assuming the spatial correlations to remain of the 2-d isotropic type, $p$ will be given by $\nu\ (2 - 2\eta) \approx 1.5$ instead of $\nu\ (2 - 2\eta + z) \approx 2.5$ (Figure 17).

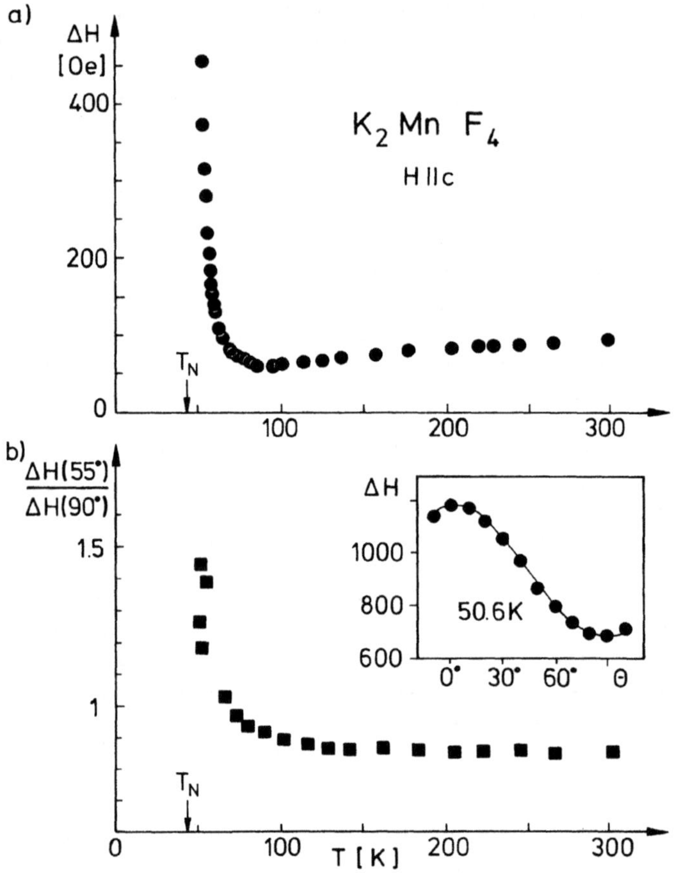


Fig. 16. (a) Critical divergence of the EPR linewidth in $K_2MnF_4$. (b) Temperature dependence of the linewidth anisotropy. Insert shows the $(a + b\cos^2\theta)$ dependence close to $T_N$. (Data from Refs. [15] and [64].

There exist still other examples of AF layer-type compounds where the experimental value of p has been reported to be close to 1.5 [70, 71]. Merely from the literature, however, it is hard to distinguish whether such exponents also correspond to some comparable cut-off mechanism of $\Gamma_{q_0}$ retaining the 2-d character, or already indicate a crossover to 3-d behaviour (isotropic: $p = 1.7$, anisotropic: $p = 1.8$), or whether they only result from an improper evaluation of the experimental data. This often turns out to be a rather delicate procedure.

Sufficiently close to $T_N$, in most of the investigated layer-type compounds with AF interplanar coupling, a crossover is observed where the increase of the linewidth becomes less drastic, and can be described by an exponent $p = 0.5$—$0.7$. There is no significant difference concerning the F or AF coupling within the

TABLE II
Experimental EPR linewidth exponents of some antiferromagnetic layer compounds.

| | range of $\varepsilon$ | $p_{EPR}$ | $\eta_{exp}$, $\nu_{exp}$* | $(3-2\eta_{exp})\nu_{exp}$ |
|---|---|---|---|---|
| $K_2MnF_4$ | 0.1—0.5 | 2.5 ± 0.2 [64] | 0.20 ± 0.05 [47]<br>0.95 ± 0.10 | 2.5 ± 0.3 |
| $Rb_2MnF_4$ | 0.1—0.5 | 2.5 ± 0.2 [64] | | |
| $K_3Mn_2F_7$ | 0.1—0.3 | 3.0 ± 0.2 [65] | 0.20 ± 0.05 [69]<br>1.1 ± 0.1 | 2.9 ± 0.4 |
| $(C_3H_7NH_3)_2MnCl_4$ | 0.1—0.4 | ≈ 2.4 [66] | | |
| $(CH_2)_4(NH_3)_2MnCl_4$ | 0.1—0.3 | ≈ 2.4 [67] | | |
| $K_2Mn_{1-x}Fe_xF_4$ | 0.1—0.7<br>0.7—1.2 | 1.4 ± 0.2 [68]<br>2.5 ± 0.2 | | |

* From neutron scattering.

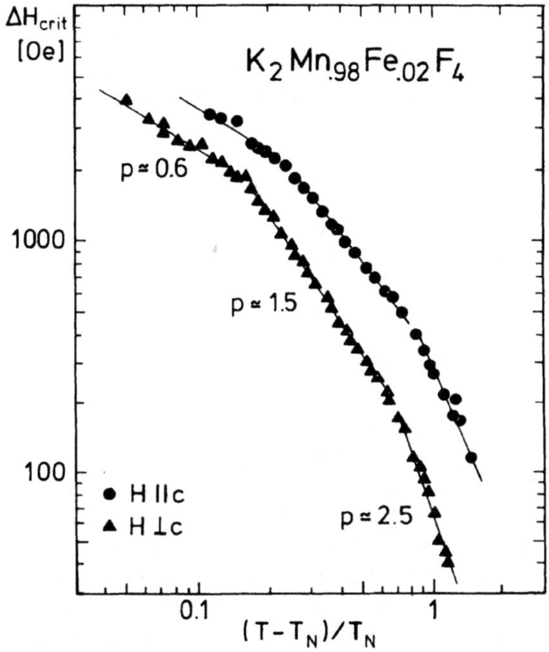


Fig. 17. Log—log plot of the critical part of the EPR linewidth in $K_2MnF_4$ doped with 2% $Fe^{2+}$ as a function of the reduced temperature. (Data from ref. [68].)

planes. Though a quantitative explanation has not yet been offered in the literature, we interpret this crossover to result from 3-d AF fluctuations, where the slowing down of the staggered mode — as in the preceding example — has been 'short-circuited' by some additional cut-off mechanism. Then the EPR linewidth merely reflects the increase of the staggered susceptibilities, yielding

$$\Delta H_{\mathrm{EPR}}^{\mathrm{crit}} \sim \kappa^{d-4+2\eta} \sim \begin{cases} \varepsilon^{-0.66} & \text{(3-d HAF)} \\ \varepsilon^{-0.59} & \text{(3-d IAF).} \end{cases} \tag{5.18}$$

In our example of $NaCrS_2$ (2-d F) the crossover-temperature takes the rather high value of 1.7 $T_N$, which corresponds to the large AF interplanar coupling, whereas in the manganese layer compounds the crossover occurs typically below 1.1 $T_N$ (compare Figure 17).

### 5.3. AC SUSCEPTIBILITY

A general problem of magnetic resonance investigations arises from the external magnetic field which — apart from causing the Larmor precession of transverse spin components — especially in ferromagnets tends to suppress the critical spin fluctuations. Existing theory, on the other hand, refers to zero-field dynamics. Therefore, one might doubt whether the EPR really represents an appropriate means to study the critical dynamics of ferromagnets in detail.

Such difficulties can be avoided by using a zero-field relaxation technique instead of resonance techniques [72]. In this kind of experiments generally the real part of the dynamic susceptibility

$$\mathrm{Re}\,\chi^{\alpha}(\omega) = \chi^{\alpha}_{\mathbf{q}=0}/\{1 + \omega^2/(\Gamma^{\alpha})^2\} \tag{5.20}$$

is measured as a function of frequency, temperature and orientation (the index $\alpha$ refers to the direction of the applied alternating field). Provided that the frequency range is comparable to the relaxation rate, which is typically of the order 1—100 MHz, $\Gamma^{\alpha}$ can be evaluated according to

$$\Gamma^{\alpha} = \omega\{\chi^{\alpha}_{\mathbf{q}=0}/\mathrm{Re}\,\chi^{\alpha}(\omega) - 1\}^{-1/2},$$

while the static susceptibility $\chi^{\alpha}_{\mathbf{q}=0}$ can be measured simultaneously by lowering the frequency down to $\omega \ll \Gamma^{\alpha}$ (Figure 18).

Since we have so far not considered the intrinsic influence of the static magnetic field on the critical spin fluctuations, the relaxation rate $\Gamma^{\alpha}$ essentially reads the same as the EPR linewidth except for the vanishing Larmor modulation of the nonsecular terms. This means that the distinction between secular and nonsecular contributions no longer remains of physical importance. Since in AC experiments the measuring frequency $\omega$ is much smaller than in common EPR, we do, however, have to include, the dipolar cut-off for each contribution. Then, for a layer-type FM, the critical part of the relaxation rate should increase in the isotropic regime, by analogy with Equation (5.13), like

$$\Gamma_{\mathrm{crit}} \sim \kappa^{-2+3\eta/2} \ln\{D\kappa^2/\sqrt{\omega^2+\Gamma^2} \qquad \text{(2-d HF).} \tag{5.20}$$

In the anisotropic regime the critical behaviour will again depend on orientation in a more complicated way. Confining to the experimental examples to be discussed

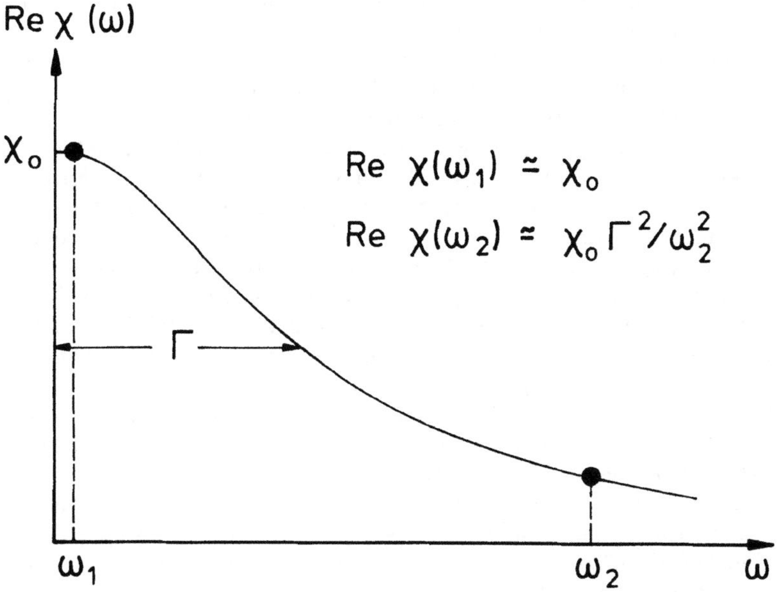


Fig. 18. Real part of the zero-field dynamic susceptibility as probed by AC susceptibility measurements.

below, we consider a spin lattice of orthorhombic symmetry where the HF field is applied along the crystal axes. The corresponding relaxation rates are given by

$$\Gamma^a = T(\chi^a_{\mathbf{q}=0})^{-1}\,|F^0_{\mathbf{q}=0}|^2\,\mathrm{Re}\sum_{\mathbf{q}} \chi^b_{\mathbf{q}}\,\chi^c_{\mathbf{q}}/(\Gamma^b_{\mathbf{q}}+\Gamma^c_{\mathbf{q}}+i\omega)$$

$$\Gamma^b = T(\chi^b_{\mathbf{q}=0})^{-1}\,|F^0_{\mathbf{q}=0}|^2\,\mathrm{Re}\sum_{\mathbf{q}} \chi^c_{\mathbf{q}}\,\chi^a_{\mathbf{q}}/(\Gamma^c_{\mathbf{q}}+\Gamma^a_{\mathbf{q}}+i\omega)$$

$$\Gamma^c = T(\chi^c_{\mathbf{q}=0})^{-1}\,|F^2_{\mathbf{q}=0}|^2\,\mathrm{Re}\sum_{\mathbf{q}} \chi^a_{\mathbf{q}}\,\chi^b_{\mathbf{q}}/(\Gamma^a_{\mathbf{q}}+\Gamma^b_{\mathbf{q}}+i\omega). \qquad (5.21)$$

Now, assuming that the $a$-axis is the easy axis, where the spin fluctuations still increase, there is always at least one dynamic correlation function of the non-critical spin components $S^b$ or $S^c$ involved in the $\mathbf{q}$-summation. The noncritical rates $\Gamma^b_{\mathbf{q}}$ and $\Gamma^c_{\mathbf{q}}$ effectively 'cut off' the critical slowing down of $\Gamma^a_{\mathbf{q}}$. Thus, all $\mathbf{q}$-summations remain constant, and only the inverse static susceptibilities occurring as prefactors will determine the critical behaviour of the ($\mathbf{q} = 0$) relaxation rates. That means, applying the HF-field in the $a$-direction, we expect a slowing down of $\Gamma^a$ resulting from the the critical term $(\chi^a_{\mathbf{q}=0})^{-1}$:

$$\Gamma^a_{\mathrm{crit}} \sim \kappa^{2-\eta} \sim \begin{cases} \varepsilon^{+1.24} & \text{(3-d IF)} \\ \varepsilon^{+1.75} & \text{(2-d IF).} \end{cases} \qquad (5.22)$$

For the $b$- or $c$-direction, the prefactors behave in a noncritical way, and the relaxation rates should remain constant.

AC susceptibility measurements on F layer-type compounds were reported by Okuda *et al.* on $Mn(CH_3COO)_2 \cdot 4\ H_2O$ [73] and on $(CH_3NH_3)_2CuCl_4$ [74]. The latter compound has also been studied in detail by Rutten [75].

$(CH_3NH_3)_2CuCl_4$ represents a well-known 2-d F, also with F coupling between the planes ($J'/J = 10^{-4}$). Its critical behaviour has been characterized by a sequence of crossovers both in spin and lattice dimensionality, i.e. from 2-d isotropic via planar to Ising like [76]. At $T_c = 8.9$ K, finally, a 3-d ordering transition occurs, where all spins line up in the $a$-direction. From temperature dependent measurements of Re $\chi^a(\omega)$ at very low frequency ($\omega/2\pi = 50$ Hz) one could directly evaluate the critical increase of the static susceptibility. Applying the HF-field along the easy $a$-axis, Okuda [74] obtained two different values of the respective critical exponent $\gamma$: $1.75 \pm 0.03$ for $0.05 \leqslant \varepsilon \leqslant 0.3$ and $1.20 \pm 0.03$ for $0.004 \leqslant \varepsilon \leqslant 0.05$. At higher frequencies (1—100 MHz) the critical increase of Re $\chi^a(\omega)$ was found to be strongly reduced, and, instead of a peak in the temperature dependence, there occurred only a narrow 'dip' at $T_c$ (Figure 19). The same effect has also been observed in $Mn(CH_3COO)_2 \cdot 4\ H_2O$ and can qualitatively be interpreted as a direct evidence for the critical slowing down of the relaxation rate $\Gamma^a$. Close to $T_c$, where $\Gamma^a$ becomes smaller than $\omega$, the dispersion signal is approximately given by

$$\mathrm{Re}\,\chi^a(\omega) \approx \chi^a_{\mathbf{q}=0}\,(\Gamma^a/\omega)^2.$$

Then, apparently, the critical increase of $\chi^a_{\mathbf{q}=0}$ is compensated by the decrease of $(\Gamma^a)^2$. Following Equation (5.22) we have

$$\mathrm{Re}\,\chi^a(\omega) \sim \begin{cases} \chi^a_{\mathbf{q}=0}, & \omega \ll \Gamma^a \quad (\varepsilon \text{ large}) \\ (\chi^a_{\mathbf{q}=0})^{-1}, & \omega \gg \Gamma^a \quad (\varepsilon \text{ small}). \end{cases}$$

Comparing high- and low-frequency data, Okuda derived an exponent for the relaxation rate of $p = 1.05 \pm 0.03$ within the whole range $0.004 \leqslant \varepsilon \leqslant 0.3$. Surprisingly, in contrast to the static behaviour, no indication of a crossover could be observed at $\varepsilon \approx 0.05$. Rutten [75] also studied the relaxation rate in the $b$- and $c$-directions. He found that even close to $T_c$ ($\varepsilon < 0.005$) $\Gamma^a$ and $\Gamma^b$ nearly coincide and show the same slowing down. By contrast, $\Gamma^c$ was found to be much smaller near $T_c$ and showed a critical speeding up.

These experimental findings appear rather contradictory and can hardly be interpreted in a consistent way. Definitely, the observed slowing down cannot be attributed to Heisenberg dynamics, in neither two nor three dimensions. The crossover observed in the static susceptibility has been interpreted as a changeover from 2-d ($\gamma = 1.75$) to 3-d Ising behaviour ($\gamma = 1.24$) [76]. The observed value of $p$ and the experimental fact that $\Gamma^a$ and $\Gamma^b$ behave in the same way, however, more likely support a planar model where the spins still fluctuate in the $ab$-plane. Though for the 2-d XY model the theoretical values for $\eta$ and $\nu$ are not known and a simple power law behaviour might be doubted, we have tried to give a rough

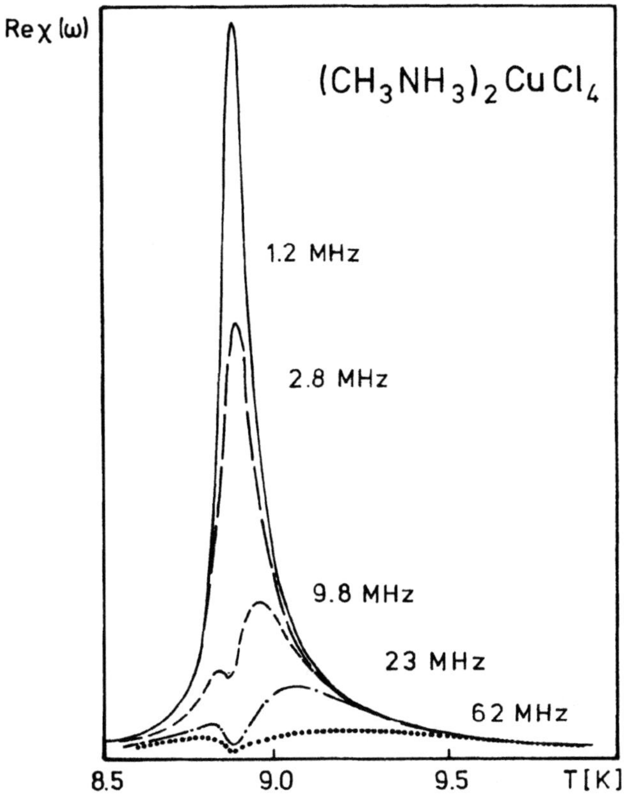


Fig. 19. Temperature dependence of the real part of the AC susceptibility of $(CH_3NH_3)_2CuCl_4$ for different frequencies. (Data from ref. [74].)

quantitative comparison of the respective models in Table III. Assuming $\eta$ and $\nu$ to be close to the 2-d Ising values (which is consistent with earlier theoretical results [77]) one could understand some of the experimental features: (i) the slowing down of $\Gamma^a$ and $\Gamma^b$ and the speeding up to $\Gamma^c$, (ii) the absolute values of the exponents $\gamma$ and $p$, and (iii) even the fact that the crossover in lattice dimensionality hardly shows up in the experimental relaxation rate. The explanation of the latter effect would be that the crossover occurring in the $\kappa$-dependence of $\Gamma^a$ or $\Gamma^b$ (from roughly 1 to 3/2) is compensated by a simultaneous crossover in $\nu$ (from roughly 1 to 2/3).

## 6. Conclusions

Both theoretical and experimental investigations of the paramagnetic spin dynamics have indicated the dominating role of spin diffusion in low-d Heisenberg magnets. The influence of lattice dimensionality shows up most distinctly in those

TABLE III

Critical behaviour of static susceptibility and linewidth; comparison with experimental data on $(CH_3NH_3)_2CuCl_4$.

| *Theory*:[1] | 2D-H | 2D-XY | 3D-XY | 2D-I | 3D-I |
|---|---|---|---|---|---|
| $\chi^a$ | $\kappa^{\eta-2}$ | $\kappa^{\eta-2}$ | $\varepsilon^{-1.32}$ | $\varepsilon^{-1.75}$ | $\varepsilon^{-1.24}$ |
| $\Gamma^a$ | $\kappa^{3\eta/2-2}$ | $\kappa^{+1}$ | $\kappa^{3/2} \sim \varepsilon$ | $\kappa^{2-\eta} \sim \varepsilon^{+1.75}$ | $\kappa^{2-\eta} \sim \varepsilon^{+1.24}$ |
| $\Gamma^c$ | $\kappa^{3\eta/2-2}$ | $\kappa^{2\eta-3}$ | $\kappa^{-5/2} \sim \varepsilon^{-1.5}$ | const | const |

| *Experiment*: | | | | |
|---|---|---|---|---|
| $\chi_a$ | $\varepsilon^{-2.5}$ $(0.5 \leqslant \varepsilon \leqslant 0.2)$ | $\varepsilon^{-1.75}$ $(0.2 \leqslant \varepsilon \leqslant 0.07)$ | $\varepsilon^{-1.25}$ $(0.03 \leqslant \varepsilon \leqslant 0.003)$ | [76] |
| | | $\varepsilon^{-1.73}$ $(0.3 \leqslant \varepsilon \leqslant 0.05)$ | $\varepsilon^{-1.20}$ $(0.05 \leqslant \varepsilon \leqslant 0.004)$ | [74] |
| | $\varepsilon^{-1.96}$ $(0.5 \leqslant \varepsilon \leqslant 0.1)$ | $\varepsilon^{-1.15}$ $(0.1 \leqslant \varepsilon \leqslant 0.01)$ | $\varepsilon^{-0.85}$ $(0.01 \leqslant \varepsilon \leqslant 0.003)$ | [75] |
| $\Gamma^a$ | | $\varepsilon^{+1.05}$ $(0.1 \leqslant \varepsilon \leqslant 0.004)$ | | [74] |
| | $\approx \varepsilon^{-0.5}$ $(0.7 \leqslant \varepsilon \leqslant 0.2)$ 'speeding up' | $\approx \varepsilon^{+0.7}$ $(0.1 \leqslant \varepsilon \leqslant 0.01)$ ↖ 'slowing down' | $\approx \varepsilon^{+0.9}$ $(0.01 \leqslant \varepsilon \leqslant 0.003)$ ↗ | [75] |
| $\Gamma^c$ | | $\approx \varepsilon^{-0.2}$ $(0.1 \leqslant \varepsilon \leqslant 0.01)$ 'speeding up' | const $(0.01 \leqslant \varepsilon \leqslant 0.003)$ | [75] |

[1] The asymptotic $\kappa$-dependence of the $\Gamma^a$ was calculated by means of the mode-coupling approach described in Section 5.2. Note that for the XY model $\chi^a = \chi^b$ and $\Gamma^a = \Gamma^b$, while for the Ising model $\chi^b = \chi^c$ and $\Gamma^b = \Gamma^c$.

physical properties which are related to $q = 0$ modes. We have pointed out that the diffusive behaviour, not only of 2-spin correlation functions, but also of 4-spin, 6-spin etc. correlation functions, represents a very general feature of a spin system with purely isotropic interactions. It is based on the fact that the respective powers of total spin operators $S^\alpha$, $S^\alpha S^\beta$, $S^\alpha S^\beta S^\gamma$, etc. are constants of motion. In NMR and EPR the diffusive long time behaviour of 2-spin and 4-spin correlation functions gives rise to several specific low-d anomalies resulting from divergences of the respective spectral densities at very low frequencies. These divergences show a characteristic shape given by $\omega^{-1/2}$ in a 1-d spin lattice and by $\ln \omega$ in a 2-d spin lattice.

Different experimental techniques have been discussed to detect these divergencies in layered compounds. In EPR experiments at room temperature their logarithmic shape was analyzed directly from the frequency dependence of linewidth, from the shape of the main EPR line and also from the shape of a satellite line occurring at half resonance field. In addition, the characteristic angular dependence of linewidth, observed in such systems, may be regarded as an indirect proof. Evidence for spin diffusion could also be obtained by NMR from the logarithmic frequency dependence of $T_1$. Each of these experimental techniques can be used for a quantitative evaluation of the respective 2-spin (NMR) or 4-spin (EPR) diffusion constants.

In contrast to the isotropic spin dynamics observed at high temperature, approaching the phase transition the critical spin fluctuations mostly become anisotropic. This is related to the fact that an ideal 2-d Heisenberg magnet cannot show any long-range order at nonzero temperature, so that the critical behaviour is sensitively affected by even a very small anisotropy in the spin-spin interactions or by weak interplane couplings. Depending on the strength of such perturbations in real layer compounds, it is possible to observe several crossovers in the critical regime $T_c \lesssim T \lesssim 1.5\, T_c$ both in spin and in lattice dimensionality. As a consequence, the critical behaviour of NMR and EPR linewidths is rather complicated and diverse. We have tried to give a representative survey by discussing the different behaviours in terms of critical linewidth exponents.

Two principal difficulties occur, however. As in the high-temperature regime, the critical spin relaxation of 2-d F's shows a pronounced non-Markovian character, that means the linewidth depends not only on the correlation length, but also on frequency, and cannot simply be described by a critical linewidth exponent $p$. Moreover, for the 2-d Heisenberg and XY-models exact predictions for the static critical exponents $\eta$ and $\nu$, which essentially determine the value of $p$, do not exist. This makes the interpretation of some experimental results rather doubtful.

A question of current interest concerns the possibility of localized excitations ('solitons') in 2-d magnets. Although the soliton concept has been well established for Heisenberg spin chains at low temperature [78, 79], little is known about the situation in 2-d lattices. For the 2-d Heisenberg magnet a static soliton solution had already been presented in the early sixties by Skyrme [80] and this has been extended to anisotropic systems very recently by Waldner [81]. However, it is not yet clear how these solutions will affect the dynamics in detail. Qualitatively, by analogy with the 1-d case, a broadening of the EPR and NMR linewidth is expected, which should be proportional to the number of solitons $n_s \sim e^{-E_s/kT}$, with $E_s = 4\pi JS^2$ being the soliton activation energy. In fact, such an Arrhenius law has been observed in the EPR linewidth of several manganese layer compounds close to the phase transition [82].

Another topical point, which has reawakened the current interest in layered compounds, concerns the percolation problem, i.e. the dilution of magnetic compounds with nonmagnetic impurities, or the mixing of different species of magnetic ions. When a magnetic insulator is diluted with nonmagnetic ions, the exchange paths are cut off at random and the spin correlation functions will decay more slowly. As a consequence, the exchange narrowing of the EPR or NMR linewidth becomes less efficient and a broadening occurs. The effect is most dramatic near the percolation concentration $c_p$ of the magnetic ions, where the coherent spin network breaks up into finite clusters. It is well known from linear spin chains ($c_p = 1$) that even very small amounts of impurities ($\lesssim$ 1 at %) will markedly increase the EPR linewidth at room temperature [83]. A similar line broadening indicating the slowing down of the diffusive spin fluctuations has been observed in square spin lattices at $c_p = 0.59$ [84]. Simultaneously, the non-Markovian character of the total spin relaxation appears more distinctly: there occurs an enhancement of the linewidth anisotropy, a change of the lineshape from a Lorentzian to an almost Gaussian shape and an increase of the $2\omega_0$ satellite's intensity by as much as two orders of magnitude [85]. The effects have qualitatively been interpreted as arising from the prevailing 1-d character of the links within a spin cluster, but they can also be understood as a general consequence of the slowing down of spin fluctuations in isolated clusters.

The percolation effect also manifests itself as a continuous decrease of the 3-d ordering temperature $T_c$ with increasing impurity concentration, until, at the percolation limit, $T_c$ reaches zero [85]. Since the correlation length depends on temperature as well as on concentration, drastic changes of both the static and dynamic critical properties are expected when the array of spins breaks up into finite clusters. For instance, no long-range order will occur any longer for $c < c_p$.

A similar situation is encountered in mixed compounds where two species of magnetic ions have competing interactions, for instance, F and AF coupling in $K_2Cu_cMn_{1-c}F_4$ [85]. In those ranges where the concentration of one of the species prevails ($c \leqslant 0.42$ and $c \geqslant 0.72$) the long-range order of the major fraction develops as in the pure compounds, but with an ordering temperature decreasing continuously to zero at the limiting concentrations. In between, a new magnetic phase occurs where the spins are expected to be highly frustrated and unable to show collective order. In both cases, magnetic resonance has been used as a powerful tool to study the static and dynamic spin correlations. Especially EPR measurements have contributed to elucidate the physical situation, but open questions do remain and work is still in progress.

## References


1. R. Kubo and K. Tomita, *J. Phys. Soc. Jpn.* **9** (1954) 888.
2. T. Moriya, *Prog. Theor. Phys.* **16** (1956) 23.
3. L. J. de Jongh and A. R. Miedema, *Adv. Phys.* **23** (1974) 1.
4. F. D. Devreux, J. P. Boucher and M. Nechtschein, *J. de Physique* **35** (1974) 271.
5. J. P. Boucher, *Phys. Rev.* **B 11** (1971) 3819.
6. N. Bloembergen, *Physica* **15** (1949) 386; L. Van Hove, *Phys. Rev.* **95** (1954) 1374.
7. R. A. Tahir-Kheli and D. G. McFadden, *Phys. Rev.* **178** (1969) 800.
8. F. Ferrieu, *J. de Physique/Lett.* **38** (1977) L 381.
9. R. E. Dietz, F. R. Merritt, R. Dingle, D. Hone, B. G. Silbernagel and P. M. Richards, *Phys. Rev. Lett.* **26** (1971) 1186.
10. G. F. Reiter and J. P. Boucher, *Phys. Rev.* **B 11** (1975) 1823.
11. A. Lagendijk and D. Schoemaker, *Phys. Rev.* **B 16** (1977) 47.
12. L. P. Kadanoff and P. C. Martin, *Ann. Phys.* **24** (1963) 419; H. Mori, *Prog. Theor. Physics* **24** (1965) 399.
13. J. E. Gulley, D. Hone, D. J. Scalapino and B. G. Silbernagel, *Phys. Rev.* **B 1** (1970) 1020.

14. F. Carboni and P. M. Richards, *Phys. Rev.* **177** (1969) 889.
15. P. M. Richards and M. B. Salamon, *Phys. Rev.* **B 9** (1974) 32.
16. A. Lagendijk, *Phys. Rev.* **B 18** (1978) 1322.
17. J. P. Boucher, M. Ahmed-Bakheit, M. Nechtschein, M. Villa, G. Bonera and F. Borsa, *Phys. Rev.* **B 13** (1976) 4098.
18. P. M. Richards in: *Local Properties at Phase Transitions*, Proc. of the International School of Physics 'Enrico Fermi', North-Holland, Amsterdam, 1976, p. 539; J. E. Drumheller, *Magn. Res. Rep.* **7** (1982) 123.
19. D. W. Hone, C. Scherer and F. Borsa, *Phys. Rev.* **B 9** (1974) 965.
20. F. Ferrieu and J. P. Boucher, *Proc. XIXth Congress Ampère*, Heidelberg, 1976.
21. W. Rappold and H. Benner, 1980, unpublished.
22. H. Nishihara, T. Goto, Y. Kimishima and H. Kubo, *J. Phys. Soc. Jpn.* **51** (1982) 407.
23. W. Depmeier, *Acta Cryst.* **B 33** (1977) 3713.
24. T. Morita, *Phys. Rev.* **B 6** (1972) 3385.
25. H. R. Boesch, U. Schmocker, F. Waldner, K. Emerson and J. E. Drumheller, *Phys. Lett.* **36A** (1971) 461; F. Ferrieu and M. Pomerantz, *Sol. State Comm.* **39** (1981) 707; R. D. Willet, R. Wong and M. Numata, *J. Magn. Magn. Mat.* **15—18** (1981) 717; R. D. Willet and F. Waldner, *J. Appl. Phys.* **53** (1982) 2680.
26. D. B. Losee, J. W. Hall and W. E. Hatfield, *Sol. State Comm.* **16** (1975) 389; H. Hagen, Ph.D. thesis, Univ. Zürich, 1979; J. Onderwaater, Ph.D. thesis, Univ. Leiden, 1980; O. Takizawa, R. Srinivasan and E. de Boer, *Mol. Phys.* **44** (1981) 677.
27. D. L. Huber, *Phys. Rev.* **B 6** (1972) 3180; M. S. Seehra, *Phys. Rev.* **B 6** (1972) 3186.
28. A. Lagendijk and E. Siegel, *Sol. State Comm.* **20** (1976) 709.
29. H. Benner, *Phys. Lett.* **70A** (1979) 225.
30. A. Lagendijk and E. Siegel, *J. Magn. Magn. Mat.* **15—18** (1980) 714.
31. Y. Natsume, F. Sasagawa, M. Toyoda and I. Yamada, *J. Phys. Soc. Jpn.* **48** (1980) 50; M. Tanimoto and Y. Yokozawa, *J. Phys. Soc. Jpn.* **49** (1980) 67.
32. H. Benner, M. Brodehl, H. Seitz and J. Wiese, *J. Phys.* **C 16** (1983) 6011.
33. H. Benner, *Phys. Rev.* **B 18** (1978) 319 and *J. Magn. Magn. Mat.* **13** (1979) 103.
34. K. Nagata, I. Yamada, T. Komatsubara and T. Nishizaki, *J. Phys. Soc. Jpn.* **43** (1977) 707; Y. Yokozawa and M. Tanimoto, *J. Phys. Soc. Jpn.* **44** (1978) 1741.
35. N. D. Mermin and H. Wagner, *Phys. Rev. Lett.* **17** (1966) 1133.
36. H. E. Stanley and T. A. Kaplan, *Phys. Rev. Lett.* **17** (1966) 913.
37. J. M. Kosterlitz and D. J. Thouless, *J. Phys.* **C 6** (1973) 1181.
38. M. E. Fisher and R. J. Burford, *Phys. Rev.* **156** (1967) 583.
39. B. I. Halperin and P. C. Hohenberg, *Phys. Rev.* **177** (1969) 952; P. C. Hohenberg and B. I. Halperin, *Rev. Mod. Phys.* **49** (1977) 435.
40. K. Kawasaki, *Prog. Theor. Phys.* **40** (1968) 11.
41. K. Kawasaki, *Prog. Theor. Phys.* **39** (1968) 285.
42. P. M. Richards, *Sol. State Comm.* **13** (1973) 253.
43. M. E. Fisher, *Rev. Mod. Phys.* **46** (1974) 597.
44. E. Brezin, J. C. LeGuillou, J. Zinn-Justin and B. G. Nickel, *Phys. Lett.* **44A** (1973) 227.
45. D. S. Ritchie and M. E. Fisher, *Phys. Rev.* **B 7** (1973) 480.
46. P. Bloembergen, *Physica* **79B** (1975) 467.
47. R. J. Birgenau, J. Als-Nielsen and G. Shirane, *Phys. Rev.* **B 16** (1977) 280.
48. E. Riedel and F. Wegner, *Phys. Rev. Lett.* **24** (1970) 730.
49. W. Finger, *Phys. Lett.* **60A** (1977) 165 and *J. Phys.* **C11** (1978) 1877.
50. L. van Hove, *Phys. Rev.* **95** (1954) 1374.
51. H. Abe and M. Matsuura, *J. Phys. Soc. Jpn.* **19** (1964) 1867; F. V. Bragin and S. M. Ryabchenko, *Sov. Phys. Sol. State* **15** (1973) 721; W. J. Looyestijn, T. O. Klaassen and N. J. Poulis, *Physica* **96B** (1979) 303.
52. Y. Kimishima and H. Nishihara, *Sol. State Comm.* **32** (1979) 245.
53. P. K. Leichner and P. M. Richards, *Phys. Rev.* **B 7** (1973) 453.
54. E. P. Maarshall, A. C. Botterman, S. Vega and A. R. Miedema, *Physica* **41** (1969) 473.

55. C. Bucci and G. Guidi, *Phys. Rev.* **B 9** (1974) 3053 and *Phys. Lett.* **48A** (1974) 233.
56. C. Bucci and G. Guidi, *Sol. State Comm.* **10** (1972) 803.
57. D. L. Huber, *J. Phys. Chem. Sol.* **32** (1971) 2145.
58. H. Reimann, H. Hagen, F. Waldner, W. Berlinger and H. Arend, *Sol. State Comm.* **17** (1975) 1319; H. Benner, *Appl. Phys.* **13** (1977) 141.
59. P. M. Richards, K. A. Müller, H. R. Boesch and F. Waldner, *Phys. Rev.* **B 10** (1974) 4531.
60. P. R. Elliston, *J. Phys.* **C 7** (1974) 425.
61. I. Yamada, I. Morishita and T. Tokuyama, *Physica* **115B** (1983) 179.
62. J. Wiese and H. Benner, 1979, unpublished.
63. D. L. Huber, *Phys. Rev.* **B 6** (1972) 3180.
64. H. W. de Wijn, R. L. Walker, J. L. Davis and H. J. Guggenheim, *Sol. State Comm.* **11** (1972) 803.
65. C. M. J. van Uijen and H. W. de Wijn, *Phys. Rev.* **B 24** (1981) 5368.
66. J. Onderwaater, Ph.D. thesis, Univ. Leiden, 1980.
67. H. Hagen, H. Reimann, U. Schmocker and F. Waldner, *Physica* **86—88B** (1977) 1287.
68. A. F. M. Arts, H. van der List, J. G. M. van Miltenburg, C. M. J. van Uijen and H. W. de Wijn, *J. Magn. Magn. Mat.* **31—34** (1983) 1181.
69. C. M. J. van Uijen, E. Frikkee and H. W. de Wijn, *Phys. Rev.* **B 19** (1979) 509.
70. E. F. Riedel and R. D. Willet, *Sol. State Comm.* **16** (1975) 413; H. Hagen, Ph.D. thesis, Univ. Zürich, 1979; A. G. Anders and S. V. Volotski, *J. Magn. Magn. Mat.* **31—34** (1983) 1169.
71. R. Boesch and F. Waldner, in *Local Properties at Phase Transitions*, Proc. of the International School of Physics 'Enrico Fermi', North-Holland, Amsterdam, 1976, p. 642; F. V. Bragin, P. V. Vodolazskij and S. M. Ryabchenko, *Sov. Phys. JETP* **43** (1976) 803; G. A. Petrakovskii, L. S. Yemelyanova and V. V. Velichko, *Sol. State Comm.* **48** (1983) 647.
72. J. Kötzler, G. Kamleiter and G. Weber, *J. Phys.* **C 9** (1976) L361; Kötzler and W. Scheithe, *Phys. Rev.* **B 18** (1978) 1306.
73. Y. Okuda, M. Matsuura and T. Haseda, *J. Phys. Soc. Jpn.* **44** (1978) 371.
74. Y. Okuda, M. Matsuura and T. Haseda, *J. Phys. Soc. Jpn.* **47** (1979) 773 and *J. Magn. Magn. Mat.* **15—18** (1980) 1027.
75. W. L. C. Rutten, Ph.D. thesis, Univ. Leiden, 1979.
76. L. J. de Jongh, *Physica* **82B** (1976) 247.
77. D. D. Betts, C. J. Elliott and R. V. Ditzian, *Can. J. Phys.* **49** (1971) 1327.
78. H. J. Mikeska, *J. Phys.* **C11**, (1978) L29 and *J. Phys.* **C13** (1980) 2913.
79. M. Steiner, *J. Magn. Magn. Mat.* **31—34**, (1983) 1277; P. Boucher *et al.*, *Physica* **120B** (1983) 241; H. Benner, H. Seitz, J. Wiese and J. P. Boucher, *J. Magn. Magn. Mat.* **45** (1984) 354.
80. T. H. R. Skyrme, *Proc. R. Soc. (Lond.)* **262** (1961) 237.
81. F. Waldner, *Proc. XXII Congress Ampère*, Zürich (1984) 147.
82. F. Waldner, *J. Magn. Magn. Mat.* **31—34** (1983) 1203.
83. P. M. Richards, *Phys. Rev.* **B 10** (1974) 805; E. Siegel, J. Ibrügger and A. Lagendijk, *J. Phys.* **C 15** (1982) 4583.
84. H. Takano and Y. Yokozawa, *J. Phys. Soc. Jpn.* **42** (1977) 1059; W. M. Walsh, R. J. Birgeneau, L. W. Rupp and H. J. Guggenheim, *Phys. Rev.* **B 20** (1979) 4645.
85. I. Yamada, S. Shimoda and S. Nagano, *J. Magn. Magn. Mat.* **31—34** (1983) 645.

# FIELD-INDUCED PHENOMENA IN TWO-DIMENSIONAL WEAKLY ANISOTROPIC HEISENBERG ANTIFERROMAGNETS

H. J. M. DE GROOT AND L. J. DE JONGH

*Kamerlingh Onnes Laboratory, Leiden University, P.O. Box 9506, 2300 RA Leiden, The Netherlands*

## 1. Introduction

The field-dependent behaviour of weakly anisotropic antiferromagnets is unique in that the anisotropy may be tuned and transformed continuously. The symmetry of these systems can be changed from Heisenberg into weakly Ising-type, from Heisenberg into weakly XY-type, or from weakly Ising-type into weakly XY-type, and back. This is achieved simply by varying two parameters, the temperature and the applied magnetic field. In the present chapter we focus on the various phenomena that may occur in magnetic fields. In particular the field-induced anisotropy crossovers only recently have been recognized [1—3] and exploited [1—7]. Since the magnetic properties of 2-d Ising-type, 2-d XY-type and 2-d Heisenberg systems are widely different, the field-induced anisotropy crossovers will give rise to important changes in the thermodynamic behaviour.

In the next section the explicit dependence of the anisotropy on the applied field is treated. The concept of effective, field-dependent anisotropy is introduced and explained. This concept will play a central role throughout the whole chapter and may be considered as a basic principle, necessary to understand the explanations we give for various field-induced phenomena, which will be treated in subsequent sections. Firstly the field-induced variation of the ordering temperature $T_c$ is discussed in Section 3. Comparisons between the theory and various quasi 2-d experimental examples are presented. In Section 4 the importance of domain walls and random fields is discussed, whereas Section 5 deals with the spin-flop phenomenon. It appears that the experimentally observed spin-flop transition in quasi 2-d systems is very broad and is therefore not of first order, as it is in 3-d systems. The broadening is attributed to small random fields arising from the number of defects naturally present in the crystals used in the experiments. Finally, the bicritical behaviour is briefly discussed in Section 6. Differences are found between 3-d and quasi 1-d systems. One may argue that these differences also arise from the small random fields present.

Since this chapter is part of a book on 2-d magnetism, we shall restrict ourselves to quasi 2-d examples. However, field-induced effects are by no means restricted to 2-d antiferromagnets. More information for the 3-d and quasi 1-d systems can be found in [1—9].

## 2. Effective, Field-Dependent Anisotropies

The basic principle for the interpretation of field-induced effects in weakly



*L. J. De Jongh (Ed.), Magnetic Properties of Layered Transition Metal Compounds* 379—404.

anisotropic Heisenberg antiferromagnets is the concept of effective, field-dependent anisotropy [1, 6], which originates from the competition between the dipolar anisotropy and the Zeeman energy of the moments in the applied magnetic field.

Let us consider a fairly general classical interaction Hamiltonian of the form

$$\mathcal{H} = -2J \sum_{(i,j)} \mathbf{S}_i \cdot \mathbf{S}_j - \sum_k (D_x S_{kx}^2 - D_z S_{kz}^2) - \sum_k g\mu_B \mathbf{H} \cdot \mathbf{S}_k \tag{1}$$

with the summation $(i, j)$ over nearest-neighbour pairs. This Hamiltonian contains firstly an isotropic (Heisenberg) exchange term, describing the antiferromagnetic ($J < 0$) exchange interaction between nearest-neighbouring atoms with magnetic moments $\mathbf{S}$, secondly an orthorhombic (biaxial) anisotropy term with $D_x$, $D_z \geqslant 0$ and finally a Zeeman term, with $\mathbf{H}$ being the applied field. The anisotropy term contains two contributions: a planar (XY) type $D_z$ term and an easy-axis (Ising) type $D_x$ term. Thus the orthorhombic anisotropy in Hamiltonian (1) establishes the $XY$-plane as the easy plane for the magnetic moments with the $X$-axis as the easy axis within this plane.

With the summation $(i, j)$ over the appropriate number of nearest neighbours $z$, the Hamiltonian (1) is thought to be a reasonable model for weakly anisotropic quasi 2-d antiferromagnets, at least at low temperatures, where $k_B T \approx 2S|2zD_x J|^{1/2}$, corresponding to the magnon gap due to the Ising type anisotropy.

In the weakly anisotropic Heisenberg antiferromagnet a spinflop transition occurs at a critical field $H_x = H_{sf}$. There the average polarization of the magnetic moments changes from parallel ($H_x < H_{sf}$) to perpendicular to the easy axis ($H_x > H_{sf}$). The spinflop finds its origin in the large difference between the magnetic susceptibility parallel ($\chi_{\parallel}$) and perpendicular ($\chi_{\perp}$) to the easy ($X$-)axis, which is typical for an antiferromagnet. At low temperatures one has $\chi_{\parallel} \ll \chi_{\perp}$. This is illustrated in the chapter by Arts and de Wijn, where the differential susceptibility of a single crystal of $K_2MnF_4$ is shown. $K_2MnF_4$ consists of weakly coupled ($J'/J \approx 10^{-6}$) antiferromagnetic layers, and is an excellent example of a quasi 2-d system. For $T < 50$ K one has $\chi_{\parallel} < \chi_{\perp}$, and $\chi_{\parallel}$ vanishes completely for $T \to 0$. For $H_x > H_{sf}$ the difference in Zeeman energy, $1/2(\chi_{\perp} - \chi_{\parallel})H_x^2$, which tends to polarize the moments perpendicular to the applied field and thus *perpendicular* to the the $X$-axis, exceeds the anisotropy energy $D_x S^2$ that tends to polarize the moments *along* the $X$-axis. At low temperatures the small $\chi_{\parallel}$ can be neglected and the effective net anisotropy, denoted by $D_{\text{eff}}^x(H)$, equals the difference between the anisotropy energy $D_x S_2$ and the Zeeman energy $1/2\chi_{\perp} H_x^2$, yielding after some manipulation:

$$D_{\text{eff}}^x(H_x) = D_x(1 - H_x^2/H_{sf}^2). \tag{2}$$

For $\mathbf{H} = H_z$ or $\mathbf{H} = H_y$ the spin-flop transition is obviously absent, since the Zeeman energy and anisotropy energy do not compete but enforce one another. In that case one may also define a net anisotropy, which, however, equals the *sum* of the anisotropy energy $D_x S^2$ and the Zeeman energy $1/2\chi_\perp H^2_{y,z}$, yielding

$$D^x_{\text{eff}}(H_{y,z}) = D_x(1 + H^2_{y,z}/\mathrm{H}^2_{\text{sf}}). \tag{3}$$

In fact we have now incorporated the last term of Hamiltonian (1) into the anisotropy term so that we may model the system by the effective Hamiltonian

$$\mathcal{H}_{\text{eff}} = -2J \sum_{i,j} \mathbf{S}_i \cdot \mathbf{S}_j - \sum_k (D^x_{\text{eff}}(\mathbf{H}) S^2_{kx} - D_z S^2_{kz}), \tag{4}$$

i.e. the anisotropy is field-dependent and various symmetries are possible dependending on the magnitude and direction of the applied field and the values of $D_x$ and $D_z$.

The various field-induced anisotropies are visualized in Figures 1 and 2 for systems which, in *zero* field, have uniaxial (i.e. $D_z = 0$) or orthorhombic anisotropy, respectively. In these figures the easy axis and easy plane are indicated for several configurations corresponding to various applied fields. As an example we shall discuss in some detail what will happen for $\mathbf{H} = H_x$ since this is the most interesting case for the subsequent sections.

In case of uniaxial anisotropy the $X$-axis is the easy axis for fields up to $H_x = H_{\text{sf}}$ (cf. Figure 1a) since, according to Equation (2), one has $D^x_{\text{eff}} > 0$. For $H_x >$

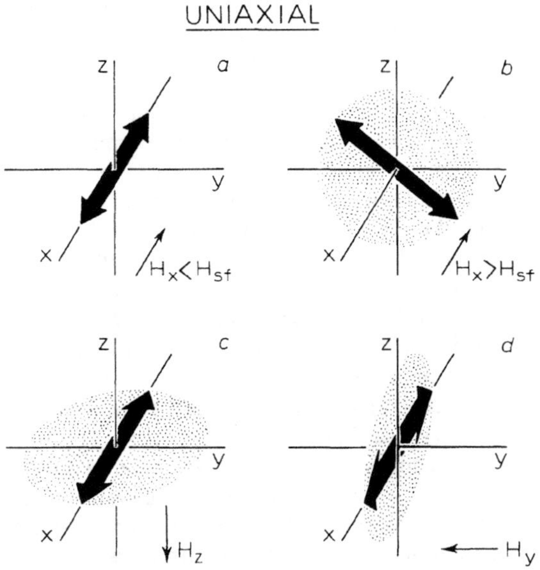


Fig. 1. Effective, field-dependent anisotropies for an antiferromagnetic system with weak uniaxial anisotropy. (a), (b) $\mathbf{H} = H_x$, (c) $\mathbf{H} = H_z$ and (d) $\mathbf{H} = H_y$. The heavy arrows indicate the easy axes and the shaded areas show the easy planes for the various cases. The small arrows indicate the directions of the applied field.

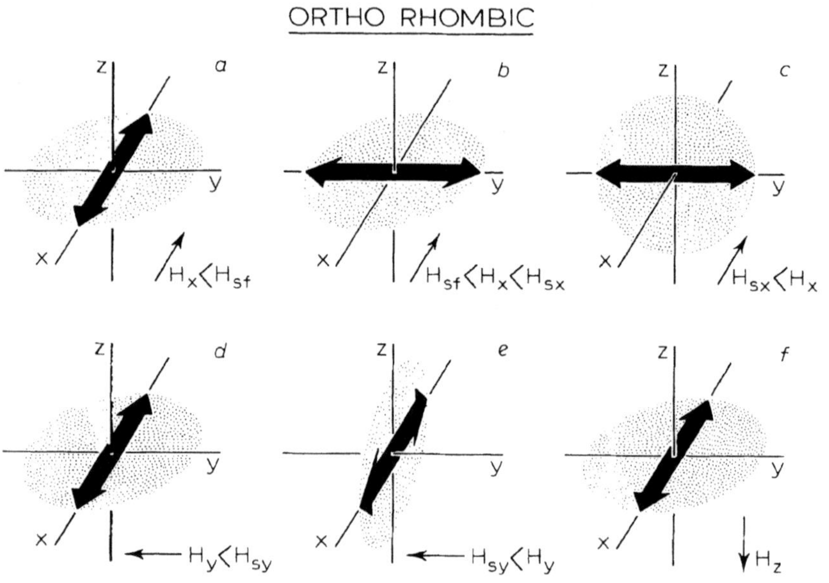


Fig. 2. Effective, field-dependent anisotropies for an antiferromagnetic system with weak orthorhombic anisotropy. (a), (b), (c) $\mathbf{H} = H_x$, (d), (e) $\mathbf{H} = H_y$ and (f) $\mathbf{H} = H_z$. The heavy arrows indicate the easy axes and the shaded areas show the easy planes for the various cases. The small arrows indicate the directions of the applied field.

$H_{sf}$ on the other hand, the $YZ$-plane is the easy plane and the $X$-axis the hard axis (Figure 1b), since in that case $D^x_{eff} < 0$. For orthorhombic anisotropy the situation is changed in that there are now two anisotropy crossovers. For $H_x < H_{sf}$ (cf. Figure 2a) $D^x_{eff} > 0$ and the $X$-axis is the easy axis. The difference with the uniaxial case is that here the symmetry in the $YZ$-plane is broken by the anisotropy $D_z$. Therefore the $Z$-axis is the hard axis, the $XY$-plane the easy plane and the $Y$-axis the intermediate axis. The anisotropy crossover in the $XY$-plane at $H_x = H_{sf}$ occurs since for $H_x > H_{sf}$ (but $H_x < H_{sx}$) one has $D^x_{eff} < 0$ and therefore the $Y$-axis is the easy axis in the easy $XY$-plane. However, the main difference between the uniaxial and orthorhombic case is that when in the latter the field is further increased, a second anisotropy crossover occurs for $D^x_{eff} = -D_z$, that is for $H_x = H_{sx} = (2(D_zS^2 + D_xS^2)/\chi_\perp)^{1/2}$. At this second anisotropy crossover the hard axis changes in direction, from the $Z$-axis to the $X$-axis.

Evidently the symmetry properties of weakly anisotropic system vary strongly with the applied field. Comparable anisotropy crossover effects occur for $\mathbf{H} = H_y$ and $\mathbf{H} = H_z$, both for uniaxial and orthorhombic systems (cf. Figures 1b, c and Figures 2d, e, f, respectively). Summarizing, one may have for the uniaxial system: (i) uniaxial Ising-type anisotropy ($H_x < H_{sf}$), (ii) fully isotropic Heisenberg behaviour ($H_x = H_{sf}$), (iii) XY-type behaviour ($H_x > H_{sf}$) and (iv) orthorhombic (Ising-type) anisotropy ($H_x$, $H_z$). For the biaxial system one may have: (i) orthorhombic (Ising-type) anisotropy ($H_x \neq H_{sf}$, $H_x \neq H_{sx}$, $H_y \neq H_{sy}$), (ii) XY-type behaviour ($H_x = H_{sf}$) and (iii) uniaxial Ising-type anisotropy ($H_x = H_{sx}$, $H_y = H_{xy}$). Since the magnetic properties depend very strongly on the underlying symmetry, widely different behaviour is expected, and indeed observed, in particular for 2-d systems.

## 3. The Phase Diagram

The magnetic phase diagram that would result from a system described by Hamiltonian (4) is sketched in Figure 3. We shall concentrate first on the uniaxial case (Figure 3a). For $H_x \uparrow H_{sf}$ one has a 2-d Heisenberg system with vanishing Ising-type anisotropy. Therefore the I-P boundary $T_c^{\mathrm{I}}(\mathbf{H})$, where the transition between the paramagnetic state and the 2-d Ising ordered state occurs, is a line of second-order transitions. Recently, theoretical predictions for the ordering temperatures of weakly anisotropic, Ising-type systems have been obtained using Monte Carlo techniques [10]. Applied to the present case, where the degree of anisotropy can be tuned by the field, these Monte Carlo predictions yield the relation

$$T_c^{\mathrm{I}}(D_{\mathrm{eff}}^x(\mathbf{H})) \propto (D_{\mathrm{eff}}^x(\mathbf{H}))^n, \tag{5}$$

where $n$ is as small as 0.0—0.05, so that the field-dependence of $T_c$ is almost logarithmic [10, 11]. The phase boundary according to Equation (5) is sketched schematically in Figure 3a for $H_x < H_{sf}$.

According to Equation (5) one has $T_c(\mathbf{H}) \downarrow 0$ for $H_x \uparrow H_{sf}$, i.e. for vanishing anisotropy. Such a behaviour should be expected since for $H_x = H_{sf}$ the system is

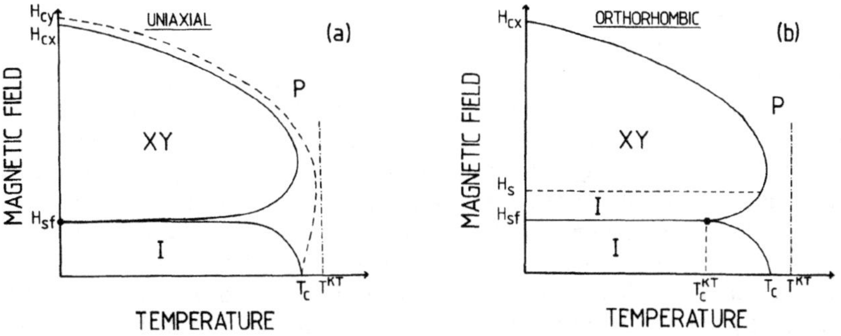


Fig. 3. (a) Schematic phase diagram of the 2-d Heisenberg antiferromagnet with weak uniaxial anisotropy. Solid curves correspond to the phase diagram for $\mathbf{H} = H_x$, whereas the dashed curve yields the phase diagram for $\mathbf{H} = H_\perp$. The dot-dashed line indicates the Kosterlitz—Thouless transition for the pure 2-d planar rotator (XY-)model. (b) Schematic phase diagram for $\mathbf{H} = H_x$ of the 2-d Heisenberg antiferromagnet with weak orthorhombic anisotropy (solid curves). The dashed line for $\mathbf{H} = H_x$ indicates the field value for which the switching of anisotropy axis occurs. The dot-dashed line indicates the Kosterlitz—Thouless transition for the pure 2-d XY-system. According to the effective anisotropy model the phase transition at the bifurcation point ($T_c^{\mathrm{KT}}$) would be of the Kosterlitz—Thouless type.

fully isotropic and it is commonly accepted that the 2-d Heisenberg system does not order at $T \neq 0$ [12].

For $H_x > H_{sf}$ the symmetry is planar. This part of the phase diagram should correspond to that of the isotropic system in a field [7]. Also for this problem theoretical predictions have been evaluated with Monte Carlo calculations [13], and the phase boundary is given schematically in Figure 3a. For the ideal theoretical case the XY-state is not long-range ordered. Instead there is topological order, in terms of strongly coupled vortex—antivortex pairs. Therefore the XY-phase has much in common with the low-temperature state predicted by Kosterlitz and Thouless [14] for the classical XY model, and the phase boundary XY-P should correspond with a line of Kosterlitz—Thouless transitions $T^{KT}(\mathbf{H})$. The transition temperature rapidly increases for $H_x > H_{sf}$ from $T = 0$ towards the Kosterlitz—Thouless temperature $T^{KT}$ for the plane rotator model. The latter corresponds with the limit $D^x_{eff}(\mathbf{H}) \rightarrow -\infty$ in the context of the present work. Estimates for $T^{KT}$ have been calculated [15], yielding $T^{KT} \approx 1.7|JS^2/k_B|$, and obtained with Monte Carlo techniques [13], which give $T^{KT} \approx 1.45|JS^2/k_B|$. However, before reaching $T^{KT}$ the curve will bend backwards due to saturation effects occuring in high fields. One has $T^{KT}(\mathbf{H}) \rightarrow 0$ for $H_x \rightarrow H_{cx}$, the critical field being given by

$$H_{cx} = 2H_e - H_{ax}, \tag{6}$$

where

$$H_e = 2zJS/g\mu_B \tag{7}$$

is the exchange field and

$$H_{ax} = 2D_xS/g\mu_B \tag{8}$$

represents the anisotropy field.

For $\mathbf{H}$ perpendicular to the $X$-axis, $\mathbf{H} = H_\perp$, the uniaxial Ising-type character of the system becomes more pronounced as the field is applied. The whole phase boundary, indicated by the dashed line in Figure 3a, formally corresponds with a line of second-order Ising-type transitions. However, for high fields the field-induced anisotropy will be much stronger than $D_x$ so that the symmetry in high fields will be approximately of the planar type and the phase transition may correspondingly alter its character from Ising-type to Kosterlitz—Thouless type. Starting from $H_\perp = 0$ the transition temperature will first increase, according to Equation (5). As for $\mathbf{H} = H_x$ the phase line bends back in high fields and will approach the Monte Carlo prediction $T^{KT}(\mathbf{H})$. Finally, for $H_\perp = H_{cy}$ with

$$H_{cy} = 2H_e - H_{ax}, \tag{9}$$

the transition temperature vanishes.

For the orthorhombic case the phase diagram is sketched in Figure 3b, for $\mathbf{H} = H_x$. In the low-field region ($H_x < H_{sf}$) the behaviour is the same as for the uniaxial case. The phase boundary I-P is given by Equation (5) and is of second order. However, for $H_x = H_{sf}$ the situation is different since the symmetry is now of the

XY-type and thus topological order will establish at temperatures below the Kosterlitz–Thouless temperature corresponding to the magnitude of the hard-axis anisotropy $D_z$. Therefore the two second-order lines should coalesce into the spin-flop line at $T^c_{KT}$. For $H_x > H_{sf}$ the phase boundary is again given by Equation (5), until $H = H_{sx}$, where the switching of the hard axis occurs. Ultimately, for the highest fields, the symmetry will approach the XY-type, since $D^x_{eff}(\mathbf{H}) \gg D_z$ in that case. Furthermore, the phase line will bend backwards and reach $T = 0$ for $H = H_{cx}$.

In the experimental examples 2-d ordering processes always entail 3-d order. This is because the increase of the 2-d correlation length will enhance the interactions between the planes. However, since the onset of such 3-d ordering is still governed by the 2-d processes, the phase boundaries that are observed should reflect the underlying 2-d character.

A few detailed studies of the phase diagrams of quasi 2-d systems have been performed in the past. In Figure 4 we show the phase diagrams of the compound $K_2MnF_4$, after de Jongh *et al.* [16], whereas in Figure 5 the phase diagram for $Rb_2MnCl_4$, taken from Rauh *et al.* [17] has been reproduced. The relevant parameters for both compounds are given in Table I. Both systems consist of weakly coupled ($J'/J \approx 10^{-6}$) antiferromagnetic layers, and are excellent examples

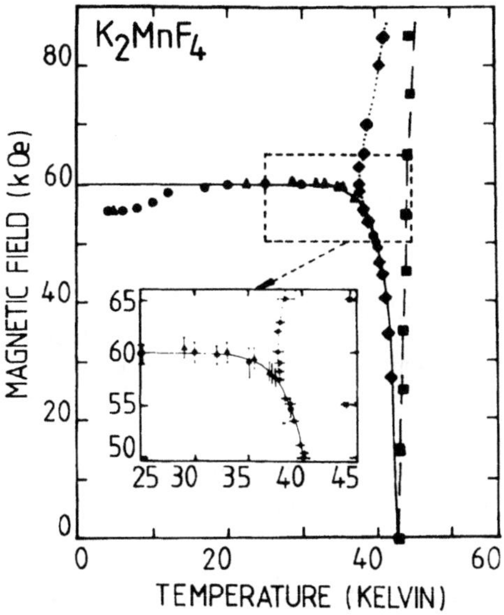


Fig. 4. Phase diagram of $K_2MnF_4$ for **H** parallel (●, ▲, ◆) and perpendicular (■) to the easy *c*-axis. The data are from [16, 18]. The solid and dashed lines are the theoretical predictions according to the effective, field-dependent anisotropy model (Equation (5)). The dotted line is a guide to the eye. In the insert the anomalous bifurcation behaviour is shown in detail.

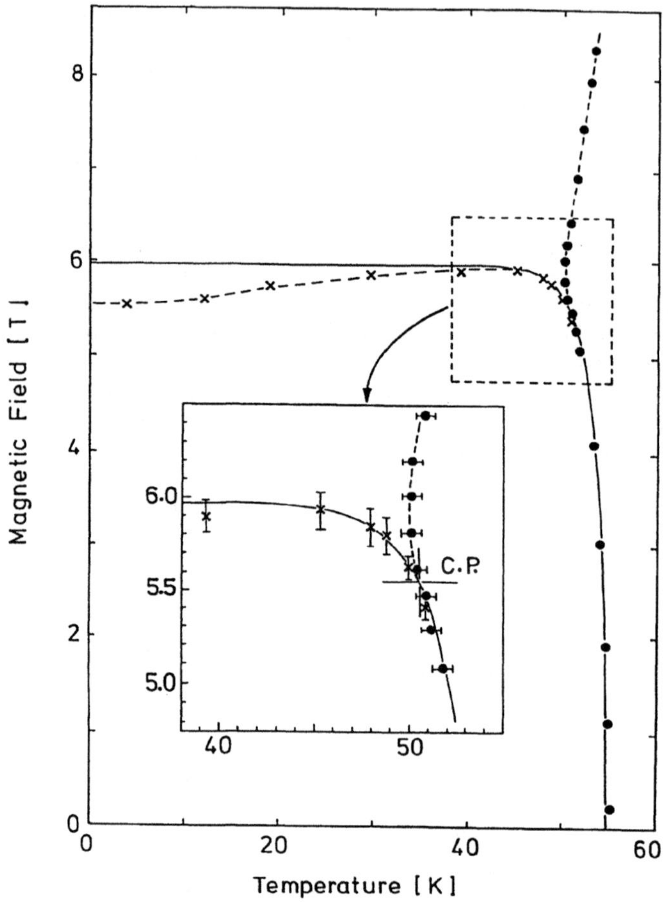


Fig. 5. Phase diagram of $Rb_2MnCl_4$, after Rauh *et al.* [17]. Data are obtained from neutron experiments in magnetic fields $\mathbf{H} = H_x$, parallel to the $c$-axis. The solid lines are the predictions from the effective, field-dependent anisotropy model (cf. Equation (5)) and the dashed lines are guides to the eye. The bifurcation behaviour is shown in detail in the insert.

of quasi 2-d systems. The easy-axis anisotropies $D_x$ have been calculated from the value of the spinflop field, extrapolated to $T = 0$, using the mean-field expression

$$D_x = g^2\mu_B^2 H_{sf}^2/32JS(S+1). \tag{10}$$

The anisotropy is of dipolar origin and is very weak, $D_x/J \approx 0.01$. It is nearly uniaxial, due to the in-layer quadratic symmetry. For $Rb_2MnCl_4$ the rhombic ($D_z$) term has been determined with antiferromagnetic resonance, and it has been reported that $D_z/D_x \approx 0.1$ [17]. Since the exchange field is quite large, only the low-field part of the phase diagram could be measured. For both $Rb_2MnCl_4$ and

TABLE I

Spin values $S$, exchange constants $J$ and $J'$, anisotropy constants $D_x$, $D_z$, and the predicted values for the Kosterlitz—Thouless temperature $T^{KT} = 1.45|JS(S+1)/k_B|$, for the experimental examples of quasi 2-d weakly anisotropic antiferromagnets $K_2MnF_4$, $Rb_2MnCl_4$, and $Mn(HCOO)_2 \cdot 2\,H_2O$.

| | $S$ | $J/k_B$ (K) | $D_x/k_B$ (K) | $D_z/D_x$ | $T^{KT}$ (K) | $J'/J$ |
|---|---|---|---|---|---|---|
| $K_2MnF_4$ | 5/2 | −4.20 | 0.046 | — | 53 | $\approx 10^{-6}$ |
| $Rb_2MnCl_4$ | 5/2 | −5.85 | 0.034 | $\approx 0.1$ | 74 | $\approx 10^{-6}$ |
| $Mn(HCOO)_2 \cdot 2\,H_2O$ | 5/2 | −0.30 | 0.003 | — | 3.8 | $\approx 10^{-3}$ |

$K_2MnF_4$ this has been done with neutron scattering, whereas for the latter compound also magnetic susceptibility measurements [18] are available.

Except for the behaviour around the bifurcation point (as shown in detail in the inserts in the two figures) the general features of Figure 3 are nicely confirmed by these two experimental examples. The solid lines give the Monte Carlo predictions from Equation (5) with $n = 0.04$ [16, 17]. The differences between experimental data and theory for $T < 40$K can be attributed to the small variation of $D_x$ with the temperature through its dependence on $\chi_{\parallel}$, since the latter has been neglected in the evaluation of the effective, field-dependent anisotropy. The dashed line in Figure 4 represents the fit to Equation (5) for $\mathbf{H} = H_{\perp}$. In that case too the agreement between theory and experiment is very good.

According to our predictions one would expect that for $H_x > H_{sf}$ the transition temperature should approach $T^{KT}$. For $K_2MnF_4$ one has $T_{KT} = 53$K and for $Rb_2MnF_4$, $T^{KT} = 74$K (see Table I). Obviously, the magnetic fields applied in the experiements are not strong enough to induce sufficient planar anisotropy in order to realize the full planar behaviour.

The phase diagram for another quasi 2-d antiferromagnet, $Mn(HCOO)_2 \cdot 2\,H_2O$ is shown in Figure 6. The experimental data were taken from Schutter *et al.* [19] and Koyama *et al.* [7]. $Mn(HCOO)_2 \cdot 2\,H_2O$ is a reasonable approximation of a 2-d system, with ($J'/J \approx 10^{-3}$). Since the antiferromagnetic coupling is an order of magnitude smaller than for $Rb_2MnCl_4$ and $K_2MnF_4$ (cf. Table I) the entire phase diagram is accessible. In Figure 6b the low-field part is shown, with the solid and dashed lines the theoretical predictions from Equation (5), this time with $n \approx 0.02$. Note that for this system too the bifuraction is anomalous. In Figure 6a and 6c the high-field data are given, for $\mathbf{H} = H_x$ and $\mathbf{H} = H_{\perp}$, respectively. The solid lines represent the Monte Carlo theory for the Heisenberg antiferromagnet in a magnetic field. We want to stress that these high-field predictions are entirely determined by the value of the exchange constant $J$, i.e. the theory is fitted to the data without adjustable parameter.

For $Mn(HCOO)_2 \cdot 2\,H_2O$ the Kosterlitz—Thouless temperature is calculated as $T^{KT} = 3.8$K. It appears that for $H > H_{sf}$ the experimentally determined transition temperatures approach $T^{KT}$ before they bend back due to saturation effects, both for $\mathbf{H} = H_x$ and for $\mathbf{H} = H_{\perp}$. As was argued above, the symmetry of the system

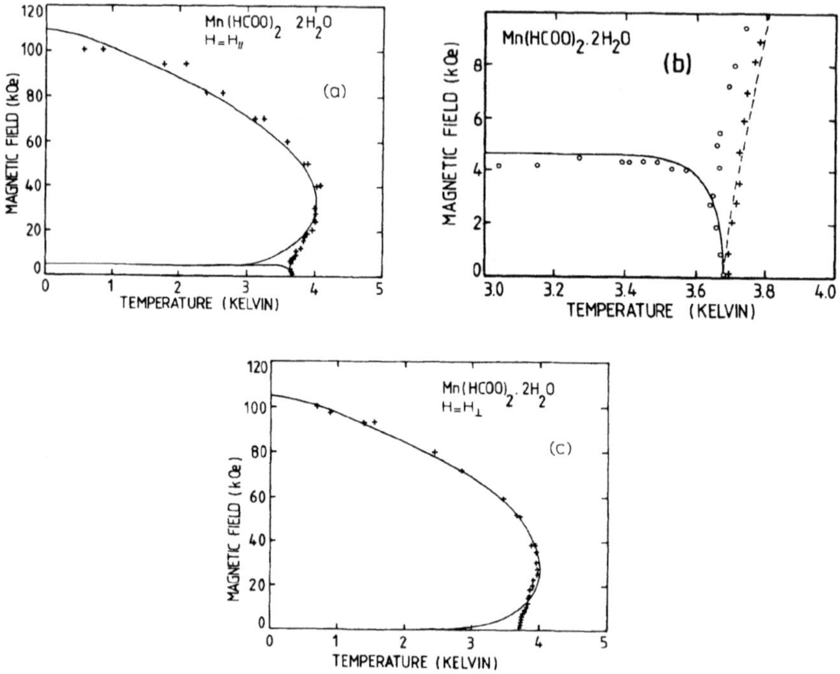


Fig. 6. Phase diagram of $Mn(HCOO)_2 \cdot 2H_2O$. Data are from [7, 19]. (a) Diagram for $\mathbf{H} = H_x$. Solid curves yield the Monte Carlo predictions according to Landau and Binder [13] ($H_x > H_{sf}$) and Equation (5) ($H_x < H_{sf}$). (b) Enlargement of the low-field region. Data are shown for **H** parallel (○) and perpendicular (+) to the easy axis. Solid and dashed curves are the theoretical predictions from Equation (5). (c) Phase diagram for **H** perpendicular to the easy axis. The solid line represents the theory from Landau and Binder [13].

should become XY-like in high fields and therefore the phase transitions should be of the Kosterlitz–Thouless type. A detailed search for Kosterlitz–Thouless behaviour has been performed by Takeda and Koyama [7]. They have measured the magnetic heat capacity in a field of 20 kOe, which is in the region where the transition temperature is almost equal to $T^{KT}$. Their results, for $\mathbf{H} = H_{\perp}$, are reproduced in Figure 7. The parameter $J^{XY}$ equals $2|J|S(S+1)$. For $k_B T/J^{XY} > 1$ there appears to be good agreement between theory and experiment. This provides strong evidence that the underlying symmetry in the quasi 2-d system is of the XY-type. For $k_B T/J^{XY} < 1$, however, important deviations occur, which are most probable due to the onset of the 3-d order induced by the 2-d topological ordering.

In conclusion it appears that the concepts of an effective field-dependent anisotropy is quite useful for understanding the various kind of phase transitions that may occur for the quasi 2-d weakly anisotropic antiferromagnet. Phase transitions of a widely different character can be put in a unified framework by this

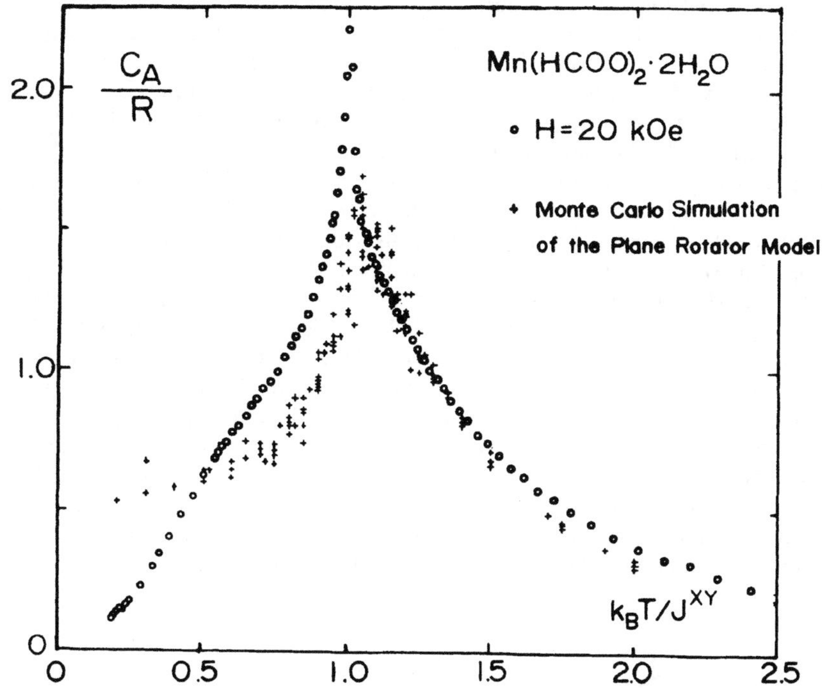


Fig. 7. Magnetic heat capacity versus reduced temperature for $Mn(HCOO)_2 \cdot 2\,H_2O$ in an applied field $\mathbf{H} = H_\perp$, after Koyama *et al.* [7]. The experimental data (○) are compared with theoretical predictions for the planar rotator model (+), as obtained with Monte Carlo techniques [49, 50].

simple concept, as is evidenced by the apparent agreement between theory and experiment.

## 4. Random Fields and Domain Walls (Solitons)

It has been argued recently that 2-d systems are very sensitive to random fields [20]. Even infinitesimal random fields may already drastically change the thermodynamic behaviour. In particular the long-range order that would occur in an undisturbed Ising-type 2-d system below $T_c$ [21] will become unstable with respect to domain formation as soon as a random field is present. This was first recognized by Imry and Ma [20], who proposed a criterion for this instability:

$$E_D = \sigma R^{d-1} - H_r R^{d/2} \tag{11}$$

i.e. the energy of a domain $E_D$ amounts to the difference of the cost in wall energy and the gain in random-field energy. In Equation (11) $\sigma$ represents the surface tension of the wall, $R$ the average radius of the domain, $H_r$ the rms random field and $d$ the dimensionality of the system.

It follows that for $d < d_{lc}$, where $d_{lc}$ is the lower critical dimensionality, $E_D$ will become always negative for large $R$, no matter how weak $H_r$ is. As a consequence, the random-field Ising system is unstable towards domain formation for $d < d_{lc}$. From the Imry—Ma criterion it follows that this instability will only occur for $d <$ 2. However, more recently a rigorous proof has been given by Imbrie [22] that it also happens for the 2-d case and thus $d_{lc} = 2$.

This has important consequences for the behaviour of the quasi 2-d systems. Since infinitesimal random fields, due to lattice defects and the like, will always be present, true long range order in these systems will be virtually impossible.

It is of interest here to consider in some detail the domain wall, since it appears that for weakly anisotropic systems its properties, the surface tension and the wall width, depend on the applied field through the field-induced anisotropy. Consequently the random-field effects will be field-dependent too, and therefore weakly anisotropic systems offer quite promising novel possibilities for the study of random-field behaviour and the associated metastability effects [6, 8, 9].

In Figure 8 a schematic representation of the domain wall pattern in a 2-d system with randomly distributed strong defects is given. Each of these defects favours one or the other domain configurations, which are separated by the domain wall that corresponds to the region where the two antiferromagnetic sublattices are interchanged. The average distance between the walls is denoted by

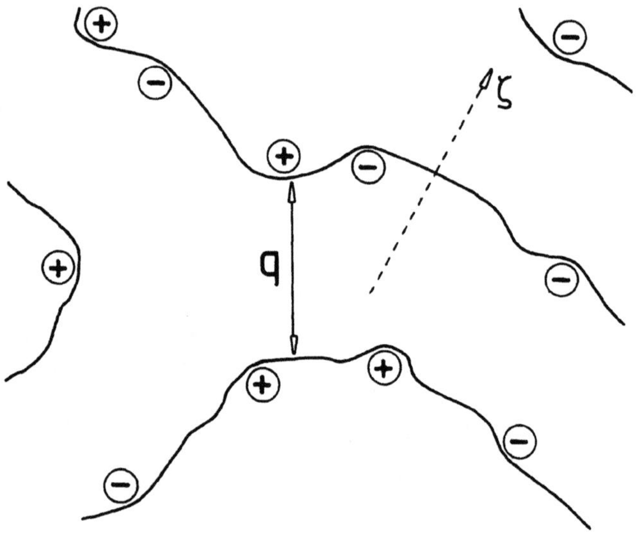


Fig. 8. Schematic representation of the domain wall pattern in a 2-d system with randomly distributed defects that may be considered to act as local sources of strong random fields. $q$ is the mean distance between the walls and $\zeta$ is a direction perpendicular to the wall.

$q$. In Figure 9 is sketched what one would find if one goes along the axis $\zeta$ in Figure 8 through the domain wall. The average polarization of the magnetic moments is rotated through an angle $\pi$. In Figure 9 a distinction has been made between $H_x < H_{sf}$ and $H_x > H_{sf}$, and furthermore the situation for orthorhombic anisotropy is given. In that case the $X$-axis is the easy axis for $H_x < H_{sf}$ whereas the $Y$-axis is the easy axis for $H_x > H_{sf}$ (cf. Figure 2).

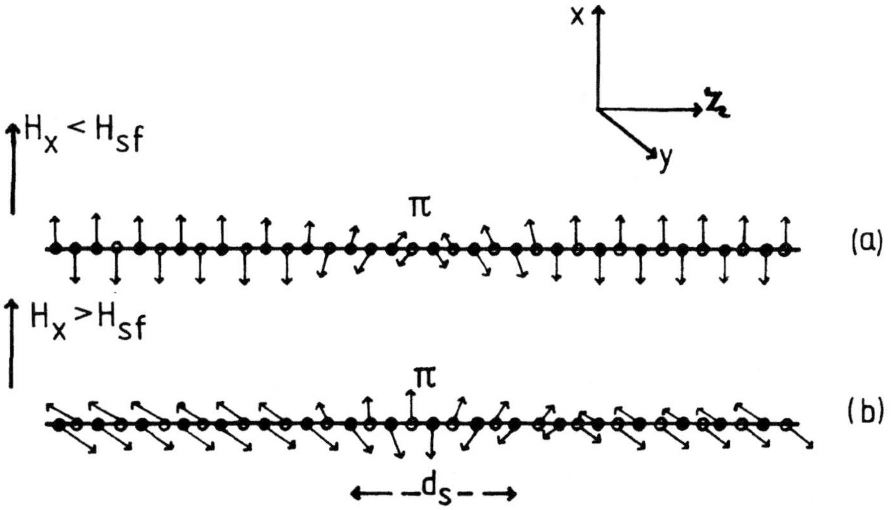


Fig. 9. Schematic representation of the domain wall along the axis $\zeta$ in Figure 8, for $H_x < H_{sf}$ (a) and for $H_x > H_{sf}$ (b). In the wall the average polarization of the magnetic moments rotates through an angle $\pi$. The wall-width is denoted by $d_s$.

Apart from the surface tension $\sigma$ the magnetic domain walls are also characterized by their width $d_s$. The variation of $\sigma$ and $d_s$ with anisotropy strength may be evaluated quite easily using arguments that were first applied to the 3-d ferromagnet and can be found in elementary textbooks on solid state physics [23]. According to such arguments the wall properties may be estimated from the parameters $J$ and $D^x_{\text{eff}}$ $(H)$ in Hamiltonian (4). The surface tension is simply the sum of the contributions due the exchange and anisotropy energy, yielding $\sigma = \pi^2|JS^2|/d_s + d_s D^x_{\text{eff}}(\mathbf{H})S^2$. Stability requires $\sigma$ to be a minimum and after minimization with respect to $d_s$ one finds $\sigma = 2\pi S^2|D^x_{\text{eff}}(\mathbf{H})J|^{1/2}$ and $d_s = \pi|J/D^x_{\text{eff}}(\mathbf{H})|^{1/2}$. Although these estimates are approximate they give the right order of magnitude and the correct functional dependence of $\sigma$ and $d_s$ on the exchange and anisotropy.

For weakly anisotropic systems a better description of the domain wall can be obtained from 2-d sine—Gordon theory. Hamiltonian (4) may be approximated [6] by the continuum description:

$$\mathscr{H} = 2|JS(S+1)| \int_{-\infty}^{\infty} \mathrm{d}x\mathrm{d}y \left\{ 1/2 \left( \frac{\partial \phi}{\partial x} \right)^2 + 1/2 \left( \frac{\partial \phi}{\partial y} \right)^2 - \right.$$

$$\left. - (D_{\mathrm{eff}}^{x}(\mathbf{H})/4J) \cos(2\phi) \right\} \tag{12}$$

which is the classical 2-d SG-Hamiltonian. In principle, Hamiltonian (12) gives the description for a 2-d *ferro*magnet. A formal transformation from an antiferromagnet into a ferromagnet is hidden in the field $\phi$ [6].

The domain walls correspond to the soliton solutions [24] of Hamiltonian (12). The wall profile of the SG-wall (along the axis $\zeta$ in Figures 8, 9) is expressed as:

$$\phi(\zeta) = 2 \tan^{-1} \exp(\zeta/d_s). \tag{13}$$

It may be seen that for each individual sublattice the field $\phi$ corresponds to the variation of the angle between the easy axis and the average polarization of the spins (cf. Figure 9). Here $d_s$ is the soliton width which corresponds to the width of the domain wall for the magnetic system. It is expressed as

$$d_s = |J/D_{\mathrm{eff}}^{x}(\mathbf{H})|^{1/2}. \tag{14}$$

Furthermore, the surface tension $\sigma$ is given by

$$\sigma = 4(S(S+1)\,|D_{\mathrm{eff}}^{x}(\mathbf{H})J|^{1/2}. \tag{15}$$

With the expressions (14) and (15) the concept of effective, field-dependent anisotropy may be used to quantitatively describe the field-induced variation of the domain wall properties.

It follows that application of a magnetic field $\mathbf{H} = H_x$ influences $\sigma$ and thus changes the balance between the random field energy and the wall energy cf. Equation (11) through the *first* term. The field-induced effects are most strongly present at the spinflop transition, when $H_x \to H_{\mathrm{sf}}$. In that case $\sigma \to 0$ and $d_s \to \infty$, since $D_{\mathrm{eff}}^{x}(H_x) \to 0$ (cf. Equation (2)). Therefore the wall-density $q^{-1}$ should increase for $H_x \to H_{\mathrm{sf}}$, provided that the system stays at equilibrium.

An interesting consequence of this argument is that, in order to study random field effects in weakly anisotropic antiferromagnets, there is no strict need to make use of the Fishman and Aharony device [25]. They were the first to remark that an external field $H_x$ that is applied to an antiferromaget diluted with nonmagnetic impurities will give rise to a random staggered field $H_r^x \propto H_x$. This allows the variation of the *second* term in Equation (11). In case of strong anisotropy or low fields and a fairly large amount of nonmagnetic impurities, the variation with $H_x$ of the second term should predominate over the first term. However, for weak anisotropy and $H_x \approx H_{\mathrm{sf}}$, the first term should become predominant. Furthermore, for nominally pure systems it is expected that the random field hardly depends on the applied field since the very small $H_r$ that is present should be mainly due to lattice defects and the like.

It should be emphasized that the random field seems to be a necessary prerequisite for the appearance of solitons or domain walls for the 2-d case. This is in strong contrast with the quasi 1-d systems, where solitons are present because of entropy considerations [26]. In the pure SG-system there are two contributions to the free energy associated with a soliton. Firstly the wall energy, which is simply the product of $\sigma$ and the length of the wall, and secondly the entropy due to the meandering of the walls [24, 27]. For the pure 2-d case the cost in wall energy for a domain wall extending throughout the whole system is always more than the entropy gain, and therefore the ordered state is stable.

Furthermore, there could be another way of getting domain walls in the system, namely if they form small loops or droplets. One may assign a topological charge to such excitations, corresponding to the degree of mapping of the field $\phi$ onto the unit circle [28]. These topological excitations [29] or soliton strings (cf. Figure 10a, b) are comparable to the vortex—antivortex pairs proposed by Kosterlitz and Thouless [14] for the XY system (cf. Figure 10c).

Since the soliton strings contain a finite energy they could be thermally excited. Similar excitations have been investigated numerically by Kosevich *et al.* [28]. They have found that the minimum energy needed for the excitation of any stable droplet, with or without topological charge, amounts to $4\pi|JS^2|$. Therefore the droplet density may be estimated as $n_d \approx \exp(4\pi|JS^2|/k_B T)$, which is entirely negligible at low temperatures, considering that $|J/k_B|$ is of the order of 1—10 K.

On the other hand, at higher temperatures, in particular for $T \approx T_c$, droplets may be easily excited and play an important role in the thermodynamics of the 2-d system, since $\sigma$ effectively vanishes at $T_c$ [30, 31]. This is in agreement with current

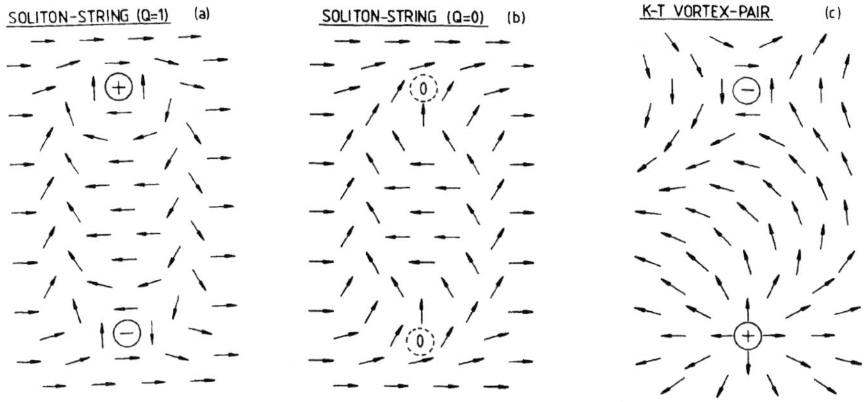


Fig. 10. Small excitations (droplets, soliton strings) for weakly anisotropic systems. Strings with (a) topological charge 0, (b) topological charge 0. (c) Vortex—antivortex pair from the Kosterlitz—Thouless theory. For reasons of simplicity the ferromagnetic case, corresponding to only one of the two sublattices in the antiferromagnet, is shown.

ideas about the mechanism of the second-order phase transition [32]. Experimental observation of droplets at $T \approx T_c$ has been claimed by Waldner *et al.*, from ESR measurements on several quasi 2-d systems [33].

## 5. The Spin Flop Transition

The first observation of a spin flop transition in a quasi 2-d system is probably by Breed, who studied the compounds $K_2MnF_4$ and $Rb_2MnF_4$ [34]. His data for $K_2MnF_4$, obtained at a temperature of 4.2 K and in pulsed magnetic fields, are shown in Figure 11, for the pure compound and for systems with part of the manganese ions replaced by nonmagnetic magnesium. The spin flop manifests itself as a jump in the magnetization $M$ at $H_{sf}$ = 55 kOe. In Figure 12 the differential susceptibility $\chi = (\delta M/\delta H)_T$ is shown, after Mulder *et al.* [18]. Here the peaks correspond to the spin flop transitions.

It is interesting to compare the spin flop in $K_2MnF_4$ with the same phenomenon in a 3-d antiferromagnet. An example is given in Figure 13, for $MnCl_2 \cdot 4\,H_2O$ [35]. The difference with the 2-d case is quite striking. For the 3-d compound the spin flop is narrow, ≈ 300 Oe, and the width may be fully explained taking into account demagnetizing effects. The magnetization for $H < H_{sf}$, which is given by $M = \chi_{\parallel} H$, is negligible compared to the magnetization for $H > H_{sf}$, given by $M = \chi_{\perp} H$. This is due to the fact that for the antiferromagnet $\chi_{\parallel} < \chi_{\perp}$. The spin flop in $K_2MnF_4$, on the other hand, extends over a region of approximately 10 kOe, i.e.

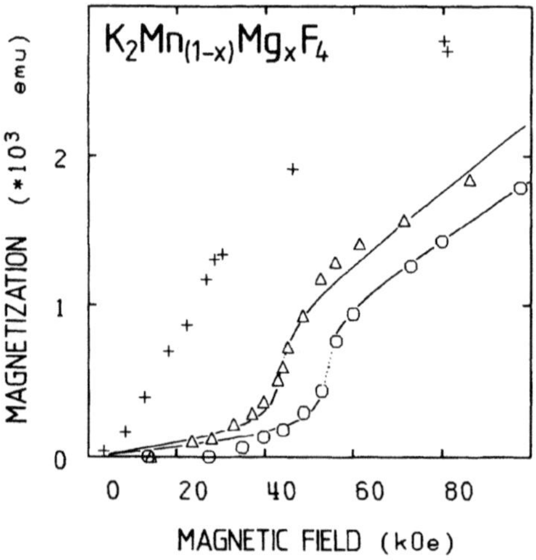


Fig. 11. Spin flop transitions for the weakly anisotropic antiferromagnetic system $K_2Mn_{(1-x)}Mg_xF_4$. The magnetization data are for the pure compound (○) and two diluted compounds with $x$ = 0.07 (Δ and $x$ = 0.24 (+). The experimental data have been obtained in pulsed fields by Breed *et al.* [34]. The solid curves are fits to Equation (17, 18) with $q^{-1} = 10^{-3}$ and with $\chi_{\perp}$ = 0.022 emu/mole and $\chi_{\perp}$ = 0.026 emu/mole for $x$ = 0 and $x$ 0.07 respectively.

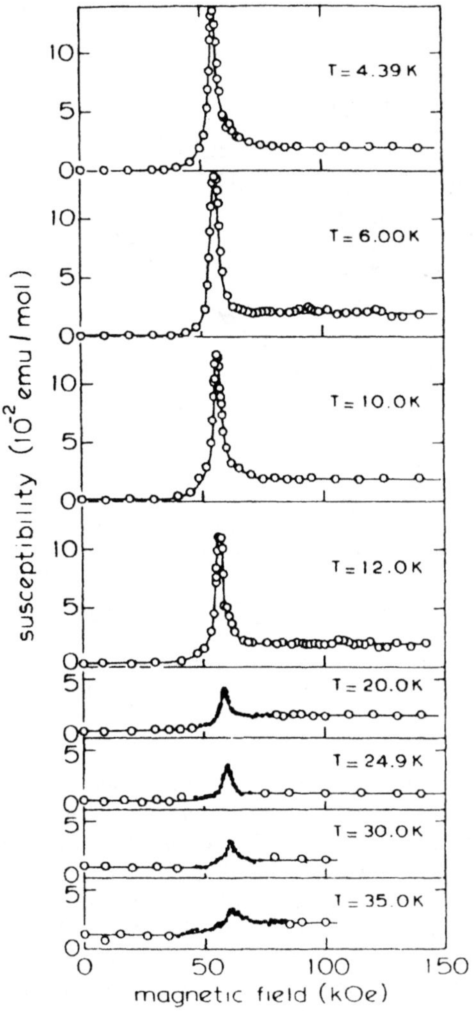


Fig. 12. The differential susceptibility of $K_2MnF_4$ as a function of the external magnetic field $\mathbf{H} = H_x$ for various temperatures, after Mulder *et al.* [18]. The drawn lines are guides to the eye.

two orders of magnitude larger. Moreover, there is substantial magnetization in low field, much more than would correspond with $M = \chi_{\parallel} H$, since also for $K_2MnF_4$ it has been determined experimentally [18] that at $T = 4.2$ K, $\chi_{\parallel}$ is two orders of magnitude smaller than $\chi_{\perp}$ (see the chapter by Arts and de Wijn).

In Figure 14 and 15 three other experimental examples of spin flop transitions in quasi 2-d systems are given, namely for $BaMnF_4$ [36], $K_2NiF_4$ [37], and $Rb_2NiF_4$ [37]. In all cases the transitions are broad. The data on $BaMnF_4$ are from

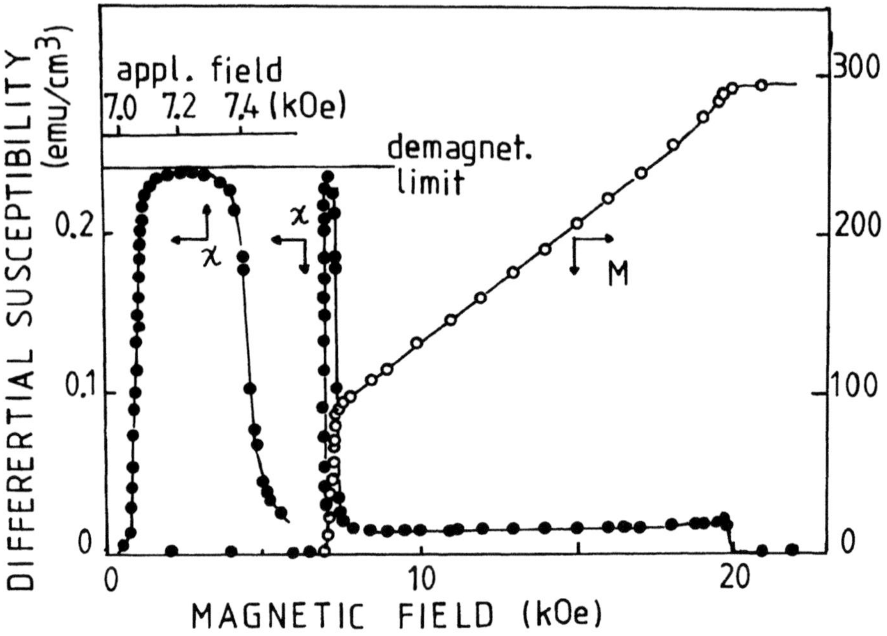


Fig. 13. Differential magnetic susceptibility (●) and magnetization (○) of the 3-d compound $MnCl_2 \cdot 4H_2O$ at a temperature $T = 0.297$ K and in magnetic fields $\mathbf{H} = H_x$, (after Rives and Benedict [35]).

Holmes *et al.* [36]. They report that in order to explain the width of the transition they had to assume a misalignment of the crystal of $\approx 3°$, which is quite large. Several other examples of broad spin flop transitions have been described in the literature, notably for $(CH_3NH_3)_2MnCl_4$ [38] and $Rb_2MnCl_4$ [17].

In our opinion the anomalous width of the spin flop transition for quasi 2-d systems should be atrributed to domain walls. As is clear from Figure 9, the domain wall for $H_x < H_{sf}$ may be considered as a small fragment of the flopped phase in the low-field phase, while for $H_x > H_{sf}$ it is a fragment of the low-field phase in the flopped phase. For the weakly anisotropic quasi 2-d systems one has $d_s \approx 5$—10 lattice units in zero field. Since, in principle, $d_s$ should diverge for $H_x \to H_{sf}$, the spin flop transition should be a gradual and continuous transition for any system with domain walls, and in particular for the experimental examples of quasi 2-d systems. Therefore the present model agrees well with the experimental observations.

Recently, Lemmens *et al.* [39] have calculated the contribution of the spins in a static domain wall to the magnetization as

$$M_s = 8S^2\chi_\perp H_x \left[ \frac{2Jd_s}{\pi k_B T} \right]^{1/2} q^{-1} \tag{16}$$

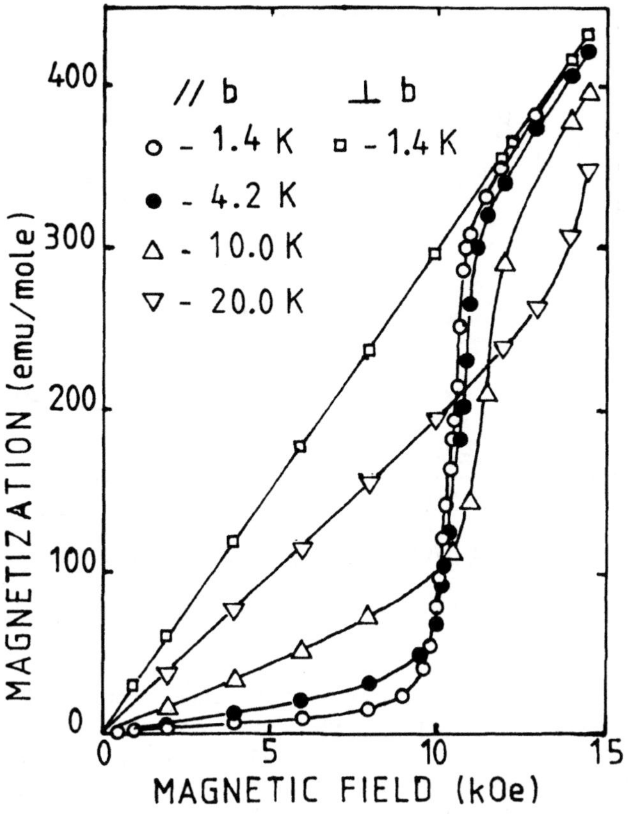


Fig. 14. Magnetization versus magnetic field applied parallel ($\mathbf{H} = H_x$), and perpendicular ($\mathbf{H} = H_\perp$) to the easy $b$-axis for $BaMnF_4$ at various temperatures. (After Holmes *et al.* [36].)

Their expression may be used [6, 9] to calculate theoretical predictions for the magnetization curves of the quasi 2-d systems according to

$$M = M_s \qquad \text{for} \quad B_x < B_{sf} \quad \text{and} \tag{17}$$

$$M = \chi_\perp H_x - M_s \qquad \text{for} \quad B_x > B_{sf}. \tag{18}$$

For $K_2MnF_4$ these predictions are given by the solid lines in Figure 11. The field-dependent expression for $d_s$ cf. Equation (14) has been used and the wall density was taken as $q^{-1} = 10^{-3}$, both for the pure and the diluted compound, and was kept constant. The initial (i.e. in zero field) wall density should be roughly the same for the pure and for the diluted compound, since nonmagnetic impurities do not lead to domain formation in zero field [25] and therefore the domain structure should be due to defects only [6, 8, 9]. Furthermore, the wall density is expected to remain constant during the magnetization experiment. Although the energy balance, Equation (11), will change upon application of a field, for the pure

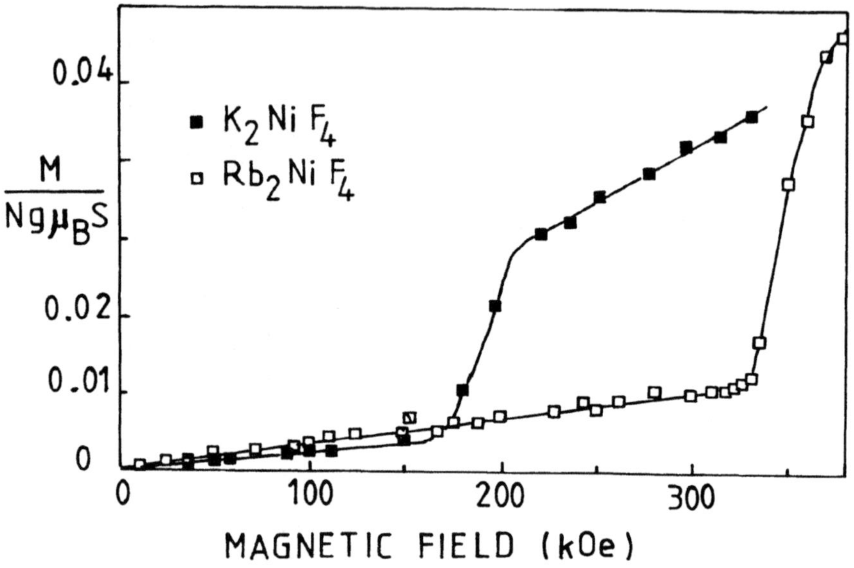


Fig. 15. Magnetization curves for $K_2NiF_4$ (■) and $Rb_2NiF_4$ (□) for $\mathbf{H} = H_x$ at a temperature $T$ = 4.2 K. (After Matsuura *et al.* [37].)

compound through the first term and for the diluted compound through both terms, this will not result in a change of $q^{-1}$, at least, not on the time scale of the pulsed-field experiment ($\approx 0.1$ s.). This is because domain patterns in random field Ising systems are metastable [40], since the walls should be strongly pinned by defects. Therefore relaxation towards thermal equilibrium may take very long times [41]. Nevertheless, the effect of the increasing $d_s$ alone is already enough to cause substantial broadening of the spin flop transition.

However, for weakly anisotropic systems the walls may become very broad ($\approx 100$ lattice units), since $d_s$ diverges for $H_x \to H_{sf}$ (cf. Equation (14)). In that case one would expect relatively short relaxation times, of the order of minutes (although still much longer than the characteristic time of the pulse experiments discussed above). An experiment that has been performed on such a time scale is shown in Figure 16. There the rotation of the average staggered magnetization **m** has been detected by following the variation of the intensity of the (3/2, 1/2, 0) Bragg reflection in a neutron diffraction experiment [8, 16]. The same experiment has been performed recently for another quasi 2-d system, $Rb_2MnCl_4$, yielding comparable results [17]. The temperature was 4.2 K, which is very low compared to $T_c = 42.5$ K and since $\chi_\| \ll \chi_\perp$ at this temperature the effective anisotropy model should be fully applicable. In Figure 16 are shown $\langle m_\|^2 \rangle$ and $\langle m_\perp^2 \rangle$, the components of **m** parallel and perpendicular to the easy axis. Again the transition is very broad and the value $H_{sf} = 54.8$ kOe agrees well with the result of the magnetization experiments.

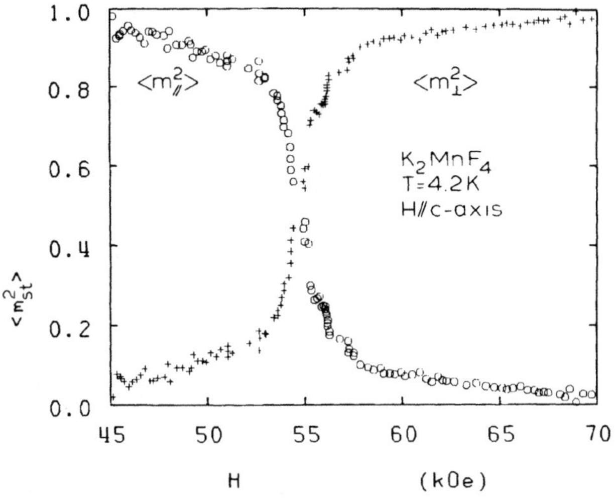


Fig. 16. Rotation of the staggered magnetization in applied fields $H_x$ for chemically pure $K_2MnF_4$ at $T = 4.2$ K. Circles: $\langle m_{\parallel}^2 \rangle$, crosses: $\langle m_{\perp}^2 \rangle$.

From a theoretical point of view there is a close correspondence between the spin flop transition for the weakly anisotropic antiferromagnet and the commensurate—incommensurate (CI-)transition that occurs in a noble gas monolayer on a solid surface when the gas pressure is varied. Both systems may be modeled by the classical 2-d SG-Hamiltonian. The CI-problem has attracted much attention lately and has been extensively studied [42]. It has been argued that the results from these studies may be applied to explain the broadening of the spin flop in the quasi 2-d system [6, 8].

The random field CI-problem has been investigated *inter alia* by Villain *et al.* [43]. They considered a modified 2-d Ising-type system with a preferential direction for walls. In such a system the walls will be long and almost straight (cf. Figure 17). In case of a weakly anisotropic antiferromagnet there is no such preferential direction but, to our knowledge, the model of Villain *et al.* is still the best approximation available. The model predicts a domain structure at low temperatures, in agreement with the proof of Imbrie [22] that $d_{lc} = 2$. The wall density is expressed as

$$q^{-1} \propto \exp\{-\gamma(\sigma/\sigma_0)^{4/3}\} \tag{19}$$

and is field-dependent through $\sigma$ (cf. Equation 15). Here

$$\gamma = 2(\sigma_0/h_r)^{4/3} \quad \text{with} \tag{20}$$

$$\sigma_0 = 4S^2|D_x J|^{1/2} \quad \text{and} \tag{21}$$

$$h_r = g\mu_B H_r/k_B. \tag{22}$$

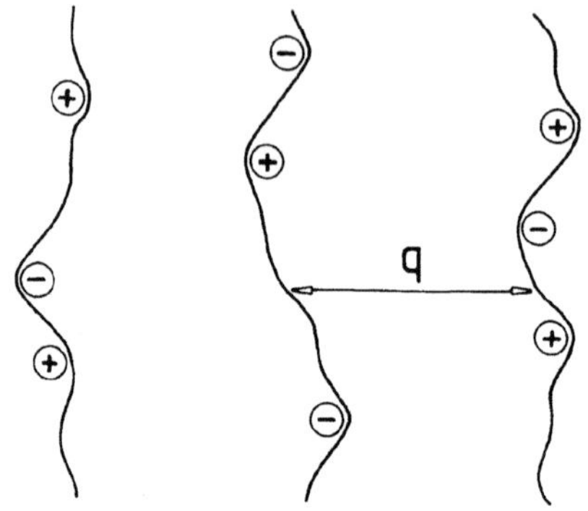


Fig. 17. Schematic wall pattern in a 2-d system with randomly distributed defects, and a strong preferential direction for walls.

This may be applied in an attempt to analyze the neutron experiment of Figure 16. From symmetry considerations it follows that $\langle m_{\perp}^2\rangle$ for $H_x < H_{sf}$ is equal to $\langle m_{\parallel}^2\rangle$ for $H_x > H_{sf}$ and we put

$$\langle m_{pl}^2\rangle \equiv \langle m_{\perp}^2(H_x < H_{sf})\rangle = \langle m_{\parallel}^2(H_x > H_{sf})\rangle. \tag{23}$$

It can be argued [6] that $\langle m_{pl}^2\rangle \propto q^{-1}$, yielding an exponential law for $\langle m_{pl}^2\rangle$. In Figure 18 the comparison with the neutron data is shown. From the slope of the curve we obtain $h_r = 8$ K. Assuming an acceptable concentration of defects for a pure system, $c \approx 10^{-3}$–$10^{-4}$ per lattice unit, one calculates a random field strength $h_0 = h_r/\sqrt{c}$ of about 500 K. This seems a reasonable value, since it is of the same order as the droplet creation energy [6, 28] $4\pi|JS^2| \approx 460$ K, as calculated from the 2-d SG theory.

## 6. The Bicritical Point

In Section 3 we have concentrated on the phase diagrams of quasi 2-d systems. An important aspect of these diagrams remains unexplained and is in contradiction with the predictions from the effective anisotropy model, namely the behaviour around the bifurcation point (cf. the inserts in Figure 4 and 5.) One may argue that the anomalous bifurcation is also a random field effect. An indication for this may be that the same type of bifurcation also occurs in the random field 3-d weakly Ising-type system $Mn_{(1-x)}Zn_xF_2$ [44], but *not* in the pure 3-d system $MnF_2$ [45].

A possible explanation can be extracted from the theoretical work on the CI-problem. For the pure 2-d SG-system (cf. Hamiltonian (12)) there are no solitons at low temperatures unless a chemical potential $\mu$ for walls is introduced [42]. Although such a chemical potential is purely hypothetical in the case of a weakly anisotropic antiferromagnet, since it would correspond to a 'twist field' that may never be applied in any experiment, it may be useful to consider for a moment its consequences. The phase diagram ($\mu - T$), as calculated by Haldane *et*

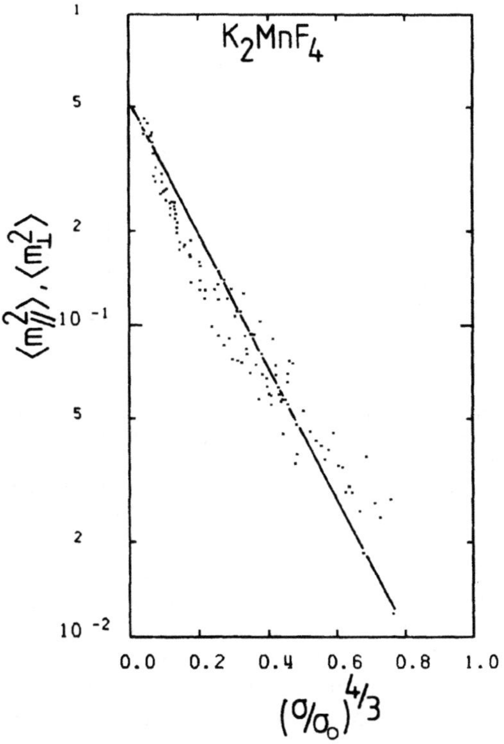


Fig. 18. Semi-log plot of $\langle m_p^2 \rangle$ versus $(\sigma/\sigma_0)^{4/3}$. The solid line is the fit to the exponential law (see text).

*al.* [46], is sketched in Figure 19a. There are three phases, the commensurate or ordered (C-)phase, the disordered (D-)phase and finally the incommensurate (IC-)phase, where the system contains long, meandering, and mutually repelling solitons. At $T = 0$ the IC and the C-phase touch each other for $\mu = \mu_c$, the critical potential, which is equal to the surface tension $\sigma$ of the solitons. For $\mu \downarrow \mu_c$ the wall density vanishes. Furthermore, at $T > 0$ C and IC-phase are separated by the disordered phase.

The corresponding phase diagram for the random field counterpart has not yet been calculated, but from the paper of Villain *et al.* [43], and other theoretical works [22, 47] for systems at $T = 0$ K, we learn that there should be walls in the low-$\mu$, low-$T$ phase, which is therefore not commensurate but rather pseudo-commensurate. In Figure 19b we give a schematic sketch of what the phase diagram might look like if a random field becomes involved [6, 9]. In that case the wall density should decrease but not vanish for $\mu \downarrow \mu_c$ and hence the IC—D phase line should join the PC—D line at some point $(T_2, \mu_2)$.

This may be applied to the spin flop transition for the weakly anisotropic

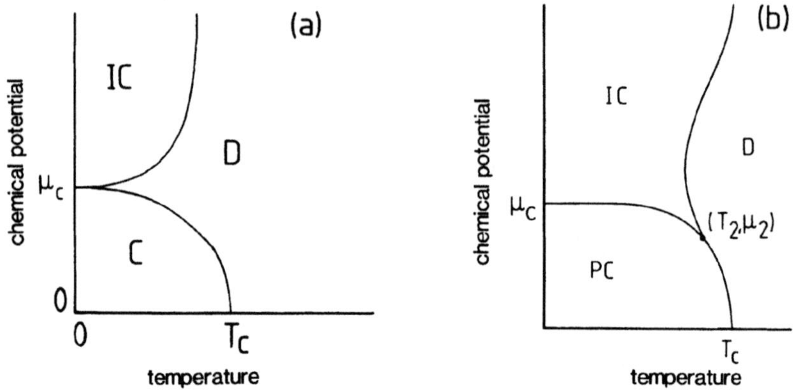


Fig. 19. (a) Phase diagram ($\mu - T$) for the 2-d SG-system in a 'twist field' $\mu$ (schematic). The various phases are: disordered (D), ordered commensurate (C) and floating incommensurate (IC). Only in the IC-phase there are domain walls in the system. (b) Proposed phase diagram 2-d SG-system in an effective 'twist field' $\mu_r$ due to a random field $H_r$. In the 'pseudocommensurate' (PC) phase the system contains (nearly) frozen walls.

antiferromagnet. For $H_x \rightarrow H_{sf}$ one has $\mu = 0$ and $\mu_c \downarrow 0$. The small random field in a chemically pure system that is almost field-independent will give rise to domain wall formation an may therefore be considered as a constant effective chemical potential for walls $\mu_r$. If $H_x \rightarrow H_{sf}$ one has $\mu_c \downarrow 0$ *and* $\mu_2 \downarrow 0$. Hence for $H_x \approx H_{sf}$ there should be a phase with thermally meandering walls between the ordered phase and the paramagnetic phase, and the bifurcation point may be a Lifshitz-type point.

One of the problems with the above explanation is that it relies heavily on the 2-d character of the system, whereas the same kind of bifurcation phenomenon has been observed in the diluted 3-d systems $Mn_{(1-x)}Zn_xF_2$. Moreover the character of the spin flop phase remains unclear and it is not explained why there is no Ising-type phase boundary for $H_x > H_{sf}$. Another problem is that in general the spin flop transition and the bicritical point are quite sensitive to hidden variables, so that multicritical behaviour may occur. Furthermore it has been argued [48] from both molecular field theory and spin wave theory that for the spin flop transition supercooling and superheating may occur. In that case the value of the spin flop field would be substantially different for increasing and decreasing field. Although these first-order effects have not been observed in pure 2-d antiferromagnets, they might be triggered by the slightest amount of impurity and give rise to domains of flopped and unflopped phases. Finally, Landau-type free energy expansions show that the occurrence of the bicritical point should be very sensitive to the presence of higher-order anisotropies, next-nearest-neighbor interactions, and other weak thermodynamic variables besides the field and the temperature. Taking all this into account, one should admit that it is impossible at present to give an unambiguous explanation of the bifurcation behaviour of the quasi 2-d system, and more study is called for.

## 7. Concluding Remarks

In the present paper we have reviewed the various field-induced phenomena that may occur for the weakly anisotropic quasi 2-d antiferromagnet. Many of these can be well understood taking into account the anisotropy crossovers that are caused by the magnetic fields applied in various directions. This holds in particular for the phase transitions, as there are the order—disorder transition and the spin flop transition. Furthermore, interesting connections exist with the random field Ising model and other problems in physics, such as the commensurate—incommensurate transition. We hope that this work will stimulate more effort along these lines and will contribute to a better understanding of the many problems that are still open in 2-d physics.

## Acknowledgements

This work is part of the research program of the 'Stichting voor Fundamenteel Onderzoek der Materie' (Foundation for Fundamental Research of Matter) and was made possible by financial support from the 'Nederlandse Organisatie voor Zuiverwetenschappelijk Onderzoek' (Netherlands Organization for the Advancement of Pure Research).

## References

1. J. P. Boucher, *Solid State Comm.* **33** (1980) 1025.
2. I. Harada, K. Sasaki and H. Shiba, *Solid State Comm.* **40** (1981) 29.
3. D. Bloch, J. Voiron and L. J. de Jongh, in: *High Field Magnetism*, M. Date, Ed. North-Holland, Amsterdam, 1983, p. 19.
4. L. J. de Jongh and H. J. M. de Groot, *Solid State Comm.* **53** (1985) 731.
5. L. J. de Jongh and H. J. M. de Groot, *Solid State Comm.* **53** (1985) 737.
6. H. J. M. de Groot and L. J. de Jongh, *Physica* **141B** (1986) 1.
7. K. Takeda and K. Koyama, *J. Phys. Soc. Jap.* **52** (1983) 656; K. Koyama, K. Amaya and K. Takeda, *J. Magn. Magn. Mat.* **31–34** (1983) 1196.
8 H. J. M. de Groot and L. J. de Jongh, *Solid State Comm.* **56** (1985) 7.
9. H. J. M. de Groot and L. J. de Jongh, *Physica Scripta*, **T13** (1986) 219.
10. K. Binder and D. P. Landau, *Phys. Rev.* **B13** (1976) 1140.
11. D. R. Nelson and R. A. Pelcovits, *Phys. Rev.* **B16** (1977) 2191.
12. See e.g. J. M. Kosterlitz and D. J. Thouless, in: *Progress in Low-Temperature Physics*, Vol. VIIB, D. F. Brewer, Ed. North-Holland, Amsterdam, 1978, Ch. 5; V. L. Pokrovski, *Adv. Phys.* **28** (1979) 595.
13. D. P. Landau and K. Binder, *Phys. Rev.* **B24** (1981) 1391.
14. J. M. Kosterlitz and D. J. Thouless, *J. Phys.* **C6** (1973) 1181: J. M. Kosterlitz, *J. Phys.* **C7** (1974) 1046.
15. J. Villain, *J. de Physique* **36** (1975) 581.
16. L. J. de Jongh, L. P. Regnault, J. Rossat-Mignod and J. Y. Henry, *J. Appl. Phys.* **53** (1982) 7963.
17. H. Rauh, W. A. C. Erkelens, L. P. Regnault, J. Rossat-Mignod, W. Kullmann and R. Geick, *J. Phys.* **C19** (1986) 4503.
18. C. A. M. Mulder, H. L. Stipdonk, P. H. Kes, A. J. van Duynevelt and L. J. de Jongh, *Physica* **113B** (1982) 380.
19. J. W. Schutter, J. W. Metselaar and D. de Klerk, *Physica* **61** (1972) 250.

20. Y. Imry and S. K. Ma, *Phys. rev. Lett.* **35** (1975) 1399.
21. L. Onsager, *Phys. Rev.* **65** (1944) 117.
22. J. Z. Imbrie, *Phys. Rev. Lett.* **53** (1984) 1747.
23. See e.g. C. Kittel, *Introduction to Solid State Physics*, Wiley, New York, 1953, p. 488; see also [3].
24. See e.g. F. D. M. Haldane and J. Villain, *J. de Physique* **42** (1981) 1673.
25. S. Fishman and A. Aharony, *J. Phys.* **C12** (1979) L729.
26. For recent reviews see e.g. *Magnetic Excitations and Fluctuations*, S. W. Lovesey, U. Balucani, F. Borsa and V. Tognetti, Eds. Springer, Berlin, 1984.
27. V. L. Pokrovski and A. L. Talapov, *Zh. Eksp. Teor. Fiz.* **78** (1980) 269 (*JETP* **51** (1980) 1).
28. A. M. Kosevich, B. A. Ivanov and A. S. Kovalev, *Sov. Sci. Rev. A Phys.* **6** (1985) 161. See also A. S. Kovalev, A. M. Kosevich and K. V. Maslov, *Pis'ma Zh. Eksp. Teor. Fiz.* **30** (1979) 321 (*JETP Lett.* **30** (1979) 296).
29. T. Nattermann, *J. de Physique* **41** (1980) 1251.
30. C. N. Yang, *Phys. Rev.* **85** (1952) 808.
31. J. V. José, L. P. Kadanoff, S. Kirkpatrick and D. R. Nelson, *Phys. Rev.* **B16** (1977) 1217.
32. L. P. Kadanoff, in: *Phase Transitions and Critical Phenomena*, C. Domb and M. S. Green, Eds. Academic Press, New York, 1976 vol. 5a, p. 1.
33. F. Waldner, *J. Magn. Magn. Mat.* **31–34** (1983) 1203; *ibid.* **54–57** (1986) 873.
34. D. J. Breed, *Physica* **37** (1967) 35; D. J. Breed, K. Gilijamse, J. W. E. Sterkenburg and A. R. Miedema, *J. Appl. Phys.* **41** (1970) 1267; D. J. Breed, Ph.D. Thesis, Amsterdam, 1969.
35. J. E. Rives and V. Benedict, *Phys. Rev.* **B12** (1975) 1908.
36. L. Holmes. M. Eibschutz and H. J. Guggenheim, *Solid State Comm.* **7** (1969) 973.
37. M. Matsuura, K. Gilijamse, J. W. E. Sterkenburg and D. J. Breed, *Phys. Lett.* **33a** (1970) 363.
38. W. D. van Amstel and L. J. de Jongh, *Solid State Comm.* **11** (1972) 1423.
39. L. F. Lemmens, I. Kimura and W. J. M. de Jonge, *J. Phys.* **C19** (1986) 139.
40. J. Villain, *Phys. Rev. Lett.* **52** (1984): R. Bruinsma and G. Aeppli, *Phys. Rev. Lett.* **52** (1984) 1547.
41. For recent reviews on the random field Ising model see e.g. J. Villain, in *Scaling Phenomena in Disordered Systems*, R. Pynn and A. Skjeltorp Eds. Plenum, 1985, p. 423; A Aharony. *J. Magn. Magn. Mat.* **54–57** (1986) 27; A. R. King and D. P. Belanger, *J. Magn. Magn. Mat.* **54–57** (1986) 19; R. J. Birgeneau, Y. Shapira, G. Shirane, R. A. Cowley and H. Yoshizawa, *Physica* **137B** (1986) 83.
42. See e.g. J. Villain, in: *Ordering in Strongly Fluctuating Condensed Matter Systems*, T. Riste, Ed. Plenum, 1980; P. Bak, *Rep. Prog. Phys.* **45** (1982) 587.
43. J. Villain, B. Semeria, F. Lancon and L. Billard, *J. Phys.* **C16** (1983) 6153.
44. Y. Shapira, N. F. Oliveira and S. Foner, *Phys. Rev.* **B30** (1984) 6639.
45. Y. Shapira and C. C. Becerra, *Phys. Lett.* **57A** (1976) 483: A. R. King and H. Rohrer, *AIP Conf. Proc.* 1976, 420.
46. F. D. M. Haldane, P. Bak and T. Bohr, *Phys. Rev.* **B28** (1983) 2743.
47. T. Nattermann, *J. Phys.* **C16** (1983) 4113.
48. See e.g. F. Keffer, *Encyclopedia of Physics*, vol. XVIII, H. P. J. Wijn, Ed. Springer, Berlin, 1966, **2**, p. 109; F. B. Anderson and H. B. Callen, *Phys. Rev.* **136** (1964) 1068: J. Feder and E. Pytte, *Phys. Rev.* **168** (1968) 640.
49. S. Myashita, H. Nishimori, A. Kuroda and M. Suzuki, *Prog. in Theor. Phys.* **60** (1978) 1669.
50. J. Tobochnik and G. V. Chester, *Phys. Rev.* **B20** (1979) 3761.

# INDEX OF NAMES

# INDEX OF CHEMICAL COMPOUNDS

# INDEX OF SUBJECTS

# Physics and Chemistry of Materials with Low-Dimensional Structures

*Previously published under the Series Title:*

PHYSICS AND CHEMISTRY OF MATERIALS WITH LAYERED STRUCTURES

1. R.M.A. Lieth (ed.): *Preparation and Crystal Growth of Materials with Layered Structures.* 1977 ISBN 90-277-0638-7
2. F. Lévy (ed.): *Crystallography and Crystal Chemistry of Materials with Layered Structures.* 1976 ISBN 90-277-0586-0
3. T. J. Wieting and M. Schlüter (eds.): *Electrons and Phonons in Layered Crystal Structures.* 1979 ISBN 90-277-0897-5
4. P.A. Lee (ed.): *Optical and Electrical Properties.* 1976 ISBN 90-277-0676-X
5. F. Hulliger: *Structural Chemistry of Layer-Type Phases.* Ed. by F. Lévy. 1976 ISBN 90-277-0714-6
6. F. Lévy (ed.): *Intercalated Layered Materials.* 1979 ISBN 90-277-0967-X

*Published under:*

PHYSICS AND CHEMISTRY OF MATERIALS WITH LOW-DIMENSIONAL STRUCTURES
SERIES A: LAYERED STRUCTURES

7. V. Grasso (ed.): *Electronic Structure and Electronic Transitions in Layered Materials.* 1986 ISBN 90-277-2102-5
8. K. Motizuki (ed.): *Structural Phase Transitions in Layered Transition Metal Compounds.* 1986 ISBN 90-277-2171-8

PHYSICS AND CHEMISTRY OF MATERIALS WITH LOW-DIMENSIONAL STRUCTURES
SERIES B: QUASI-ONE-DIMENSIONAL STRUCTURES

B1. P. Monceau (ed.): *Electronic Properties of Inorganic Quasi-One-Dimensional Compounds.* Part I: Theoretical. 1985 ISBN 90-277-1789-3
B2. P. Monceau (ed.): *Electronic Properties of Inorganic Quasi-One-Dimensional Compounds.* Part II: Experimental. 1985 ISBN 90-277-1800-8
B3. H. Kamimura (ed.): *Theoretical Aspects of Band Structures and Electronic Properties of Pseudo-One-Dimensional Solids.* 1985 ISBN 90-277-1927-6
B4. J. Rouxel (ed.): *Crystal Chemistry and Properties of Materials with Quasi-One-Dimensional Structures.* A Chemical and Physical Synthetic Approach. 1986 ISBN 90-277-2057-6

*Discontinued.*

PHYSICS AND CHEMISTRY OF MATERIALS WITH LOW-DIMENSIONAL STRUCTURES
SERIES C: MOLECULAR STRUCTURES

C1. I. Zschokke (ed.): *Optical Spectroscopy of Glasses.* 1986 ISBN 90-277-2231-5
C2. J. Fünfschilling (ed.): *Relaxation Processes in Molecular Excited States.* 1989 ISBN 07923-0001-7

*Discontinued.*

# Physics and Chemistry of Materials with Low-Dimensional Structures

9. L.J. de Jongh (ed.): *Magnetic Properties of Layered Transition Metal Compounds*. 1990
ISBN 0-7923-0238-9
10. E. Doni, R. Girlanda, G. Pastori Parravicini and A. Quattropani (eds.): *Progress in Electron Properties of Solids*. Festschrift in Honour of Franco Bassani. 1989
ISBN 0-7923-0337-7
11. C. Schlenker (ed.): *Low-Dimensional Electronic Properties of Molybdenum Bronzes and Oxides*. 1989
ISBN 0-7923-0085-8
12. R. H. Friend (ed.): *Conducting Polymers*. 1990 (forthcoming)

GPSR Compliance

*The European Union's (EU) General Product Safety Regulation (GPSR) is a set of rules that requires consumer products to be safe and our obligations to ensure this.*

*If you have any concerns about our products, you can contact us on ProductSafety@springernature.com*

In case Publisher is established outside the EU, the EU authorized representative is:

Springer Nature Customer Service Center GmbH
Europaplatz 3
69115 Heidelberg, Germany

**Batch number: 10815974**

Printed by Printforce, the Netherlands